Field Life

INTERSECTIONS: HISTORIES OF ENVIRONMENT, SCIENCE, AND TECHNOLOGY IN THE ANTHROPOCENE

SARAH S. ELKIND AND FINN ARNE JØRGENSEN, EDITORS

Field Life

SCIENCE IN THE AMERICAN WEST DURING THE RAILROAD ERA

Jeremy Vetter

University of Pittsburgh Press

Published by the University of Pittsburgh Press, Pittsburgh, Pa., 15260

Manufactured in the United States of America
Printed on acid-free paper
10 9 8 7 6 5 4 3 2 1

ISBN 13: 978-0-8229-4453-9
ISBN 10: 0-8229-4453-7

Cataloging-in-Publication data is available from the Library of Congress

Jacket art: Photo by William Henry Jackson (1870). USGS Photographic Library.
Jacket design by Melissa Dias-Mandoly

Contents

Acknowledgments

This book has gestated through my residence at four institutions—the University of Pennsylvania, the Max Planck Institute for the History of Science, Dickinson College, and now the University of Arizona—and I heartily thank all the faculty and students at each place for providing stimulating intellectual engagement, even when they may not have realized I was thinking about this book. First of all, I would like to acknowledge the enduring influence of Rob Kohler and the late Riki Kuklick, who served as advisors for the doctoral dissertation from which my research program has developed. Rob has remained a valued colleague and sounding board even as he has moved into an intellectually active retirement and as I have moved into my own professorial career. The graduate student community at Penn was also vital to the development of my thinking, and I warmly thank all of those who provided feedback on much earlier iterations of this work long ago, and especially Hilary Smith, Babi Hammond, Emily Pawley, and Roger Turner, who have done the most to keep those conversations going in the many years since graduate school. I also enjoyed spending a little over a year as part of the international community of scholars in Berlin, supported by a postdoc in Raine Daston's lively department. My other long-term mentor who deserves special thanks is David Cahan of the University of Nebraska, who started me out in history of science as a senior thesis advisor when I was an undergraduate, and who has faithfully maintained an interest in my work ever since.

No one deserves greater acknowledgment than my parents, Jay and Maureen Vetter, who have been unfailingly supportive throughout my life and academic career. A few years ago, they retired to the Rocky Mountains of Colorado, where I have enjoyed a dream location for a writing retreat during the hot Arizona summers—and other times of the year! Since coming out West, I have appreciated the stimulating intellectual conversations on Rocky Mountain hikes with a variety of historian colleagues, especially regulars like Mark Fiege, Adrian Howkins, Paul Sutter, Christine Talbot, David Ortiz, and whomever else I could coax up into the mountains when I was up in Colorado. When still based out East, I spent several summers out West, including some longer stays in beautiful Bozeman, Montana, where I benefited from interactions with colleagues there, especially Michael Reidy, who provided a place for me to stay on more than one trip. These visits were crucial for expanding the geographical purview of my research to include the northern Great Plains and Rocky Mountains. My sister, Molly, who lives in California, has also been an important source of all-around support since childhood. After coming to Arizona, I benefited from ongoing collaborations and discussions with many colleagues, especially my fellow environmental historians Katherine Morrissey and Doug Weiner, as well as department head Kevin Gosner, a collegial supporter of junior faculty.

Most of this book has not been published before. However, some portions of the fifth chapter and a few passages in the epilogue appeared in a previous journal article, which has been reproduced with kind permission from Springer Science+Business Media from the *Journal of the History of Biology*, "Labs in the Field? Rocky Mountain Biological Stations in the Early Twentieth Century," vol. 45 (2012): 587–611. Various pieces of the research related to this book have been presented at so many different conferences, workshops, and other venues and has benefited from conversations with so many colleagues around the world, that it is impossible to name them all. In the homestretch, however, as the book took its final shape, I benefited from the advice and assistance of those at the University of Pittsburgh Press, as well as two anonymous external reviewers, and I would like to thank Sandy Crooms for shepherding the manuscript through the editorial process.

Field Life

Introduction

In 1859 and 1860, at the tail end of the era historians have called the "Great Reconnaissance" of the West—just before the building of the transcontinental railroad—Captain William F. Raynolds of the US Army Corps of Topographical Engineers led an exploring expedition from the Great Plains to the Wind River Mountains of present-day Wyoming, part of the larger Rocky Mountain region. Despite being guided by legendary mountain man Jim Bridger, the expedition failed in its goal of reaching the fabulous Yellowstone Park area.[1] Yet the enterprise was not entirely a failure for science. In the late spring of 1860, a young Ferdinand V. Hayden, the expedition's geologist and naturalist, found something novel in those mountains that he would publish in a national scientific journal almost a decade later, in the same year that the transcontinental railroad was completed. "One day," he recounted, "I noticed a group of singular tracks on the snow which seemed different from any I had ever observed in the West, and they appeared to belong to an enormous species of hare. Descending the western slope about a third of the way from the summit, we saw a number of these animals in the little patches of pine forests, and succeeded in capturing several of them, old and young." Hayden described this new species, which he named *Lepus bairdii* after Spencer Baird of the US National Museum, not only by its physical traits and habitat but also by its fine taste: "Its meat is very white and tender, affording the most delicate food for the traveller. For tenderness and fineness of fibre, the meat of this hare not only differs from, but surpasses all others of the West."[2]

The year 1860 also marks the starting point for this study of the field sciences in the US Great Plains and Rocky Mountains of the American West. Hayden's scientific work in the Wyoming Rockies illustrates some key features of field science on the cusp of the railroad era. In the origins and funding of the expedition itself, as well as in Hayden's naming of the species, it suggests the early subordination of regional scientific development to the institutions and scientific centers of the East. But it also foreshadows the region's environmental distinctiveness, as well as the shaping of knowledge production by a challenging material context. At a more practical level, the case shows how scientists themselves used that environment's resources for their own subsistence in the field. Finally, it symbolizes the era of natural history before more rigid disciplinary divisions, when multipurpose expeditions identified new plant and animal species, minerals, geological formations, and native peoples to be reported in general-interest publications such as *American Naturalist* that appealed far beyond narrow specialists.

Now shift forward sixty years to 1920. In that year Francis Ramaley, a biology professor at a regional institution, the University of Colorado, published a paper in a disciplinary journal, the research for which involved ecological field work at a series of subalpine lakes, all in the vicinity of the university's "Mountain Laboratory," a biological field station at the small village of Tolland, located in a high mountain valley within the towering Front Range of the Rocky Mountains and conveniently accessible by railroad. "The area in which lake-shore vegetation has been studied," wrote Ramaley, "is a strip about 5 miles wide, along the eastern slope of the Rocky Mountain National Park south to Parry Peak, a distance of 24 miles."[3] In his paper, Ramaley described the formations of vegetation in the technical language of the emerging discipline of ecology. Moreover, his botanical field work at the Mountain Lab involved the prolific use of disciplinary innovations such as the quadrat, pioneered by Frederic Clements and Roscoe Pound at a neighboring state institution, the University of Nebraska, in the 1890s.[4] After more than ten years of field work in that high mountain valley, Ramaley had produced scientific knowledge using intensified and systematic methods at a site that extended the reach of one of the interior West's own institutions into its fabulous mountain landscapes.

The contrasting example of Ramaley at Tolland reveals a geography of knowledge production quite different from Hayden's discoveries on the Raynolds expedition. By the 1910s, regional institutions—state universities, agricultural experiment stations, civic museums, geological surveys, and the like—had gained a substantial foothold. They were joined in the field by powerful institutions from outside the Great Plains and Rocky Mountains, most often from the eastern United States, and including major research universities, large urban museums, and sprawling federal science agencies such as the US Geological Survey and the US Department of Agriculture. Moreover, the practices used had grown more intensive, rigorous, and systematic, with field scientists adopting the cosmopolitan standards set in distant scientific centers and applying them in the distinctive environmental context of the West. Finally, most practice had shifted from multipurpose expeditions studying the whole gamut of natural phenomena of the region—its rocks, fossils, plants, animals, weather, artifacts, and people—to research now almost always divided by disciplinary boundary lines, connecting the region's environmental resources to specialized sciences. In one sense, this book seeks to show how a region's scientific practice in the field developed from the mid-nineteenth century to the early twentieth century. Yet while change over time in field practices is a significant part of this narrative, it is also intended to reconstruct a world we have lost—a world of field science that existed during the era when railroads dominated long-distance transport. "Field life" during the railroad era was produced through intertwined environmental and technological systems, the organization of work, and a geographical division of labor. But it is the analysis of science in regions and the historical articulation of its diverse modes of practice that bind and define its subject matter.

Regions of Science, Modes of Practice

Why, one may ask, should we study the history of science in regions at all? The region has been a rare unit of analysis in the history of science; national or disciplinary traditions have received far more attention. While many excellent histories have been written about national or disciplinary science, this book is grounded in the belief that the region can be a useful scale at which to examine the environmental and working conditions of science, as well as a middle level of analysis between the local and global that does not presume the nation-state to be the only important intermediate unit.[5] One of my principal aims is to more systematically develop a regional approach to the history of science, further sharpening the analytical tools necessary to make it a compelling scale for research. Moreover, by placing greater emphasis on the natural environment and the work organization of knowledge production than some previous regional accounts in the history of science, I seek to strongly emphasize place and practice.

Regions exist in a global context, of course. Scholarly approaches to capitalism can be instructive by analogy in expressing how global processes interact with place. "Capitalism," notes geographer David Harvey, "does not develop on a flat plain surface endowed with ubiquitous raw materials and homogeneous labour supply with equal transport facility in all directions. It is inserted, grows and spreads within a richly variegated geographical environment which encompasses great diversity in the munificence of nature and in labour productivity."[6] The same could be said for the geographical development of science.[7] By the 1860s the American West was already beginning to be opened to the production of scientific knowledge. Expeditions led by the US Army, which included some naturalists, had begun to circulate knowledge of this region more widely by the mid-nineteenth century. Topographical engineers, as well as the scientific assistants they sometimes took along with them, provided valuable observations in their reports and collected specimens from a region that had hitherto been largely outside the domain of cosmopolitan science. In the late 1860s the building of the transcontinental telegraph and railroad opened a new frontier for the expansion of scientific practice, beginning the process of converting nature into knowledge that could travel to metropolitan centers.

By conceiving of the expansion of science regionally, we can also help address another problem that has vexed the history of science, and science and technology studies more generally, for the past several decades: the tension between the contextualized, empirical richness of the local case study and the explanatory, weighty allure of the "big picture" global grand narrative.[8] In examining scientific practices that extended over many local places within larger regions, we can generalize beyond the local while still preserving a fidelity to the texture and specificity of the past. One way that historians have negotiated this tension is to focus on the nation-state, an emphasis that has generated significant insights, sometimes even through the explicit comparison of "national styles" in science.[9] By focusing on regions, however, the roles of environmental context and the closely connected technologies and practices deployed by field scientists become more salient. At the same time, despite focusing on environmental regions, I have deliberately confined this study within the borders of the United States, not because tracing the story into Canada—where both the Rocky Mountains and Great Plains extend—is any less worthy, but because I prefer to sidestep making cross-national comparisons the central focus of my analysis, which would be a significant, and probably irresistible, temptation. Transnational histories of science of environmental regions are likewise well worth doing, but I have chosen to resist making *Field Life* a book about transnationalism, which could easily overshadow other important themes. In focusing on environmental regions, even within one nation-state, this book operates at the middle range, between the general and the particular. By doing so, it seeks to avoid having to "trade off abstraction and generalization with the demands of remaining faithful to the complexity of [our] empirical material."[10]

Just as a regional frame of analysis permits an analytical balance between the local and the universal, and between abstraction and empiricism, so too does it help locate a middle ground between what people were physically doing on the ground and their cognitive goals. One way to accomplish this is to focus on midlevel patterns in the practice of science—in other words, what I shall refer to as *modes of field practice*. Each differing mode indicates the organization of both people and techniques and thus is simultaneously social and material, and at once practical and methodological.[11] By defining modes of knowledge production in the field as they emerged historically and regionally, this history articulates a taxonomy of scientific practice that is intentionally close to the ground. While most histories of scientific practice have started with specific disciplines and ideas, this book starts from the practices themselves, spanning diverse disciplines and theoretical perspectives. In particular, it focuses on the four most prevalent new modes of field science that were emerging during the railroad era: lay networks, surveys, quarries, and stations, each of which will be treated in a separate chapter.[12] As an analytical strategy, dividing "field practice" into four distinct modes has the advantage of highlighting the diversity of ways that scientific work was accomplished in the field, while at the same time making a serious effort to generalize and seek larger patterns beyond local case studies.

By putting practice at the center of the analysis, this book approaches science as a form of *work*. Scientific work is partly mental, of course, but it also involves moving through the environment, making observations, taking specimens, and sometimes transforming the earth itself. And like the history of miners, farmers, or factory workers, the work history of field scientists is also a form of social history. A central phenomenon of social history has been the emergence of status levels, some of which have been called classes. The closest analogy to a class division in the production of knowledge is probably the line separating those who publish or otherwise culturally display it from those who participate in making knowledge—as field assistants or local collaborators—without gaining such rewards. Similar to class-based exploitation under capitalism, scientific appropriation across the class lines of knowledge work may or may not be regarded as such by participants.

But just what is scientific "field work"? Here I adopt an operational definition that emphasizes its distinction from laboratory work.[13] While work in the laboratory is intended (if not in actuality) to take place in a controlled, placeless setting—"simplified and standardized, stripped of all context and environmental variations," laboratories "are places apart from the world," writes historian Robert E. Kohler—field work is always tied in some way to the specific geographical place in which it is done.[14] In other words, the environmental and locational features of place are explicitly part of reports of the knowledge produced, rather than remaining unmentioned. Often field work has involved the publishing researcher actually traveling into the field—for example, on an organized expedition—but some field work has relied on surrogates in the field who communicate with and send specimens to a scientist located elsewhere. Either way, however, field work is more or less place bound.

Historical work on scientific practice is, like all human practice, constrained by the availability of documentary evidence about everyday activities and interactions. Laboratory work, for its part, produces very little permanent written record of the face-to-face conversations among those who participate in it.[15] Archival records of correspondence in field science, on the other hand, provide access to communication between field agents and their metropolitan supervisors, which can reveal abundant, rich detail about scientific practice. Moreover, key insights about field work are often included in reports and publications, as well as more fully in programmatic statements and purposely written reflections. In other words, a wide range of primary source materials is available to reveal the texture and development of scientific field work. In this book, then, my task is to analyze the historical emergence of these four material modes of field practice during the railroad era. The exposition will combine a synchronic, historically ethnographic approach, seeking to reanimate the social and material world of scientific field practice, with some attention within each chapter to diachronic transformations within each mode, between the 1860s and 1910s. Finally, each chapter will feature two distinct case studies to propel the narrative forward and provide flesh-and-blood texture to the analysis. These case studies have been chosen to represent some of the diversity of subject matter within each mode of practice, and their presence, along with many other comparative examples, is intended to prevent the conflation of each mode of practice with any particular example.

An Envirotechnical Infrastructure

The title of this book indicates that it takes place during "the railroad era." In part, this is simply a convenient shorthand for a period that could also be labeled as the Gilded Age and Progressive Era, or the period from the Civil War to World War I. But this technological label also conveys something important about the American West from the late 1860s to around 1920: to understand the practice of field work requires diligent attention to the infrastructure of transport and communication. The railroad not only provided for rapid movement into and out of the field, as well as within it; the railroad often defined the very point of entry into the field.[16]

Yet, as Richard White has observed, while the railroads "made the far near . . . they did so unevenly and chaotically."[17] Despite the dominance of railroads in the long-distance transport infrastructure, for more localized movement it was still in many respects a "landscape for horses," in the words of historian Ann Norton Greene, "with living machines and steam machines each occupying a unique niche in nineteenth-century industrialization."[18] The botanist A. Isabel Mulford expressed this transport bifurcation succinctly in 1894, reporting that for summer travels in the Rocky Mountains of Idaho she had "traveled by railroad to the principal points, and by stage, carriage, or horseback to the more distant ones." (In common with most other field scientists, she also "walked a good deal.")[19] The railroads undergirded a new travel geography of the West that was as critical to the practice of science as to any other activity.

In communications, the most important technological infrastructure was provided by the US Post Office, which by then generally followed the railroad, though with a significant supporting network of horse-drawn mail stagecoach lines and postriders to reach outposts beyond the rail. Postal historian Richard R. John has convincingly argued for the early bureaucratic development of the post office prior to the 1830s—a "communications revolution" before the advent of rail or steam. In his view, "it was neither the railroad nor the telegraph nor even the commercialization of fossil fuel, but rather the stagecoach and the postrider that liberated Americans from their crippling dependence on the vagaries of geography, wind, and water." Moreover, steamboats of this era played a very limited role, even where there were pliable waters, given their unreliable schedules, in part due to the perpetually "changing river conditions."[20] At the same time, historian David Henkin has aptly described the period from 1840 to 1860, too, as a "revolutionary era . . . when a critical mass of America began reorganizing their perceptions of time, space, and community around the existence of the post," and he highlights the "substantial reductions in the cost of sending a letter" in 1845 and 1851. The number of items posted during this time accordingly exploded "from about 27 million to about 161 million," which meant a tripling of the per capita rate, even when adjusted for population growth.[21] Moreover, while the increasing movement of mail by railroad, which began in 1831, may not have changed all aspects of postal practice at first, by the 1860s the advent of "railway mail," based on "continuous sorting on specially outfitted railroad cars," signaled further transformation.[22]

Supplementing the postal network was the expanding telegraph, which provided a more rapid (and expensive) means for communication into the field.[23] As Elliott West has noted, "the West was acquired, conquered, and largely consolidated into the nation coincident with the greatest breakthrough in the history of human communication . . . the telegraph."[24] While we must be cautious not to overdramatize the role of the telegraph—after all, most scientific messages, forms, circulars, and material specimens still traveled through the mail—it is nonetheless significant that a form of nearly instantaneous communication became available during the railroad era at some locations in the Plains and Rockies, beginning with the first transcontinental telegraph route, which opened in late 1861. The telegraph could be a powerful tool or instrument of science, just as much as for any other activity, when time was of the essence.

There has been much written more generally on the roles of tools, technologies, and instruments in the history of science. In considering the field sciences especially, the key "instruments" need not be solely conventional artifacts that might be displayed in science museums or hidden away in the back rooms of scientific institutions, although these were obviously important. Tools, technologies, and instruments may always have some material component, of course, but these might range quite broadly, even in laboratory and theoretical science, as other historians of science have shown—for example, legal technologies reflected in the demarcation and enforcement of property boundaries, or the example of paper and pencil techniques in mathematical physics or Berzelian chemical formulae as "paper tools."[25] Tools in the field deriving from the natural history tradition, and often carried over into the field sciences, also included various "written categorization procedures" such as lists and reports and "material collection techniques" such as boxes and capsules.[26] Other technologies and instruments in the field sciences that developed during the railroad era were quite distinct, as we shall see. Often these were integrated into large technological systems linking everything from a transcontinental railroad line to a single paper form delivered by railway to be held in one person's hand.

Moreover, shifts in infrastructure need not be related merely to technological hardware, such as railway or telegraph lines. In the case of the post office, for example, simple changes in policy or fare structure—what we might call the "software"—have had significant implications for the practice of science. An instructive example of such a change is noted by Martin Rudwick in his study of early nineteenth-century gentlemanly geology in Britain. He observes that these "gentlemen of science" inhabited a social world "in which spontaneous and fluent letterwriting was a routine accomplishment." Yet at first the high cost of postage beyond London—a cost paid by the recipient, not the sender—gave correspondents about geological matters a strong incentive, even a social "obligation," to provide "good value for [recipients'] money, to fill the sheet with interesting and worthwhile material." While such reliable and rapid communication was in fact made possible by a relatively well-developed postal infrastructure, compared to earlier times and other places, it was still a cultural-technical system requiring considerable economy and care. "All this changed," Rudwick points out, "with the introduction of the cheap prepaid penny post throughout Britain in 1840; as soon as a low and uniform cost was incurred by the sender, the informative quality of long-distance scientific letterwriting fell off abruptly."[27]

While the transport and communication infrastructure receive top billing in *Field Life*, they do not constitute the entire material basis of scientific field work. Some other technologies were crucial to particular modes, such as the paper reporting form in lay networks, camping gear for surveys, property control (legal technologies) for quarries, and laboratory buildings for stations. Other technical means and tools have been important to particular sciences, such as farm equipment for agricultural science, thermometers and rain gauges for meteorology, and indeed the whole gamut of specialized measuring tools for each field discipline. While these more specific technologies underlying more specialized systems of production will not be my primary focus, they will come up from time to time, and they are certainly important in case studies of the particular systems of knowledge production that flourished within each discipline.

Thinking about the tools of science in the field purely as technology is not enough, however, even when it incorporates both hardware and software. The actual use of technologies of transport, communication, and everything else that we might consider important was deeply intertwined with the particular natural environments in which those technologies were deployed. For thinking about this relationship, a conceptual framework from two fields closely related to the history of science—environmental history and the history of technology, or, more precisely, from the intersection between them—can be useful: the "envirotechnical system." One leading envirotech historian, Sara Pritchard, has defined *envirotechnical systems* (deliberately pluralized) as "historically and culturally specific configurations of intertwined 'ecological' and 'technological' systems, which may be composed of artifacts, practices, people, institutions, and ecologies."[28] Note that "enviro" and "tech" are not merged into complete indistinguishability: they are "intertwined." As another influential envirotech historian, Tim LeCain, has argued, "the goal of envirotechnical analysis . . . is to demonstrate how this system that is both human and nonhuman, artificial and natural, technological and ecological, does actually exist even if our culturally constructed ideas and words often keep us from recognizing it."[29]

Thus, without denying that technology and the environment are to some extent distinct and analytically separable elements, historians of science can immediately recognize, in agreement with the envirotech historians, that the natural environment and human technology were so closely interconnected in scientific practice—especially but not only in the field—that it makes sense to view them as elements of the same system. To take just one example, the railroads of the interior American West constituted a powerful, even dominant, technological infrastructure for the practice of field work during this period, but they were always used and understood in relation to the environmental landscapes in which they were located, built, maintained, used, and embedded. Railroads could, of course, be built in a wide variety of natural places (though not everywhere), and environmental regions such as the Great Plains have been inhabited by humans with a variety of technological systems, from horses and tepees to gasoline-powered vehicles and center-pivot irrigation equipment—both before and since the railroad era. Yet it is only by attending to the envirotechnical system as a relational whole that the conditions of practice can be fully understood, including the conditions of scientific practice.

Microdivision of Labor: The Organization of Work

Complementing my emphasis on the envirotechnical basis of scientific field work—and closely overlapping with my attention to the relationship between the environment and scientific bodies in the field—another significant focus in this book is on divisions of labor, including both those within the day-to-day practice of scientific field work at ground level and those of the larger geographical relations between regions of cosmopolitan science. For the first of these, while the history of science has traditionally focused on lead research scientists, this book aims to uncover the hierarchical division of labor that made up rigorous scientific field work, from leaders to subordinate contributors, and crossing the conventional boundaries that have divided publishing scientists from their technicians, field hands, and other assistants. Another important aspect of field work, which has distinguished it from the laboratory with its access restrictions, has been the involvement of lay people in the process of knowledge production, which I also examine in this book—for example, the reliance of survey parties and quarry prospectors on their knowledge of where to find promising natural phenomena and objects.

One conventional way of characterizing the division of labor in science—if truth be told, more in rhetoric than in actual practice—is to distinguish between head and hand work.[30] As Mary Louise Pratt argues, "natural history set in motion a secular, global labor that, among other things, made contact zones a site of intellectual as well as manual labor, and installed there the distinction between the two."[31] Some of the field work described and analyzed in this book descended directly from the natural history collecting of the eighteenth and early nineteenth centuries that Pratt analyzes, and such a division of labor carried over to the modes of field production that I describe. Indeed, within each of these modes, as the several chapters of this book will show, field researchers employed or voluntarily organized in their work collaborators who engaged in relatively more manual labor than they themselves did.

Nevertheless, this conventional formulation fails to tell the whole story: practical realities in the organization of field work often meant that researchers would be engaging their hands, while those whose status they attempted to subordinate sometimes contributed vital mental skills and experiential knowledge. Other historians have recognized that the grubby, hands-on character of field work extended across the lines of social class, so that Britain's gentlemen geologists, for instance, loved to recount examples in which their gentlemanly status went unrecognized in the field.[32] Even when assisted by a myriad of guides, cooks, packers, hired workers, and other proletarians, field work routinely involved an engagement with the natural world that belied any simple exclusion of manual tasks from the activities of naturalists and other scientific researchers. Likewise, the local or experiential knowledge of lay people and hired laborers in the field, along with the mental habits that systematic field work often required of them, was a crucial component of its success. The chapters that follow will offer numerous examples of the mental aspects of subordinate labor in the field.

The organization of scientific work, again like other types of work, has been accomplished through hierarchical structures of authority. These go well beyond the simple division between research authors and their assistants. Just as managers and laborers in industry are further stratified, so too are scientists divided by myriad status distinctions. For example, museum directors have authority over department curators, who in turn guide field staff, who in turn direct assistants in the field. Different status levels pervade the "work history" of science. Besides status distinctions, field scientists are also divided socially by career status, institutional affiliation, sex, race, and geography. Overlapping significantly with all these status distinctions is the development of careers based on field science work. Some participants in knowledge making in the field established scientific careers, while others were involved only temporarily (e.g., students) or as an unpaid adjunct to another career (e.g., doctors, clergymen, and businessmen). During the period of this study, career makers gradually came to dominate, although some noncareer participants remained visible into the twentieth century.

Historians have learned to treat the issue of scientific identity with some historical sensitivity. In fact, I have focused deliberately on the emergence of scientific careers in part to avoid relying on the more conventional category of professionalization. It is not so much that I want to continue flogging the category of professional, so much under attack during the past several decades—and, to be sure, I do sometimes use the terminology of *amateur*, which implies its professional antonym—but it is important to avoid teleological assumptions about what particular identities were coveted by the makers of natural knowledge in the nineteenth century. In this case, the most questionable assumption is probably that scientists in the nineteenth century sought to become professionals in the twentieth-century mold. As numerous historians have demonstrated, knowledge makers did not necessarily aim to become professionals, or even scientists.[33] Rather, the task here is to display the historical emergence of scientific careers or life paths through field work based on changing practices and identities.

An important related trend was the displacement of independent field scientists by those with institutional affiliations at museums, universities, and government agencies. Nearly all field scientists at all status levels during the railroad era were men, although there were a handful of women, many of them conspicuous in part due to their sex. Nearly all of them were white, with the crucial exception of American Indian field assistants, especially in anthropology. Low-status participants such as cooks and packers were sometimes nonwhite, although the rarity of extended comment on such workers by scientific writers makes it difficult to say with any precision how many. Finally—and with great significance for the scientific development of the region—field scientists could be from inside or outside the West, and by the 1890s the early overwhelming dominance of Easterners was challenged by researchers from within the region.

Despite these consequential differences among the field scientists in the American West, they also had a great deal in common. All were involved in the production of knowledge out of Western nature, by converting their observations and material specimens into symbolic expressions—oral or written—that could circulate in a wider scientific community beyond the locality where they began. As such, these migrant knowledge makers were not fundamentally different from the many other new producer groups who came to the West in the late nineteenth century, such as miners, farmers, and ranchers. Like them, scientists subsisted on the natural resources offered by the land, attempting to earn a living through their own sweat and toil. Some carved out niches as old-middle-class-style independent producers, in the specimen business, for instance.[34] But as institutions such as universities and government agencies came to dominate employment of scientists, most would be better described as members of the new middle class of college-educated white-collar professionals and managers.[35]

Although the producer groups provide the dominant metaphor underlying this project, we can also recognize a second comparison group, the tourists or vacationers. Scientific field work in the West shared a great deal with the emerging middle-class style of vacationing in the late nineteenth century, which Cindy Sondik Aron calls "working at play." Middle-class vacationers sought to avoid the old upper-class style of vacationing, which they saw as luxurious, extravagant, and fraught with idleness. Instead, they sought self-improvement.[36] Participating in a scientific field expedition might well be called the extreme version of a middle-class, working-at-play vacation, in which tourists saw a lot of interesting sites—especially at popular field work locations like Yellowstone and the Colorado Rockies—but were doing scientific work too.

While spatial hierarchy in knowledge making has endured over time, the specific types of labor have shifted. As Henrika Kuklick has pointed out in her study of British social anthropologists, the nineteenth-century "division of disciplinary labor" between a recognized elite of armchair scholars and their collectors in the field reflected the community's constitution, giving "distinct roles for foot soldiers who collected data and the scholars who interpreted them." Yet such a division "dissolved" with the expansion of the university system, which opened up new career opportunities for researchers and made field work an "essential" activity for gaining prestige in anthropology. By the First World War—near the endpoint of this study—personal involvement in systematic field work was crucial for all different kinds of field sciences.[37] Scientists in the center still controlled the production and circulation of knowledge, but they did so increasingly by going into the field themselves.

Even while largely eschewing the amateur-professional distinction, as noted above, this book foregrounds another distinction with which it is often conflated: the lay-expert divide. While experts and professionals are often thought to be the same people today, the leading experts in a given field of study in the nineteenth century might or might not be professionals. Arguably, the professional identity was increasing in significance, especially by the turn of the twentieth century, but we need not place a great deal of weight on the category of the professional in order to recognize the analytical importance of the growing divide between lay people and experts. Of course, as Paul Lucier has reminded us, the term *expert* itself underwent a significant set of shifts during the nineteenth century.[38] Nonetheless, while it may be somewhat difficult to pin down the "expert," the social role of the "lay person" in the production of knowledge, as an analytical category, is in my view unavoidable in examining field work during this period. The involvement of lay people—usually, locals who lived and worked near the field site—in knowledge production was simply too important, and sometimes contested, for the dynamics of such participation to be elided because of a fear of using anachronistic terminology.

Indeed, the lay person is a constitutive social category for understanding one of the period's most important modes of knowledge production—what I have called the lay network—and lay people were nearly always important in other modes as well, acting as guides to survey parties, as well as hired laborers for many field stations and quarries. The lay person is a crucial social category for understanding the status hierarchies and patterns of work organization discussed in this book.

While I only rarely use the terminology of *expert*, leaving such a role implicit, it will thus prove to be crucial to talk a lot in this book about lay people. In some cases, for precision, it will be preferable to refer to amateurs instead, or to amateur and lay together, when the people involved possessed a clear avocational interest in the subject matter, or if some of them did, in order to convey the social and epistemic relations as closely as possible.

Experiential and Cosmopolitan Knowledge: The Epistemic Rift

To situate field practice in its context, it will also be essential to take seriously the experiential knowledge possessed by lay people and amateurs in the field, which often complemented what we might call the "cosmopolitan" knowledge of field scientists who encountered them, collaborated with them, and occasionally conflicted with them in the practice of field science.[39] The distinction between experiential and cosmopolitan knowledge, and what we might call the "epistemic rift" between them, is one of the most important conceptual frameworks for understanding the practice of field science.[40] All modes of field science have attempted to bridge the rift between the knowledge of experience based on living, working, or playing in a particular place and the cosmopolitan knowledge that travels well between places and enables the facts from any one place to be situated in larger taxonomies, categories, and comparisons. For instance, while a farmer might know where and when to find various animals that live nearby, along with their locally peculiar habits, a visiting scientist with access to specialized knowledge and distant collections would be better equipped to classify and map those animals in terms of cosmopolitan scientific knowledge. We might furthermore think of this distinction as directly parallel to that between use value and exchange value in the material economy: the local use value of experiential knowledge within everyday life and work, likewise, contrasts with the long-distance exchange value of cosmopolitan knowledge within science.

The distinction between experiential and cosmopolitan knowledge, as well as the possibility of bridging the epistemic rift between them, not only is important for the historian's analytical toolkit but was also sometimes present in discourse during the railroad era itself. During the late 1860s, for example, as the first transcontinental railroad was in progress, a new periodical called the *American Naturalist* began publication, with a stated aim to go beyond an audience of scientists and expert naturalists and become "equally welcome to the Farmer, Gardener, and Artisan," by emphasizing "the practical benefits resulting from the study of nature." However, its editors pledged to offer readers content "drawn from original sources," including "the most recent discoveries of general interest from the English, German, and French reviews and Journals,—for science is cosmopolitan."[41] A quarter of a century later, near the midpoint of the railroad era, the naturalist C. Hart Merriam drew further attention to the distinction between different types of knowledge, writing in the pages of *Science* that he "would rather have the farmer's boy who knows the plants and animals of his own home than the highest graduate in biology of our leading university." Yet he also recognized a "chief disadvantage in manufacturing naturalists in this way": in Merriam's view, "they lack the education possessed by college-bred men," which would be "sorely felt" over time. "To be well equipped for this work," he concluded, "a naturalist or biologist needs a college education," thus admitting the importance of cosmopolitan knowledge alongside the knowledge of experience possessed by the farm boy.[42]

The adjective *experiential* is only one of many options for making this key distinction, and the term overlaps considerably with other commonly used terms such as *local, vernacular, indigenous,* or *folk,* which also identify this type of knowledge to varying degrees.[43] The term *local* is perhaps the easiest to use, and I will use it often, but we must be careful about the hierarchical assumptions built into the antinomy of local and global, which often implies that the one is geographically subsumed under the other. (In such a formulation, local knowledge is sometimes thought to have validity only insofar as it can be aggregated into global knowledge.) However, I do use the term *local* frequently in this book because of its simplicity and aptness for many specific examples, and occasionally I also employ other alternatives such as *folk, indigenous,* or *vernacular* when the connotation seems especially appropriate.[44]

More broadly, as shall become clear at various points in this book, people of lower status levels—field collectors, amateurs, local collaborators, and the like—did possess some leverage that softened the more exclusionary, professional identity then emerging in the cosmopolitan scientific community, especially among laboratory researchers.[45] Perhaps one reason why so many researchers took to the field at the end of the nineteenth century—when both infrastructure and academic positions began to enable them to do so—was to exert a greater degree of control over the production of knowledge in the field, which had hitherto been under the partial authority of peripheral collaborators. Still, except perhaps for the example of intensive ethnographic field work, most scientists did not live permanently in the field and had to rely on people who did.[46]

Macrodivision of Labor: Cores and Peripheries

As this discussion of long-distance collaboration indicates, a patterned division of labor existed not just within the local practice of regional field science but also between larger geographical regions. In the end, whether in any part of the Americas, Africa, Asia, Australia, or even peripheral zones of Europe itself, those living and working in regions connected to the cosmopolitan science of Europe and eastern North America faced the challenge (and opportunity) presented by the geographical division of scientific labor between core and periphery. One of the underlying aims of this book is to explore the historical unfolding of that system of spatial hierarchy. A geographical division of labor has been crucial to the practice of science, just as a similar division of material labor has underpinned the spread of capitalist production and trade. Beyond the organization of the field work itself, then, this book also examines the geographical division of labor that characterizes the larger world of cosmopolitan science. As a new periphery for scientific practice in the nineteenth century, the Great Plains and Rocky Mountain regions were at first dominated, in the production of knowledge about them, almost completely by outsiders. Over time, regional institutions and researchers struggled to redirect the epistemic profits of research to the region itself, with partial success, so that by the mid-twentieth century it had arguably gained at least semiperipheral status. By discussing the division of labor both within research enterprises and between geographical regions, this book aims to highlight the role of hierarchies of labor and work in the history of science.

Such hierarchies were evident to contemporary observers. On the occasion of the American Association for the Advancement of Science (AAAS) meeting in Denver in 1901—its first meeting in the region—an editorial report that appeared in *Science* advanced a suggestive historical comparison. "During the first hundred years," it noted, "the nation was in scientific matters somewhat in the relation of a colony to Europe. Our students went abroad for study; we depended on Europe for our journals and books, and did not contribute our share to the work of the world." Yet, in the last quarter century, "great progress" in American science led to a new set of relations: "The central and western states have been in the colonial relation to the Atlantic seaboard that it had previously held to Europe. Students have come to eastern universities, and the scientific men for the central and western states have been drawn from the east."[47]

East-West hierarchical relations in American science—especially as manifested in interactions between the Western states and the federal headquarters in Washington, DC, along with such metropolitan centers as New York, Boston, Chicago, and Pittsburgh—were in some respects similar to the relations of European capitals such as London and Paris with their peripheries.[48] Moreover, many of the dynamics and tensions discussed in this book were felt in other parts of the United States. But there was also an environmental dimension to East-West relations. Regional characteristics like the semiaridity of the interior West made it a different kind of continental periphery than the Midwest, the South, or northern New England. Such environmental differences between East and West—mediated through the environmental and technological infrastructures of transport, communication, supply, and provisioning—reinforced and even strengthened the more general geographical relations of core and periphery.

For the author of the 1901 report in *Science*, however, such a hierarchy was not unchangeable: further mobility could be imagined. The editorial report therefore concluded that "it is perhaps worth noting that the fiftieth meeting of the Association in the first year of the new century should celebrate an epoch in the development of science in America."[49] Was the first meeting of the AAAS in the Rocky Mountains and Great Plains region a harbinger of a future liberation from the constraints of scientific colonialism? How would scientific work be organized to accomplish scientific development? In the end, rhetorical enthusiasm for progressive frontier scientific development must not be assumed to describe the actual reality; rather, it requires a clear-eyed confrontation with the historical evidence. It is the task of the rest of this book to supply that evidence and to interpret it in its larger context.

At its most basic level, this book is a historical reconstruction of the practice of scientific field work in the Plains and Rockies during the railroad era, oriented around a few simple questions: What was it like to do science in the field? How were natural environments and human tools and technologies brought together? Who was involved in field science and how was their work organized? How did they interact with others outside the West? This book aims to give readers a ground-level perspective on scientific practice in the field during an especially intriguing time and place in the history of science—the American West during the late nineteenth and early twentieth centuries. The title, as many readers may have surmised by now, builds on the classic science studies book by Bruno Latour and Steve Woolgar, *Laboratory Life* (1981). Like *Laboratory Life, Field Life* is meant to provide a window onto the everyday practices involved in the production of scientific facts. There are a few key differences, however. First, I emphasize the diversity of ways of organizing the production of knowledge in the field, rather than identifying "field life" with any single mode. Second, my approach is consciously transdisciplinary in its subject matter, bringing together evidence from a wide range of field science disciplines, including paleontology, botany, zoology, ecology, archaeology, meteorology, geology, and agricultural science. Third, I have deliberately situated the field sciences both historically and geographically, so that the effects of period and place are a crucial part of the story. Finally, as suggested above, I aim to point out analogies and connections between science and the developing material economy even more emphatically, in order to suggest how field science has been in dynamic interaction with the larger capitalist political economy.

After an opening chapter on the making of the field as a place for doing science, which also considers the field in relation to the laboratory and the natural history tradition, the four remaining chapters are each oriented around thick description and analysis of a particular mode of field practice in the Great Plains and Rocky Mountains. Beginning with lay networks, followed by surveys, quarries, and stations, I consider the practice of science during the railroad era under each of the modes of field work. In each chapter, I consider the natural environment and technological systems, the organization of work, the ground-level division of labor, and the larger East-West relations, aiming to provide a synthesis of science practice across a range of disciplines, from the earth and physical sciences to the human and life sciences. Throughout, I pay close attention to how these configurations of work, environment, and technology bridged—selectively, distinctively, and partially—the epistemic rift between the cosmopolitan knowledge of science and the knowledge of experience. Finally, in the epilogue, I briefly consider the transformation and persistence of the field sciences after the end of the railroad era. By the end, I hope to demonstrate that "science in the field" in this period was a diverse, polyglot assemblage of practices, but that those practices followed common patterns we can trace.

Chapter 1

Making the Field

Two distinct kinds of field scientists inhabited the interior American West during the railroad era: travelers and long-term residents. Both groups depended on the environmental and technological system built upon the railroad. First, consider a traveler: Born in Massachusetts, but spending his youth and college years on the Ohio farming frontier, Ferdinand V. Hayden was one of the most successful and upwardly mobile traveling scientific entrepreneurs in the American West, who experienced firsthand not only a dramatic social rise (and partial fall) within science but also the shift from limited steamboat and arduous overland travel to a railroad-dominated geography of science. He traveled in the 1850s along the Missouri River fur trading and army exploration corridor, then benefited from the dramatic remaking of the field after the building of the transcontinental railroad in the 1860s, which enabled him to build his massive government-funded survey science empire. This regime lasted until the formation of a unified US Geological Survey in 1879, when, in his ill-fated attempt to secure the directorship, he lost out to fellow "great survey" leader Clarence King. In that quarter century, Hayden participated as much as anyone else—certainly as much as any traveler—in the making of the field in the Great Plains and Rocky Mountains (see figure 1.1).[1]

Next, consider a long-term resident: An immigrant from England, first arriving in Colorado in 1887, Theodore Dru Allison Cockerell came from much farther away than Hayden did, but he ultimately made his permanent home in the border zone between the Great Plains and Rocky Mountains. After shorter stints in West Cliff (now Westcliffe), Colorado, and in Las Cruces and Las Vegas, New Mexico, he eventually pursued a life and career at the University of Colorado, Boulder, which is about as close as anyone can get to the precise junction of the Plains and Rockies. Cockerell was an astonishingly prolific scientific author whose working style unabashedly perpetuated the tradition of the well-rounded, autodidactic naturalist: his professional specialization

FIGURE 1.1. Ferdinand V. Hayden with a field party in Wyoming in 1870. Original caption reads: "Wyoming. Group of all the members of the Hayden Geological Survey of 1870, made while in camp at Red Buttes at the junction of the North Platte and Sweetwater Rivers. Standing left to right: John 'Potato John' Raymond and 'Val,' cooks; Sanford R. Gifford, landscape painter; Henry W. Elliott, artist; James Stevenson, assistant; H.D. Schmidt, naturalist; E. Campbell Carrington, zoologist; L.A. Bartlett, general assistant; William Henry Jackson, photographer. Sitting left to right: C.S. Turnbull, secretary; J.H. Beaman, meteorologist; Ferdinand V. Hayden, geologist in charge; Cyrus Thomas, agriculturist; Raphael, hunter; A.L. Ford, mineralogist. 1870. U.S. Geological and Geographical Survey of the Territories (Hayden Survey)." Credit: US Geological Survey. Photo by William Henry Jackson (1870). USGS Photographic Library, http://library.usgs.gov/photo/#/item/51dc99ece4b097e4d383a799

was in zoology, particularly the study of insects, but he also published as a self-described amateur in botany, especially the study of flowering plants.[2] He also participated centrally in building up scientific higher education within the interior West, first at various institutions in New Mexico and then for most of his career at the University of Colorado. Moreover, as he helped make the field, he also helped to remake his own body, since he had chosen this region for health reasons. By the time he arrived, the railroad had already expanded its tentacles throughout much of the region, but he relied upon it as much or more than Hayden did. Cockerell's scientific career is as characteristic of the second half of the railroad era as Hayden's was of the first half.[3]

As the successful and prolific careers of Hayden and Cockerell demonstrate, even as the laboratory rose to its dominant position in science between the mid-nineteenth century and the early twentieth century, scientific practice in the field also changed quite dramatically. Indeed, far from declining, science in the field expanded and intensified. It also became ever more systematic, highly structured, and spatially coordinated. With the rise of the laboratory, "the field" came to be increasingly defined in relation to it. But it was more than simply an abstract place on a conceptual landscape: field science was produced in material environmental places whose construction was, in turn, shaped by both nature itself and the human technologies and cultural assumptions that transformed it. By taking an envirotechnical approach that draws attention to the deep intertwining of the natural and the technological in particular environmental regions, we can more fully understand the making of the field as a place for science.

This chapter begins with an overview of how historians have understood natural history practice through the middle of the nineteenth century, followed by a discussion of the making of the field in the Plains and Rockies from then onward, emphasizing its technological and environmental dimensions. Natural history practice comprised both long-distance expeditions and local collecting, with communication thus based on travel diaries and journals, as well as on correspondence. The earliest natural history in the US West mainly came from long-distance expeditions, which were dramatically affected by the shift from the very slow and cumbersome prerailroad travel infrastructure to a region threaded—though much less densely, compared to the eastern United States—with railroads. Complementing this emphasis on railroads and other industrial-age technologies, such as telegraph lines, I then present an overview of the natural environment itself as a shaping context, emphasizing the distinctive features of the Great Plains and Rocky Mountains. Such an envirotechnical approach should not, however, be imagined as collapsing technology and nature into one another, a full blending that would render the terms *technology* and *environment* meaningless as separate analytical categories. On the contrary, like the term *technoscience*—which has often suffered a similar slide from legitimate claims of intertwining into full conceptual blurring—it will be important to consider, in turn, technology, environment, and both together, keeping in mind that such emphases are always choices made for analytical clarity of expression and argument, not because they can ever be fully separated.

Finally, this chapter also considers the rise of the laboratory as a privileged place for the production of knowledge in the modern world, including some brief discussion of what an envirotechnical perspective might reveal about even the lab itself. The rise of the laboratory had profound ramifications for the development of all scientific practice, and this chapter concludes with an examination of some responses within natural history and the emerging field sciences.[4] To make such a claim is not to argue that the lab ideal completely dominated the field, nor to presume that lab and field were always and everywhere in conflict. As we shall see, the relations between the two were complex and do not fit a single pattern. Indeed, one goal of this book is to emphasize how scientific practice in the field changed and developed on its own terms, and only sometimes and partially in direct relationship to the lab. Nevertheless, to ignore the effects of the rising stature of laboratory practice, and the challenges to their own credibility and rigor that it posed for scientists working in the field, would miss a key influence on the history of the field sciences. The age of the railroads was also the age of laboratory, which did so much to change the conditions under which scientific field work would be conceptualized and evaluated.

The Geography and Practice of Natural History

Prior to the rise of the laboratory there was nothing entirely like "the field" in the modern sense, since the field would come to represent the messy, complex world outside the lab. Yet important predecessor practices can be found in the natural history tradition stretching back centuries, and its various prevailing modes of practice such as local collecting and long-distance expeditions were important not only to natural history itself but to the later development of the field sciences. While some could argue, as T. D. A. Cockerell did, that an expedition "need not be to a remote country, or of long duration" as long as it had a "definite desire or purpose to make discoveries," for most people the difference between covering a small local or regional terrain within the radius of everyday life and traveling by ocean, river, or—before the railroad, very slowly—overland was highly significant.[5] Yet it is also true that naturalists widely aspired to make the categories of local experience commensurate with those of cosmopolitan science, so that many core natural history practices were shared by those with widely divergent spatial ambitions for traveling through nature.

An essential starting point for the making of the field in scientific practice, then, is the natural history tradition, when conceived not strictly as a set of ideas and theories but in practical and material terms. As Janet Browne notes, "the term 'natural history' perhaps ought to be regarded as a designating technique rather than as a subject area. To conduct a natural historical study was to collect, describe, classify, and from the seventeenth century onward, to perform minor experimental investigations into observable phenomena."[6] Accordingly, there was a wide variety of practices encompassed within the category of "natural history," and we might regard it as a composite term encompassing such disparate activities as local collecting on foot and long-distance expeditions around the globe by ship.

Not only does the natural history tradition provide substantive historical roots to the development of the field sciences, but its conceptualization by historians also contributes analytical insights to any serious attempt to grapple with the emerging field sciences of the late nineteenth century. Since at least the mid-1990s, historians have sought "to portray natural history as the product of conglomerates of people, natural objects, institutions, collections, finances, all linked by a range of practices of different kinds" and by defining the category broadly enough to include material, social, literary, bodily, and reproductive practices.[7] However, this focus on the diverse kinds of practices involved in natural history further calls out for a deepened emphasis on the environmentally situated technologies deployed by naturalists.

Collaborative practices in the natural history tradition have been studied most thoroughly in Europe, including its far-flung imperial outposts, and of course that continent was the source for much American practice. Important natural history networks in the eighteenth and nineteenth centuries were organized around not only Linnaeus and his rival Buffon but also Joseph Banks at the head of a whole variety of London institutions, André Thouin at the Jardin du Roi in Paris, and later Joseph Hooker at Kew Gardens near London, all of whose collecting empires have been examined fruitfully by historians of science.[8] Interpretive debates have often focused on the utility or disutility of an analytical framework based on Bruno Latour's geography of centers of calculation, toward which collections and observations were brought from various global peripheries. Such structured geographical models cannot plausibly be imagined as absolute, however. In the context of informal empire, for example, British naturalists and their Chinese collaborators each had some leverage, but "the distribution of power in fieldwork relations did not necessarily favor the naturalists," as Fa-ti Fan has pointed out.[9] Even in formal European colonies, collaboration was far from one-sided; despite the clear importance of hierarchical structure to "imperial relationships," historical circumstances sometimes led to "renegotiation of relations between those at the center and those on the imperial periphery."[10] While historians are surely right to warn against insufficiently nuanced structural portrayals of centers dominating peripheries, at the same time the hierarchical structure of natural history remains manifestly evident. Balanced historical analysis requires simultaneously remaining attuned to these structuring forces and recognizing that naturalists at both centers and peripheries enjoyed advantages and suffered disadvantages based on their positions. And the same balance remains important for studying the field sciences of the railroad era.

In the United States, no less than in Europe, centers and peripheries were structurally important in the practice of natural history. The largest and most important correspondence network was the one centering on Spencer Baird at the US National Museum (Smithsonian) in Washington, DC, which relied on a sprawling and diverse set of local correspondents spread throughout the country.[11] James Hall of Albany, New York, was another early example of a successful collecting network builder, focusing on fossils and other geological materials. The network he built eventually included more than 150 "amateur collectors," who sent specimens to Hall over an entire half century between 1847 and 1898.[12] To overcome the distance and isolation of collectors scattered across the American periphery, Frederic W. Putnam of Massachusetts began publishing the *Naturalists' Directory* in 1865, at first as part of the *Proceedings of the Essex Institute* and later as a separate publication. In 1877 the *Directory* was taken over by Samuel E. Cassino, who shepherded it through thirty more editions into the mid-twentieth century.[13] Putnam himself also cultivated his own network of archaeological field collectors in the Ohio Valley focusing on American Indian mounds.[14] Individual collectors in hierarchical relationship to one another thus dominated the practice of natural history.

To be sure, institutions eventually came to dominate knowledge production in the American West, but individual collectors—often in relation to other collectors in distant places—were still quite important through at least the late nineteenth century. Arising out of the gentlemanly context in Britain in the early nineteenth century to become a more widespread leisure activity for the middle class, this tradition also flourished across the Atlantic and was brought along with the westward expansion of white settlers in the United States.[15] Before moving in 1887 to the small settlement of West Cliff, Colorado, at the foot of the Wet Mountains, T. D. A. Cockerell prepared himself by searching "for literature on the Mollusca of the Rocky Mountains," using mainly a report from "one of the volumes of the Hayden Survey." Nevertheless, once he arrived, he "collected what [he] could" but found that the Rockies were "poor in snails and slugs" compared to most other places. He therefore turned his primary attention elsewhere, especially to "the very abundant and varied insects."[16] Always situating himself in relation to naturalists and collectors throughout the cosmopolitan world of science, Cockerell nevertheless shifted his practices to suit the place where he found himself.

Cockerell's arrival in Colorado coincided with a period of greater maturity in American natural history, when naturalists in the United States interacted with their European counterparts on a more equal footing, especially with respect to the nature of North America itself. Nearly always during the colonial and early national periods, by contrast, Americans had found themselves in subordinate positions in natural history hierarchies, though this was not always, or often, accepted without protest. Historians of early American natural history have emphasized the conflict between the prevailing hierarchy of natural history practice and spreading egalitarian ideals, concluding that "Americans' distaste for their subordination became more acute after independence." Conflict arose between a "democracy of facts" and an "empire of reason," and the latter included among its advocates not only European naturalists but also, increasingly, their elite American counterparts.[17] These tensions in the structure of natural history practice persisted into the late nineteenth and early twentieth centuries, yet those very practices also constituted the operational social infrastructure of natural history.

Functional collaboration coexisted with some tension inside American natural history as well. This structure was partly geographical: western naturalists were usually subordinate to their eastern and European counterparts. The most accomplished expert naturalists were nearly always based on the East Coast or in Europe, especially in the early part of the railroad era and especially in rare specialties with only a few western practitioners, such as conchology.[18] Moreover, nearly all of the largest collections within the United States were on the East Coast, and that would only gradually and partially change with the turn-of-the-century founding of such regional repositories as the Rocky Mountain Herbarium in Laramie, Wyoming, and the Colorado Museum of Natural History in Denver, Colorado. Moreover, some of the largest museums and herbaria were across the Atlantic, and the more extensive collections in Europe or on the East Coast had to be consulted in order for more systematic, cosmopolitan knowledge to be produced.

This material reality could be frustrating, or at least bemusing, to naturalists in the West, who found that they had to travel eastward or even to Europe in order to systematically situate their finds scientifically. "It seems extraordinary to have to go to Edinburgh to settle a point in the botany of New Mexico," Cockerell once remarked.[19] The necessity of pilgrimage to eastern and European repositories was only enhanced by the possession of knowledge by on-site experts familiar with the material. (The tables would occasionally be turned, as when Cockerell's knowledge of a specialized branch of entomology grew so much that "people began to send [him] scale-insects from many parts of the world, South America, Africa, Japan, The Philippine Islands etc.")[20] Likewise, in reviewing the past century of plant collecting in his state, botanist J. W. Blankinship of the Montana Agricultural College in Bozeman estimated "that four out of ten collections made previous to 1862 went to Europe along with the publication of the species contained, and that all the others, for the first three quarters of the century, remained in the eastern United States." However, he took comfort in the fact that, after 1880, "a large part of the collections made have remained, in duplicate at least, in this state," mostly in Bozeman.[21]

For his part, Cockerell found compensation in the benefits of place-based knowledge making. "The workings of nature in the west are peculiar," he pointed out in a report while working at the New Mexico Agricultural Experiment Station, "and many a farmer has discovered to his cost that what is very well in the East or in Europe, will not do at all out here." Even within the West, he thought, the challenge was to study "great differences of altitude and climate," in order "to discover the exact nature of the conditions obtaining in each locality, and adjust ourselves to them."[22] Cockerell also incorporated colorful details of the experience of collecting into his accounts, portraying it as an embodied practice deeply embedded in place: noticing something novel while sitting to rest, or wetting his finger to collect a bee.[23] The editor of the decidedly western-oriented botanical journal *Muhlenbergia* was even more emphatic. "The man on the ground," he declared, "is infinitely better qualified to do the work than one many miles away who is totally unacquainted with the field and the conditions prevailing there." Focusing on "the whole region from the Rocky Mountains west," the editorial cautioned that "a great herbarium" might be "filled with specimens, and supplemented by a fine library," but would nonetheless fail to "qualify a botanist located there to be the arbiter of everything relating to plants the whole country over. The man in the field is the one to settle problems."[24] Despite the frustrations from lack of access to distant collections and cosmopolitan expertise, then, western naturalists claimed the benefit of year-round residency in the Plains and Rockies.

This privileged access to the natural environment shared by residents and long-term travelers alike conveyed a richer combination of experiential and cosmopolitan scientific knowledge than was available to those who only studied and classified the specimens moved out of the West. An early botanical traveler to the Rocky Mountains, Charles Christopher Parry, collected for several field seasons from 1861 onward, and by 1862 was already serving as an unacknowledged "guide and botanical expert" for other traveling collectors.[25] Moreover, one of the first publishing botanists to live in the Rocky Mountains, Edward Lee Greene, experienced this divergence personally in 1872, when he hiked in the mountains of Colorado with the visiting Harvard botanist Asa Gray. "It was very, very interesting," Greene recalled, "to see him pick up something which he himself had first described and named, and ask, 'Parry' or 'Green,' 'What is this' and then after a closer examination, 'Oh, it is ——— ——— isn't it? O yes, I never saw it growing before."[26] While the availability of railroad travel into the Colorado Rockies brought both long-term residents and temporary travelers into the region, the distinction in experience between easterners and westerners persisted. As late as 1920, local residents were still cited for information about specific places where a certain species might be found and how common it was.[27] Even when authority over naming and classification remained—as it often did—with outsiders, such knowledge was more strictly cosmopolitan and thus limited in its purview.

The knowledge of experience that western residents gained from living and working in the region may be brought into sharper relief by considering how those without expert knowledge in natural history nevertheless learned a great deal about their local environments. Evidence for experiential knowledge is diffuse, both in scientific publications and more personal sources. Published accounts in scientific journals referred intermittently to the knowledge derived from lay people encountered in the American West, such as farmers, ranchers, hunters, fishermen, and miners, as well as tourists and amateur naturalists.[28] Occasionally, publishing naturalists demonstrated a high regard, even admiration, for experiential knowledge. "The common names of plants," argued two Montana botanists, "aptly portray the genius of a people":

> They express disgust at unwarranted intrusions and disagreeable dispositions and flout before us the appearance of the ugly and misshapen. They advertise their evils [*sic*] habits and warn us against their unknown dangers. . . . We also find their good qualities often mentioned, as well as their various and peculiar habits of growth or seed-dissemination, while their friendship for their late friends and admirers, the noble Red Man, and often even their Indian names, yet cling to them in our popular nomenclature, as a badge of honor and a well-earned recognition of their former services. . . . It is a common habit of an immigrant people to christen the strange plants of their new homes with the names of the loved flowers of their former land and to keep up the old names of the weeds that follow in their footsteps. Yet, the genius of a people finds its most natural expression in the new names given to these new species which bloom around them and it is here that we see the comic, the poetic, the indignant hatred, the love and the wild, free happiness of childhood and the country home so often displayed.[29]

An appreciation for the use value of the knowledge of experience, such as the local names given to plants, is perhaps the most palpable feature of this statement. However, at least as often, if not more, field scientists' reports on the experiential knowledge of lay people were transmitted through a judgmental filter, even when they recognized the existence of experiential knowledge, that denigrated such knowledge as "superstition," or at least inferior.[30]

Casual references are also ubiquitous in the personal writings of nonscientists but, as a rule, are rarely organized around their experiences in the main archives of the history of science, even in cases where local amateurs and lay people took a special interest in science. One exception to the rule was the Cook ranching family of northwest Nebraska, whose records at the Agate Fossil Beds National Monument provide an unusual set of documents organized from the perspective of local lay collaborators rather than visiting scientists. The Cooks always maintained that they had discovered the fossil hills that would eventually be visited by several regional and eastern museums, though at least one eastern museum, the Carnegie Museum, fought bitterly with the Cooks to contest this assertion.[31] In a much later autobiographical account, the ranching family's son, Harold Cook, defended his father's discovery claim by referring to the experience of regularly riding a trail from Cheyenne to northwestern Nebraska as producing the knowledge. "My father and mother knew this trail well," he insisted, "for mother traveled over it twice every summer, and father had often made the ride," sometimes in the company of his mother while courting her. While his father had first seen "scrap bone lying in the grass" as early as 1878, it was on "one of those leisurely rides" several years later that the elder Cook had "discovered fossil bones eroding out of a large conical hill east of the ranch buildings" and then "stopped to pick up and examine the fossil bones."[32]

The ranch's patriarch, James H. Cook, not only insisted on his own knowledge of fossils based on experience but also the knowledge possessed by American Indians. This knowledge was both general and specific. He regarded Red Cloud, for example, as "a most interesting student of nature" and commented that the Sioux man's "knowledge of the flora and fauna of the land over which he and his people roamed was simply marvelous." Yet Cook also recounted learning from American Horse about "the bones of the strange creatures which had once lived in the land of the Sioux," which had been "turned to stone." The Sioux man showed him a "gigantic jawbone" and "explained that it had belonged to a 'thunder horse' which had lived a long time ago," which would "sometimes come down upon the earth in a thunderstorm and chase the buffalo," thus driving many of them into the camp.[33] The valuation of American Indian experiential knowledge of nature was also evident in Gilbert Wilson's sympathetic account of Hidatsa agricultural knowledge, as told to him by Buffalo Bird Woman (Maxidiwiac), which he originally published in 1917 as *Agriculture of the Hidatsa Indians: An Indian Interpretation*. He records her as saying things such as: "We knew when corn planting time came by observing the leaves of the wild gooseberry bushes. This bush is the first of the woods to leaf in the spring." He cited local knowledge of the upper Missouri River area of North Dakota in this account as compatible with modern science.[34] Thus, some observers, such as Cook and Wilson, sought to credit highly place-specific American Indian experiential knowledge in relation to cosmopolitan science.

More broadly, however, other observers emphasized the irreducibly local and holistic character of the knowledge of experience. The ethnobotanist Melvin Gilmore noted that certain specialists in northern

Great Plains tribes led "disciples on private excursions to the haunts of the plants and there impart to them the knowledge of the characteristics and habits, ecologic relations, and geographic distribution of the plants, together with their uses, methods, and time of gathering, preserving, and preparing for medicinal use, and the proper way to apply them."[35] Another Native American in the Great Plains region recalled being brought up to value learning about nature: "to know the habits of wild creatures, to know about trees, wild plants and fruits, to be able to judge of weather, to foretell what the seasons would be, whether a cold winter or a dry summer—there were signs one might learn to read all these things by—all these things constituted a well-rounded education for the average young Indian man."[36] Thus, while some experiential knowledge might be validated by cosmopolitan scientists, it also included a great deal that might not be, which incorporated the intimate local details of place.

Another window into the world of experience of place, but for white settlers rather than American Indians, can be found in the autobiographical writings of T. D. A. Cockerell, the English naturalist introduced above who ended up settling in Boulder, Colorado. In narrating the childhood origins of his emerging interest in studying nature during his time spent at his grandfather's farm in Sussex, England, Cockerell recalled that he "used to wander about the fields and woods admiring the birds and butterflies and the wildflowers, watching the men at work, and helping in such small ways as were permitted." Such a recollection may seem utterly conventional, but he then articulated something equally common, but less often explicitly stated, which nicely expresses the value and distinctiveness of experiential knowledge in the field. "Although the country folk know nothing of science as we understand it," he pointed out, "they did know a lot about the events of the country side. They know when the birds came in the spring, where they nested, where the spreckled trout might be found in the brook, where the cowslips and lilies-of-the-valley grew."[37] They knew where things were, when things happened.

For the Great Plains and Rocky Mountains, of course, the particular plants and animals were different, but Cockerell and others found the same thing to be true of how people knew nature in Colorado. Naturalist Charles E. Aiken, who had arrived in Colorado a decade and a half earlier, near the beginning of the railroad era, described his experiences in mid-October as he was "riding past Beaver Creek, a larger and beautiful mountain stream" in the south-central part of the state.

According to Aiken, his "attention was attracted by a great twittering among the feathered tribe," which, implicitly drawing on his year-round knowledge based on experience living in the area, "seemed to be an unusually large congregation of species for this season of the year." Thus cognizant of a notable anomaly to incorporate into his experiential knowledge of place, Aiken "dismounted from [his] pony, and leaning upon the cottonwood rail-fence . . . watched the birds for nearly an hour, noting the different varieties, and observing the actions of each."[38] Similarly, a British naturalist who had become a rancher in Montana in the 1890s related his ten years' worth of experience in the region before presenting an unusual bird migration event, which he credited "Mr. H. Tusler, whose ranch adjoins mine on the south" for being "the first to observe," and for bringing three of the birds to him.[39] In both of these cases, knowledge of unusual events in nature was deeply rooted in many years of local experience.

The knowledge of experience extended beyond living organisms to other parts of nature. At the state's agricultural college, L. G. Carpenter mentioned his "mountain friends, who with long acquaintance with mountain conditions had had unusual chances for observation." From them, he sought knowledge of when particular snow cover conditions occurred, so that he could visit to study the relationship between forests and snow cover. He then received on-site guidance from "Mr. John Zimmerman, who is a native of Switzerland, and who has lived in the high elevations of Colorado for nearly twenty years" and "Mr. John McNabb, who has also had a long and intimate acquaintance with high elevations."[40] For another project involving stream measurement in the Arkansas Valley, Carpenter valued the knowledge of those managing irrigation systems at the ground level, asking a man in Pueblo, Colorado, for referrals to people who could help navigate the area. "In order to do the work expeditiously," he wrote, "we need local help from those who are acquainted with the location of the headgates and have a knowledge of the general character of the stream."[41] Whether for living or nonliving nature, Colorado residents possessed experiential knowledge based on year-round close acquaintance with the local environment—knowledge that was valued by field scientists.

When T. D. A. Cockerell arrived in West Cliff, Colorado, in 1887, he befriended local people and joined them in gaining rich and varied experiential knowledge through collecting and observing the plants and animals around them. On the way home one evening, he "looked in at the Cusacks for a moment, and Mrs. C. gave [him] some insects she had caught." Two other locals named "Regy and Miss Urquhart" he hoped would "come down again" so that Regy could show him some interesting birds' eggs, and Cockerell himself caught birds and observed flowering violets that "look very lovely under the trees by the creek." Cockerell expressed the holistic, integrative quality of local experience vividly in a letter back to England: "The quaking asps are well forward, the young leaves being apple-green. The weather is warm, with thunder and showers—good for the plants."[42] He also began working hard to collect for distant naturalists interested in receiving specimens of the "insects, plants, shells, fossils, and some birds" in return for knowledge about them. "The experts," writes William Weber, "jumped at the chance of having an alert observer and correspondent in this remote area," which enabled Cockerell to form "long-lasting ties with his future colleagues."[43] Yet he also pleaded with his naturalist colleagues to permit the naming of "slight varieties" he had discovered near West Cliff with internationally recognized names to prevent their passing unnoticed in cosmopolitan science.[44]

Cockerell assisted his new neighbors in connecting their experiential knowledge of that place to the wider cosmopolitan systems of knowledge that he brought with him from England. He helped a fellow immigrant, "Mrs. M. E. Cusack, an elderly English lady who had always longed to study botany," to build her own herbarium based on "Coulter's *Manual of Rocky Mountain Botany*," which allowed both of them to "acquire a fair knowledge of the local flora" by reframing her lengthy (and his shorter) experiences with local plants in the light of cosmopolitan taxonomy.[45] A local hotel keeper even paid Cockerell "65 cents and a little chunk of silver" to name the plants in his wife's "seven little books of pressed flowers," based on Coulter's manual. Such earnings may have provided Cockerell with a convenient rebuke to two local men named Vetter and Samse—"miserable creatures," he called them, "with no ambition in life except to make sufficient money to take things easy"—who were perplexed that Cockerell could "work at Science, when (as they suppose) 'there is no money in it.'"[46]

Such local detractors notwithstanding, in bridging the epistemic rift between the local knowledge of experience and the cosmopolitan knowledge of science, Cockerell was carving out a niche for field work as a distinct and valuable category of practices. More pointedly, a decade after his arrival in the American West, Cockerell explicitly distinguished between "collectors and specialists," the former offering "field data" revealing the "biological" knowledge of place—thereby integrating the experiential with the scientific—and the latter focusing more exclusively on the "systematic" knowledge of cosmopolitan science. The problem, he complained, was that too many publishing specialists were neglecting to credit or express the knowledge of field collectors.[47] While Cockerell's conception of the "biological," which he envisioned as "field data" that could be incorporated into scientific publications, was undoubtedly selective and somewhat reductively extracted from the full experiential knowledge of place that he and others living in Colorado would have possessed, it nevertheless drew upon it. Knowledge derived from local experience of nature has been potentially important everywhere, yet has everywhere been distinctive and particular. In the changing geography of natural history, the role of such experiential knowledge has remained relevant to new modes of practice.

Industrial-Age Technological Systems in the Field

Expanding cosmopolitan science encountered the knowledge of experience long before the railroad era. But with the advent of an intertwined industrial-age system of environment and technology, specific modes of practice emerged, linking places in the field to the larger world in new ways. The emerging technologies of the field in the mid-to-late nineteenth century—most significantly, the railroad itself—were foundational. As Cockerell pointed out, the railroad made it "much easier to travel about than formerly" for naturalists, especially for places "which were a few years ago reached only with great trouble and expense." Other field scientists, as we shall see, questioned Cockerell's claim that the proliferation of railroad lines made it unnecessary "to wander far from established routes," and some of them would likely have demurred to his contention that "the progress of road-making is much faster than the progress of students of the fauna who travel on the roads."[48] More broadly, the railroad, the telegraph, and other field technologies were part of a much larger process of industrialization during the nineteenth century, involving a historically unprecedented dramatic increase in production output, a more highly structured division of labor, and an energy revolution, focusing on the shift from an advanced organic economy to fossil fuels.[49]

By just about any conceivable measure, during this industrial age the output of scientific knowledge production was increasing quite steeply between 1860 and 1920, not only in the laboratory sciences but also in the field. It may be impossible to pinpoint a single explanation for this dramatic surge in scientific activity, since it was due to a confluence of institutional, political, economic, cultural, and social factors in industrial society. In the field, however, especially in the interior American West, the railroad was a key catalyst. This was recognized at the time and eagerly anticipated. In Ferdinand V. Hayden's view, "a very large portion of the West" was still remaining to be "explored by the geologist," and he dismissed all previous work before the railroad as "of necessity very superficial." Soon, however, he thought, "the facilities which will soon be afforded for travel through those wild regions, on the completion of the Pacific Railroads, must give a very great impulse to explorations," and he "hoped that not many years will elapse before a sufficient number of facts may be gathered together from every portion of the west, to enable the geologist to work out the general plan of its geological structure with some degree of completeness."[50] Hayden's vision of science in the railroad era, echoed by many of his contemporaries, would become reality, as field work, not only in geology but in other field sciences as well, proliferated on the Great Plains and Rocky Mountains after the late 1860s.

However, we should also consider industrialization not just as a quantitative expansion in scientific work but also as a qualitative change in the division of labor. In other words, just as textile and iron manufacturers replaced an artisanal mode of production with a hierarchical separation between managers and workers, so too did scientific field work become an increasingly hierarchical enterprise between 1860 and 1920, as large institutions such as museums, government agencies, and their university-based collaborators came to dominate the production of environmental knowledge. In many ways the domination of field work by such large institutions was even more marked in the field sciences than in the laboratory, where some university-based researchers (for a while, at least) maintained relatively small-scale, on-campus operations. By contrast, the field sciences in the American West by the turn of the twentieth century had become dominated by a handful of large institutions, such as the US Geological Survey, the US Department of Agriculture (with its many departments), the American Museum of Natural History in New York, and the Carnegie Museum of Natural History in Pittsburgh.

Many of these institutions vertically integrated during the railroad era, bringing the entire process, from field collecting—previously dominated by independent collectors—to subsequent processing, analysis, and scientific publication, under their own authorities. The American Museum, for example, initiated "a policy of exploration rather than of purchase for the acquisitions of collections" starting in 1881, which by 1912 had led them to "every continent on the globe, except Australia" and funding thirty-five field parties in total that year.[51] State institutions within the region, such as land-grant college–based agricultural experiment stations and geological surveys, were much smaller, but they also integrated and coordinated the activities of multiple scientists. Individual field researchers were still at work too, especially at colleges or universities, but even they were increasingly associated with these large national institutions, or at least their state-level counterparts.

A sharpened division of labor in scientific field work was becoming more evident, not just between large institutions and their teams of subordinates fanning out over the western American landscape but also within field parties. It is true that labor had long been divided—particularly on long-distance expeditions with especially large organizational structures and (often) military leadership—but the routine deployment of specialty scientific parties in the field entailed a rearticulation of that internal division in labor, in which lead investigators were accompanied by scientific assistants as well as subordinate "technicians of field": cooks, packers, guides, and other laborers. The production of scientific knowledge in the American West, as elsewhere, would depend on a permeable but definite boundary between those in charge of field parties, who were entitled to reap the scientific rewards of credit through publication, and those who labored with and for them. Much like the middle managers of business corporations, many field scientists—geologists working for the US Geological Survey, agricultural scientists working for the US Department of Agriculture or state experiment stations—were entitled to receive some of the epistemic profits of the larger institutions that supervised them. Recognizing the increasing importance of multiple divisions of labor, then, is crucial to understanding the industrialization of science in the field between the 1860s and 1910s.

The proliferation of field research and accompanying changes in the division of labor were undergirded by a third aspect of the industrialization of field work: the energy revolution that accompanied changes in transport and communication infrastructures. The shift to fossil fuel sources was significant just about everywhere, but it was particularly important in the interior American West, where limited organic fuel resources and the insufficiency of waterways for transportation through the elevated, mountainous terrain meant that the region would be dramatically transfigured with the exploitation of coal and the building of railroads.[52] It was primarily along those railroads that personnel and supplies moved in and out of the field, and they formed an increasingly fossil-fuel-based foundation for the practice of field science.[53]

The transport infrastructure was also largely the communication infrastructure, since correspondence between field parties and their metropolitan supervisors, or with lay collaborators, traveled almost entirely through the US Postal Service, which utilized the railroad network alongside connecting animal-powered mail stages.[54] Communication by mail was supplemented by occasional, more urgent or time-sensitive (and expensive) messages sent through the telegraph network, which often paralleled the railroads. To be sure, not all railroads, especially early ones, were powered by fossil fuels, but in the West it was ultimately a coal-based railroad infrastructure that carried the load. In sum, the industrial-age infrastructures for communication and transportation, as well as fossil-fuel-based energy systems, were decisive in how the field sciences were transformed between the 1860s and 1910s.

The technology of the industrial age was especially significant in remaking the practice of science in the field, across all disciplines, through its influence on transportation. The history of expeditions on land and sea offers an important prehistory to the transformation of field practice in the railroad age, a shift whose study reveals both continuities and important changes.[55] As Richard Sorrenson has emphasized for the eighteenth century, when ships were so fundamental to scientific practice that they even served as key instrumental technologies themselves, scientific travel has long been closely shaped by the means of transport that underlie it.[56] And that was certainly true of natural history practice and other scientific practices in the field during the nineteenth century. Expeditions were an important mode of practice in the investigation of nature; they enabled scientific travel rather than just investigating local nature. Although long-distance expeditions by horse and wagon—or by boat in favored partial access corridors such as the Missouri River—played a key early role in the practice of natural history in the American West, especially those of the US Army, the greatest catalyst to scientific travel was the building of the transcontinental railroad, along with its accompanying postal service and (often) nearby telegraph lines. Just as transportation and communication lines facilitated capitalist expansion, so too did they open up the region to bringing observations, specimens, and research results to the cosmopolitan scientific community.

For a large land area such as the interior West, with mountains at its center—and thus no rivers for through-travel—railroads became the key conduits of travel, but they were really just part of a larger age of steam that also included steamboats in large river valley regions, such as the Ohio and Mississippi, and oceangoing steamships that linked empires and commerce around the world. "In the eighteenth century," observes Paul Farber, "naturalists might be out in the field for years at a time, with their collections ultimately destroyed by pests or shipwreck. In contrast, nineteenth-century railroads, steamships, canal systems, road building, and harbor construction all meant that more people traveled, goods were cheaper to ship, and shipments followed dependable timetables."[57] Before the railroad, by contrast, naturalist travelers and their scientific cargoes depended on the same limited opportunities as everyone else. Some observers explicitly likened it to oceanic travel in the pre-steam age of sail. "Prairie travel resembles that by sea," remarked William Ludlow, "which indeed the landscape not unfrequently suggests. The compass is the guide, the direct course is not always the best, and the probabilities of finding wood, water, and grass, and a good road, compare with those of obtaining favorable and moderate winds and a smooth sea."[58] Despite the varied purposes of Euro-American exploration of the West before the 1860s—including mountain men, military reconnaissance, and more systematic mapping by Army topographical engineers—all endured very slow transportation and poor communication with the East. The expense and difficulty of reaching the West in the first half of the nineteenth century virtually guaranteed that few travelers would have access to the region.

The easiest westward passage before the railroad, though limited in its reach and terminating below the Continental Divide along the crest of the Rocky Mountains, was the Missouri River, the same route favored by Lewis and Clark. Fur trading companies operated steamboats up through what would become Nebraska Territory—organized in 1854, with key areas later becoming part of Dakota Territory and Montana Territory—thereby making the region more accessible to scientific travel as well. While this Missouri River corridor only skirted the northeastern edge of the US Great Plains and Rocky Mountains, the cost and time advantages of steamboat travel were significant. Estimates provided by historian William Lass suggest that steamboats on the Missouri could travel between about 3.5 and 7.5 miles per hour in the 1840s, and over 8 miles per hour by the 1850s, whereas wagons pulled by oxen or mules averaged only 2 or 2.5 miles per hour, respectively. And while steamboats could transport two hundred tons for $13,000, a standard wagon freight train moved only seventy-five tons for "as much as $35,000."[59] Such transport efficiencies were a double-edged sword for scientific travelers, since steamboats improved access but made it more difficult to stop for collecting and observing.

In the 1850s, scientific collecting expanded along the Missouri River corridor, including plant, bird, and mammal collecting—as well as ethnographic observations of the Sioux and other Upper Missouri groups—in 1850 by Thaddeus Culbertson, who had connections with the American Fur Company.[60] Ferdinand V. Hayden followed the same fur trading river route.[61] Because of Hayden's later fame as a "great survey" leader, his expeditions of the 1850s up the Missouri River have been amply chronicled, and his work may thus serve as an illustrative example of field practice on the cusp of the railroad era. Hayden's first expedition up Missouri River was in 1853, with collaborator F. B. Meek, followed by later voyages with American Fur Company and US Army support.[62] Hayden's letters from his first three years in the field, mainly traveling the Missouri corridor, reveal an environmental vision dominated by the river itself and surrounding bluffs—in fact, observations of bluffs and geological cross-sections exposed by the river constitute a major portion of his written comments. While approaching the Fort Union in July 1854, Hayden lamented the steamboat's "unusually quick trip, not giving me as good a chance to collect as I could wish." Not having to stop for fuelwood harvesting, adverse wind, or encounters with American Indians, Hayden noted, they were never more than six hours in one place, thus leaving him with "rather small" collections. "As soon as I can get free from this boat," he declared, "I propose to spend 3 or 4 weeks ranging up and down the river," visiting army outposts along the way.[63] He would stay in the river corridor—it was his lifeline for supplies and security, after all—but hoped to free himself from the steamboat's constraints.

Despite the vital hospitality and security upon which he depended at army posts along the Missouri River, by the late 1850s Hayden did not have such harmonious relations with US Army officers leading expeditions. His most significant conflict was with army officer G. K. Warren, who led the northern Plains expeditions that Hayden accompanied in 1856 and 1857.[64] The compatibility of scientific goals and military goals on a single expedition and the apportioning of credit for discoveries seem to have been the primary points of contention. Hayden's frustration with military leadership was also shared by others, such as fossil collector George Bird Grinnell, who traveled with George Armstrong Custer's famed 1874 expedition to the Black Hills and complained of "not securing a larger collection of fossils." The expedition's captain of engineers, William Ludlow, attributed this to "the facts that our marches were necessarily rapid, the time being limited to sixty days, and that the country traversed was nearly barren of the objects of his search."[65] This litany of problems offers a glimpse of the potential tensions that might exist between civilian scientists and their military collaborators.

At the same time, others depicted military objectives and civilian science as compatible, or even symbiotic. George M. Wheeler, who was a military leader of explorations that included civilian scientists, was predictably one of the most enthusiastic. In a synopsis of the work of the nine field parties under his command in 1874, Wheeler described the co-operation of military officers with "professional gentlemen drawn from civil life" who were successfully investigating a wide range of "cognate branches of scientific research, geology, paleontology, mineralogy, natural history, etc." In Wheeler's view, this "partly military, partly civilian" effort was a model of collaboration, and he portrayed them as "working always in harmony."[66] Moreover, field scientists found ways to adapt to their circumstances that sidestepped the tensions inherent in having differing objectives. Hayden, for his part, would often free himself, in the company of Meek or other companions, to venture away from the river corridors in order to examine intriguing geological features such as the Dakota Badlands and the Laramie Mountains, on various expeditions through the late 1850s and into the 1860s. But he always had to return eventually to the river-based system of army posts on which he depended.[67] While the privileged riverine access to the West provided by numerous US Army expeditions made cooperation of civilian scientists with the military very attractive at first, it thus also generated many tensions in the structure of expedition management.[68]

However strong these tensions were, Hayden managed to present and publish his results at the Academy of Natural Sciences in Philadelphia, including a map of geological systems in the northern Plains. Hayden's mapping in the 1850s was dominated by a geography of rivers, which remained the most prominent features for travel and orientation in the prerailroad age. Rivers also shaped Hayden's environmental perception, much as railroads would just a decade later. "The river bottoms, which are quite extensive," Hayden wrote in one report based on his 1857 travels with Warren along the Missouri River, "possess an inexhaustible fertility and sustain a luxuriant vegetation, while the broad upland prairies, clothed with grass and flowers of great variety and beauty, meet the eye of the traveller on every side."[69] He also published results from his Missouri River journeys in various national scientific periodicals, such as *American Naturalist* and the *American Journal of Science*.[70] Joseph Leidy published a study based on an analysis of specimens collected by Hayden on the Warren expedition of 1857.[71] Thus, the opening of the Missouri River corridor to fur trading traffic did open up at least the northern part of the region to a more substantial outward flow of knowledge.

But the enduring catalyst for opening the Plains and Rockies was the transcontinental railroad, carved out in a central corridor through the heart of the region. The Union Pacific linked with the Central Pacific at Promontory Point in northern Utah in May 1869, completing an unprecedented access corridor through the West. Moreover, the impact of the railroad extended beyond the initial route in subsequent decades to include several other (predominantly east-west) routes through the region, including the Kansas Pacific, the Atchison, Topeka, and Santa Fe, the Burlington and Missouri River, the Northern Pacific, the Great Northern, and others. Just as the railroad played a crucial role in the extension of capitalist frontiers, so too did it enable the expansion of scientific field work.[72] The transportation revolution in the West was preceded by a quantum leap in communication with the building of the transcontinental telegraph. The Pacific Telegraph Act of 1860 granted an annual stipend for the building of a transcontinental telegraph line, completed in October 1861.[73] This telegraph line was designed to run from Missouri's western boundary to San Francisco, and the transcontinental railroad would later follow much of its route.[74]

The extension of industrial-age technologies, especially the railroad and telegraph, into interior western environments during the nineteenth century produced a new hybrid system of environment and technology in the practice of science. At the most basic level, this not only enabled a much greater proliferation of recorded field work in nature but also served as a pathway for environmental perception and reflected the increasing differentiation of field work into ever more specialized domains of expertise.[75] Upon his first arrival in the American West in July of 1887, Cockerell first encountered the yellow flowers of the buffalobur, butterflies, beetles, sunflowers, and prairie dogs, as well as the "Rocky Mountains rising high in a long chain in the distance—in places with snow still on them" from the windows of the train or around the railway depots where they stopped on the way across the plains of Nebraska and eastern Colorado.[76] Over time, specialists increasingly encountered and recognized each other on trains and waiting at depots. After meeting on the streets of Albuquerque and discovering their mutual interest in collecting snails, for example, Cockerell and his colleague Rev. E. H. Ashburner, whom Cockerell judged to be "the only other student of snails at that time resident in New Mexico," next saw one another "on a train," where they exchanged information from the field and worked together on common problems.[77] The field technologies of the industrial age were not just practically expedient; they also channeled social interaction among field scientists, much of it cooperative.

At the same time, however, the newfound accessibility of the field by railroad could also generate frantic competition among rival naturalists, who could feverishly telegraph their scientific results to gain priority over one another. Among the emerging specialized disciplines where fierce competition was the rule, vertebrate paleontology was probably the most eminent (and notorious) to develop rapidly in the interior American West, as rival paleontologists O. C. Marsh of New Haven and E. D. Cope of Philadelphia deployed teams of rival fossil collectors across the region. At the beginning of this shift, during the early 1870s, Marsh and Cope themselves traveled the rails in a frantic struggle for the fossil riches of the West. Given their distance from scientific centers of the East—and their crucial publishing outlets, which provided the basis for allocating priority in naming new species—the rivals often dispatched articles directly from the field for rapid publication. In some cases, the reports of new fossil species were even transmitted directly by telegraph, as when Cope sent an August 17, 1872, telegram from Black Buttes, Wyoming, to the secretary of the American Philosophical Society in Philadelphia. Yet the use of telegraphy as a technology for securing priority, with its great speed advantage, also brought potential deficiencies in clarity. As the recipient noted, "the telegram was so badly transmitted by the operators as to be read with difficulty, and the precise forms of the specific names could not be certified until the return of Prof. Cope from the field."[78] Nevertheless, it is significant that telegraphy permitted a kind of nearly instantaneous publication of field research despite the long distance.

Over time, the field practices of Marsh and Cope became further adapted to the changing transport and settlement geography. Marsh, for instance, may have begun his western field work by relying on large parties and army escorts departing from the single transcontinental railroad line existing in the early 1870s during a series of Yale scientific expeditions led by himself.[79] But within a few years, he too shifted to a much more geographically dispersed network of "local collectors and small parties," only occasionally visited by Marsh himself. The "constant stream of vertebrate riches" that "flowed into the Yale University Museum from the Rocky Mountain region" made an impression on late nineteenth-century observers of the scientific scene. "The rapid settlement and development of the West," noted Marsh's obituary in the *American Journal of Science*, "rendered it no longer necessary to fit our expensive expeditions, especially as many of the localities were easily accessible by railroad."[80] Cope, for his part, despite traveling for longer periods in the field himself than Marsh did, also cultivated a dense network of collectors who relied on the expanding railroad, telegraph, and settlement infrastructure of the West. As these networks of collectors for Marsh and Cope became more geographically extensive and hierarchically structured between eastern experts and local collaborators, they increasingly took on the characteristics of lay networks, and for this reason will be revisited in the next chapter.

Yet while natural history collecting was transformed by the railroad and other industrial-age technologies, there were some who raised concerns about the negative effects of traveling by rail on collecting practice, especially how it limited the routes that could be taken and how it encouraged haste. The French-born botanist Julien Reverchon, who settled in Texas in the late nineteenth century, expressed these reservations pointedly in the *Botanical Gazette*, based on his experiences in botanizing across the state, from April until the end of July in 1885. "By 'botanizing,'" he declared, "I do not mean taking a railroad and stopping at such and such a station, taking a ramble or two in the neighboring hills, or sometimes jumping from the cars at a coal station, tempted by some tantalizing plant, and running back with only the top of said plant, at the call of the imperious whistle, and after that running maybe a hundred miles before stopping again." Reverchon disavowed such a mode of practice—"not my way," he sniffed—on the grounds that "railroads do not pass exactly where many nice things are found, and I don't care to be in a hurry." Instead, Reverchon and his small field party, which also included his wife and an assistant named Robert Freeman, who was "a fine fellow, exactly fit for driving, hunting, fishing, and other duties invaluable on such a trip," traveled by "covered wagon, drawn by a good team, [which] was packed with provisions, drying papers, arms, etc." In contrast to the fixed path followed by the railroad, Reverchon emphasized that his wagon and team could "travel any length of time, and over any extent of country."[81] Such flexibility for travel would not be replicated by fossil-fuel-era transport until the arrival of the automobile as an affordable field technology in the early twentieth century.

Since Reverchon was a resident of Texas, he could depart by wagon and team directly from his Dallas County home, but most field naturalists in the late nineteenth century, except when collecting locally, followed the railroad as far as possible, even when most or all of the work was accomplished off the rails. Indeed, many naturalists adopted the opposite approach from Reverchon, using the railroad itself as a time-saving and efficient pathway through nature. The entomologist Morgan Hebard, for example, collected specimens in August 1904 across a wide swath of the West, "extending from eastern Montana to Salt Lake City, and from that locality to east-central Colorado." Unlike Reverchon, Hebard had only a single month, and without the railroad it would have been impossible to visit so many sites in so little time. "Collecting at some localities was limited to a few minutes search permitted by the stoppage of the train," while "at other localities several days were spent," Hebard and his senior coauthor James Rehn wrote. Notably, Hebard stopped for longer periods at Yellowstone Park and Pikes Peak, two of the West's most popular tourist sites, which underscores the close overlap between natural history and tourist travel during this period.[82] More typically, scientific travelers simply used the railroad depot as the key embarkation point into the field, spanning great distances to put distant field sites close at hand, as Baylor University museum curator John H. Strecker did when collecting animals in northwestern Texas with a group of local teachers in 1910. They had departed "from Goodnight, on the Fort Worth and Denver Railroad, on a trip through some of the outlying arroyos and draws forming a portion of the great Paloduro system of plains canyons," which constituted their extended field site.[83]

As these examples of naturalists with diverse specialties indicate, even as the practice of science was transformed by new envirotechnical regimes, not all field work used the railroad infrastructure in exactly the same way. But for all of them, even in such rare examples as the botanist Reverchon who invoked the railroad as a foil, the spatial organization of scientific practice was profoundly reshaped (for good or ill) by the new transportation geography of the industrial age. In the end, it is difficult to see the railroad era as anything other than a massively expansionist era for natural history, and that was in significant part due to the new possibilities opened up by the railroad and other, closely associated, technologies, embedded as they were within the natural environment of the Plains and Rockies, which had presented so many challenges to access before the railroad era.

An Environmental History of the Field

This synopsis of technological change in the mid-to-late nineteenth-century American West has already made the role of the natural environment quite evident, if indirectly. But by focusing more closely on the environment itself, we can more fully assess its constitutive role in the making of the field. After reviewing some of the key natural features of the Rocky Mountains and Great Plains, along with their many complex internal variations—as they became evident to scientific travelers and other observers during the railroad era—we can then consider how the practice of science was shaped by the regional particularities of the Plains and Rockies. This influence was felt not only in the uneven distribution of resources for scientific field work to study but also in the material basis of practices in everyday experience, including subsistence and bodily health. It was this combination of environmental content for scientific knowledge production and the knowledge of everyday experiences on the ground that provided the basis for the distinctive synergies between cosmopolitan and experiential knowledge in the field sciences.

Considering the history of science from the perspective of the natural environment itself has, on the whole, been less thoroughly attempted than focusing on instruments, techniques, and infrastructures that form the technological basis of scientific practice. Declaring that the sciences—even the field sciences—have an environmental history that might reframe the whole of the history of scientific practice, and perhaps the entire history of science, distinct from the more widespread acceptance of the history of the environmental sciences as a topical niche, might even seem implausible in light of the postconstructivist sensibilities that, for good reason, now dominate the history of science. Yet the task is worth contemplating, for it promises to afford historians a means to reincorporate the influence of the natural environment in our analyses, even if not in the direct sense of providing a sufficient historical explanation of how scientists have come to know what they know about nature.

The interior American West is an environmental region of arid or semiarid plains, but with isolated islands of greater rainfall and elevation on the slopes and peaks of great mountain ranges—especially the Rocky Mountains, whose imposing presence impressed nineteenth-century gold and knowledge seekers as much as it does the early twenty-first-century observer. In 1883, Ferdinand V. Hayden, along with the Canadian national geological survey's director A. R. C. Selwyn, described the entire western half of North America as "one great plateau, crested by the numerous ranges which collectively make up the Cordilleras of North America."[84] However, as Hayden had already pointed out in a richly illustrated address before a "large audience" at the American Geographical Society of New York in April 1874, including views by "photographer's camera and the stereopticon" of Yellowstone and the Colorado Rockies, early suppositions about a single mountain chain were incorrect. "Our present knowledge," he declared, "reveals to us the fact that the name 'Rocky mountains' is only a general, rather indefinite term, including an almost limitless series of ranges or folds like the waves of the sea."[85]

Hayden's growing appreciation for the complex mountain geography of the West, and the insufficiency of earlier perceptions remind us that environmental regions are historically constructed divisions of nature. This process of delineating regions for scientific inquiry involved both field scientists who spent ample time in the interior West and their counterparts based on the US East Coast and in Europe, as a few early examples illustrate. Hayden, for his part, at the very beginning of the railroad era had defined a "Rocky Mountain region" that included a "combination of mountain and prairie," and which tracks closely with what would come to be known as the Rocky Mountains and Great Plains, though delineating the region more expansively on its eastern flank: "Nearly all the vast area west of the Mississippi may be divided into mountain and prairie," he observed, "for very soon after passing westward from Leavenworth"—in eastern Kansas along the west side of the Missouri River—"there is very little timber to be seen except that which skirts the streams."[86]

Hayden's colleagues Asa Gray at Harvard and Joseph Hooker in England adopted his "Rocky Mountain region" in their own work on plant geography several years later, referring to it even more broadly as "the wide central tract which lies between the Atlantic United States and those which border on the Pacific." Yet they clarified that their "Atlantic" region extended as far west as the states on both sides of the Mississippi River, ending only at "the great woodless plains," with the region in question encompassing "the gradually elevated plateau which flanks the eastern base of the Rocky Mountains" and "equally elevated district or plateau" to the west. The Rockies themselves, they thought, extended as far west as the Wasatch Mountains of Utah.[87] With the exception of the arid region west of the Wasatch, such definitions coincide with what

FIGURE 1.2. The Rocky Mountains and Great Plains as physiographic regions situated within the western United States as a whole. Detail from hand-drawn map insert in Nevin F. Fenneman, "Physiographic Boundaries within the United States," *Annals of the Association of American Geographers* 4 (1914): 84–134.

would later become widely known as the Great Plains and Rocky Mountains (see figure 1.2).[88]

The Rocky Mountains of Montana, Idaho, Wyoming, Colorado, Utah, and New Mexico have harsh, cold winters, so human habitation has been primarily seasonal, with dominant activities shifting over time from hunting to mining, then (in many places) to tourism and recreation.[89] Valleys near mountain ranges that provide ample snowmelt, such as the mountain parks of Colorado, also became, and have remained, popular areas for ranching.[90] The most fertile area along the eastern front of the Rockies, especially in Colorado where geographers call it the Piedmont, has long been home to more extensive occupation and is now a dense urban and suburban corridor that economically dominates much of the region.[91]

East of the Rockies are the grasslands of the Great Plains, which stretch from Canada south to Texas and arose about five to seven million years ago.[92] The Plains were formed in relation to the Rocky Mountains, as researchers already believed by the early twentieth century. As Hayden and Selwyn put it, the Plains on "the eastern slope of the great Cordilleran plateau have no well-defined eastern limits, but shade imperceptibly into the prairies of the Mississippi Valley."[93] This description was amplified in a soil survey report in 1911, which described the Great Plains as "a vast plateau sloping away from the Rocky Mountains. Its surface formations, except where they have been removed by erosion, are made up of sediments brought down from the elevated region to the west during and since late Tertiary time and deposited by the streams in thick strata of fan-like aprons, so that it may be regarded as the outwash débris removed from the Rocky Mountains."[94]

The close relationship between the Rockies and Plains has not been viewed solely in geological terms, however. Nebraska botanist Charles Bessey thought of the Great Plains portion of his state, especially its trees, as "really an extension of the eastern foot-hills of the Rocky Mountains."[95] The predominant native vegetation was drought-resistant short grass, with bison playing a crucial maintenance role.[96] The semiarid climate of the Great Plains, along with its grassland vegetation, results from continental air circulation patterns dominated by dry westerlies from the Rocky Mountains, which capture most of the remaining moisture, rather than the snowy Arctic or moist tropical Atlantic air masses that more strongly affect the rest of North America east of the Rockies.[97] A useful analogy—similar to one deployed by many nineteenth-century travelers—is to liken the interior West to a vast sea of relatively semiarid, grassy plains, punctuated dramatically by isles of mountain country.[98]

Local details have been important, too. The Great Plains region is crosscut by major rivers that drain the snowmelt of the Rockies eastward to the Mississippi—most notably the Milk, Missouri, Yellowstone, Powder, Platte, Arkansas, Red, and Canadian, but also including other smaller tributaries such as the White, Niobrara, Loup, Republican, Smoky Hill, and Cimarron. These conduits across the dry plains were important not just for farmers and ranchers but also for scientific field workers, and not just for the subsistence of people and pack animals. Outcroppings of sedimentary strata along these rivers, for instance, constituted a veritable treasure trove for paleontologists and other fossil collectors. Any overall picture of the topographical diversity of the Plains and Rockies must include such smaller-scale variations, which had as much to do with shaping the geography of scientific work and settlement as did larger-scale factors such as semiaridity.

Due to its high altitude relative to the Mississippi and Ohio Valleys—with a gradual slope downward from west to east—the central part of the region has often been called the High Plains. At the same time, the southern portion in Texas and New Mexico has become known as the Llano Estacado, or Staked Plains. Lacking abundant rainfall, and more distant from mountain snowmelt, agriculture has depended on securing enough water from the handful of rivers flowing west to east or pumping underground water from the Ogallala Aquifer. American Indian occupation of this resource-poor region was traditionally seasonal or—after the introduction of the horse—nomadic. In areas such as northeastern Colorado, early hunter-gatherer subsistence depended on the utilization of small patches or islands of high-density vegetation.[99] Rivers, such as the Niobrara in western Nebraska, also provided vital "resource corridors" for subsistence in the seasonally challenging plains environment.[100]

The vast area north of the High Plains consists primarily of the Missouri Plateau, spanning much of central and eastern Montana, northeastern Wyoming, and the western halves of both Dakotas. The Missouri Plateau is much like the High Plains but exhibits more surface topographical variation, deeply cut by the Missouri River and its tributaries and with more small bluffs and localized mountainous areas. The far northern and eastern edges of the Missouri Plateau in Montana and the Dakotas have been glaciated and thus form a modified surface with moraines and glacial deposits, as well as a convenient boundary to the region. Finally, though not considered part of the Great Plains proper, much of central and western Wyoming between mountain ranges consists of even more arid plains, where large-scale, extremely extensive uses such as hunting and ranching have prevailed, except where mineral resources have been found.

Within the region are certain smaller, distinctive areas. The Black Hills, which are mostly in the southwestern corner of South Dakota, are a great island of outlying Rocky Mountain landscape, with a familiar economic pattern of hunting, forestry, mining, and tourism, although American Indian access claims there have remained more visible and openly contested than in many other areas of the Plains and Rockies.[101] The area was likewise especially prominent in scientific field research. In 1895, C. Hart Merriam proclaimed the Black Hills to be "of more than ordinary interest to the naturalist," due to its biogeographical distinctiveness. "Perhaps no area of equal size in the United States," he estimated, "has been more closely studied by geologists and paleontologists, and it has been visited more than once by zoölogists and botanists."[102]

Environmental variability has also been related to the presence of river corridors and equally distinctive landforms. North and east of the Black Hills, for example, the wide Missouri River bends its way through the northern Great Plains and then along the region's eastern edge, providing a riparian landscape that is much different from the surrounding Plains.[103] And concentrated in north-central Nebraska are the sublime Sandhills, the largest area of sand dunes in the western hemisphere, stabilized by grass and providing ample range for bison—now mostly replaced by cattle—as well as significant numbers of birds in a multitude of tiny, shallow lakes (see figure 1.3). Over the long term, the abundant underground water and mixture of warm- and cool-season prairie grass species produced less severe climatic fluctuations than in other parts of the Plains.[104] Nevertheless, human population density in the Sandhills has remained among the lowest anywhere in the United States, with the aridity and scarcity of trees eliciting unusual cultural responses, such as building houses out of baled hay.[105] Other smaller subregions along the eastern edge of the Rockies include the Pecos Valley and the Raton Uplift.[106]

An environmental history of the scientific field work in the Plains and Rockies suggests multiple productive avenues for reframing our view of scientific development and change. The predominance of semiaridity and high altitudes in the region defied its settlement by native-born Americans and immigrants alike based on the food-producing tradi-

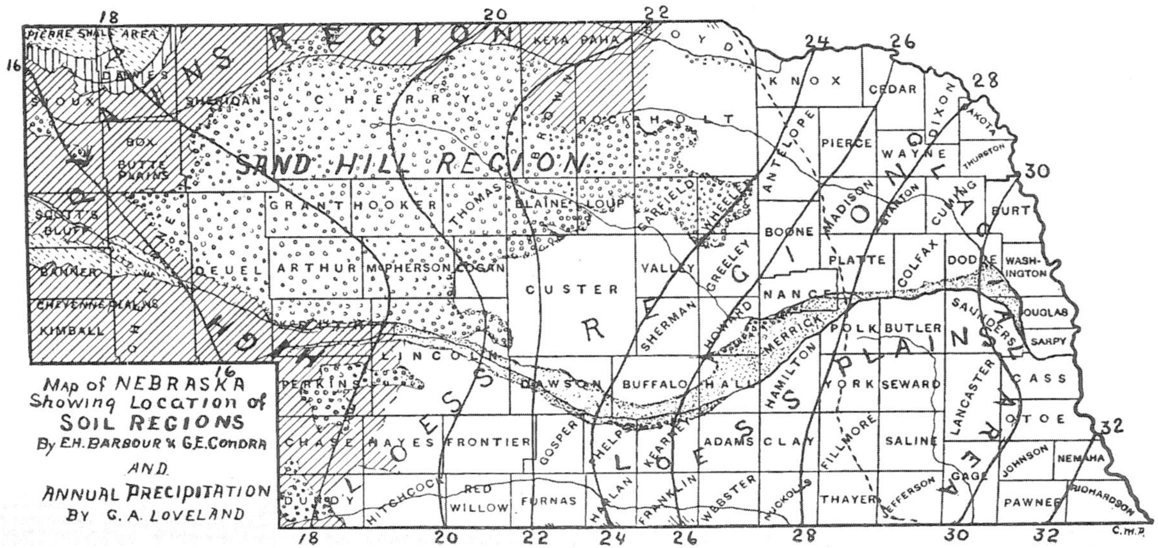

FIGURE 1.3. Subregions within the Great Plains state of Nebraska, including the distinctive Sandhills. W. W. Burr, *Storing Moisture in the Soil*, Bulletin of the Nebraska Agricultural Experiment Station, no. 114 (1916): 4.

tions of Europe and the eastern United States. As environmental historian Donald Worster contends, climatic limits—and the challenges they posed to creating "democracy in the desert"—are central to understanding the development of the interior West.[107] As part of what two agricultural scientists of the time called the great "Arid and Semiarid Region" of the country, the interior West's environment was largely hostile to more productive economic activities. "Ranching is a pastoral and retrogressive kind of farming," they lamented, "but it is the only kind suitable to the unirrigated portions of this great region."[108] High altitudes only exacerbated the disadvantage of aridity for settlers in much of the Great Plains region. To confront these difficulties, state-supported science and technology played a more visible role in making settlement possible.

Likewise, as mining historian Rodman Paul argues for the case of Colorado, the minerals of the Rocky Mountains region resisted extraction with simpler techniques, thus stimulating the unprecedented application of specialized science and technology.[109] Ovando Hollister, who wrote about Rocky Mountain mining in the late 1860s, described how mineral resources in Rocky Mountains were different from those in California. While "in the early settlement of Colorado, the Foothills furnished many gulch, placer, and bar mines," they "were limited in extent compared with those of California" because of "Colorado's greater distance from the sea" and its relative aridity. "Her mountains are approached by inclined planes six hundred miles long from which they rise only about six thousand feet," he observed, "while the Snowy Andes of California rise almost from the sea-coast to a height of twelve thousand to fifteen thousand." Hence a syllogism: "Where there is little water there will be slight wash, and where there is little wash there will be few surface gold mines."[110] In the view of Hollister and other early observers of Rocky Mountain mining, then, just as a distinctive environment challenged agricultural development of the Plains, so too did it challenge the mineral development of the Rockies.[111] For both agriculture and mining, the environmental characteristics of the interior West elicited an unusually high demand for scientific and technological expertise.

What was hardship for the settlers of the West, however, was a windfall for scientists. The environmental features of the Plains and Rockies presented exceptional opportunities for knowledge production. Even apart from the practical demand for the assistance of science and technology in the region's economic exploitation, within the larger scientific community, these same challenging natural features also proved to be excellent resources for solving interesting intellectual problems. Semiaridity may have impeded the economic goals of settler farmers, threatening bankruptcy during dry years, but it was a cornucopia of abundance for a knowledge maker. In the lexicon of economic historians, semiaridity was a natural resource endowment for scientists. The same could be said for high altitudes, complicated geological stratigraphy, and other natural features. By harnessing these resources, field scientists who had access to the region enhanced the value and significance of their research work. Other resource endowments of the interior West included the prevalence of eroded geological formations available for fossil prospecting, its distinctive flora and fauna—much of it contrasting vividly with eastern species, or forming complex boundary lines with them—and the naturalized human inhabitants confined over time on American Indian reservations. The natural and human resources of the Plains and Rockies thus provided the material basis for more extensive and valuable field science.

Intraregional variability has markedly affected the presence of scientifically valuable resources, such as fossils. Since paleontology was one of the signature specialties that flourished in the interior American West, achieving early international recognition for the field scientists who worked in the region, it may serve as an instructive example to consider. Fossils could, of course, be found scattered throughout the area. But many places were unusually rich in fossil resources, such as the badlands, foothills, and arid buttes near the boundary between the Rocky Mountains and Great Plains. The vegetative covering also influenced the accessibility of fossil resources. As one member of a Nebraska paleontological family, Caroline Barbour, pointed out, most of her state was covered with prairie grass instead of "tuft grass," which was "a distinct characteristic of the extreme arid region" of Wyoming and the South Dakota badlands. "In such regions," she observed, "it is rather a simple matter to find exposed on the surface between the tufts of grass such bones as may be washed out by storms and freshets; however, in Nebraska it is a much more complex matter to find such remains, because of the covering of grass which practically grows over the whole state, save in the very bad lands."[112]

Where prairie grass covered the ground, on the other hand, the frontier expansion of railroads, highways, farms, and irrigation works was often crucial to the uncovering of fossil resources on the Plains. Fossil hunters of the time recognized their dependence on such earth-transforming processes. Barbour noted that fossil bones "have generally been brought to light by the plow and scraper in preparing some cut or fill for the railroad or public highway or for an irrigating ditch." One exceptionally nice set of mammoth tusks was uncovered by a railroad cut in Gosper County of southwestern Nebraska. In addition, settler farmers sometimes found bones around prairie dog holes. For example, a farmer in Buffalo County in central Nebraska found a mammoth tusk along with other bones "by following the windings of a prairie dog burrow for nearly forty feet through the loess formation."[113]

Despite the fossil riches sometimes uncovered by road cuts, irrigation ditches, and prairie dog burrowing, the best places to look for fossils were, as Barbour suggested, in the badlands. What were these "badlands" and where could they be found? In the drier western part of the Great Plains, this was land that had been severely eroded by rivers into formations that looked strikingly different from the flat, featureless plains around them. To one Princeton-based observer, the "bad lands" were "a great cemetery of extinct animal remains."[114] The best-known region of badlands, explored by travelers associated with the fur trade along the Missouri River, was the area sometimes known more specifically as the White River Badlands in the southwest part of Dakota Territory (in South Dakota, from 1890). Sometimes simply called *the* Badlands—some of them are protected within Badlands National Park today—this area witnessed a steady onslaught of fossil collectors, including Ferdinand Hayden, who delighted in the tremendous volume of fossil materials that the continuously eroding formations revealed.

After traveling across the monotonous expanses of the Plains, the sudden arrival at the badlands formations could be a memorable event. One fossil collector recounted his first experience, in the early 1890s, of the badlands of the Hat Creek Basin just across the border in Nebraska, to which access had been just opened up by the building of a new railroad to the northwest corner of the state:

> At last, as we rode along, one of the party, who had been there before, told us not to look up until he spoke. Three minutes passed, and then the signal came. We gazed on the most desolate spot I ever saw. For miles it was all the same. The names Bad Lands and its French equivalent, Mauvaises Terres, need no defense. Not a bit of green,—nothing but that creamy-white, calcareous, clayey rock; and this was not level and flat, but eroded into the most irregular surface one could imagine. Ridge and gully, ridge and gully succeeding each other for miles,—the summits of the ridge as sharp as the roof of a house, while the gullies

> in most instances were not wide enough to allow the passage of anything larger than a wheelbarrow. . . . The light reflected from the creamy ground was very trying to the eyes, while the heat on a warm day was oppressive. Not a breeze finds its way into these narrow valleys. The walls sometimes rise at the angle of forty-five degrees; at others they are all but perpendicular.[115]

Not only does this description make clear how strikingly different the badlands landscape was from the rest of the Great Plains, it also vividly illustrates how the particularity of quite local places produced the finely textured environmental context for scientific field work, alongside the more familiar, common features of the Great Plains and Rocky Mountains.

The influence of the region's environmental features on science was not just felt directly. The indirect effects of the economic development of the Plains and Rockies provided the basis for modest local patronage, and it also shaped the most prominent directions of research conducted. The economic activities of the region varied across space: mining, dry farming, irrigation-based agriculture, and ranching dominated in different places. With considerable variation among states, and even among regions within them, related scientific work grew up as a subsidiary to these economic sectors. Schools of mines, for example, with their associated research and testing activities—including field geology—grew up predictably adjacent to the precious metals mining zones of the Black Hills in South Dakota, along the Colorado Front Range, and near other similar places.[116] The University of Wyoming's School of Mines, on the other hand, focused its bulletins almost entirely on place-specific studies of the state's suddenly emerging petroleum districts in the late 1890s.[117] At the same time, some kinds of inquiry, such as insect collecting, were virtually universal and could be pursued just about anywhere. "In the course of an hour," remarked T. D. A. Cockerell, without identifying any particular place, "one may make a collection of insects, only part of which can be named even by a very learned specialist."[118] Even in such cases, however, the results were place-specific.

Finally, the environment of the Plains and Rockies provided the material conditions for the practice of field work in the region. Like other settlers or visitors who tried to make a living in the West, scientists faced the same embodied constraints for human survival: weather conditions, water supplies, food sources, and the need for shelter. Animals, too, especially the horses and mules brought when traveling away from railway lines, required food and water. Thinking about the experience of field work in this way also situates science in the same analytical framework as other practices implicating the relationship between health and the environment, an especially rich area of historical examination in the past few decades.[119]

Due to the high altitude and sometimes mountainous terrain, field work in the interior West was usually seasonal, with the snow and cold of winter as the major limiting factor. The heat of summer also limited productivity in some places, although it rarely stopped field work altogether. Whatever the season, the availability of water for both human and livestock needs was a primary constraint on movement in the field. Access to food varied considerably depending on the availability of wild game and fish, as well as the presence (or absence) of nearby towns and ranches for provisioning. The low settlement density of most of the interior West made it an especially challenging region for research work. Grazing lands for horses and other draft animals, or the proximity of hay suppliers, could also be decisive in the practical decision making of field scientists. For shelter, researchers often relied on tent camping, although as settlement expanded boarding at ranches or in towns became increasingly common—a significant shift within the railroad era. These practical features of scientific field work may seem mundane, but they were also the most direct way that the natural environment shaped science in the Plains and Rockies.

Scientific travelers and settlers also experienced the western environment as something connected to their own bodies. Indeed, those naturalists and field scientists who moved into the Great Plains and Rocky Mountains often did so in part because of a desire to live and work in western environmental conditions. This was not true for everyone, of course. When Wallace Craig, "after having lived in many beautiful places," arrived in Valley City, North Dakota, in the early twentieth century to teach for two years at the state normal school, "his first view of the new home gave him a shock even worse than he was prepared to receive." In particular, he "keenly" experienced "the monotony of the vast level expanse, and the awful barrenness of a landscape without trees." The "barren" flora, which was "never green like the grass of the East, but only a greenish gray," compounded how the "severe winter" compromised the "impulse to enjoy nature." He lamented "the low temperatures, the dangerous blizzards, the snow-blocked roads, and the almost complete disappearance of wild life," which rendered "a walk in the country uninviting even to an enthusiast."[120] Such experiences of nature were most extreme on the northern Great Plains but extended across the region to varying degrees.[121]

More often, though, expectations were positive. New arrivals were interested in improving their own health, as well as that of their collaborators and assistants. "As a sanitary retreat during the summer months," declared botanist Charles Christopher Parry, the Rocky Mountain region was "unexcelled in the purity and coolness of its atmosphere, the clearness of its flowing streams, and its picturesque extended views."[122] The practice of science in the Plains and Rockies was shaped by the interactions of bodies and environments, which were simultaneously real in their material effects and culturally represented and constructed within the context of health beliefs and knowledge of this historical period. Dry climates, as a rule, were favored for their recuperative powers, including but not limited to respiratory ailments such as tuberculosis (consumption). Moreover, as advocates of medical climatology such as Charles Denison emphasized, higher altitudes were thought to best protect against consumption, with an "altitude of approximate immunity" for Colorado estimated "at 6,000 feet."[123] In neighboring New Mexico, an entire "laboratory of pathology and bacteriology was established in the late 1890s," which aimed "to undertake a series of investigations into the climatology of the plateau with especial reference to the influence of climate upon disease."[124]

Consequently, the regional climate's restorative health potential, particularly for the southern Rocky Mountain states of New Mexico and Colorado but more generally for the arid and semiarid West as a whole, was a significant part of the region's history of knowledge.[125] To be sure, the Great Plains region, as a rule—except for places adjacent to the Rocky Mountain front—was much less often sought out for its healthful qualities. A few special places, such as Waconda Lake in north-central Kansas, held special significance for healing purposes, for both American Indians and white settlers, but mainly it was the Rocky Mountain health resorts, especially those in the southern Rocky Mountain states of New Mexico and Colorado, that attracted the most attention.[126]

Many field scientists, like other settlers, moved to the area explicitly for health reasons, including Cockerell, who made most of his career in the American West. Cockerell relocated first to southern Colorado, then New Mexico, and ultimately to northern Colorado—where he served for much of his long and unusually productive career as a professor at the University of Colorado in Boulder—in order to improve his health. Cockerell suffered from "a mild case of tuberculosis of the lungs," and moved into the region as part of a massive wave of tuberculosis sufferers moving to the arid West, with some also believing in the merits of high altitude. "I was to live in the mountains of Colorado, well known for their health-giving climate," Cockerell later recalled, "and I think that today I may stand as a good advertisement for my adopted state."[127] Cockerell was one of the most prolifically publishing naturalists in the Plains and Rockies, and his choice of field sites for collecting and observing in Colorado and New Mexico proved to be desirable both for his own bodily health and for the accessibility of novel phenomena and objects in the natural world. At least as common, however, were instances of health migration later in life, such as geologist W. G. Tight, who spent most of his life and career in Ohio but moved to New Mexico near the end of his life for health reasons.[128] While health migrants were not the majority of field scientists moving into the Plains and Rockies, they were a significant minority.

More commonly, as shall be evident intermittently throughout this book, both material effects of the environment on the body and guiding concerns about the intertwining of environments and bodies permeated the practical experiences of travelers to the region, including field scientists. Of course, not all health expectations and bodily experiences were strictly positive. The astronomer Samuel Pierpont Langley, for example, when traveling to Colorado in order to view the solar eclipse of 1878 as it passed "directly over one of the loftiest mountains of the country, Pikes Peak," recognized that while "pleasure tourists" might enjoy a brief ascent, "it is one thing to stay there for an hour or two, and another to take up one's abode there and get acclimated" to the fourteen-thousand-foot altitude. He complained that "to do the latter we must pass through the horrors (not too strong a word) of mountain-sickness." Nevertheless, even Langley appreciated the distinctive observations made possible by "the extreme altitude and the clearness of the air" that he experienced on Pikes Peak.[129]

Regardless of whether health considerations were centrally important, naturalists often located their work headquarters and their own bodies with the surrounding natural environment uppermost in their minds. Within the interior West, particular locations might be chosen for their local advantages. For instance, the Colorado naturalist Martha Maxwell set up a site for her taxidermy—processing materials collected in the field nearby—on Boulder Creek. In her view, the place was "an admirable one for a naturalist, as it was visited by animals from both the mountain ranges and the plains."[130] Such a location not only allowed Maxwell to collect efficiently within a close radius of her home base, it also permitted her to address very specific questions about environmental factors directed to her by metropolitan naturalists such as Spencer Baird.[131] Even when establishing temporary homes away from home, naturalists chose their collecting locations with an eye toward both experiential and productive environmental qualities. Such collecting "camps," as they were often called, could be just as carefully scouted as were permanent headquarters. "Last spring, while on a collecting trip in North Dakota," wrote bird collector Gerard Abbott in one typical description, "the writer was encamped for a month on a narrow neck of land, surrounded on three sides by a chain of lakes. This point was covered with a small growth of timber, mostly poplars, and was an ideal spot for a camp."[132]

Long-distance expeditions, like short-term camps and permanent migration to the interior West for Cockerell, Maxwell, and others, could also offer opportunities for bodily regeneration. But expeditions also presented special challenges due to unexpected experiences with the natural environment of the interior West. On July 16, 1892, at "Camp San Juan" in Utah, for example, published reports from an expedition sponsored by the *Illustrated American* magazine narrated how the party broke camp while the San Juan River's waters rose and abruptly flooded their campsite. Even in carefully edited accounts for a general audience, such an experience was revealed as traumatic and threatening to comfort, safety, and well-being.[133] To be sure, these narratives of harrowing experiences could also be—and more often were—construed as heroic adventure tales, but they nonetheless reveal material realities that were as poignant in their pain and unease as other experiences of the healthful qualities of western travel and settlement were enjoyed for their positive qualities.

The evidence from long-distance expeditions can also help us trace the increasingly intertwined worlds of nature, human settlement, and infrastructure for provisioning in the field at both short and long distances, which were becoming evident by the last decade of the nineteenth century. In the early years after the building of the transcontinental railroad, wild animals were still quite easy to obtain, as George Bird Grinnell experienced during his time as a student assistant to paleontologist O. C. Marsh in western Nebraska in 1870. "The country was full of game," he recounted, "antelope being by all odds the most abundant." In his view, the lack of settlement on the Great Plains by white Americans meant that "antelope were still nearly in their primitive numbers" and elk along the Loup River "were found and killed, and fresh meat for the party was never lacking."[134] At the same time, Marsh and his counterparts in the field also benefited from innovations in food preservation and canning, which had expanded in the early 1860s during the US Civil War. With the fortuitous expiration of Gail Borden's initial patent in 1861, scores of competing food companies such as Van Camp, Armour, and Swift entered the market, and their products designed for army camps and naval voyages became thereafter available to traveling field scientists as well.[135] The invention of a "steam retort" in 1873, based on "the same principles as a steam engine," helped ensure that "low-acid, high-protein products like baked beans and soup" were properly heated to kill bacteria without destroying the food's taste. But few canners had the resources to install such equipment, and canned food remained fairly rare until the 1910s, except for a few special situations.[136] The long field excursion was one of those situations, although their suppliers may have relied on older methods.

By the late nineteenth century, then, travelers depended on the developing infrastructure of processed and canned food, as well as the burgeoning western cattle industry—and occasionally also American Indian sheepherders—as the experience of the *Illustrated American* expedition's daily life from "Camp F. W. Putnam" near Bluff City, Utah, suggests. According to its report, "The cook is astir at six o'clock in the morning, and frequently at five. The teamsters arise at the same hour, to gather in the stock and see that it is fed." The human participants typically consumed "bacon, coffee, and hot bread" for breakfast and "dried fruit, Armour's canned beef, potatoes, corn, coffee, and bread" for dinner. "Occasionally," in place of beef, the cook might serve "mutton, which we buy of the Navajos, or fresh beef purchased at the ranches along our route."[137] Around the same time, an 1895 University of Kansas fossil-hunting expedition in Wyoming experienced the shift from hunting to canned food. One journal entry, from J. P. Sams, commented that, despite some hunting efforts "in a wild uninhabited country" with "lots of game and fish," their most plentiful "game" was "hame [*sic*] and bacon from Armors Packing House" and fish was "salmon and sardines well packed in cans and boxes."[138] These depictions of routine camp life hint at the transition to field work in a context of more settled infrastructure and an economy rapidly becoming integrated with the wider world.[139]

All these bodily and local provisioning experiences of scientists and their collaborators collectively remind us that the environment is always locally variable, even within well-defined regions such as the Rocky Mountains and Great Plains. They also illustrate the translocal geography of field practice, perpetuating and transforming the existing geographical relations of natural history and scientific travel. But the changes in field work in the interior American West during the railroad era did not take place in a vacuum. Knowledge making in the field during the second half of the nineteenth century proliferated in the growing shadow of the lab. The same industrial-age technologies that made this great expansion of natural history and new modes of field science possible also fostered the emergence and proliferation of new urban and suburban modes of scientific practice, including most notably the laboratory.

The Laboratory Challenge

Today we often associate science with laboratories. Yet before the mid-nineteenth century, laboratories were not really all that common. During the second half of the nineteenth century, the laboratory rose from relative rarity to its high status as a premier site for scientific practice by the turn of the twentieth century. Notably, pure science research was not the sole driving force. In British physics, for instance, the rise of the lab was closely connected with its role in university and college settings, for teaching and the related aim of precision measurement for the practical economic sphere, especially for telegraphy.[140] We have some hint of a similar pedagogy-driven trajectory in American science, including at least one assertion that the "laboratory system" for studying plants began around 1870 "in connection with the teaching of botany."[141] Nonetheless, the rise of the laboratory occasioned a significant shift in the landscape of knowledge production as well. German examples were commonly credited as the models, but the laboratory as a mode of knowledge production soon spread elsewhere in the scientific world, including Britain and the United States.[142]

By the 1880s American scientific observers were conscious that a dramatic shift had taken place. "The material circumstances under which scientific discovery is prosecuted have been completely revolutionized during the last forty years," proclaimed one commentator in the pages of *Science*. "Forty years ago there were very few, more properly no laboratories which we of to-day would consider even tolerable. Now every university of importance and high repute, the world over, has large suites of rooms for each department of science, and often numerous great buildings within whose walls thousands and thousands of students are daily brought face to face with the facts and laws of nature."[143] Similarly, an 1895 address on the rise of the laboratory by Johns Hopkins University pathologist William H. Welch contended that "the systematic study and advancement of any physical or natural science . . . requires trained workers who can give their time to the work, suitably constructed workrooms, [and] equipment with all of the instruments and appliances needed for the special work"—in other words, "a well-equipped and properly organized laboratory."[144]

Such views were widely shared. Indeed, by the turn of the twentieth century laboratories had become more prevalent, prominent, and prestigious not only at universities and colleges but at government agencies such as the US Geological Survey and the US Department of Agriculture, as well as at many large business corporations. In specific disciplines, even those with long-standing traditions of field work, the laboratory was greeted with enthusiasm. In 1885, for example, J. C. Arthur described the lab as "one of the most powerful auxiliaries to the advancement of botany in this country," focusing especially on its contribution to "a wider range of study and a more intimate and thorough acquaintance with the structure and life of the plant." Moreover, the lab's "tendencies are toward habits of careful investigation," not only in a comparison of the potential of labs in botany to other fields such as zoology but also of US labs to their German counterparts.[145] As places for scientific knowledge production, laboratories were heralded for providing more highly structured environments in which control of the variability of nature and the regulation of human activities, such as the control of access and the deployment of skilled human labor in the operation of specialized instruments, prevailed.

The work organization of the laboratory has never been monolithic—there have been many variations in how scientific labor was allocated among various roles, for example—but one common characteristic has been the hierarchical division of labor. The same *Science* article that hailed the laboratory revolution also specified three levels of personnel for a "good scientific laboratory," with at least one person at each level: (1) the "highest officer . . . a responsible director,—a man of superior ability, extensive attainments, and prolonged experience"; (2) "an assistant, preferably a young devotee of science, who will be fitted by his experience as an assistant to ultimately become himself director of a similar laboratory"; and (3) a "laboratory keeper (*diener*), who needs must be a man of some mechanical skill, so that the precious instruments may be safely intrusted to his care." The latter role, which would come to be known in the English-speaking world as the laboratory technician, "should be something more than a servant, and less than an assistant."[146] Notable in this three-part division of labor is the distinction between two classes of subordinates: the scientific assistant with the chance for status mobility to become a director, and the keeper or technician, who typically did not have such an opportunity. A similar class divide would be evident in several modes of field practice, as shall become clear in later chapters, such as in the field survey party or field station staff.

While the historical emergence of the laboratory—what one 1884 article in *Science* called "the most remarkable and influential creation of science in our time"—has mostly been highlighted in metropolitan centers of the Europe and the US East Coast, the rise of the laboratory was also visible in the Great Plains and Rocky Mountains regions.[147] New laboratory buildings were celebrated in the national scientific media, as occasions for advertising the rising prestige of even institutions on the western frontier. At the University of Nebraska, for example, the construction of a chemical laboratory in 1887 was lauded in *Science* with a full spread of architectural drawings and photographs.[148] Only a few years earlier, Nebraska's botanical lab was first opened in April 1885 (though subsequently moved to the new chemical lab building) with at least one lavish description in a national scientific journal of the instruments already available for teaching and research work, especially microscopes.[149] In some cases, visiting scientists were invited to deliver major addresses upon the dedication of such laboratory buildings, as did University of Michigan chemist S. Lawrence Bigelow in 1904 at the opening of Palmer Hall at Colorado College. Bigelow used the opportunity to highlight the dramatic rise of the laboratory as a place for doing science, noting that "we might well marvel that such a complete revolution of conditions could occur in fifty years. It would be harder to find a university without moderately good laboratories to-day than it was to find one with them in 1850. And they are increasing in numbers and size, through the munificence of individuals and of legislatures and governments, at a surprising rate."[150]

Likewise, in retrospective chronicles of regional scientific development written toward the end of the transformative period examined in this book, the rise of the laboratory was often given top billing. This was acknowledged even by field scientists. The paleontologist Samuel W. Williston, for example, who was educated in Kansas, emphasized the role of laboratories in his 1918 account of the "progress of science" as exemplified by the first half century of the Kansas Academy of Science. After noting the pioneering roles of early frontier naturalists, and the tremendous growth in world science as measured by publications and careers, he singled out as local touchstones the "first chemical laboratory in the state of Kansas . . . organized at the Kansas Agricultural College in 1873," and seeing his "first compound microscope . . . in the same year."[151] This association of laboratories and instruments was crucial, for it was the controlled conditions of the laboratory that made possible the rigorous deployment of such material technologies in scientific practice. Similarly, laboratories often also housed collections of specimens, and many of them traced their lineages to museums and natural history collections. A final crucial linkage, perhaps less often remembered, was between laboratories and libraries available in close proximity to provide access to scientific publications, and this connection was made explicit in many discussions of laboratories, by Williston and others.[152] The same built environment that housed the university, government, or corporate laboratory also accommodated rapidly expanding libraries of scientific books and journals. These adjuncts to laboratory life—instruments, collections, libraries—were crucial to its success as a mode of knowledge production, and the various modes of scientific practice in the field would have to compete with these advantages too, just as they would the laboratory's emphasis on control and rigor.

Laboratories, like field sites, were also constituted by the melding of the natural environment with material technologies. Indeed, a fully envirotechnical history of the laboratory would itself be a worthy endeavor. While historians have long recognized the importance of instruments and techniques to the history of the lab, its environmental context has been less fully recognized.[153] Typically situated in an urban or suburban setting, labs were supplied by modern technological systems for electricity, fuel, and a dense transportation and communication infrastructure.[154] They were also, like other urban institutions such as museums and office buildings, situated in places where workers—scientists, assistants, and technicians—could rely on housing, food, and other daily needs supplied by the modernizing city.[155]

At the same time, urban-based labs also suffered the disadvantages of the noise, mechanical vibrations, pollution, and other complications of their surrounding environments. Delicate experimental apparatuses in labs were vulnerable to disruption. This was especially true of labs that maintained high levels of instrumental precision, which could be disturbed by the electrical fields and vibrations of streetcars, for example.[156] Even the human body of the scientific worker could easily be compromised by excess indoor heat or inadequate ventilation systems. Laboratory materials themselves could prove injurious, or even fatal, to scientists or their assistants working indoors. In one rare case, for example, the chemist Robert N. Hartman at the Colorado School of Mines "was asphyxiated in the chemical laboratory" in 1903 shortly before he was to take up a new position in New Mexico.[157] It is true that fatalities also occasionally occurred in the field, as when geological assistant Henry A. Newton fell ill and died during the 1877 field season, or when an oologist named E. S. Bryant drowned in North Dakota's Sweetwater Lake in March 1895.[158] And of course other field scientists suffered nonfatal injuries or illnesses. But the point is this: we should not assume that it was only the field scientist working outdoors who faced environmental challenges or threats of bodily harm.

Some idea of the more typical local environmental perils faced by lab scientists and other indoor knowledge workers in urban settings, by contrast with field scientists, can be gleaned from a remarkable set of complaints that appeared in the 1911–1912 annual report of the US Geological Survey. The opening section of the report focuses not on the content of the survey's work but rather on the dire need for a larger building with more space and better working conditions. The Hooe Building at 1330 F Street in Washington, DC, was "described without exaggeration as in large part comparable in crowded and insanitary conditions to a sweatshop. Treatment of wage earners that would excite severe criticism if a corporation were the offender is tolerated in the case of these employees of the Federal Government."[159]

Many of the specific complaints about the US Geological Survey headquarters revolved around the material and environmental conditions of the building, including lack of space, as well as a lack of fresh air and light, which were explicitly (and unfavorably) compared to the conditions that "field men" enjoyed during the summer field work. What the report called a "violation of practical hygiene" was mitigated, they argued, by "the fact that a considerable number of the members of the Survey are field men, whose work during the summer is performed in the open air where they have opportunity to recuperate each year." Such "open-air work" for the five-month field season was thought to "protect certain members of the force from lung disease and defective vision," but this only brought into sharper relief their plight back in the "congested conditions" of Washington, DC, for the other seven months. Field men, in a memorable phrasing, suffered a dramatic compression from a space of "1,000 square miles" to "an allowance of 100 square feet of floor space."[160]

By linking bodily health to the indoor urban workspace, the report was not only drawing attention to the contrast between indoor work and field work but also providing a rare descriptive account of some of the ecological perils facing those working in urban buildings of all kinds, whether in office or lab work. That high-temperature conditions affected indoor work in general as well as specific aspects of lab practice is evident. "The top floor in the main building and the loft in the annex becomes so hot in the summer," the report observed, "that at times it is impossible for the employees occupying these floors to perform their duties, and indeed they frequently have to be assigned to work elsewhere." Accordingly, it was mostly "field men" who were assigned to work on the uppermost floor, since they "usually are absent from Washington during the hot weather." But the same was not true of the loft annex, "an old building in an alley," where "the summer temperature runs as high as 110° F, and neither the employees nor the solutions and emulsions used in the photographic laboratory can be depended on for high-grade results under these conditions."[161] Beyond the unfavorable comparison with the conditions of field work, then, the report also reveals many of the specific difficulties that scientists working in urban buildings faced due to indoor environmental conditions.

Besides excessive heat inside, the report also discusses fire risk and the impairment of efficiency, as well as dirt and noise from outside. "Not only do the neighboring chimneys pour out soot and fine cinders that keep the books, files, and records covered with dirt," observed the report, "but on three sides the building is bordered by a street and alleys, which are a constant source of additional dirt and also a source of noise." The ensuing description affords a vivid description of the perilous conditions of urban laboratory life at their most crowded extreme:

> These alleys, which afford access to a theater, two newspaper offices, a hotel, an office building, and several saloons, necessarily have an almost constant traffic of heavy wagons, and the noise of these passing up and down the cobblestones is deafening; the only variations are the rehearsals of a summer opera company and the chorus of yells by newsboys who congregate at the rear of the Survey building awaiting the various editions of an afternoon paper.
>
> The street noises in the summer, when the windows must be opened, are so great and annoying that it is at times impossible to use the rooms on F

Street for dictation to stenographers, and to use the telephones in those rooms it is often necessary to close the windows temporarily. The almost constant passage of heavy cars along F Street, in front of the building, sets up vibrations so great that at times scientists wishing to use delicate instruments have been obliged to do such work at intervals after midnight, and even then the effect of vibration was not wholly avoided.[162]

Such "street noises" and "vibrations" thus not only were a source of bodily discomfort to the scientists and their assistants working inside but also directly threatened the viability of office practices (e.g., stenography) and highly sensitive laboratory practices (e.g., use of instruments). Of course, not all urban lab and office buildings presented the same level or scope of environmental challenges as the Geological Survey headquarters, but this example at least suggests the range of possibilities and reveals more explicitly than usual the potential perils.[163]

More broadly and positively, as the survey's reference to the delicacy of lab instruments hints, the technologies of the laboratory, in their material and environmental settings, were also the foundational bases for the construction of knowledge across widely separated locales. Historians of science have come to recognize the significance of new instruments to the development of whole new areas of science, including the "instrumental communities" that grew up around them, as well as the experimental systems that have developed around standardized laboratory organisms.[164] Instruments not only were important technologies for the transformation of nature (if usually on a fairly circumscribed scale) but also have been manufactured out of the same raw materials as everything else in industrial society, or indeed in any society. By thinking of the laboratory in its urban or suburban setting as an envirotechnical system, we can illuminate its own history and also establish a basis for comparison with the various modes of field practice.

Field Response

In the wake of the laboratory's rise to the pinnacle of scientific practice, field science evolved—and, in many respects, expanded and proliferated—rather than disappearing into the sunset. The field response took several different and overlapping forms, including direct rebuttals to claims of the laboratory's epistemic superiority, articulations of complementary relationships combining lab and field in the research process, the persistence of natural history practices despite their lower status, and, finally, the spread and development of new modes of scientific practice in the field. The rest of this book will be organized around the latter, and I will examine how these emerging modes of field practice in the railroad era—lay networks, surveys, quarries, and stations—evolved and solidified in response to the laboratory challenge. For now, however, it may be useful to elaborate briefly on the other three responses.

First, consider explicit rhetorical challenges to the emerging hegemony of the laboratory. Direct rebuttals tended both to attack the exclusive association of modern scientific practices with the laboratory and to emphasize the advantages of field work as a means for engaging with the real world of nature. The contrast between lab and field—and the need for researchers to recognize the value of both—was broached, for example, in 1901 by Edmund B. Wilson, who applauded what he perceived as the reconciliation of the lab and field after some divergence and invidious distinction. At an address to the American Society of Naturalists, he lamented that some experimental morphologists based in laboratories had contended "that the scientific study of nature is possible through experiment and through *experiment alone*."[165] The soil scientist George N. Coffey offered a more forceful perspective, contending that "although the laboratory may be the only place where conditions are under control[,] things act differently when this control is removed."[166] Across a variety of field science disciplines, then, the advocates of field work pointed out how it could remedy the defects of laboratory-produced knowledge.

One of the foremost advocates of field work in the early twentieth century was the botanist and ecologist Frederic Clements, who took this argument to perhaps its greatest extreme. "As botany becomes broader and surer in its progress," he predicted, "the relative position of laboratory work and field work will be changed. The field will take the first place. The laboratory, though indispensable, will become secondary in that its chief use will be to assist the interpretation of field facts and field experiments."[167] Elsewhere, he argued that field-based ecology was "as natural a rebound from the intensive laboratory research of the last twenty-five years as this was a logical reaction from the more diffusive studies of natural history."[168] Although Clements's prognostication may now seem naïve, it nevertheless indicates the defiance with which at least some field scientists greeted the laboratory challenge. Echoing Clements in the realm of animal life, though more evenhandedly admitting that both "field and laboratory methods each possess advantages," Walter P. Taylor of the US Biological Survey argued that "the possibility of a complete control of environmental factors under laboratory conditions has sometimes led to undue enthusiasm for this type of research." Especially for mammals and birds, he thought, the laboratory induced "a loss of normality of response on the part of the organism," thus requiring "direct observations of the animal in a natural state," since "many of the most important details of animal life-histories can never be learned in the laboratory."[169]

Not all field scientists portrayed the relationship between lab and field practices in such agonistic terms. As an alternative, many of them articulated connections between lab and field work as complementary parts of a single, unified research process. Examples of complementarity, too, will be evident throughout the rest of this book, but a generalized example up front may be useful: consider the work of the US Geological Survey, where mineral specimens routinely moved from collection in the field to analysis in the lab. Both steps were recognized as essential, and both field collectors and laboratory analysts were credited in publications, often as coauthors, with environmental place and specific location explicitly indicated as integral components of the knowledge that was produced.[170] Still, it is true that many, perhaps most, of these apparently seamless linkages between lab and field tended, in practice, to subordinate field to lab in the sense that the lab became the final arbiter of scientific judgment, rather than the other way around—in other words, to concentrate most of the value added in the production of knowledge in the lab rather than the field. Even in more complex cases, such as through the field surveys of the US Department of Agriculture's Bureau of Soils, the judgments of field men in the process of mapping, while given a great deal of deference, could be overruled by supervisors in charge of "soil correlation" in the bureau's Washington office, more often than not reliant in large part on lab determinations.[171]

If much of the scientific practice in the field was related in some more direct way to the hegemonic laboratory mode, either antagonistically or as its complement, it was nevertheless also true that a significant quantity of scientific work continued with natural history practices. As this chapter has emphasized, natural history was transformed in the Great Plains and Rocky Mountains during the railroad era, proliferating and expanding far beyond all previous levels. This persistence—or even growth and development—of natural history was not at all unique to the frontier regions of the American West, though of course the opportunities were somewhat greater there for the discovery of new species and revealing their distributions, compared to areas that had already been more closely studied.

To some commentators, perhaps more conscious of differentials in scientific status, natural history can barely be said to exist at all in the world of modern scientific specialties and disciplines—this despite, or perhaps because of, its high level of public participation. "It is paradoxical, but no real surprise," observes James Secord, "that natural history is today both hugely popular and at the bottom of the scientific hierarchy."[172] Those scientists who study the amazing diversity of the natural world have tended to gain prestige by associating themselves with more conventionally professionalized scientific disciplines such as population ecology or numerical taxonomy. Science with too much of the broader public actively involved has made it low status and hazardous to the scientific career-making aspirations of its practitioners. Recently, there have been signs of a possible reinvigoration of public participation in science, but so far these efforts have not yet overturned the late modern disciplinary hierarchy.

And yet, whatever we think of its more recent status, it is now quite clear that natural history flourished around the turn of the twentieth century. Large urban museums expanded and transformed themselves through the "new museum idea" into great centers of both education and research. Field expeditions were launched to all corners of the earth. Institutionally—and in the number of new species discovered—it was a golden age.[173] Of course, many of those expeditions undertaken by urban museums and other powerful institutions would partake in the emerging, more intensive modes of field practice discussed throughout the rest of this book, such as surveys and quarries. But even more traditional collecting expeditions would persist too, often in close association with the larger public. College expeditions for collecting natural history specimens for campus collections, for example, while increasingly viewed as unsystematic and not on the cutting edge of scientific rigor, continued as beloved traditions at many institutions. Moreover, some expeditions were arguably more concerned with adventure and discovery than with adhering to the latest standards of systematic rigor in survey practice. Such examples never disappeared entirely, and the public appetite for such science has sustained it, just as new transportation and communications technologies have made it easier to undertake.

Still, as scientific disciplines such as biology both strengthened their hold on the production of knowledge and diversified into numerous specialties, natural history found itself pushed to the margins of cutting-edge research. Lynn Nyhart sums up the paradoxical situation with an agrarian metaphor. "If in 1835 natural history was the crop cultivated by most students of living things," she observes, "by 1900 this was no longer the case. Simultaneously, however, the overall landscape under cultivation had grown. This combination of factors left natural historians with a territory that at least in institutional terms was larger, but that also constituted a relatively smaller portion of the total area under cultivation. In this sense, natural history was *both* growing *and* declining in the early twentieth century." Proponents of new research areas combining aspects of traditional natural history with elements of the new experimental modes of practice, such as "life history" studies, as Nyhart shows, found themselves competing for funding, prestige, and jurisdiction over research territory.[174]

This paradox of simultaneous growth (in absolute terms) and decline (in relative terms) characterized the field sciences in general, not just the life sciences Nyhart examines. Even sciences that preserved their commitment to field work as the heart of scientific practice during the railroad era, such as geology, were transformed by the rise of the laboratory, in which determinative analyses and experiments were increasingly undertaken. Despite significant differences in how field work was defined and practiced from one scientific discipline to the next, one argument of this book is that the commonalities in field practice indicate patterns crossing diverse disciplines, by virtue of their shared ways of organizing the work of human workers in relation to material technologies and environments. Thus, having now explained the natural history background of field practice and its transformations during the railroad era, as well as the epistemic challenge posed by the rise of the laboratory, the rest of this book will focus on the four most visible modes of field practice that proliferated during this period, each of which enabled field science to become more systematic, rigorous, and tightly organized.

Ferdinand V. Hayden and T. D. A. Cockerell were very different kinds of people, and very different field scientists too. Whereas Hayden built his career by traveling repeatedly from east to west (and back again) in the early part of the railroad era, helping the US government extend its epistemic authority over a vast terrain—from the Plains of Nebraska to the heights of the Rockies—Cockerell settled later in the railroad era at a series of specific places in the West along the border between mountain and plain, becoming intimately familiar with each, before settling permanently in Boulder, Colorado. Yet both Hayden and Cockerell participated in making the field; rooted in the natural history tradition, transformed by new industrial-age envirotechnical systems of the railroad era, and innovating new responses in field practice alongside the rise of the laboratory. Moreover, foreshadowing the field practices discussed in this book, Cockerell, especially, bridged the epistemic rift between the cosmopolitan knowledge of science and the experiential knowledge of local place. For Cockerell this was perhaps most evident during his early years in West Cliff, Colorado, when he lived in a tight-knit village community with other new residents becoming acquainted over a period of years with the place they now inhabited, through everyday life and work. But he would carry this orientation to the particularity of local and regional experience with him for the rest of his career in New Mexico and Colorado.

Beyond Hayden and Cockerell, and in the wake of the laboratory's rise, this chapter has considered how the field sciences responded and changed: how the field was made. Yet the natural history tradition—that great project of naming and classifying the natural world—continued, even as some naturalists were transformed into scientists, and much of natural history into the field sciences. More broadly, one might say that the modern field sciences were historically constituted by their transformation during the railroad era. By examining the technological and environmental characteristics of the field as it was made in the Plains and Rockies, we can trace some overarching patterns evident in the specific technologies and their relation to the natural environment that were common among many diverse modes of field practice. Whether in lab or field, the practice of science can be most productively conceptualized as both environmental and technological together.

Well beyond the initial observations of nature that accompanied expeditions to locate railroad lines, the long-term consequence of railroad development for the field sciences was the creation of a new spatial geography and landscape of natural knowledge production. These railroad lines, along with accompanying material technologies, embodied the union of technology and environment. As pathways to nature, they blended the industrial-age technology of steam power with the treeless open expanses of the Great Plains and the exceptionally rugged topography of the Rocky Mountains. The rapid proliferation of field science work in the Plains and Rockies occurred in the wake of this shift, and it enabled not only an increase in traditional natural history practice but also the more systematic modes of knowledge production in the field that are analyzed throughout this book. Throughout the second half of the nineteenth century, the building of railroads across the Plains and Rockies remade the field for scientific practice. In later chapters we will consider how field practices under the newly proliferating modes—lay networks, surveys, quarries, and stations—were transformed under the partial influence of industrial-age technologies in conjunction with those material environments.

Hayden, for his part, would later parlay his initial experiences in the western field into a position as head of one of four "great surveys" that preceded the US Geological Survey. Hayden's federally funded operation began in Nebraska upon its attainment of statehood in 1867, extended into Wyoming Territory in 1868, then became "a reconnaissance along the eastern edge of the Rocky mountains from Cheyenne, Wyoming Territory, to Santa Fe, New Mexico" in 1869. By the early 1870s Hayden's field enterprise would expand still further into a more systematic "Geological and Geographical Survey of the Territories," complete with subdivision into multiple field parties under his direction, "each provided with a geologist, topographer, meteorologist, naturalists, assistants, collectors, and laborers."[175] By then Hayden had moved into a full-fledged survey mode of practice. But Hayden's success at attracting government patronage in the earliest years of the railroad era was unusual. While other contemporaneous "great survey" leaders—Clarence King, John Wesley Powell, George Wheeler—followed a similar strategy, most other naturalists had to join one of the leaders or secure their own travel funds. But sometimes they adopted a different strategy: find some way to connect with people who were traveling or living there.

Chapter 2

Lay Networks

Vernon Bailey was a quintessential field man. Over several decades, he would become as intimately familiar with the animal life across the entirety of the American West as anyone else of his generation—quite possibly of any generation. Yet in his early years working as "chief field agent" in the early 1890s for the US government agency that would later become the Biological Survey—with Bailey rechristened as "chief field naturalist"—he found himself depending on a heterogeneous network of collectors and observers. Some of them had scientific titles and claims to expertise, but most did not. Still, since Bailey could not immediately travel everywhere in the West himself, these collaborators were crucial for his attempts to map animal distributions across the vast interior of the continent. For example, in mapping prairie ground squirrels, or sphermophiles (seed lovers), as he more precisely delineated them, Bailey relied on responses his agency received from two mass-distributed messages that were circulated in the late 1880s. "Each locality from which specimens were examined was marked on a large map of the United States devoted to the species," he noted. "Between the positive localities a multitude of reports of occurrence cover the ground," thereby suggesting where "specimens should be obtained" in the future in order to solidify and verify this knowledge.[1]

While ground squirrels were tangible physical creatures that could not only be observed but sometimes also collected as scientific specimens, other natural and social phenomena remained solely in the nonportable realm of observation and data collection, such as the weather. As the organizers of weather-observing networks worked to gather and map information obtained across a wide geographical area in order to turn it into expert-based weather forecasts, they may have wished that they could transport the weather itself, but they had to be content with standardized and instrument-based data. Often this information traveled through the mail, but in certain prioritized cases—the grain market–making forecasts during the growing season in the Central

Plains beginning in the early 1900s, for example—it traveled by telegraph too.[2] The US Weather Bureau set up branch offices in several locations throughout the country, attempting to tie its work as much as possible to local agricultural needs.

One of these branch offices of the Weather Bureau was in Kansas City, Missouri, a town on the eastern edge of the Great Plains, which was initially settled in the 1830s and incorporated in 1853, finally booming in the late nineteenth century during the railroad era. By the early twentieth century, Kansas City's meteorologist-in-charge Patrick Connor, who typically signed his letters with the title "Local Forecaster," guided two networks of lay weather observers scattered across Kansas and Missouri. These groups were designated as the Corn and Wheat Region Observers, who recorded daily measurements such as temperature and rainfall from April to September, and the River Observers, who were located along major rivers in the region subject to flooding. Some of the most important places for lay observers were in the central and western part of Kansas, including the High Plains, since most weather systems moved from west to east. These weather observations were then aggregated in Kansas City, supporting forecasting operations there, which were closely watched not just by farmers themselves but by a variety of city businesses as well and especially by the boards of trade, which were concerned about the effects of weather conditions on the future production levels of these key grain commodities. Monthly paper reports of weather observations were crucial to the observational record generated by Connor's lay networks but were of little use in short-term forecasting, and the daily production of weather knowledge in Kansas City depended on the telegraph. Members of both teams of observers took their measurements with standard instruments, proceeded to the local telegraph station as soon as possible after 7:00 a.m., and then telegraphed the data using a code system.[3]

As these examples make clear, whether physical specimens were involved or not, much of the knowledge produced from the field during the railroad era was grounded in lay or amateur networks. More broadly, then, naturalists and scientists wanting to produce knowledge of the vast Great Plains and Rocky Mountains region overcame their lack of access to it by forming networks with people who lived there. The lay networks that emerged during the railroad era were shaped by the new envirotechnical infrastructure of transportation and communication, which enabled them to become more systematic and standardized. While networks of lay people, however carefully coordinated, could rarely approach the level of top-down direction and guidance that the supervisors of survey scientists could demand, they did become increasingly distinct from prerailroad natural history practice in the field. And unlike some other modes of field production such as quarries or stations, which concentrated research at single locations, lay networks by design aggregated observations and specimens from many different places. As was true for all field research, place-based information—including not just location in space but also other environmental data such as altitude, topography, and climate—constituted an essential component of scientific reporting.

Those participating in lay or amateur networks followed in the long tradition of natural history networks stretching back to the early modern period.[4] Prior to the development of the infrastructure of railroads, telegraphs, and rapid postal delivery, many natural history collectors communicated with each other on a much longer time horizon, often spanning months or even years. The ease of communication in a matter of minutes or days made it possible to recruit lay or amateur collaborators much more quickly and widely, and it enabled their work to be coordinated more closely and systematically. In turn, responses by lay people in the field could also be transmitted much more rapidly, and the shipment of specimens by train from networks of collectors in the field was also much quicker and more reliable. The railway lines of the Great Plains and Rocky Mountains were not, as a rule, as densely built as their eastern counterparts, and the vast spaces between railway stations constituted not only the domains of survey field work but also the local perambulations of lay collectors and observers. It is thus especially important to place networks in the environmental context of these particular regions to see how the natural world shaped the development of networks.

Such networks in the field operated above all through the trust and goodwill of lay people, which presented distinct challenges. Without some modicum of trust of local collaborators by scientists, there is no way that a network could function at all to produce reliable knowledge. The validity of the knowledge was thus ultimately dependent on the credibility of people who were usually neither trained and disciplined as part of institutional structures nor, in most cases, specially educated in the sciences. Local participants in networks were usually—though not always—assumed to be reliable in reporting basic facts and observations, but the proper interpretation of physical objects and aggregated observations was more commonly claimed by credentialed scientists at museums, universities, and government agencies. (This did not preclude lay and amateur collaborators from developing their own interpretations, however.) Furthermore, ever-increasing standards of rigor in the field sciences also inspired efforts at standardizing lay collecting and observing, as well as mechanisms for maximizing trust, through rules, guidelines, and formal structures for exercising authority.[5]

As a mode of field practice in the West, much like in other parts of the United States and the world, the lay network harnessed the observational and collecting powers of geographically dispersed people and linked together those who lived, worked, and played in diverse places. The participants in networks were generally lay people rather than career scientists, even if those who made their careers in one field might also collect and observe nature in other domains avocationally. Whatever their occupational status, those who lived in the field possessed their own experiential knowledge of place, which could be selectively appropriated by scientists who lived elsewhere. The knowledge of experience could be a powerful resource, but it posed a problem of translation for all the field sciences, as many historians have fruitfully discussed.[6] The construction of lay networks in the field during the railroad era is a promising arena for further developing these lines of inquiry.

What become especially visible, then, in examining the lay network as a mode of field practice are the roles of other participants.[7] By its very nature, the field as a place has been less bounded than the laboratory: access is difficult to restrict. For a lay network, moreover, widespread participation in knowledge making was not just a noteworthy feature; it was central to its functioning. Career field scientists benefited from this permeability of their knowledge-making domain to outsiders, through the experiential knowledge, specimen gathering, and distributed observational and collecting power of people on the ground. At the core of this chapter will be the revealing interactions between career scientists and members of other social groups that were essential to producing field science.

As a means of work organization transforming nature into knowledge, the lay network was also a powerful way to connect geographically extensive natural environments to scientific nodes and centers. But places did not report themselves. The distribution of human settlement in the American West undergirded the production of knowledge in lay networks. Their building required the cooperation of local residents and travelers who were not only sufficiently interested—whether through a desire to participate in science, make money, or both—but also regarded as trustworthy. At the same time, American Indian inhabitants, who might be relied upon as guides and whose local knowledge was often used with minimal attribution, were rarely trusted as observers or collectors in networks, with the conspicuous exception of anthropological research informants. The history of producing knowledge through lay networks thus largely coincides with the history of westward expansion of white settler society. Early lay networks were based on existing institutions that preceded large-scale civilian American settlement, especially the outposts of the US Army. With the arrival of larger waves of immigrants making a living from mining, ranching, and agriculture, lay networks incorporated some of these new arrivals. Many of these networks were based on specimen collecting while others were designed mainly to circulate observational data, or both physical specimens and observations. After discussing the transition to a more densely settled landscape, this chapter will examine examples of both specimen and information networks. Ultimately, the range of environmental phenomena studied by field networks through both collecting physical specimens and aggregating observations from diverse locations was staggeringly wide.

Environmental Change and Human Migration

Human interactions with the environment through lay networks were fundamentally shaped by the geography of settlement—which, during the railroad era, was exceptionally dynamic. In establishing the envirotechnical basis of scientific knowledge production through lay networks, then, it is essential to sketch the contours of settlement geography during this period. Yet while the movement of human beings on the settlement landscape may have been the obvious and determinative shaping force, we should not overlook the dynamism of the natural world itself. To be sure, the enduring environmental features of the Plains and Rockies such as aridity or semiaridity, high altitude, and diverse topography grounded the environmental history of scientific practice. But we can also discern environmental factors that reflect the dynamics of change over time, including natural resource depletion, grassland and woodland fire, and powerful climatic cycles—as well as more distinctive episodic events, such as the remarkable Rocky Mountain locust invasions that spread across the Great Plains in the late 1870s, or later attacks by potato beetles and spotted-fever ticks.[8] In addition, we can look back into prehistoric time to acquire a sense of the natural origins of the Plains and Rockies, as well as human attempts to make a living as revealed through archaeology.

The historical climate of the region, especially for the Great Plains, has been a perennially contested issue. Undoubtedly, significant changes have occurred over time that have altered the conditions of human habitation. Archaeologists conventionally divide the long span of human occupation into three periods: the Paleoindian period, dominated by big game hunting; the archaic period, including a time of great drought called the Altithermal from about 6,500 to 4,500 years ago; and the modern period of the last two thousand years or so.[9] Despite the "extremely adverse climatic conditions" during the Altithermal period, the northern Plains, at least, may have supported some human occupancy through communal bison hunting.[10] The southern Plains, likewise, were seriously affected by warmer and drier climatic conditions, with human adjustments such as the digging of deeper wells.[11] Toward the latter part of the archaic period, about five thousand years ago, humidity seems to have increased, as indicated by the pollen records in a Nebraska Sandhills lake.[12] During more humid times in the modern era (such as AD 1000–1500) sedentary American Indian villages appeared along the rivers of the Great Plains, partly drawn by, but not heavily reliant on, the bison.[13] Evidence from fossils, soils, and archaeology suggests that the climate could sometimes change dramatically even over short time spans, so that severe periods of drought such as occurred from the twelfth to the fourteenth century pushed settlement back east.[14] White settlers hoped to transform the region's environment, but through all the cyclical climatic fluctuations of the past century or more, the lack of rainfall has remained a stubborn environmental constraint. Despite the increase in productivity made possible by intensive irrigation, the Plains have steadily lost population.[15]

Climatic changes have also occurred on a shorter time scale. For example, some early nineteenth-century observers claimed that the climate was relatively arid then, perhaps even a Great American Desert. This may have been true in the popular imagination, but apparently not on the ground. According to tree ring evidence, the first half of the nineteenth century was "considerably wetter" than other half-century periods of the past four centuries, with a "higher precipitation effectiveness" due to the colder temperatures of the Little Ice Age.[16] So while it is clear that conditions have varied over time, historians must be cautious in making inferences on the basis of anecdotal information from travelers. And despite considerable cyclical variation, the key general features of the environment of the region have usually remained somewhat constant over time: dry and high, yet with islands of exceptional height the most likely to offer some reprieve from the region's aridity. Moreover, both the forested, mountainous Rocky Mountains and the grassy Great Plains experienced wildfire, which reconfigured and restored landscapes.[17]

American Indian societies migrated into the region and gained knowledge of its environmental conditions over the course of many centuries.[18] By the mid-nineteenth century, even after considerable traffic on cross-country paths such as the Oregon Trail, the entire Plains and Rockies area was still mostly devoid of Euro-American settlement. The first great event to change that was the Mormon migration to present-day Utah in the late 1840s. This earlier wave of American settlement along the Wasatch Front on the western edge of the region relied on the communal institutions of Mormon society.[19] With the exception of this crucial migration pattern along its western boundary, most of the Euro-American settlement in the region was concurrent with, and strongly shaped by, the knowledge frontier of science and technology. This wave also reshaped how people living in Utah related to their environment, although perhaps less completely than for the rest of the interior West.[20]

The patchy, sparse settlement pattern of Euro-American civilians was strongly influenced by the distribution of environmental resources. Geologically rich mining areas in the Rocky Mountains and irrigable river valleys across the Plains attracted greater clustering of settlers. Concentrated mineral resources in the Rocky Mountains, in particular, made that part of the region a zone where active collecting of scientific objects took place early in the railroad era. Strong reciprocal relations between officers and experts of the mining industry and geological scientists, for instance, stimulated the production of knowledge about mineralogy and related subjects. As they dug into the mountains, mining companies uncovered new and interesting mineral specimens that often circulated between private collections and the museum collections studied by scientists in urban centers and universities. Although mineral collecting lost much of its visibility during the twentieth century as the frontier of capitalist expansion in the United States shifted away from hard-rock mining, when the Rocky Mountains were first settled by Euro-Americans in the second half of the nineteenth century, mineral resources loomed large in natural science as much as in the material economy.

Much of the interior West outside the mining centers remained sparsely inhabited by white settlers into the 1900s. Yet there were some parts of the region that experienced substantial white civilian settlement by that time. These included the towns along the transcontinental railroad corridor in Nebraska, Wyoming, and Utah, as well as subsequent lines in other states, starting with Kansas and Colorado, the Colorado Piedmont region around Denver, and the Mormon settlements along the Wasatch Front in Utah. During the relatively humid 1880s, Great Plains agricultural settlement expanded westward across Kansas and Nebraska, reaching even into the semiarid western halves of those states. Although this wave of expansion subsided during the drier 1890s—before returning with a fierce intensity in the early 1900s—this significant movement of people began to gradually populate the region with field naturalists, mostly amateurs, in some parts of the region by the end of the nineteenth century. Additional waves of agrarian settlement—in Montana and the western Dakotas, as well as the Texas Panhandle—swept over the dry farming regions of the Plains after 1900.

Over the first few decades of American settlement in the Plains and Rockies, patterns of land use would change significantly, as those settlers became more aware of the advantages and limitations of the material environment. By the late nineteenth century, settler farmers on the Plains, such as in southwest Nebraska, would find through hard experience that adjusting agricultural practices to the aridity or semiaridity of the West was a necessity.[21] In a relatively less humid environment like the Great Plains, wherever mineral resources or intensive irrigation was not possible, extensive ranching or farm-based livestock grazing would typically become the default economic activity.[22] Despite the common perception that the Plains were mostly plowed up for agriculture, vast portions of the region have remained as grasslands supporting cattle and other grazing animals.[23] For those areas that were farmed with crops, the land also supported animal populations that settlers found troublesome, such as the prairie ground squirrels identified by field naturalist Vernon Bailey as agents of "immense damage" and thus the appropriate targets of field research on "their food habits" and methods "for their destruction." Likewise, as historian Susan Jones has noted, prairie dogs became key pests on the Great Plains rangelands and, accordingly, objects of research investigation.[24]

For naturalists who recognized the changes taking place, collecting native plants and animals became all the more urgent. People living in this area during its early settlement years were thus not only few in number but also highly coveted as collectors and observers with whom to be in correspondence. As elsewhere, residents often simply reported what they experienced in their day-to-day work or on recreational rambles. Sometimes these facts on the ground were reported in small notices by amateur collectors, such as one written by George P. Anderson of Dannebrog, Nebraska, in the *Proceedings of the Nebraska Ornithologists' Union*. "One day last summer, while taking a ramble along the Middle Loup River, I found a nest," his typical local amateur's report began.[25] Even when people did not live along the fuzzy yet fascinating boundary line between eastern and western species, they were still often along the edge of the range of some species that interested metropolitan naturalists. In Garfield, Utah, for example, a high school senior named Guy Hart dug up some honey ants, which he presented to A. O. Garrett, his biology teacher in Salt Lake City, who in turn sent them on to nationally known myrmecologist William Morton Wheeler on the East Coast. They were the first specimens found in Utah, and Wheeler identified them as "the farthest north and farthest west of any locality for any species or variety of the honey ant."[26]

Compared to the East, the interior West had a far lower population density and, accordingly, far fewer people interested in participating in scientific networks. Constructing large-scale networks was made even more difficult by the relative scarcity of cultural and educational institutions that could foster enthusiasm about the formal study of science. Moreover, the interior West was a region with comparatively few scientific institutions and lacked the dense web of small colleges that characterized other regions of the country such as the Midwest and Northeast, so scientists had to work unusually hard to develop and maintain their networks of correspondents in the region. Furthermore, the much more widely spaced railroad and telegraph networks in the Plains and Rockies, compared to the eastern United States, channeled flows between network nodes, selectively favoring certain corridors while rendering vast swaths of territory either inaccessible or marginal. This was especially true for the Kansas City–based weather observing networks, for which the network of telegraph lines "constituted the *de facto* locational realm of possibility for the geographical distribution of daily weather data reporters on the frontier."[27] On the other hand, the abundant environmental riches of the outdoors increased the attractions of observing and collecting work in the Plains and Rockies. Likewise, from the field scientist's perspective, the value of specimens and information from the West for producing new knowledge was often quite high. A common thread in all the diverse examples discussed in this chapter is the intense interest in the natural resources that the environment of the interior West had to offer despite problems of gaining access to the region and the selective limitations on knowing nature that the lay network entailed as a mode of practice during the railroad era.

From Army Networks to a New Geography of Settlement

Collecting and observing by residents in the interior American West were at first largely organized around the system of US Army forts that provided protection and supplies within the region. With the exception of a small number of nascent towns and cities, the Plains and Rockies region as a whole was only minimally settled by Euro-Americans. Army naturalists, best known for collecting mammals and birds, operated in the West even before the Civil War. In the first decades after the war, the army through its western operations continued to have the advantage of access to remote field sites, producing such notable figures as Elliott Coues, whom one historian views "without doubt" as "the most prominent military naturalist of the post-war era," and Edgar Alexander Mearns, who collected specimens all over the West despite his focus on the Mexican boundary region in his own publications.[28] Through at least 1876, frequent wars with American Indians made access to the region difficult for anyone not affiliated with the US Army, and even those independent traders and travelers who did venture into the interior West during the time of most violent conflict often took advantage of established forts and military escorts for protection.

Of the people stationed at military forts who participated in collecting and observing nature, the most common participants were post surgeons.[29] Coues himself was an assistant surgeon, as was the bird and mammal observer W. J. Hoffman (better known as a folklorist and ethnologist) in the Dakota Territory in the early 1870s. Another army surgeon active in field collecting was George M. Sternberg, the older brother of independent fossil collector Charles H. Sternberg. The elder Sternberg was assigned to various posts in Kansas in the late 1860s, including Fort Harker, Fort Riley, and Fort Hays, and traveled with expeditions such as Major General Philip H. Sheridan's. "Spending a great deal of time on the march," one biographer observes, "he saw much of the country at first hand and had opportunities to study the geologic formations along the way." Sternberg also collected American Indian artifacts, birds, mammals, and fossils, sending the latter to Philadelphia scientist Joseph Leidy for inclusion in his reports on vertebrate paleontology, as well as some American Indian skulls to the Army Medical Museum.[30] As this last example indicates, some surgeons extended their collecting to human bodies. A surgeon at Fort Laramie in Wyoming in the mid-1860s stole a Sioux chief's body from its burial scaffold before being ordered to return it to avert a conflict. Likewise, a surgeon at Fort Harker in Kansas decapitated six Pawnee bodies to procure the skulls for the Army Medical Museum in 1867.[31] Whether on expedition or staying at a frontier post—and despite the brief tenure common to such positions—army surgeons combined access to the field with a medical background that better prepared them for scientific collecting than most others.

But post surgeons were not the only collectors and observers of nature in the US Army. Others associated with military forts and campaigns in the West, including officers, also participated in scientific collecting. General Custer, for example, collected fossils and animal skins during the Black Hills expedition of 1874 and also sent live animals to the zoo at New York's Central Park, while General George Crook collected birds and butterflies in Wyoming's Bighorn Mountains.[32] In accordance with the great interest in American Indian languages among military officers, Colonel Phillipe Régis de Trobriand put together "an extensive Lakota Sioux vocabulary while stationed in Dakota Territory, though he never tried to publish it."[33] Even the US Weather Bureau, well before it established the telegraph-based weather observing networks based in Kansas City to track the movements of weather systems across the Plains of western Kansas in the early twentieth century, began its life as the US Army Signal Corps, using the web of army posts across the country, including the interior West, as the basis for its information-gathering activities. These military roots were institutionally preserved through the continuity of personnel and operational procedures during the changeover to civilian control in the early 1890s.

Other collaborators resided in or traveled through the towns that grew up around military posts to supply them. Drs. J. Van A. Carter and Joseph K. Corson, for example, were friends of Philadelphia paleontologist Joseph Leidy who lived around Fort Bridger in southwestern Wyoming in the early 1870s.[34] And Fort Bridger continued to be important as a frontier town connected to metropolitan networks with ample fossil beds around it into the 1880s. During the Princeton expedition of 1885, for example, paleontologist William Berryman Scott, while staying "at the little hotel at Fort Bridger," encountered a Salt Lake City dentist named Dr. Whytock. "While I was still at the hotel, packing up our baggage," Scott reported, the dentist "came in from one of his tours, having travelled from Henry's Fork by the mail buckboard, and showed a large fragment of fossil bone evidently freshly broken from a skull."[35] Mission-

Table 2.1. Years of transition from territorial status to statehood

1845	Texas
1861	Kansas
1867	Nebraska
1876	Colorado
1889	North Dakota
1889	South Dakota
1889	Montana
1890	Idaho
1890	Wyoming
1896	Utah
1907	Oklahoma
1912	New Mexico

aries, too, were important in scientific networks, especially given their privileged access to American Indian reservations. Missionary members of the American Folklore Society in its early years in the late 1880s and early 1890s included Episcopal Bishop William Hobart Hare of the Dakotas, among others.[36] Whether generals, townspeople, traveling naturalists, or missionaries, early residents of the interior West contributed valuable observations and material specimens from their unusual vantage points on a frontier still populated by more American Indians than white settlers.

From a landscape of military outposts, the transition to a new settlement geography can be traced both through the shift from territorial status to statehood and through trends in population growth. Except for Texas, all the states of the Great Plains and Rocky Mountains passed through a territorial period before achieving statehood. Three states in the central zone—Kansas, Nebraska, and Colorado—became states near the beginning of the railroad era, while five states were established around its midpoint in 1889–1890, and three states came later (see table 2.1). Rapid population growth transformed the settlement geography of the Plains and Rockies between 1860 and 1920, as statewide population numbers suggest (see table 2.2). Starting slowly in the 1860s, settlement grew rapidly in the wake of railroad building, beginning with the central corridor of the first transcontinental railroad and later expanding to the northern and southern corridors, which received their own rail-

Table 2.2. Population by state (or t = territory), US Census

State	1860	1870	1880	1890	1900	1910	1920
North Dakota	—	—	—	182,719	319,146	577,056	646,872
South Dakota	—	—	—	328,808	401,570	583,888	636,547
Nebraska	28,841(t)	122,993	452,402	1,058,910	1,066,300	1,192,214	1,296,371
Kansas	107,206(t)	364,399	996,096	1,427,096	1,470,495	1,690,949	1,769,257
Oklahoma	—	—	—	—	—	1,657,155	2,028,283
Texas	604,215	818,579	1,591,749	2,235,523	3,048,710	3,896,542	4,663,228
New Mexico	—	—	—	—	—	327,299(t)	360,350
Colorado	—	39,864(t)	194,327	412,198	539,700	799,024	939,629
Wyoming	—	—	—	60,705	92,531	145,965	194,402
Montana	—	—	—	132,159	243,329	376,053	548,889
Idaho	—	—	—	84,385	161,772	325,594	431,866
Utah	—	—	—	207,905	276,749	373,351	449,396

Source: US Census, Historical Census Browser

roads by the 1880s and 1890s. Mining booms were responsible for some of the influx, especially in the Rocky Mountains and its outposts such as the Black Hills of western South Dakota, but much of the enduring settlement in these states consisted of those living on farms and ranches, as well as in small towns.

By the early twentieth century, a few urban centers were also beginning to emerge within the region, such as Denver (founded in 1858 during a gold rush), which was located on the boundary between the Plains and Rockies, but most cities were on the eastern edge of the Plains, including Omaha, Nebraska; Kansas City, Missouri; and Lubbock, Texas. The populations of these states—cities, towns, and rural areas together—formed the pool of potential lay and amateur collaborators for field science networks, and the vast increase in population thus played a key role in enabling these lay networks to grow. The overall population density of the Plains and Rockies remained quite small compared to the United States as a whole, but it was nonetheless a significant, and much more widely distributed, increase over the sparse population at scattered outposts of the US Army and occasional missionaries.

The establishment of a new geography of settlement in the Great Plains and Rocky Mountains did not result in a static landscape or constant set of environmental conditions, however. Just as climatic cycles were important to the movements and activities of American Indians and early settlers, so too did they continue to play a role throughout the railroad era. Into the early twentieth century, variations from one year to the next continued to shape the environmental experiences of farmers, ranchers, miners, local naturalists, and other lay or amateur collaborators in networks to produce knowledge in the field. When a Nebraska agricultural scientist set up a series of alfalfa experiments with about three dozen cooperating farmers in several distinct subregions of the state in 1910, for example, he found that for all of them, "weather conditions were very unfavorable and it was thought advisable to repeat the tests in the spring of 1911," when "conditions were more severe" and farmers suffered from "extremely dry weather and a large number of grasshoppers."[37] Thus, while the human interactions and power relations between scientific experts who organized networks of local people to produce knowledge and their lay and amateur collaborators were central to the practice of lay networks, we should not underestimate the enduring influence of the natural world itself, which remained a strong shaping force, even for lay networks that depended on the participation of settlers determined to transform the environment.

ORGANIZING THE NETWORK

At the simplest level, the lay network operated as a more rigorous, systematized version of the networks that had long existed in prerailroad natural history practice, which were characterized by the hierarchical flow of materials or information toward a center. The practices of the modern field sciences—including a modernized version of natural history itself—developed in part from those of the natural history tradition, and that is especially true of the lay network in the railroad age. Sketching some of these key practices in the gray area between older natural history practices and emerging lay networks can reveal key patterns characterizing the earlier practices out of which lay networks developed. Above all, examining the geography of natural history indicates the continuing role of center-periphery relations while emphasizing the advantages and disadvantages of being in each place within such a hierarchical system.

While acknowledging the strong structuring influence of long-distance core-periphery relationships, the practice of field science in lay networks should also be seen in its local and regional context. As historians have emphasized, local collecting has a lengthy pedigree in defining motivations and activities for naturalists.[38] Likewise, local observation has been important as a complement to collecting, as well as an activity undertaken for its own sake in domains such as meteorology.[39] By studying the places in which they lived, lay or amateur field naturalists were strongly defined in their practices by a local orientation, even as they were sometimes also connected to distant metropoles. Within the US Great Plains and Rocky Mountains, unlike areas that had been more densely settled for centuries, local naturalists were at first thin on the ground. American Indians and early white settlers were keen observers and sometimes even collectors of the natural world, but only a few were in close communication with distant scientific centers in the 1860s and 1870s. By the end of the nineteenth century, however, and certainly in the opening decades of the twentieth century, a critical mass of local collectors and observers who were practicing naturalists, or at least lay people who were in communication with distant experts, had emerged. These people were simultaneously linked with cosmopolitan naturalists and immersed in their own local contexts of everyday experience of nature.

A close analysis of the proliferation of naturalists' activities in the Great Plains and Rocky Mountains in the late nineteenth and early twentieth centuries affirms that natural history relied on a hierarchical division of labor between publishing researchers at central locations and their correspondents spread throughout the field. Historians writing about natural history networks have long relied on a geographical language of center and periphery that traces crucial hierarchies in the production of knowledge. Bruno Latour's emphasis on "centers of calculation," which emphasizes how facts are assembled to produce more powerful knowledge that describes and applies to a larger geographical area, has been widely debated.[40] However, as Steven Shapin has argued, Latour's picture is missing a "conception of normative order." To transcend accounts that are "all natural fact and no moral fact," he has urged an emphasis on trust relationships.[41] In the practice of natural history, it seems clear that maintaining trust between collaborators across geographical distance was central. As we shall see below, insisting on the strong structuring role of center-periphery relationships is just part of the story, and a full understanding of the practice of natural history requires attention to the texture of conflict and cooperation that characterized the human linkages between core and periphery.[42]

In long-term historical context, centers in core regions were large associations of those who studied the natural world; the Royal Society of London of the early eighteenth century provided one such institutional base. Andrea Rusnock identifies a "scientific method of correspondence" for which the society "acted as a critical centre of collection" under Secretary James Jurin in the 1720s.[43] But most were based in institutions, such as museums and botanical gardens, with large collections of specimens. In practice, then, one of the greatest advantages of being at the center was simply having access to large and diverse collections. Centers typically also conferred cultural power. At the Jardin du Roi, for example, in E. C. Spary's analysis, "problems of social and natural historical order were to be solved by the good management of people and specimens," on ever-increasing scales from France to Europe to the entire globe.[44]

Metropolitan centers in the United States were at first secondary satellites to European capitals such as London and Paris. By the middle of the nineteenth century, however, American centers were becoming more important, especially the Washington, DC–based correspondence network of Spencer Baird, as noted above. Other US centers emerged around specific institutions and individuals in such eastern cities as Albany, New Haven, Boston, and Philadelphia. When combined with westward frontier expansion, the structure of American natural history thus increasingly took on a strong West-East dichotomy. While the southern states, as well as the hinterlands of the northeastern states, were also peripheral, when focusing on the Great Plains and Rocky Mountains regions, the articulation of a core-periphery relationship as between East and West was ubiquitous. And for naturalists in the eastern United States, the western states and territories loomed very large and to a great extent were viewed as the most desirable field in which to operate, at least until the burgeoning interest in Alaska and the American tropics around the turn of the twentieth century. (Even then, the American West remained a significant region for American naturalists.) As specimens collected through western expeditions flowed into Washington, DC, and other eastern centers, the advantages of eastern US scientific centers came more and more to resemble those of their European counterparts.

While tacit cooperation between rival metropolitan centers often reigned in natural history collecting, there were notable exceptions, such as the well-documented "fossil war" between O. C. Marsh of New Haven and E. D. Cope of Philadelphia in the 1870s and 1880s, introduced in chapter 1.[45] Effective organization for Marsh and Cope mostly took the form of directive correspondence—usually by regular mail, with the occasional telegraph dispatch for urgent matters—supplemented by occasional on-site visits or field excursions, particularly for Cope. Our history thus far has tended to be uneven and skewed toward a small number of atypical yet better-known collaborators who later developed their own scientific careers, such as John Bell Hatcher.[46] It is difficult even to identify all of the collectors who were in Marsh's and Cope's networks. But each significant collecting location was usually associated with one or more local collaborators who agreed to join one rival team or the other. A crucial feature of these competitive collecting networks was that they accelerated a shift from voluntary to paid collaboration in paleontology, as Marsh and Cope struggled against each other to secure the best fossil material.[47] Rather than relying only on local scientific enthusiasts who might be encouraged to send in material in return for identification, naming, and the pleasures of intellectual association, the competing paleontologists enlarged their networks to include anyone who found (or could find) valuable material and was willing to sell it.

Initially, the collecting networks of Cope and Marsh developed in a more conventional fashion. Like so much early field work in the West, collecting localities at first centered on US Army forts. Post surgeon Theophilus P. Turner, who was stationed at Fort Wallace in Kansas in the late 1860s, found a fossil skeleton "in a ravine near an old Indian camp" while hunting. He sent the specimen to Cope, who hailed its "unusually good condition" and named it *Elasmosaurus platyurus*.[48] Likewise, at Fort Washakie in Wyoming, assistant surgeon Charles K. Winne shipped a box of "little fossil shells" to Marsh in August 1879, "asking him whether they were salt or fresh water shells," and a reply encouraged him to look for more shells by going camping in some nearby canyons.[49] Over time, and as civilians began to expand in greater numbers across the Great Plains and along the Rocky Mountain front, the competing networks of Cope and Marsh gradually came to include other settled areas outside the military forts.

It might be expected that local residents with some degree of at least amateur expertise would be the most likely to communicate to scientists elsewhere and to send them material specimens, and this was undoubtedly true. But there were many others who possessed little or no special expertise, yet whose lay status did not necessarily prevent them from participating in such interactions. The variability and particularity of local environments made these lay people dispersed across the Plains and Rockies highly desirable as correspondents and collectors, despite their lack of formal knowledge. This was especially true for objects such as meteorites, which could fall anywhere. There was some speculation that certain places, such as western Kansas, had produced more known meteorites because it had soil "especially favorable to the finding of meteorites, as it contains few terrestrial rocks," as well as a dry climate "tending to the preservation of meteorites," though these advantages were offset "by the scantiness of the population."[50] Much like any other specimen of scientific interest, the meteorite reached the field scientist through the collaboration of local finders and property owners. But unlike the typical mineral specimen, the meteorite was an object of great rarity and high scientific value because of its extraterrestrial origin. Outdone perhaps only by the great dinosaur skeletons of the West, the meteorite was one of the most highly sought scientific objects. This potent combination of tremendous scientific value and recurring local agricultural hardship led to an intermittent series of headlong rushes to locate and exchange such specimens, typically following directly on the heels of major falls. Some specimens were sold directly from their finders, while others had already been exchanged locally before making their way into the hands of meteoritic experts.[51]

Collecting actual physical objects like fossils, animals, birds, minerals, and meteorites was the focus of most natural history field work, with subsequent shipment of specimens to recognized experts elsewhere, and these practices were continued even as the expert-lay boundary line became more definite and as the lay networks become more systematized. For scientists located at museums, botanical gardens, government agencies, and other central institutions, collecting actual specimens continued to have the epistemic advantage of delivering the material evidence directly from the field. While collaborators were obviously trusted to report (whenever possible) on where exactly specimens were obtained, as well as to deploy their skill in preserving them—to say nothing of using their skills and local knowledge of experience to find such objects in the first place—dependence on the cognitive labor of amateurs or lay people in the field was reduced by the very materiality of the objects that were collected. In other words, the main work of transforming natural objects into scientific knowledge on paper took place under the direct control of the distant field scientist.

The material infrastructure that made possible these exchanges of physical specimens was a seamless union of transportation and communications networks. Railroads, supplemented by animal-drawn conveyances, moved the mail through the vital interconnecting information web of the US Post Office, and in a few cases also its foreign counterparts. The movement of these objects constituted both an act of material transport and an act of scientific communication. This dual reality usually passed without comment, but its centrality to networks linking scientists with lay and amateur collaborators, as well as with each other, occasionally surfaced at times when that system came under threat. One such time was in the early 1890s, when the post office disclosed that the "existing provisions" of international postal agreement—governed by the Universal Postal Union, which was established in the 1870s and still exists today—did not permit "objects of Natural History" to be sent "as samples of merchandise and under the rates of postage therefor." Yet, as two officers of the Academy of Natural Sciences in Philadelphia pointed out, the regular rate for letters was "ten times that required for samples of merchandise," and they worried that "such a rate for specimens of Natural History is virtually prohibitive."[52]

An attempt by the US Post Office to change the international rules "so that such specimens might be received into the mails at the same rate as samples of merchandise" had already been rebuffed by seventeen other countries, including such scientific powers as Great Britain and Germany. As a consequence, the Philadelphia naturalists' letter to the editor of *Science* implored "various scientific bodies" around the world to lobby for the change.[53] In reply, however, a Canadian colleague redirected "the blame" upon the United States itself, attributing the problem to absence of parcel post. "There is no difficulty in transmitting specimens from Canada to the most remote countries," he pointed out, "but the United States by their policy make it impossible to receive or to send them." He therefore urged US naturalists to lobby for "more advanced and liberal postal arrangements" in their own country.[54] Nevertheless, four years later, American naturalists were still working through the International Congress of Zoology to facilitate the sending of natural specimens at merchandise sample rates.[55] Among the many intriguing aspects of this debate—the invited comparison of natural history specimens to their sample counterparts in the world of commerce, for example—is that it underscores the close reliance of these scientific exchange networks, including those that linked scientists with their lay and amateur collaborators, on a merger of transport and communications systems for moving physical objects from periphery to center, or back out again.

An increasingly important technological innovation for representing material objects in the field by the turn of the twentieth century was photography. In the spectrum of practices for knitting together lay or amateur networks, photography serves as an intermediate case, since it produced material artifacts that were distinct from the physical objects themselves. In other words, when material specimens were not available, a photograph could be transmitted as evidence of visual appearance. Photography was also a new tool physically under the control of lay or amateur collaborators. Yet even when photographic evidence was available, it could be less certain than material specimens themselves, as the US Biological Survey's director C. Hart Merriam and chief field naturalist Vernon Bailey experienced in replying to Colorado naturalist E. R. Warren in 1912. "Bailey and I have examined your Yellowstone Ground Squirrel photographs from Mammoth Hot Springs with considerable care," Merriam noted, "but [we] do not feel entirely certain whether they are *armatus* or *richardsoni*."[56] In practice, then, even the visual realism of photography was not immune from the interpretive challenges that all field scientists faced when assessing the value and credibility of the vast amount of observational information that came from their lay or amateur collaborators in the field.

While the movement of material objects, or occasionally photographs, was important for all natural history practice, including when conducted through increasingly systematized networks of lay people dispersed across the landscape, one striking characteristic was the rising importance of controlling and systematizing information flows in addition to or instead of the movement of physical objects. In the main, these flows consisted of verbal or numerical reports. The example of meteorite collecting from lay people, however, also highlights the need for securing observational information distributed through geographical space in conjunction with the scientific objects themselves. In order to produce knowledge about meteorites, scientists studying them typically possessed a strong desire to secure observations from people who had observed meteor falls. Even more than the procuring of physical specimens, which might be exchanged locally with others who had greater knowledge or interest, acquiring observational knowledge unavoidably meant relying in many cases on lay people who possessed no cosmopolitan expertise. It is time, then, to bring the significance of information flows, including for meteorites, into sharper focus.

Information Flows

The collection and transport of physical specimens were not the only activities that knit together lay networks—far from it. More and more, scientists studying the environment of the Great Plains and Rocky Mountains relied on verbal and numerical reports from lay people. These could involve the observational information that researchers sought in order to describe or locate specimens, samples, and other collected objects, or even to understand the ecological, stratigraphic, or cultural context of the objects collected—a desire that, when strong enough, motivated scientists to undertake field work in other modes such as conducting surveys or establishing stations. Often, however, the limits of funding and logistics meant relying on contextual information that lay people provided. While some lay networks, especially in the agricultural sciences, began with the flow of specimens such as seed varieties outward through the mail for testing and experimenting, in exchange for a promised return flow of information, elsewhere the local knowledge was solicited wholly in the absence of any material specimens. Or, information gathering from lay people could involve observations about things that could not be physically moved, such as the weather or the observed paths of falling meteors. In all these cases, facts were constructed through the flows of information in lay networks.

No investigator of nature in the field could be everywhere at once, and certainly no one could be everywhere through all times of day, all seasons, and throughout the duration of longer-term climatic cycles.

Surveys, in particular, despite their comprehensive geographical scope and sometimes even multiple teams in the field simultaneously, as we shall see, could never fully replicate the experience of living and working in a place. In the history of scientific observation, the experiential knowledge of the environment possessed by lay people has constituted a complementary asset to the cosmopolitan knowledge of scientific experts. Early scientists working in the newly founded universities of the Plains and Rockies recognized this as well as or better than anyone. Charles Bessey, the eminent Nebraska botanist, writing in *Science* in 1906, considered everyday observation of the environment—in his words, "observation of the plants and their environment which one makes when living much in the forests, the prairies, the swamps, the fields, and the gardens"—to be "the simplest kind of field work, and on many accounts the most productive and helpful."[57]

Moreover, Bessey distinguished the knowledge of everyday experience from scientific knowledge, but with a special appreciation of the former over the latter. "The man who lives 'in the open' and learns of plants and about plants while living with them," he remarked, "obtains a mass of most valuable botanical knowledge even though it may not be formulated in botanical language."[58] Nebraska forestry students in the first decade of the twentieth century, to whom Bessey was a leading inspirational figure, developed a tradition of summer vacation excursions to the forests of the Rocky Mountains, where they would "hire out as ordinary day laborers with the result that they are found engaged in a wide range of work such as building roads, constructing camps, carrying supplies, fighting fires, planting trees, cutting timber," or even "the somewhat more dignified position of forest guard." Doing such everyday work activities in the forest, "the men are kept very busy during the daily routine," but if they paid attention they had the chance for "interesting observations . . . which, if made, might be of inestimable value in their future career as scientific foresters."[59] Likewise, when botanist Alice Eastwood scientifically described a plant that her trusted local collaborator, the rancher Al Wetherill, had found "abundant" in the San Juan River canyons of southeastern Utah, she also reported on the environmental conditions in which he told her it was commonly found: "under shady cliffs where the rocks and soil are always moist from the oozing alkaline water."[60]

Both Bessey and Eastwood, as field scientists oriented toward cosmopolitan knowledge, situated the experiential knowledge of their lay and amateur collaborators alongside much more systematic and larger-scale projects. Indeed, Bessey's remarks were delivered at the tenth anniversary of the University of Nebraska's "Botanical Seminar," which engaged in extensive survey field work, as we shall see in the next chapter. As Bessey's ambitious and energetic students traveled across the Nebraska landscape during summer vacations and other times, it would not have escaped his notice that they could never fully capture all the insights of the locals they encountered and with whom they were sometimes in correspondence. However, the Botanical Survey of Nebraska, as its name suggests, was deliberately designed as a survey enterprise, relying on trained travelers, so Bessey's admiration of local knowledge may have been part of a wider effort to demarcate the knowledge of science from the knowledge of everyday experience. Tellingly, it was not just survey science to which Bessey was comparing lay experiential knowledge. "Many an unlearned Thoreau," he added, "knows much more about the habits of plants than the laboratory botanist who is a stranger to the wild plants in their native haunts."[61] For Bessey, speaking and writing after several decades of laboratory ascendancy, the legitimation of the knowledge of lay experience in the field could also serve as an indirect affirmation of the practices of field scientists who sought to tap into the rapidly expanding pool of experiential knowledge, such as through the lay networks that redirected the flow of information, even if such efforts were sharply distinguished from the work of the Botanical Survey itself.

Operating in a continental, rather than a global, context of expanding empire in the nineteenth century, researchers in the United States developed ways of routinizing the information that could be gathered by observers scattered throughout the field. Most commonly, systematic information-gathering networks were designed and led by agents of the state and federal governments. One crucial instrument of standardization was the paper reporting form, which was the key material technology for transmitting information for Patrick Connor's weather observing networks in Kansas. In addition to the daily observations telegraphed to Kansas City, local collaborators also mailed in written monthly synopses to confirm their results and to provide a permanent record. These monthly reports were submitted "upon the first day of the succeeding month" on particular forms provided by the Weather Bureau.[62] In the off season, beginning in October for Corn and Wheat Region Observers, monthly reports were instead sent to the weather offices in state capital cities.[63] Archives bulge with filled-out copies of these forms, which constituted the ongoing inscription record of the local observations in the field.

Scientists at agricultural experiment stations also became especially prolific organizers of lay networks to gather information in the field. Despite conducting much of their research in field station mode, as we shall see, the substantial expenses for building, operating, and staffing more intensive research sites located in particular environmental places limited the number that could be supported at any one time. By conducting trials or experiments in cooperation with numerous farmers and ranchers who were spatially dispersed, agricultural scientists, like other organizers of lay networks, aggregated results from across the wide range of environmental variation that prevailed in the states of the Plains and Rockies. Similarly, state biological and geological surveys extended their own observations in the field, limited by time and finances, by harnessing the distributed observational powers of people already living there. Information about the movement of large-scale natural phenomena, such as meteor falls and weather patterns, was also gathered by state agencies and other institutions.

Typically, the most systematic attempts to gather information took the form of circulars, an increasingly popular networking technology that was widely deployed in order to collect responses from local naturalists, farmers, ranchers, mine operators, and other lay people.[64] In order to fully understand the broader context of mailed circulars, it is important to consider that they were mainly used for purposes other than information gathering by scientific agencies. "Thousands of them entered the mail each day," historian David Henkin points out, "especially after 1852, when the postage for any unsealed circular weighing three ounces or less and traveling anywhere in the country was set at one cent."[65] Circulars were also used to collect government statistical data, in much the same way that they would be used for scientific research, and as such constituted a basic form of knowledge gathering. In preparation for his 1870 report on mining in the western United States, for example, Rossiter Raymond, who was the national commissioner of mining statistics, widely circulated forms asking for information, to supplement his own "23,000 miles" of travel in 1868 and 1869. He sent the blanks to various local and county officials, newspaper editors, mine and mill superintendents or owners, and other prominent individuals, but the outcome was "a great disappointment," since very few "were received properly filled." Many mines "declined giving any information whatever," presumably for reasons of commercial secrecy, if not lack of interest in Raymond's project.[66]

Some of the wider historical context of the printed circular in nineteenth-century everyday life may help to explain Raymond's disappointing results. In general terms, circulars were "printed letters or advertisements posted or distributed to numerous recipients" and served numerous functions: they "promoted political causes, announced social events, solicited donations, and advertised goods and services." Henkin tellingly emphasizes the association of circulars with "the increasingly ordinary business of advertising, fund-raising, and propaganda" and points out that "they were used to perpetrate frauds, swindles, and schemes that took advantage of both the anonymity of the mail and the size of the postal network to reap considerable profits." Indeed, while sometimes perceived as "timely and interesting to their recipients, more often they were regarded as a paradigmatic form of junk mail."[67] This broader context may help to explain why circulars, even for information-gathering purposes, were ignored by so many, and it certainly suggests the expectations and associations we can expect that the material format of the circular may have evoked in recipients.

Nevertheless, circulars remained an important, and often at least partially effective, material technology for the collection of information from dispersed local observers. In circulars that were used for scientific information-gathering purposes, some of the questions posed were open-ended, but often the circulars asked for basic factual information or solicited numerical information. The pervasiveness of circulars for collecting information from lay people can hardly be overemphasized. As a technology of long-distance observing and for remotely organizing dispersed work in the field, the humble circular deserves more recognition than it has hitherto received from historians of science. It was a communications tool that at first glance seems detached from the natural world, but which on closer reflection presents something rather more interesting from a material and physical point of view. The circular was a mass-replicated printed form, card, or other piece of paper sent through the mail, usually through the federal postal service to a large number of people—or perhaps to newspapers for further dissemination, especially for initial appeals.[68] While some circulars were sent only to a select list of a few dozen recipients, the distribution lists of many circulars tended to be quite large, frequently numbering in the hundreds. Other circulars were sent to entire existing mailing lists, such as the membership of a state horticultural society.[69] In the Great

Plains and Rocky Mountains, as elsewhere, these pieces of paper typically traveled over the railroad to local depots, and then in some cases by wagon to more remote locations. Across the vast spaces of the Plains and Rockies, these circulars—followed by the responses to them—journeyed through the sprawling envirotechnical infrastructure of the vast spaces of the American West that the field scientists themselves did not have the time or money to travel through in person.

But a piece of paper is obviously quite different from a trained scientific observer. It must persuade lay people to supply information in a standardized format, and generally of a factual nature that can be trusted by a distant scientific authority. The genius of the circular is that its very design compels compliance with the expectations for more reliable knowledge production, through the wording and ordering of questions. Of course, actual compliance rates were something else entirely—many recipients failed to return any information at all, returned information that did not properly answer the questions posed, or failed to provide information in a form that could be aggregated to derive scientifically rigorous results. This was especially true during the earlier years of settlement and institutional development. As state agricultural research stations in the Plains and Rockies started organizing lay networks of farmers and ranchers in the 1890s, they often experienced disappointment at first. A cooperative seed test in Wyoming in 1895, for example, which involved "about one hundred and thirty farmers and ranchmen of the State," proved to be "not . . . satisfactory, as the reports are so meager and general in their nature as to be of no scientific value."[70] A few years later, in Nebraska, a series of cooperative experiments produced results that were "not . . . very satisfactory as comparatively few of those taking seed have rendered a report of their results."[71]

Over time, the efforts of network organizers to aggregate results from lay people such as farmers and ranchers would become better refined and more effective, on the whole, but the problem of ensuring compliance and proper data collection always weighed heavily on their minds. In the early years of the twentieth century, for example, Montana's agricultural experiment station's chemist F. W. Traphagen expressed "disappointment" at receiving returns from "only thirty farmers" despite sending out "over 500 samples of seed" for sugar beets.[72] The next year, his colleague F. B. Linfield likewise received only 93 replies out of 204 requests, many of which were "incomplete or very general in character."[73] And as late as 1908, at least one state agricultural expert in Wyoming lamented that only 75 replies were received from 400 circulars that had been mailed out, though he was pleased that many of the replies "went into the question very fully" even though no seeds or other materials were provided.[74]

Another crucial aspect of the circular was its geographical dispersion. The long reach of the federal postal network made it possible to reach widely distributed respondents in a variety of environmental settings. By tying each response to its physical location, the responses could be used to represent—or in some cases even map—the territory in much the same way as a survey, yet at a much lower expense.[75] Replies received from places with widely varying altitudes, soil types, topography, geological formations, and other environmental factors could be brought together in a central location and compared with one another. They could also produce unitary generalizations across a wider territory, such as when the "majority of answers from leading farmers and horticulturists of Oklahoma included crabgrass as one of the worst of weeds," despite this not being the case in other places, like Florida.[76] If objects conveyed geographically distributed environmental knowledge through their very movement through space, then replies to circulars constituted a more complex, but no less material, counterpart. The greater complexity arising from the need to simultaneously discipline and trust observers rendered the paper reply to the printed circular a source of great frustration, at times, to scientists of the field but also a source of great power for assembling environmental knowledge.

Consider once again the example of meteoritics, and in particular the question of how to gather information about a meteor fall. One such "fireball" occurred on December 7, 1900, in the skies of northwestern Colorado. Hearing the news, the director of the Chamberlin Observatory at the University of Denver "prepared circulars of inquiry, containing a number of questions, which were lavishly distributed over the region from which the meteor was visible." He also disseminated queries through local newspapers. In this case, local observation depended more on serendipity than on special skills or robust experiential knowledge, and so the net was cast as widely as possible. While a few particular observers with especially noteworthy reports were mentioned with their occupational titles, the aggregation of data from multiple observers itself was the primary means by which the reports were verified or corrected. Approximately 150 replies were received, and these observational reports were combined in digest form for the observatory's director by a student at the university identified as Miss Stingley, thereby revealing the likely path of the meteor.[77] In a similar example in Texas in 1917, geologist J. A. Udden compiled a map of observational reports obtained through a statewide appeal, and the map clearly reveals that while some observations seem anomalous, a pattern consistent with the general trend of the reports can be easily discerned from the practice of mapping them all together.[78] Thus, by combining and plotting enough observational reports, field scientists studying meteorites hoped to generate trustworthy and reliable general patterns, even when the information was gathered from lay people who possessed no special knowledge of meteorites.

Despite the interest attached to spectacular and unpredictable events such as meteor falls, the most obvious type of information to seek from lay people in the field through printed circulars was gathered through everyday experiences in practical activities such as agriculture, horticulture, ranching, and mining. While those who engaged in such pursuits possessed a great deal of experiential knowledge through their work in nature, the individual pieces of factual information extracted from that body of experience and wisdom were typically most desired. Much of this information was sought by state government agencies, such as agricultural experiment stations, boards of agriculture and horticulture, and geological surveys.[79] For such state agencies, building networks for gathering field information also meant cultivating ongoing relationships with the constituencies those agencies were intended to serve. In a representative early example, the Kansas State Board of Horticulture sent out circulars from 1880 through 1887 "to its leading members in each county, inquiring about existing forestry conditions and the progress of their own efforts during the preceding year."[80] State boards of agriculture, state geological offices, and other specialized agencies likewise collected information from farmers across their territories, based on their experiences with diverse local environments.

Managing Environmental Variability

Whether through physical objects or information about them, the knowledge of experience gained through working in agriculture, ranching, mining, amateur collecting, or other practical uses of the environment offered an enticing prospect for building bridges across the epistemic rift between experiential and cosmopolitan ways of knowing. Occasionally, this meant the more or less complete and holistic appropriation of knowledge acquired through experience. For example, when agrostologist (grass scientist) H. N. Vinall began his studies of the Nebraska Sandhills region—an unusual and distinctive environmental subregion of the Great Plains, about which local residents often knew more than scientists—he initially conducted "a study of the practices of the more successful farmers" in order to better understand what they had already learned.[81] Likewise, in Montana, agricultural scientists at Bozeman, located in Gallatin County, mailed a "detached card . . . to each owner of farm lands in this county"—nearly a thousand in total—and thereby acquired from the farmers themselves "an intimate knowledge of previous farm results," which they regarded as "very valuable in many ways."[82] Such inquiries were more typical in the early stages of research, and even those field scientists who continued to produce knowledge through lay networks nearly always shifted later to more tightly focused inquiries designed to elicit specific pieces of factual information.

Yet since lay or amateur collaborators were distributed spatially across the Great Plains and Rocky Mountains, managing the regions' environmental heterogeneity was a paramount concern. Even within more narrowly circumscribed subregions, researchers recognized the importance of constructing lay networks that embraced as much of that variation as possible by identifying dispersed local collaborators. In experimental forestry work initiated across the Nebraska Sandhills in early 1905, for example, despite having shifted to the more tightly controlled strategy of distributing jack-pine seedlings and requesting later reports, foresters recognized the need for identifying "ranchmen in widely scattered localities in the sand-hill country" in order to maximize the applicability of the results. Moreover, in summarizing the "authentic reports from nearly all of these plantations," the foresters pointed out that, while "the great majority" succeeded, the "two failures" were probably "due to unfavorable local conditions."[83] Similar tree-planting work was attempted through a lay network in Kansas, enabling tentative conclusions about the diverse environmental zones within the state (see figure 2.1). Thus, even when field scientists directing lay networks did not emphasize or explicitly discuss the distinctive experiential knowledge possessed by their rancher collaborators regarding local conditions, which they might incorporate into their work, they still recognized the importance of spatial distribution in diverse environmental conditions to the results.

More broadly, agricultural scientists and other field researchers appreciated the variability in environmental conditions—soil, climate, topography, altitude—that challenged the production of knowledge that would be valid throughout the vast land areas of their states. As Wyoming irrigation specialist Herbert T. Nowell declared in a 1908 re-

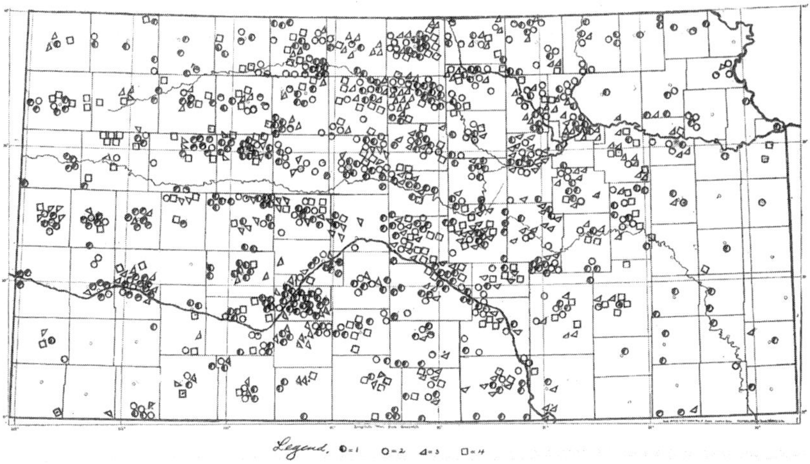

FIGURE 2.1. Map showing the distribution of locations studied through a typical state-based lay network. The reported results for this study included tables by year and species indicating the percent of living trees reported in different parts of the state. The extensive original caption reads: "The above map of Kansas above shows the distribution of trees sent out from the state nursery during the years 1912, 1913, 1914, and 1915, as indicated by the several characters: 1 shows the points to which shipments were made in 1912. 2 shows the points to which shipments were made in 1913. 3 shows the points to which shipments were made in 1914. 4 shows the points to which shipments were made in 1915. This chart, furthermore, shows the sections of the state in which tree planting is a factor in farm and home development." Chas. A. Scott, *Trees for Kansas,* Circular of the Kansas Agricultural Experiment Station, no. 55 (1916): 7.

port, "the localities devoted to farming" in his state were "somewhat scattered" and the "conditions in different parts of the state differ quite widely," thus leading him "to find out from actual correspondence with the farmers of the state what crops were most profitable in the various localities," among other things.[84] To solve this problem of producing knowledge that would be applicable in diverse places, they often resorted to circulating crop varieties to willing farmers in diverse locations. In many cases, the farmer collaborators were selected proportionally to represent different sections or counties, as when South Dakota's agricultural experiment station arranged in 1911 "to send ten plants free to the first ten applicants in each county" for its cooperative tests of alfalfa varieties brought from Russia.[85] By the early twentieth century, such intentionally systematic geographic distribution was becoming more common as a necessary condition for experimenting in the context of maximum environmental variability within state or regional boundaries.

A rare counterexample to the strong impulse to distribute varieties for testing in lay networks that encompassed diverse environmental conditions was at Fort Lewis College in Durango, Colorado. In 1920, Fort Lewis's School of Agriculture worked in cooperation with the main agricultural experiment station in Fort Collins to test out varieties in the area's "high altitude and arid conditions," in response to the "insistent demands" of local farmers to try them out.[86] In that case, the agricultural scientists at Fort Collins advised the man at the Fort Lewis School of Agriculture to "pick farmers having similar soil and irrigation conditions, if possible," and who were "living near together," so that they would each grow "one variety" that was "grown under comparable conditions" in the field. "You know the local conditions," he wrote.[87] It seems likely that Fort Lewis was exceptional only when viewed in isolation, however, for it was charged with producing valid knowledge in its own immediate local vicinity, whereas the main station in Fort Collins—and, even more emphatically, the US Department of Agriculture in Washington, DC—viewed Durango as just one locality within a much greater universe of places with contrasting environmental conditions.

The apparent enthusiasm among local farmers does, however, suggest that, by 1920 if not earlier, adapting knowledge to local conditions through lay networks had become a known and expected practice among farmers and other lay people, even in the most distant, remote corners of the Rocky Mountains and Great Plains. Indeed, almost three decades earlier, certain farmers were already so enthusiastic that they engaged in experimental variety testing for many years. Kansas farmer Elam Bartholomew began seeding a winter wheat plot under US Department of Agriculture direction in October 1891, "treating the various plots in different ways for the prevention of rusts." His diary entry for October 9 indicates his sense of pride and self-perception: "I have been appointed a special agent for the prosecution of this work. It is a very difficult piece of business and will require some skill and much patience as well as some very fine work."[88] He planted another rust-prevention experimental plot the following year for spring wheat as well as oats.[89] The third year, however, witnessed the beginning of the mid-1890s drought across the Plains. As Bartholomew recorded in his diary one day in August 1893: "Worked in the a.m. at making my report to the Sec'y of Agriculture on the Rust prevention experiments carried on here this year, which were a complete failure on account of drought."[90] Despite this difficulty, federal researchers continued to collaborate with Bartholomew. In 1907, he planted some experimental varieties of "Central American corn" as well as "24 plats of alfalfa 9x200 feet in size."[91] The next year, his cooperative work expanded to include thirty-eight varieties of corn and "113 plots of various strains and well-marked varieties of alfalfa."[92] Then in 1909 Bartholomew even added "a small tract of experimental cotton" to his US Department of Agriculture cooperative work.[93]

Yet while certain farmers were eager to cooperate in testing new seed varieties, they were not always willing to participate in sufficient numbers for all the disparate kinds of studies that field researchers wanted to do, especially in the earlier years. When Nebraska entomologist Lawrence Bruner decided to study an outbreak of chinch bugs in 1893, he looked to affected farmers for assistance. His plan was a kind of biological control: to develop a line of diseased bugs to send out to farmers that would infect other chinch bugs. But it was difficult to build a large enough network of cooperating farmers sending in live bugs. The chancellor at the University of Nebraska had asked that Bruner not "go out into the field personally to secure the necessary [chinch] bugs" because of the high cost of travel. "He seems to think that if the farmers of the state want any help in this line they should want it bad enough to send in the necessary live bugs for keeping the disease alive in the laboratory," explained Bruner in a letter beseeching a farmer to help. "We have the disease going now, but if no fresh bugs are received in a day or two I am afraid that it will die out on us."[94] Several days later, Bruner was still begging for farmers to send in more chinch bugs.[95] Despite the difficulty of building a network of cooperating farmers, the report on the work several months later indicated some useful results. "We sent out in all about 200 lots of the infected bugs to 164 different persons," wrote Bruner.[96]

Besides the state agricultural scientists, there were other government agencies that used lay networks to gather information in the field, such as the US Weather Bureau. Consider once again Patrick Connor's network of weather observers who sent reports by telegraph and mail to Kansas City, from across the plains stretching westward to the far reaches of the state. The deliberate geographical dispersion of this lay network was based on a somewhat different kind of environmental variability. Each place in the network had its own distinct weather conditions, of course, but it was only by mapping those geographically distributed observations together that the larger patterns of moving weather systems could be discerned. At the center of the network, Connor "particularly needed reliable observers in the sparsely populated western part of Kansas, where weather systems typically arrived first, before they reached the heart of the Midwestern grain belt."[97] In the case of the weather, variable local environments were parts of a larger whole, which was key to succeeding at day-to-day management of weather forecasting.

Whatever observations or physical objects were gathered, and however unevenly distributed lay and amateur collaborators may have been across the landscape of the Great Plains and Rocky Mountains, lay networks enabled researchers to receive physical objects and information flows from disparate locations. In other words, most kinds of knowledge made through networks of information gathering have been more or less spatially comprehensive depending on the availability of suitable local collectors and observers. Some special cases, such as weather-observing networks, on the other hand, required the closer control of more evenly distributed observers so that large-scale forecasting could be undertaken. While all lay or amateur networks involved the creation and maintenance of social relationships with local nonscientists, new technologies of the field during the railroad era were more often constructed to reinforce existing social hierarchies of scientific interpretation, not undermine them.

Power and Hierarchy

As most of the examples presented so far have suggested, the work organization of the lay network mode of knowledge production in the field was usually characterized by a durable structure of status hierarchy, organized across geographical space. But how were the connections in the lay network made in the first place? And how were they maintained over time? In plant, animal, and fossil collecting that persisted from the natural history tradition, the ties of identification and naming through correspondence played a strong role in establishing relations between local people and distant authorities. These hierarchical relations were dependent on acceptance of an unequal relationship by both parties. Finding good new local collectors was a matter of serendipity, cultivating promising correspondents, and conveying tokens of appreciation.[98] Even in the most strictly routinized field networks, such as the early twentieth-century US Weather Bureau telegraphic observers based in Kansas City—with regular participation that was reinforced through monetary payments—securing dependable lay observers could be challenging.[99] Moreover, this system erased the local knowledge of experience in favor of standardized, measurable facts that could aggregate more easily in cosmopolitan science. "To ask a local observer to merely take simple observations with instruments and transmit them to a distant center of calculation," as I have argued, "could thus be interpreted as a markedly top-down alternative to any local ways of knowing an environment."[100]

And even when local collaborators accepted metropolitan authority over interpreting their finds, there was still often conflict between eastern scientists and their western collaborators as subordinates acquired their own ambitions within the world of cosmopolitan science. This was especially evident in the case of the fossil warrior O. C. Marsh due to his authoritarian and controlling style of network leadership. As historian Ronald Rainger shows, Marsh encountered "serious trouble" with both his collectors and his preparators in the 1880s. On the field-collecting side, paid local collaborators, such as Wyoming's William H. Reed, "complained that Marsh failed to pay them promptly or give them credit in his publications."[101] Each of these groups brought different skills to collecting, but Marsh and Cope directly paid their collectors and expected them to follow directions.

The hierarchical relationship was explicit and tightly controlled. Early in 1875, Marsh issued a formal set of instructions to his collectors. These fifteen rules specified the practices that were to be followed in collecting and preserving vertebrate fossils. These fifteen numbered directions were indicators of the state of the art in mid-1870s paleontology, to be sure, but they were also important in managing the social relations at a distance that characterized the dispersed field network as a way of organizing knowledge production. First of all, they reinforced the power relation between Marsh and his collectors: he had the authority to specify how the collecting was to be done. But second, they also suggest that managing collectors at a distance rather than in person is inherently problematic, making explicit rules necessary from Marsh's perspective. Third, they mediated between the potentially differing expectations of Marsh and his collectors about how the fossils would be used and interpreted. As a mixture of detailed practical instructions ("Pack fossils in boxes of moderate size, and made of inch boards") and proverbial guidelines ("Haste makes waste in collecting"), Marsh's instructions attempted to standardize both the collecting practices themselves and the general approaches taken to the work by his collectors.[102] Of course, away from Marsh's direct supervision in the field, collectors often took their own initiative, subverting or augmenting his rules. One celebrated innovation was the development of a new bandaging method for fossils.[103]

While the rivalry between Cope and Marsh was the most well-known example of competition between field networks that drove rapid scientific colonization of a new frontier, it was not the only one. The archaeological networks of the Peabody Museum and the Smithsonian, especially for mound excavation in the Ohio Valley, further east, also competed with one another in the field.[104] To a certain extent, virtually all collecting networks were "competing" with others. Even such apparently hegemonic networks as the federal government system of Spencer Baird at the US National Museum (Smithsonian) at midcentury had rivals such as Louis Agassiz's at Harvard. But few collecting networks were in such heated competition over the same field territory and natural resources as Cope and Marsh were in the 1870s and 1880s. Their rivalry thus furnishes a provocative example of the heightened significance of authority and hierarchy in the work organization of field networks.

Classic nineteenth-century collecting rivals, such as Cope and Marsh, were not the only ones concerned to delineate a division of scientific labor; a more general vision of cooperation was beginning to emerge. An 1880 article in *Science*, for instance, explicitly posited a division of labor separating "scientific men" from amateurs. "From the fact that some of our best collectors are not scientific men," the article contended, "it would seem that a convenient division of labor is here suggested, namely: the amateur who collects, and the scientist who describes and classifies." At the same time, if metropolitan centers maintained their access to diverse collections, the western field was not simply a powerless periphery. As the article continued, it acknowledged the local and experiential knowledge of amateur collectors: "Some men possess special qualifications for collecting, they discover localities which others would never dream of, and they have a happy knack of always finding what they are in search of." Recognizing such knowledge as its own kind of "expertness," perhaps "due to constant practice," the article nonetheless sympathized with the "professional scientist [who] finds himself tied to his study and laboratory, and has not time for these constant perambulations; he must therefore largely rely on the amateur collector for his material, as the latter must depend on him, for his technical and scientific knowledge."[105] While the scientist's paradoxical dependence on collaborating collectors probably best describes the more knowledgeable amateur collector, the key point is more generalizable: power relations in lay networks, though structured and hierarchical, were neither unidirectional nor absolute.

This observation matches what has been found in other parts of the world where scientific cores related to collecting peripheries during the same period. Indeed, historians of science such as Jim Endersby have reconceptualized the power relations of global and imperial natural history by pointing out that power did not lie solely in the center or metropole. Thus, in interpreting London-based Joseph Hooker's relationship with his colonial collectors Ronald Campbell Gunn and William Colenso, Endersby argues that it was "not reducible to a one-way power relationship in which metropolitan gentlemen overrule colonial collectors." He emphasizes in particular the leverage that colonists in the field had based on their access to plants that Hooker did not have.[106] While we must be careful not to overstate the extent to which peripheral advantages outweighed the hierarchical relations inherent in the geography of natural history, it is an undeniably crucial point that naturalists on the periphery were advantaged by their access to novel and interesting plants, animals, rocks, minerals, artifacts, and environmental observations.

The relations between core and periphery were fraught with tensions but were often effectively managed through careful attention to human relations of trust and collaboration. Metropolitan naturalists typically had to offer something in return. In the case of collecting natural history specimens, a return flow included their proper identification according to cosmopolitan classificatory standards as well as occasional eponymous rewards. This basic exchange formula could be summed up as identification and naming in return for sending specimens for research publication from otherwise inaccessible places in the field. For instance, in January 1889 Kansas farmer Elam Bartholomew sent "about 100 specimens of parasitic fungi" to Prof. W. A. Kellerman at the state agricultural college so he could identify them. The reply came a month later: "Returns received from the package of fungi sent away the latter part of last month prove to be very gratifying as a number of them were found to be new to science. One of the species—Diplodina Bartholomi—being named in my honor as its discoverer."[107] Likewise, when a young Charles N. Gould, future Oklahoma geological survey leader, was still a "country boy" and amateur collector, he was elated to learn that "a new species of fossil reptile" he found had been named *Plesiosaurus gouldii* in his honor—the "first of my scientific progeny," he fondly recalled—he was "sure no little boy with his first red-top boots was ever prouder."[108]

Naming a species in honor of the field collector was one obvious and flattering way to cultivate good relations, but in everyday practice there were many other, less dramatic means as well. State geological surveys, for instance, sought facts extractable from the knowledge of experience from those in the mining sector by promising cosmopolitan scientific knowledge in return. James E. Todd of the South Dakota Geological Survey, also professor of geology at the University of South Dakota, began correspondence "as early as December 1892 with geologists, well drillers, prospectors, quarry operators, travelers, and anyone else able to furnish data. He created a spirit of cooperation," remarked the survey's historian, "by offering to answer inquiries and to analyze any specimens sent to him. This type of mutual inquiry and service became a vital function of the Geological Survey."[109]

The same was true of career naturalists collaborating with amateur naturalists, who offered their own expert guidance in return. Paying close attention to the particular words and phrases used in correspondence reveals many subtle examples of the maintenance of status hierarchy alongside positive words of encouragement. The University of Wyoming botanist Aven Nelson, for instance, by the 1910s had acquired a reputation as a leading authority on Rocky Mountain plants, and thus novice collectors throughout the region would write to him and send sets of plants for identification and guidance.[110] His replies reflect an acute sensitivity to the maintenance of collaborative relations. For just one illustrative example, consider the letter he wrote in October 1912 in reply to a young man named Edwin E. Payson, who was then at Montrose, Colorado. "I have had very much pleasure in examining your plants," he began. After judging Payson's plants to be "quite interesting and admirably prepared," Nelson offered to name one of them ("the first time that genus has been found in Colorado") in honor of the young collector. He continued: "It is certainly a great pleasure to get acquainted with one who is so much interested in the plants of the west, and who has begun to study them in so painstaking and discriminating a way." In this letter—typical of natural history correspondence in its mixture of subtle reinforcement of structured authority while at the same time offering a positive tone of flattery, encouragement, and collaboration—the offer of a possible naming honor is present but it is clear that the relationship being cultivated is much broader than that. This was no simple, one-time exchange transaction. In fact, Nelson was so encouraging in this particular case that he concluded with an offer for Payson to come study with him at the University of Wyoming, as well as to pay expenses for the young naturalist's summer collecting work.[111]

As this example shows, the personal relations that prevailed in Aven Nelson's botanical network suited well the mobility of amateur collaborators into scientific careers. However, when those networks became much larger and managed more strictly—and, not coincidentally, when the social status of participants was much closer to lay than amateur—the social distance between the network organizer and its local collaborators widened. For a sharp contrast to the warm relations of Nelson and Payson, consider again Patrick Connor's lay network of weather observers in Kansas. At the lay network's center in Kansas City, Connor was especially strict about the monthly paper report forms. When Royal Dunham, the observer in Russell, Kansas, filled out his July data on a Form 1009 instead of a Form 1005, he received a terse reply from the metropole: "While it may not make difference, if any, in the data," Connor admitted, "uniformity requires the use of forms 1005."[112] Only one month earlier, Connor had already scolded Dunham for requesting a government check without returning signed receipts.[113] Such trust-verifying technicalities characterized the ubiquitous and mundane communications that accompanied any institutional bureaucracy by the early twentieth century—government or private, scientific or nonscientific—and arguably they are inevitable in the management of large organizations. But when lay networks grew large and impersonal enough to require them, it is hard to miss the shift toward a cold, impersonal tone and an emphasis on strict standardization of procedure.

Despite their stark differences, however, both the interactions between corresponding naturalists and those with dispersed weather observers shared something crucial that was common to most lay and amateur networks: the people involved were at different locations and occupied distinct positions in the hierarchy of power relations, yet interacted voluntarily and with some mutuality. In a few cases, however, field network organizers could rely on local officials who might be compelled by law to participate. Despite this obligation, or perhaps because of it, researchers at the center sometimes encountered difficulties with compliance. Consider the efforts by Prof. H. S. S. Smith at the state university in Lawrence, Kansas, to organize a network of county surveyors in his state to measure magnetic declination across Kansas. Despite their official status, the county surveyors offered only spotty cooperation with Smith, which irritated him. "It is required by the law of this State," he remonstrated, "that each county surveyor shall compare the magnetic with the true meridian at least four times a year." Yet, he complained, "up to the present year, this law has been, as far as I know, a dead letter." As a result, slightly less than half of Kansas counties had submitted reports in October. "It is to be regretted that the discussion is so partial," he concluded.[114]

Far fewer yet submitted responses during the other three quarters of the year, which, in Smith's view, "render[ed] any conclusions for those months next to worthless." Of the reporting counties in October, only fifteen were in the western half of the state, and most of those were near the center of the state. Thus, the more sparsely settled far western counties were the least compliant. In addition to his difficulty in getting any report at all from over half of the state's counties, he also found the accuracy of many of their observations doubtful. The results differed substantially from what J. E. Hilgard, superintendent of the US Coast Survey, had estimated for the state. "If, then, the reports of the county surveyors are correct," Smith mused wryly, "there is here an interesting field for investigation." Instead, Smith attributed discrepancies to errors and measurement problems, suggesting that "we will get much nearer the truth by making a different supposition, viz.: that the reports are not all correct." In particular, he cited imperfect measuring instruments, carelessness (but, tactfully: "By carelessness, I do not mean hasty and poor work"), and defective meridian marks.[115] Thus, even when knowledge-producing networks consisted of public officials, their systematic observational practices could not be taken for granted.

An even more fraught example of a metropolitan expert questioning the work of a local participant in a lay network emerged at one of the newly founded agricultural colleges of the Plains and Rockies region, where scientists sometimes covered a wide range of subject matter, sometimes beyond their established domains of expertise. Such researchers might be thought of possessing a lay or amateur status outside their specialties, and their position could accordingly become more like a participant in a field network. Wyoming's Prof. J. S. Niswander, primarily an entomologist, had published an official bulletin about "ground squirrels" in the early 1890s, in order to recommend methods for killing them to farmers and ranchers, but his taxonomic judgments were sharply questioned by mammologist Vernon Bailey of the US Biological Survey. As part of his larger project aggregating collection and observation of ground squirrels by local people, Bailey warned readers that Niswander had "mistaken the southern form of Richardson's Spermophile . . . found in this locality, for Franklin's . . . a widely different animal and one that does not reach Wyoming at all." He further criticized Niswander for including "two full-length illustrations" of other spermophiles, "neither of which species occurs in Wyoming."[116] In this case, cosmopolitan knowledge quite publicly trumped the local observer's beliefs, even when the local person was a scientific expert in a different field.

The case of Bailey and Niswander also highlights the distinction between topics that might be considered within the judgment of lay people based on the knowledge of experience, and those topics where expertise held greater sway over judgment. Niswander himself may have relied upon "evidence from several correspondents as to the injuries caused by spermophiles in the vicinity of the station at Laramie, Wyo.," and thus to have acted effectively as the organizer of his own, more local, lay network for matters relating to the experiential knowledge of ground squirrels.[117] Yet within Bailey's larger network, as we have seen, his judgments on more technical subjects, such as the proper classification of distinct species within cosmopolitan taxonomy, might not be so easily accepted. Thus, whether interacting with a local amateur, a public official, or even a scientific researcher in a different field, experts who coordinated networks made sure to assert their judgment in accordance with their superior position in the structure of epistemic power relations.

Evaluating and Improving Lay Contributions

More than a decade before Bailey's report on the spermophiles for the government agency that would evolve into US Biological Survey—indeed, several years before Merriam's division even existed—some other agency heads in the agricultural department were already confronting the opportunities and challenges of collecting information in lay networks through circulars. Botanist George Vasey received 350 responses to his circular about "grass-culture in the South and West," including reports from Idaho and Montana, Utah, New Mexico, and Texas, many of which, however, "were not so full and complete as desired." Vasey acknowledged that some respondents to the circular would write "more fully on the subject, giving the value of particular grasses for grazing and hay, and their comparative merits, together with some of the causes which have operated to produce failures." Such holistic and detailed, yet heterogeneous, knowledge of experience, would have presented some challenges to use simply because it was not standardized for easy comparison or aggregation with other responses.

But the problem that most concerned Vasey was that many of the respondents had difficulty in situating the knowledge in cosmopolitan scientific taxonomies. "As a general thing," he remarked, "the correspondents were not acquainted with the botanic or technical names of the grasses, and gave the common or local name where there was one." This was a big problem, because either "the same grass will have different local names even in places not far remote from each other," or "the same name will be applied to grasses very unlike." Moreover, many of the grasses lacked a "common name" and would instead were referenced with such generic labels as "wild grass" or "wood grass." In some cases, this "perplexing difficulty" was mitigated by requesting material specimens, though many did not send them or the samples never arrived.[118]

The same problem confronted Montana botanists J. W. Blankinship and Hester F. Henshall. Even as they appreciated the value of local knowledge from experience, they urged Montana residents to learn the "hard scientific names." This was necessary, they pointed out, because "the same common name is often applied to two or more very different species in different parts of the state and to still other species in other countries or states, so that any discussion of flowers by their common names would scarcely be understood outside of the limited locality where the paper was written." On the other hand, there would be only "one scientific name, expressed in Latin, for each species in every language and in all countries," thereby permitting an "accuracy of expression otherwise unattainable." Expressing neatly the distinction between experiential and cosmopolitan knowledge, they contended that "common names may be very helpful in the local study of a flora" but could not replace "their Latin equivalents in papers intended for general circulation."[119] This divergence between local experiential and cosmopolitan scientific taxonomies was mirrored in cooperative research in lay networks with farmers who tested varieties in their own local environmental conditions. "The making of new experiments, and testing and verifying the reliability of results," commented one Oklahoma report, "is a slow, tedious work, for which very few farmers are either fitted or equipped."[120] Whether in plant taxonomy or field experimentation, the epistemic rift between experiential and cosmopolitan knowledge was often left gaping wide.

As the comparable frustration of Vasey and his state-level counterparts in Oklahoma suggests, this divergence in how nature was understood and described was not just a sign of miscommunication between two social worlds but also a serious barrier to appropriating that knowledge for use in larger scientific studies. More broadly, the problems of when to trust local observational judgment and how to evaluate credibility were perhaps the central epistemic issues faced by scientists at the center of field networks. In part, the core issue was simply one of assessing the reliability and truthworthiness of individual local collaborators. But it ran much deeper than that. In most cases, it was crucial to systematize and standardize observations from disparate collaborators in many places. Local observers often had priorities in their lives other than scientific work, making a living, as almost all did, through some other kind of activity, such as teaching, business, or farming. Moreover, even when lay people did take a special interest, this might actually work against strict accuracy, if material considerations were at stake, especially for high-value items like meteorites. As Oliver Cummings Farrington complained, "it is probable that full credence can not always be given to the statements of a finder who may wish to conceal the exact location of the specimens, either in the hope of obtaining more or to avoid all question as to their ownership."[121]

Another problem was the divergence in purposes between experiential knowledge for practical use and cosmopolitan knowledge for long-distance exchange. In agriculture, for example, farmers were often eager to try out new crop varieties and species, but they sometimes had different ideas about the aims of such work. As botanist O. F. Cook put it, there was "no lack of those who are willing to plant new seeds, but a warning is necessary, to the effect that to be of real value this class of work must be looked upon as purely experimental and should be undertaken only by those who have an interest in the study of crops, as well as a practical knowledge of at least the locally cultivated varieties." The US Department of Agriculture's Division of Botany, he reiterated, wanted to guarantee that "an actual experiment will be undertaken, that the results from the imported seeds will be brought into comparison with those obtained from other varieties grown under similar conditions, and that an honest and intelligible report will be made."[122] This set of desiderata nicely outlines the exchange value of cosmopolitan knowledge in distinction to any local use value of new seeds for testing. The same epistemic rift existed across the field sciences, and researchers such as Cook felt they needed to take proactive steps to bridge it.

Given these difficulties, disciplining observers in a network was essential to ensure that the observational data they supplied could be aggregated at the center to produce reliable higher-order knowledge. One way to evaluate local testimony was to judge the character of the observer. Sometimes character judgments were made explicit in the scientific reports that relied on them for important facts. For instance, when Kansas professor G. E. Patrick presented the observational details about an intriguing spring in the western part of the state, he had to rely on a local man's character. "Upon this subject my knowledge is all at second hand," he admitted, "but I am fortunate in having the following statement from one of the earliest settlers of Mitchell county, now a resident of Cawker City, a man whose character vouches for the truth of his words."[123] But when local people made unlikely or novel observations, their personal character might be brought into question. When US Biological Survey head C. Hart Merriam received notice of a specimen of *Bassariscus* supposedly obtained near Blue Springs, Missouri, he decided to write to Prof. L. L. Dyche, who was nearby at the University of Kansas: "Do you happen to know anything about the reliability of the man who wrote you? If the beast did not escape from confinement, its occurrence in that region would be most astonishing. I think our nearest record is in the Red River country of western Texas."[124]

The need to judge the value of incoming observations in a lay network was even more complicated when very large numbers of lay people across the country were involved. It could not always be an individual judgment of character when so many people were part of the same network.[125] In his 1893 report for the US Bureau of Biological Survey on the geographical distribution of spermophiles based in part on circular data, for example, Bailey had to resort to judging the quality of data received on the basis of his appraisal of the factual knowledge itself. "In response" to the first circular in 1886, he commented, "hundreds of letters were received, and while many of them contained matter of little value or statements that were evidently incorrect, the greater number were simple statements of fact that added much new and original information to the previous knowledge of the habits and distribution of North American mammals." Likewise, in evaluating responses on the follow-up circular in 1888 and 1889, he simply stated: "All doubtful records have been discarded."[126] While judging the quality of data solicited through a circular by referring to the content of the data itself may seem like circular reasoning, the problem of judging the trustworthiness of incoming data was a dilemma faced by any scientist or naturalist sitting at the hub of a vast network of local informants spread throughout the country, not all of whom could be scrutinized thoroughly and individually to determine their character and reliability.

Another strategy that organizers of lay networks might use to evaluate and improve the quality of lay observations was to stress the distinction between objective data and contestable interpretation—or, as Nebraska horticulturist Fred W. Card more baldly put it, between "facts" and "opinions." In seeking to aggregate the knowledge of local tree growers' experiences in diverse environmental contexts across his own state and other states further east, Card solicited responses under both headings. Under "fact," each grower was requested "to report effects, either beneficial or injurious, which he himself had observed in connection with windbreaks," and only then to offer "opinions." Card cautioned that some of them "were apparently not altogether successful in keeping the two distinct, though in the main the two lines were fairly well defined."[127] By inviting what he regarded as both facts and opinions, even if the distinction might not always be perfectly upheld, Card seemed to recognize that many lay people, especially those who possessed experiential knowledge of the natural world through their work as growers, would be disposed to offer interpretations. Yet by also asking them to report direct observations separately, he hoped to improve the credibility and rigor of those claims designated as factual.

Sometimes the information-gathering operation focused mostly or entirely on the gathering of quantitative data and a reliance on standardizing measuring instruments, strategies that also promised to separate the scientifically objective or factual from the subjective lay interpretation. The regional weather-observing networks led by Patrick Connor and based in Kansas City—with their River Observers and Corn and Wheat Region Observers scattered westward across Kansas—were among the more extreme examples of a network organizer attempting to standardize observational practices in the lay network mode of practice. In a circular to his River Observers in April 1905, Connor laid out the observing practices that needed improvement among "some river observers" who "render[ed] monthly reports" that were "not altogether satisfactory." Invoking the authority of the chief of the weather bureau, he called for "improvement along the following lines: Neatness and legibility; care in entering special readings on form 1006 and the postal card; times of beginning and ending of precipitation must be entered in all cases in accordance with instructions" in a circulated pamphlet. "When rain begins and ends after 7 a.m.," he clarified, "the measurement should appear on the following date, for the reason that it is the total for the 24 hours ending at 7 a.m. Several observers make the mistake of entering the amount on the day it ended whether it was after 7 a.m. or not." In other words, if the weather observations from geographically dispersed localities were to be combined together, they had to be collected in exactly the same way everywhere. He even instructed the observers on proper numerical notation: "When readings, including 24-hour changes, are less than a foot, the figure '0' must precede the decimal point."[128]

Connor managed the network not only by distributing circulars that detailed the standard procedures for observers but also by responding individually to observers who violated the rules and standards. In the small town of Beloit along the Solomon River in central Kansas, for instance, river observer Jared Collar submitted a monthly report for February 1905 that Connor found triply deficient. First, Collar had recorded twelve inches of snow on the ground on February 15 but did not show it falling on any previous date. Second, he had entered the "depth of rain and melted snow" as ".3," which Connor considered to be ambiguous between .03 and .30. Third, he had failed to give any explanation of the cause of high water on February 25 and 26 and its sudden fall soon after.[129] By the end of March, Connor was already seeking a replacement for Collar, whom he found "unsuited to the position" due to a lack of "sufficient intelligence to comprehend all the duties." His reports were "not at all satisfying" and showed "no improvement."[130] Within a few weeks, the Kansas City forecaster had asked Collar to resign; he then offered the river observer position at Beloit to Jesse Thompson, whom he instructed to begin his duties on April 11.[131]

While various problems developed between the Kansas City office and individual observers—occasionally even over the monthly written reports, such as ambiguities in how precisely the precipitation levels were recorded—the daily telegraph reports seemed to cause the most trouble. Sometimes the observers failed to follow the rules dictated from the Kansas City office. Messages might not be received in Kansas City as early as expected, leading Connor to investigate whether the observer or telegraph station was at fault. The Abilene station was one that encountered this sort of problem. By mid-April, the reports from Abilene were coming "with great irregularity," often received "from thirty minutes to one hour too late."[132] After local observer John H. Sherman apparently protested that he submitted his reports promptly at 8:00 a.m., when the telegraph station opened, Connor had to inquire with the station's manager.[133] These delays were of great concern to the Kansas City office not merely because of the overly meticulous, bureaucratic personalities who staffed it (although one gets the sense from reading Connor's letters that he was such a person) but because the station reports were compiled into weather bulletins that very morning. Without data from all of the widely distributed stations, the meteorological knowledge embodied in those reports would be less comprehensive.

In his communications with his River Observers and Corn and Wheat Observers, Connor "worked diligently to maintain common recording standards and reporting times."[134] As noted above, he instructed his local observers through circulars, and he seems to have followed up with direct letters in every case where the guidelines were not strictly observed, even going so far as to enforce the precise placement of decimal points. He also admonished his observers to telegraph the reports in a timely manner, and always to use the bureau's efficient code system, intervening in specific cases. Once, for instance, he criticized river observer John H. Sherman in Abilene, Kansas, for sending a "long, cumbersome telegraph message" rather than the code words devised to keep transmission costs down and to insure fidelity in transmitting measurements.[135] The focus on numbers rather than qualitative description, while in part due to the need to aggregate data from disparate locations, was also surely related to the epistemic authority that such evidence provided.[136] Just as quantitative evidence was important for its persuasive power in the public sphere, it was also crucial to the production of knowledge through lay networks that relied on trusting local observers.

At the same time that weather observation network organizers like Connor were distributing circulars in order to standardize and improve the quality of the return flow of information from lay collaborators, other scientific experts were using circulars for standardizing and improving the material specimens that were submitted from local people in the field. Numerous circulars were issued by state and federal scientific agencies that provided instructions for collecting and materially processing objects of nature before sending them. One typical example, *Directions for Collecting and Preserving Insects and Plants*, was issued in 1901 by the Oklahoma botanist and entomologist E. Mead Wilcox, a Harvard PhD who sought to "establish a corps of voluntary observers and collectors of our native flora and insect fauna with representatives in all parts of the territory." He aimed for both geographical comprehensiveness and higher standards for lay collaborators, and he hoped that his *Directions* would put "in the hands of citizens of Oklahoma sufficient information to enable them to intelligently co-operate" so that he and his coworkers could "build up . . . a complete collection of the insects and plants of Oklahoma." Like many similar circulars issued in the Plains and Rockies, as elsewhere, Wilcox aimed his *Directions* not so much at skilled, knowledgeable amateur naturalists but at novices. As he put it, he hoped "to encourage many persons to undertake, in their own locality, the more serious study of their common plants and insects."[137] While the desire of field scientists in Oklahoma such as Wilcox for better instruction and recruitment of lay participants in collecting and observing may have been especially keenly felt—since Oklahoma was not yet even a state and had comparatively few adept amateur naturalists—the desire for greater state comprehensiveness to cover as much environmental variation as possible meant that such efforts to instruct and guide lay people were common in all states of the Plains and Rockies.

By the 1910s strategies for improving the scientific rigor and quality of lay contributions to the production of knowledge had become much more refined. Though they sometimes still relied on the same basic infrastructure of circulars and more individualized messages traveling through the railroad-based postal delivery system, supplemented in some cases by telegraphs, lay network organizers came to adopt some even more supervised and coercive practices. For just one of many examples, consider the work of Nebraska agricultural scientist C. W. Pugsley, who began a multiyear series of cooperative alfalfa tests in the winter of 1909–1910. For these tests, he recruited farmer collaborators more selectively through "correspondence and by conferences at the various short courses held during that year." As part of these trials, "cooperators were asked to sign a statement promising to carry out all instructions given the Station and to keep a record of all the field operations," and those instructions included a long list of specifications, including "seeding, measurement of the ground, application of culture, soil and manure, and such other details as were necessary." This use of a formal, signed, and printed agreement, reflecting the state scientists' research design, constituted yet another material technology of paper that further refined the less fully coercive guidelines that were contained in printed circulars. These details were important for the rigor and validity of the scientific research produced through lay networks. Control of environmental conditions was especially important. "The plots were located in such a manner," reported Pugsley, "that there would be no wash from inoculated to uninoculated plants and in such a position that the soil would be as uniform as possible."[138]

To further ensure compliance, Pugsley and his colleagues, as many other lay network organizers were doing by this time, also included site visits for guidance and inspection in the routine practice of the lay network. "Wherever it was thought advisable," he noted, "a man was sent from the Experiment Station to assist in locating the plots and in applying the various treatments."[139] Similarly, when foresters at the US Department of Agriculture wanted to experiment through a lay network in the summer of 1899, they sent a circular offering "expert tree planters to examine the land," which they deployed later that year, and with more than nine of ten of the sites in the "treeless" Great Plains states.[140] Moreover, in the case of Pugsley's lay network in Nebraska, initial site visits to the farms of many of the collaborating farmers were then followed up later each season with further visits to all of them. These personal visits were intended to make sure that proper scientific records were kept, by having agents of the station itself perform this part of the practice of field research. "Each field man" sent out to represent the station "carried a notebook in which he collected data concerning the soil, preparation of plots, and seeding."[141] These site visits constituted a vital part of the practice of science in the lay network mode—ensuring not only that the data were collected but that they were collected in a uniform and standardized fashion from one farm research site to the next—and they were likely also the occasion for accountability by the farmers involved. The added expense and effort of traveling around to see each farm and to make proper records were significant but could improve the scientific rigor of the final results.

Some coordinators of research in the lay network mode tried the reverse strategy: bringing farmers to meet with them, instead of going to their farms. In December 1913, E. W. Harrison, who was based on the small town of Spur in the semiarid northwest part of Texas, helped to organize the Dickens County Experiment Association. "We station men," he reported to the Texas Farmers' Congress, "gave the proposed organization all the publicity we could and notified all the men we could reach that were going to organize." Despite a perceived lack of strong support from the business community, they attracted around thirty members and held a half-dozen meetings with them "on the station farm," distributing seeds, and initiating cooperative research with them. They stressed the importance of "maintaining pure strains of each particular kind of seed" and were pleased with the initial progress made through this organizing model.[142] By combining on-site meetings with membership in the association, they could improve the quality and rigor of cooperative research. Whichever direction the people traveled—lay people to the center, or scientists to the periphery—the key insight was that close personal instructions and human bonds could increase the value of lay participation in research.

Lay Responses and Perceptions

Compared to those at the center of these lay networks, it is more difficult to find concrete source evidence of how the relationships and activities were experienced from the perspective of the local observer or collector. Moreover, the attitudes of settlers who acted as local observers varied as much as the people themselves—and those who coordinated their networks. The greatest availability of evidence tends to be for amateur naturalists who possessed some scientific knowledge, and who therefore partially bridged the gulf between expert and lay, rather than for those who were more strictly and clearly in the lay social category. While many of these local amateurs acknowledged their relative lack of expertise and authority in an era when the categories of expert and professional were increasingly overlapping, some of them nonetheless published in scientific periodicals oriented toward naturalists. Their responses and perceptions may be indicative of the potential tensions in how more highly structured lay networks operated out of larger institutions during this period.

Some local amateur naturalists accepted their subordinate positions in the hierarchy of knowledge production. The Rev. J. M. Bates in the small town of Long Pine, Nebraska, along the northern edge of the Sandhills, for example, participated in both state and national networks. "It falls to the lot of some, through lack of time or ability," he admitted, "to confine themselves to the elementary work, in natural science, of observing and reporting." As "unfortunately one of those who are thus limited," he modestly observed, he could "offer, after eleven years of interest in the birds of Nebraska, nothing that indicates intense study." Bates regretted that he had been "unable to make the subject anything more than an incident in my life and a side issue of the nature study I love so well."[143] Even if his lay status was only relative—after all, he did possess knowledge and interest, just not as much as others—Bates accepted his role in producing ground-level facts for the use of scientists in distant cities, whether in Nebraska itself, national centers such as Washington, DC, or even international scientific centers such as London. Notably, however, the sharpening hierarchy of expertise, for Bates and others like him, coexisted with an enduring interest in and enthusiasm for the study of nature.

Such cooperative sentiments were also quite often perceived in lay network participants by their expert organizers, though of course they had their own reasons to portray relations as mutual and agreeable. In the view of one professor at Washburn College in Kansas, for example, finding weather observers was a simple matter of harnessing local interest. "There are always to be found in every community willing observers of the weather," he predicted. By "utilizing the universal interest in the subject," network coordinators could "obtain a multitude of willing observers of weather phenomena, each of whom will help in establishing the basis of fact out of which all valuable theories grow and on which they rest."[144] Far from being an impediment to credible knowledge production in a subordinate position, local enthusiasm for studying nature was probably essential to making the network function at all.

At the same time, no amount of enthusiasm for the study of nature could keep some of the local collectors and observers from at least occasional, or more frequent, resentment of the imposed authority of those who attempted to direct their work from afar, nor from chafing against the hierarchical relations of the lay network. Hints of such reactions can be gleaned even in the scientific publications that resulted from collaborations. For example, when state animal scientists organized a group of local stockmen in 1909 as part of a "breeding study of dairy cattle" in a twelve-mile radius of the small town of New Salem in southwestern North Dakota and imposed a set of guidelines for record keeping and procedures for the actual breeding trials, "the chief difficulty" was "to get the members to appreciate the value of good care and to give their cattle the attention required for best results." Unlike research conducted by animal scientists themselves, or by subordinate staff members under their control, they had to persuade lay people, with only partial success, to accede to the demands of their study design. "There is a tendency on the part of the farmers," the study's authors lamented, "to resent being dictated to concerning the care and disposition of their stock. It is in some cases hard to get them to follow instructions." Even site visits by a field superintendent employed by the researchers and stationed at New Salem could not guarantee compliance from the stockmen.[145] That such critical comments made it into print suggests that the lay resistance to the guidance of expert animal scientists must have been very significant, given the more usual tone of mutuality and appreciation that often characterizes publications that depend on local collaboration.

Even more, the enthusiasm and interest of local naturalists who interacted with larger institutions could not prevent their relations from becoming strained by the early twentieth century, at least in those cases where they perceived increasing structural inequalities that worked against them. The North Dakota botanist Joel Lunell, who was born and trained in Sweden and a prolific contributor to *American Midland Naturalist* in the 1910s, complained vociferously about how local collectors were treated by the increasingly powerful organizations that were coming to dominate all the field sciences. While Lunell had undertaken advanced academic study in botany, he made his living around Leeds,

North Dakota, as a frontier physician, and according to his obituary, "during these strenuous times [was] also president of the village council, alderman, coroner, U.S. examining physician, and postmaster." Although he "devoted his leisure time" to local and regional botany, he was in many regards the quintessential amateur naturalist in a world of scientific career making and institution building.[146]

Lunell's complaints against large institutions and the career makers who staffed them spilled out in a 1915 article that was ostensibly about the vascular plants of North Dakota. He prefaced his substantive material with a protestation that "the large public or private institutions," or at least an unnamed "some of them, often commanding almost unlimited funds," were "in the habit . . . of sending circulars to botanists, requesting them to 'give up' what they have on hand or to collect the desired plants for them." The use of printed circulars not only to solicit information flows from geographically dispersed localities but also to direct the movement of material specimens from many places at once undermined the authority and pride of some local amateur naturalists, and it seemed to replace the reciprocity and mutuality of personal relationships with a more highly structured and unequal exchange. "In the majority of cases," he noted, "the botanist so addressed will 'for the best interests of science' yield, and some will even find the proposition immensely flattering to themselves!" While "perhaps a few" local collectors might be willing to part with specimens, if the larger and wealthier institutions would pay for the "cash expenses necessarily connected" with collecting them, in Lunell's view they were taking advantage of local naturalists' generosity—"why pay for something that can be had for the asking?"—offering only token gratitude, if any. "Some institutions make their requests for material most valuable to them with 'thanks,'" he allowed, but "others do not even send acknowledgment when they have received the goods." To make matters worse, he claimed, some institutions failed to maintain reciprocity by sending any plant duplicates in return, or "use the occasion to unload the most repugnant and worthless trash," with only the "most generous of the institutions" treating their local collectors fairly. Finally, he criticized the increasingly prevalent institutional practice of "incorporating and matriculating the received specimens as gifts," scoffing at the pretense of mutuality when they would "immortalize the donator by adding his name in a special column . . . thus substituting stones for bread!"[147]

As a local naturalist who resented the loss of his own authority to these larger institutions, and the diminution of a more equal reciprocity that seemed to accompany the changes, Lunell is perhaps an extreme example of adverse reaction, due to his status as a highly trained amateur botanist who nevertheless found himself at the receiving end of the directives of those who were organizing more hierarchically structured lay networks. In all likelihood, few local collaborators in these lay networks, especially among those with much less expertise and scientific training, felt as aggrieved as Lunell. Yet his articulation of these grievances is revealing of the potential structural tensions of lay networks, and it is instructive to consider how such atypical examples may illuminate the experiences and perceptions of at least some lay and amateur participants in these networks, if in exaggerated form. Indeed, the efforts of some local naturalists to organize their own more systematic networks for producing knowledge in the field may in part have arisen out of a desire to maintain more of their own collective control over their activities.

Collective Collecting and Observing

Ongoing relationships between career researchers with greater expertise at the center and their distributed local collaborators were typical for the lay network as a mode of knowledge production. Just about all of the cases I have considered thus far have involved some version of this status hierarchy, along with all the issues of systematization, trust, and practical organization that it involved. Yet an alternative, more egalitarian variant of how to make knowledge through lay and amateur networks did emerge over time: collective collecting and observation through state scientific associations. To be sure, through the state societies people interested in the natural world could gather together to do many things, including share their results or build up enthusiasm for science, both among the members and in the tax-paying public at large, in addition to engaging in collaborative research. But here I will focus on the societies' collective efforts at knowledge production, especially to the extent that they attempted to make their activities more systematic and better coordinated.

For some key examples of systematic efforts to engage in collaborative research through lay networks, we can turn to the state scientific societies in some of the earlier territories of the Plains and Rockies to achieve statehood, located at the center of the region: Kansas (1861), Nebraska (1867), and Colorado (1876). One of the earliest organizations, the Kansas Academy of Science, included some efforts to organize the collective production of knowledge by its own members. In theory, these networks were organized through the committees of the academy, which would enable members to compile their knowledge collectively, in a more egalitarian fashion. In reality, however, the prevailing organizational patterns were similar to traditional natural history networks in that particular individuals compiled data gathered by distributed local observers and collectors. Frank H. Snow's bird catalog, for example, acknowledged the assistance of "a few friends of ornithology in Kansas" (who did not actually all live in Kansas) including Edwin A. Popenoe of Topeka, George F. Gaumer and Nelson J. Stephens of Lawrence, and T. M. Brewer of Boston.[148] Similarly, Popenoe presented and published a list of Kansas butterflies, with contributions from Snow, George P. Cooper of Topeka (who also collected in western Kansas), and T. B. Ashton of Tonganoxie.[149]

A more explicitly collective approach was adopted by Kansas University's Professor Carruth, who began giving annual botany reports on behalf of the entire academy at several meetings in the 1870s, including his "Report on Botany" (1874) and his "Centennial Catalogue" (1876), so named in recognition of the nation's hundredth anniversary. Carruth's compilations were based on the local work of people throughout the state. "I have not traveled widely," Carruth admitted, "but have been greatly aided by the observations of others." This collective network of observers began, of course, with species already identified in publication by those who had traveled through Kansas. Carruth's network included career scientists like his colleague Professor Snow, who, "with the help of his students and correspondents, has furnished more than twenty" additions to the list of Kansas plants, as well as amateurs such as Miss P. Wright of Burlington, who provided Carruth with "specimens of three—two of them common at the east, but not hitherto found in Kansas." Another woman with the same surname, Miss E. C. Wright, who lived near Irving in Marshall County, sent Carruth a list of forty to fifty observed species names "of which she [felt] certain," although the material specimens had been "accidentally spoiled." Other contributors included Popenoe, who provided a list of forty plants he found around his home county and during a trip to western Kansas, and medical doctors L. Watson of Ellis, who in 1874 "sent [Carruth] plants all summer—more than [he had] found time to analyze" for a total, two years later, of four hundred to five hundred species, and W. H. Saunders of Lawrence, whose offerings were less prolific. Other local collaborators included J. Henry from Salina, W. S. Newlon and Mrs. Jewell from Irving, James Wilson and Professor Wherell from Leavenworth, and Professor Benjamin Mudge from Manhattan.[150] Thus the contributors not only were a mixture of career scientists and amateurs but also differed in whether they sent material specimens or simply lists of observations. In the latter case, Carruth, like all network organizers, had to judge the reliability of the data, in this case the "certainty" that local collaborators such as the second Miss Wright were willing to express.

One important subgroup within the Kansas Academy was the group of students at Kansas University, whose network operated in a tighter fashion. Students working with Snow dispersed to collect butterflies around Douglas County (in which Lawrence is located) and occasionally elsewhere, to their homes and other places in the state. Of the 144 species listed in one report, almost all of them in fact came from Douglas County, along with ten from Leavenworth in the northeast corner of the state and only two from western Kansas. (Of all places across the entire region, the counties containing state universities or agricultural colleges were invariably among the most thoroughly worked environments.) When he presented the results of their cooperative effort to the academy, Snow recognized "the invaluable services of the students in my Zoölogy classes, who have searched the country in all directions for material with which to enrich their own entomological collections, and have invariably allowed me to retain for the University cabinets all species new to the Kansas list." Snow especially singled out a handful of students—Louis L. Dyche, Richard Foster, John E. McCurdy, Frank Prentiss, and Ed. G. Smith—who were "especially successful in obtaining new species."[151] The adoption of natural history collecting as a hobby by some university students is presented here as a given—it was surely stimulated by their association with Snow—but it took his network organizing and the forum of the academy to aggregate their efforts into something larger that could be published for wider circulation.[152]

While Kansas by virtue of its earlier dense agricultural settlement got the earlier start on collective scientific networks, the people of other states in the central part of the region gradually became involved as well. In Colorado, for example, the Colorado Biological Association began in 1887 as the Colorado Ornithological Association. Its first leader, Charles F. Morrison, coordinated a list of roughly 350 bird species and subspecies for publication as an "Ornithology of Colorado," relying on the association's network of local collectors.[153] In March 1887, acting on his belief "that, in ornithology, a given district could be far better investigated by a number working together, than by individual research," Morrison had "sent circulars to all the ornithologists in the state of Colorado, known to him at that time," with a "most satisfactory" result of "all, with but one exception, joining heartily in the work." Shortly thereafter, Morrison faced the challenge of deciding how far to widen the network, and he initially decided to opt for a more exclusive membership. "Many names were sent in," he noted, "but it was found necessary for our own interests to enroll only those as members who were known to be workers of repute, and upon whose word reliance could be placed." The next year, Morrison published a list of Colorado birds compiled and edited for the association.[154]

Even after changing its name shortly thereafter, the association still maintained a cohesive research agenda and was now established "for the purpose of investigation, in a detailed and systematic way, the Fauna and Flora of Colorado, both recent and fossil, and recording the results in such form as to be a permanent contribution to our knowledge of the Biology of the State." With a resident membership fee of one dollar, the association now aimed to stimulate participation across the state. An announcement suggested the collaborative network approach, indicating that the association would "heartily welcome all co-operation."[155] It proved challenging to strike the right balance between a broad geographical coverage and membership base, on the one hand, and greater selectivity for "reliability" and "reputability," on the other. However, in the end the most determinative factor in the short life of the association was probably the departure of Morrison from what he called "the perfect climate of Colorado and genial friends for the hot summers and arctic winters of Montana."[156] Morrison was not completely alone, however. A Custer County branch of the Association held a meeting on a January 1889 evening "at the West Cliff schoolhouse," organized and headlined by the Colorado naturalist T. D. A. Cockerell, who apparently even persuaded C. Hart Merriam to join the Colorado Biological Association as a show of support. Later meetings devolved into general nature education for the community and covered a wide range of topics, including placer mining, butterflies of mountain summits, the "origin of man," slugs and beetles, and charcoal from the mines.[157] This was obviously a very different plan from what Morrison had envisioned, less a systematic network for coordinated research than a field and nature club.

Around the same time, in neighboring Nebraska, only a small amount of systematic cooperative research was undertaken by networks of participants in the state's most general association of scientific enthusiasts, the Nebraska Academy of Sciences. Members from particular counties, some in the western part of the state, compiled local synopses of flora and fauna. One early example was Dr. H. Hapeman of Holdrege, who produced a list of the freshwater algae in his southwestern Nebraska county from fall 1891 to summer 1892.[158] And Edward M. Hussong of Franklin, who was superintendent of schools there, reported on the flora of his county and later reported the research of high school students in a local network of sorts under his direction in a three-year lab and field effort to examine seed distribution.[159] These local efforts were largely uncoordinated, unlike in Kansas, and so the extent to which they contributed to the production of knowledge through larger, structured networks was limited.

More systematic was the endeavor called the Nebraska Botanical Survey, initiated and coordinated by the University of Nebraska's unique scientific club, the Botanical Seminar. Although unfunded by the state of Nebraska, in many of its practices the botanical survey fit more into the survey mode (it will thus be revisited in the next chapter), but in part its work was based on the lay network mode. For example, when Roscoe Pound reviewed the progress of the survey through 1893, he noted the parts of the state visited by various student and faculty participants but also noted local work by Rev. J. M. Bates, at the time a resident of Valentine, which was, like his later home of Long Pine, also in the north-central part of the state along the edge of the Sandhills. In acknowledgment of the heavy local field work by university people around the city of Lincoln itself, just as around the University of Kansas, Pound noted that by June 1892, "Lancaster county had been pretty thoroughly worked."[160] In thanking various amateur contributors around the state, the organizers did remind them of its systematizing and centralizing ambitions, emphasizing that "no species can be reported unless a specimen is deposited in the herbarium of the Survey."[161] As the survey developed, however, it came to rely mostly on university-trained field researchers, who undertook systematic survey work by traveling planned routes across the state in addition to thoroughly exploring the area around Lincoln.

As the example of the Nebraska Botanical Survey suggests, often the most systematic efforts at collaborative field work were undertaken by specialist rather than generalist societies, whose members had enough of a common interest to sustain research projects together. The Nebraska Ornithologists' Union, founded in 1899, was another active and successful specialist society, and unlike its Colorado counterpart it continued to publish for several years without transitioning to a more generalist agenda. Following the saga of the union provides a well-documented example of how local observers came together for collective knowledge making through a lay or amateur network. Founded in 1899, it became a part of the Wilson Ornithological Club in Chicago in 1916. Despite its eventual demise, between 1900 and 1915 the volumes of the *Proceedings of the Nebraska Ornithologists' Union* regularly appeared, thereby providing a window into the collective knowledge-making projects of a state specialist society in the West, just before the US Biological Survey launched a concerted effort to direct lay and amateur observations on a more national, hierarchically coordinated basis.[162]

Some members of the union undoubtedly wanted simply to have a social association with other bird enthusiasts, while others also presented formal papers at the meetings. But there were a few members keen to coordinate a more systematic network for research in ornithology in Nebraska. One of these advocates for a more collective approach was Lawrence Bruner, whose presidential address at the first annual meeting highlighted the achievements of such collaborative programs. Recalling his experience "as Entomologist and Ornithologist, and a few other 'gists' to the Horticultural Society," where he had been "asked to get together some notes on Nebraska birds," Bruner felt that his own bird study "in a kind of dilatory way for twenty-five years" had only provided him with "some notes on the species belonging to portions of Nebraska where I happened to be at different times." In order to achieve a more comprehensive, collective ideal, Bruner wrote letters to those he knew to be interested in bird study. From this initial list of fifty-six, he gathered "the notes they had made concerning the birds of Nebraska," which generated "a larger list of birds for Nebraska than [had] been obtained in any other state in the Union."[163] The 416 species and subspecies on the Nebraska list that Bruner compiled represented the fruits of his network-based approach to compiling knowledge about birds. By combining their observations, Nebraska's ornithologists might produce something worthy of greater notice by scientists outside the state.

Bruner's program was assisted by local observers such as the same Rev. Bates who had contributed to the Nebraska Botanical Survey several years earlier, now living in Long Pine. In 1896, he had given Bruner his "notes and observations on the birds of Northern Nebraska, centering in Valentine and Cherry Co[unty], and covering eight years." In 1900 Bates offered the union the results of three or four more years of field research: a list of birds observed along with notes on their ranges to assist in compiling a statewide guide to Nebraska birds. By offering over a decade's worth of local field observation data for Bruner's (and now the whole union's) collective project, Bates demonstrated an impressive commitment to collective knowledge production without apparently seeking greater individual rewards for himself. His participation in larger scientific networks was not limited to his state project; he also reported on bird migration for the US Department of Agriculture's Division of Ornithology, which would later develop into the Biological Survey. Bates avidly studied a few scientific manuals on birds and guided a high school boy named Will Smith, who became his assistant.[164] In addition to Bates, other local collaborators in the state also offered their own observations. One young man named Merritt Cary—later to become a researcher with the US Biological Survey—tracked birds around his hometown of Neligh in the northeastern part of the state.[165] Thus the participants in the union's network for producing collective bird knowledge included both local amateurs with no career ambitions and those who hoped to make a vocation out of it.

Most of the efforts of the union's members were based on viewing alone. This focus on observational data made systematizing and standardizing field practices even more imperative. When physical specimens of birds were collected they could be examined directly by experts, though some uniformity in preservation techniques was obviously desirable. Yet the strong movement against the shooting of birds, which was a contentious issue among ornithologists in the late nineteenth and early twentieth centuries, increasingly tilted practice toward observing rather than collecting. In neighboring Kansas, at least one ornithologist wished "for the day when the scientists of this state will bring such concerted stress to bear upon our uncertain State Legislature that they may grant to accredited and genuine bird students the right to take, and to hold in possession specimens of bird-life."[166] Reliance on local naturalists raised an epistemological problem: how to standardize those observational data, not only so that the reports could be trusted but also to make it possible to combine data gathered in disparate locations. At the union's first meeting, Dr. Robert H. Wolcott suggested a solution to the problem. Consistent with the evolving practices of cosmopolitan science, he proposed particular quantitative and seasonal-length ranges for various descriptive words used in reporting such as "abundant," "numerous," "scarce," and "winter visitant."[167] By adopting more rigorous criteria for verbal reports, held in common, the members of the union would be able to combine their observations more reliably into larger knowledge claims about the birds of Nebraska.

Wolcott's crusade for making the network's bird observations more systematic extended from standardizing the meanings of descriptive terms to attempting quantitative precision. At the union's second meeting, Wolcott reported that he had tried out some new, even more rigorous observational methods around Lincoln, providing tables of numbers of birds seen and ratios per mile. In his view, meeting higher standards in the methods they used for observation would allow the Nebraska network of ornithologists to address problems of greater concern to ornithologists outside the state. Of special interest to Wolcott were environmental problems relating to the position of the state's fauna between East and West, which he thought should "be met and answered in a spirit of scientific accuracy, if our conclusions are to carry weight with those living outside our borders." He called for collective restraint: "We must know what we know and record only what we know we know." While in Wolcott's view they should remain "open at all times to conviction," they should also "subject every fact presented to the most thorough criticism." The union, he argued, "must judge kindly, but most critically, with the records presented" by its own members. According to Wolcott, the effort to standardize and systematize local observations was thus crucial for making claims in science. The prevailing epistemology of science outside Nebraska required greater rigor than many local observers were evidently supplying. Another component of making the collective project more systematic was to aim for comprehensive geographical coverage. Wolcott therefore presented lists of species that needed additional observations, especially in western Nebraska.[168]

Sentiment among the majority of members seemed to be consistent with Wolcott's agenda. Resolutions at the second annual meeting in Omaha called for a distinction between "positive" records based on an actual specimen or the observation of experienced observer, "probable" records based on observation by an experienced observer for "well marked" and "clearly seen" birds or by an inexperienced observer for common species, and "doubtful" records based on a less-strict criterion.[169] Such a taxonomy represented a sincere effort to standardize the quality of local observations so they could be combined, thereby improving the rigor of the union's collective knowledge production. It also opened up a whole host of debatable questions regarding who might be qualified as an "experienced observer," what constituted a "clearly seen" bird, or what counted as a "common species." For networks of amateurs and lay people, such epistemic issues were virtually unavoidable at some level, given the ongoing struggle to publish knowledge that would carry weight among scientists outside Nebraska. It was in a sense a variant solution to the problem of trust in network research, made more complicated by the usual lack of material specimens for inspection by higher-status naturalists.

What makes the case of ornithology in Nebraska so revealing for an analysis of field network knowledge production is that it was one of the few specialties that remained so dominated by noncareer scientists so late. Into the twentieth century, when some (though not all) other disciplines were becoming mostly career-based, observations and collections in Nebraska ornithology were still done by local residents. Occasionally, "local" observers could be found who covered such sizeable territories that their efforts approached the regional survey mode. One Nebraska bird researcher who covered an impressive radius around his home was A. R. Graves of Kearney, who worked at a whole series of locations, some of them several hundred miles away.[170] Another impressive regional traveler within Nebraska was John T. Zimmer, who made bird observations in November and December 1910 while engaged in his entomological field work around Crawford in Dawes County. The example of Zimmer also indicates how specialists in other scientific disciplines might contribute to geographical (and seasonal) diversity of sites covered by union members.[171] Over time, though, even an association as hospitable to amateurs as the Nebraska Ornithologists' Union found its published papers increasingly dominated by reports from the expeditions of professors and students at the state university.[172]

Why did the collective variant of the network mode so often seem incapable of enabling reliable and systematic knowledge production? In part, the rigorous observational criteria laid out by systematizers like Wolcott closed off participation by some local amateurs and downgraded the status of what some others observed. Moreover, despite the existence of some ambitious examples like Graves, few amateurs had the time or resources to venture systematically beyond their own immediate vicinities. Indeed, before they gained access to funds for expeditions and surveys, even university researchers focused intensively on their own surrounding countryside. To be sure, the union's network of lay and amateur observers, supplemented by heavy participation of university researchers, did generate some systematic knowledge in areas with many local participants, such as ornithology. By 1908, Wolcott had produced a synthesis of the faunal regions of Nebraska, delineating Missouri, Prairie, Sandhill, Plains, and Pine Ridge regions. In his summary, he also emphasized the larger distinction between eastern and western species rather than by specific locations.[173] Ultimately, then, another reason for the limited success of field networks of local observers was that so many of the interesting questions about the environment of the interior West could not be answered by simply investigating the local—it required travel through a larger region.

In the Great Plains and Rocky Mountains, as elsewhere, lay networks were ubiquitous during the railroad era. Vernon Bailey's informal network of local collectors of small animals at the US Biological Survey and Patrick Connor's two formal networks of weather observers across Kansas were only two of many. The diverse examples of science examined in this chapter—everything from fossil- and animal-collecting networks to meteor and weather observation networks—varied significantly in the dynamics of the social relations generated between scientific knowledge producers and their local collaborators in the field. Yet collectively they demonstrate that the work of making knowledge through field networks involved much more than the cognitive work of combining data and specimens. Without the material technology of the printed circular or report, or occasionally the telegraph, and the transportation routes to convey them quickly and efficiently—for example, gathering information from lay people for combining and aggregating with other bits of information collected from others—would have been very difficult or, in some cases, impossible.

As a mode of practice, the lay network bridged the epistemic rift between cosmopolitan and experiential knowledge in the field by linking together distinct individuals with worldviews reflective of each side of the relationship. For lay people without any considerable scientific expertise the distinction was clearest, but even for amateurs with varying degrees of knowledge of cosmopolitan science to complement their experiential knowledge, it was their possession of local knowledge based on the experience of everyday life and work in particular environmental settings that made them so valuable as network participants, since their access to nature in the Plains and Rockies was unlocked not only by their geographical location but also by their knowledge of where to find things, how to see things, and how to engage materially and bodily in transforming local nature, whether through agriculture, mining, hunting, or any of the other myriad activities they engaged in. The availability of willing and able lay or amateur participants thus served as the most significant constraining factor that shaped field science in the lay network mode and limited its reach. Nature was only visible where eyes were there to see it, and its products could be obtained only selectively based on the distribution of particular live human beings who had settled across the Plains and Rockies. With a much lower population density than the East or Midwest, or parts of Europe, the limited purview of the lay network traced its human contours, mediated by the railroad and telegraph infrastructures that allowed information and specimens to flow back and forth from core to periphery.

The social dynamics of the lay network also defined its organization of work. The most systematic field networks were typically coordinated by at least one central researcher. The basis of the network was the communication of instructions and observational information, combined in many cases with the transportation of physical specimens, predominantly although not exclusively from the periphery of the network to its center. Establishing and maintaining trust relationships, often between participants of different status levels, constituted a central problem of the network. Above all, the conversion of nature into knowledge to circulate in a wider global system—whether it be through new fossil species reported in the *American Journal of Science*, agricultural field trials aggregated in experiment station bulletins, or daily weather maps and forecasts—required the negotiation of social boundaries and reciprocal obligations at a distance. Even when the networks were constituted by state associations of local observers, such as the Nebraska ornithologists, many of the same issues of trust, standardization, and coordination arose. Although the obstacles to cooperation did often present significant barriers to knowledge production in the field, these collaborative relationships ultimately allowed scientific knowledge producers to conduct research that they would be unable to undertake without the contributions of local people. It was certainly one important means to undertake field science over vast environmental domains, but those lay and amateur collaborators were soon joined—or even preceded—in the field by traveling field scientists themselves.

Chapter 3

Surveys

It would be hard to imagine a more enviable travel appointment for a field scientist during the railroad era than to join the Yellowstone Division of the United States Geological Survey, which began its work in 1883—the same year that the Northern Pacific Railway completed its transcontinental link across the northern tier of western states—under the direction of Arnold Hague. For survey scientists of the late nineteenth and early twentieth centuries, as for travelers today, Yellowstone National Park was an unusually appealing place, and performing field work in Yellowstone had certain aesthetic and experiential dimensions that made it a special place to do research. Throughout the rest of the 1880s, and then intermittently for the next two decades, the field parties of the Yellowstone Division came to know this spectacular place intimately.

The pleasures of Yellowstone were both scientific and personal, as survey geologist Joseph Iddings, who worked under Hague, discovered. He declared, "There is probably nowhere in the United States a region so fascinating to the geologist, with such a variety of interesting phenomena" that could yet be "explored with such ease and comfort." As Iddings reckoned, "the geologists who were to spend seven summers in its thorough investigation were indeed fortunate." He later recalled feeling "unalloyed joy to ride over the elevated plateaus and open mountain slopes, often brilliant with flowers, under clear skies, breathing the invigorating air." Moreover, his survey field parties also found "wild game in the forests and alluring trouts in the rivers." Complementing this general experience of the Rocky Mountains were the more particular delights of Yellowstone itself: "the curious sights of hot-spring formations, mounds and terraces, grayish white in the dark green setting of pinewoods, the delicate structure of their coral-like substance, and the wonderful coloring of the crystal waters or underlying basins, the spouting of geysers, and the boiling of steam pools." Yet his experiences were not solely those common to all tourists but also those specific to a geol-

FIGURE 3.1. Typical high Rocky Mountain scenery, with US Geological Survey, Yellowstone Division, field party in the foreground. Original caption reads: "Crowfoot Range, Gallatin Mountains, showing upturned beds of measured section." Arnold Hague et al., *Geology of the Yellowstone National Park*, US Geological Survey Monograph no. 32 (1899), Part II, between pp. 8–9.

ogist, for "everywhere he looked there were new problems," and "each mountain and canyon in the plateau lavas invited study." All in all, he considered it "an ideal region for an enthusiastic geologist and a lover of nature and life in the wilds."[1] Put simply, Yellowstone was not only a very enjoyable place to spend the field season but also a geological wonderland (see figure 3.1). While not all field sites in the American West could offer the distinctive experiences that Yellowstone provided, survey scientists tended to find many of the same qualities—interesting topography, clear skies and air, and enjoyable flora and fauna—elsewhere in the West too.[2]

Geological survey may have dominated the field, but its counterparts in other domains—plants, animals, soils, human cultures and languages—shared many of the same characteristics of field practice. The US Bureau of Biological Survey, though much more modest in funding and personnel, was one of the clearest analogs. Beginning as a tiny agency known as the Office of Economic Ornithology in 1885 under C. Hart

Merriam, within a few decades it grew to full bureau status—a shift hailed by the *Botanical Gazette* as "the beginning of a new era in the botanical field-work in the United States"—and became heavily involved both in mapping life zones across the country and in making recommendations for predator control.[3] Vernon Bailey, introduced in the last chapter, served from 1890 as the biological survey's chief field naturalist—a title that "he cherished until his retirement in 1933," without a successor—and who collected for the survey some "13,000 specimens, a large number of them new species."[4] Often in collaboration with his wife, Florence Merriam Bailey (sister of Hart), he became one of the most celebrated field survey scientists of his generation. Bailey's commitment to field work over the long term also shows how, when resources permitted, lay networks could give way to full-fledged surveys.

The survey was a mode of field practice in which researchers, usually subordinate to a supervisor at a metropolitan headquarters, traveled into the field along a planned route that enabled the more or less systematic coverage of a territory. As J. A. Allen wrote in a review of US national biological survey work in the mid-1890s, Merriam and his collaborators, through their geographically extensive survey practices, enjoyed "opportunities . . . to bring together material from a wide area and in an abundance scarcely dreamed of" by other naturalists.[5] For survey scientists during the railroad era, the survey typically involved a two-part, sharply bifurcated itinerary, with rail transport to cover as much distance as possible to reach the field, and more deliberate and much slower movement within the field using a combination of draft animals and human locomotion. The knowledge produced through surveys, often depicted on maps, has become a durable feature of the epistemic landscape of modern society.

Surveys were also political and economic instruments: gathering knowledge in the field through survey was a tool of statecraft, of extending political authority over geographical space for capitalist economic development.[6] In the words of North Dakota geologist A. G. Leonard, surveys were "found in almost every civilized country."[7] The purpose of survey science was rarely defined simply as political administration per se but, rather, as the promotion of rationalized economic development. Morton J. Elrod, in his 1904 address to the Montana Academy of Science, called for a state geological and natural history survey with typical language: "The state's geological wealth and natural history wealth are the property of the people," he declared, "and the people know little about it. It is not good business to be possessed of valuable property without making strenuous efforts to know something about the property."[8] Most survey proponents appealed to similar sentiments in order to secure patronage. Yet while surveys often centered on a fusion of political and economic goals, their aims varied. Sometimes other justifications were also proffered for surveys, such as the promotion of public education, especially for natural history surveys. In 1913, F. C. Newcombe of Michigan argued in *Science* that scientific and economic justifications would be unlikely to maximize funding for surveys of plants and animals. He recommended that proponents of biological surveys highlight the benefits for schools, since these were more likely to motivate the educational constituency—especially if surveys were conducted on a smaller, regional biogeographic scale and involving the state academies of science as much as possible.[9]

Whatever the multiple goals of surveys, both rhetorical and substantive, they shared some key features in practice that defined the survey as a distinctive mode of knowledge production in the field during the railroad era. At their core, surveys deployed field research on itineraries from place to place across the land in order to systematically collect, observe, and map over a defined geographical space. They were made possible by the combination of a modern infrastructure of transportation and communication—dominated by the railroad, telegraph, and postal service—along with certain systematic ways of organizing labor in the field through those technologies and others. The survey thus involved travel, usually seasonal, into a geographical territory by scientists and their support personnel. By covering an extensive area comprehensively, science in the survey mode aimed to go beyond the exploratory expedition to a more thorough, rigorous, and all-inclusive knowledge of a field region, through a more systematic set of practices, sometimes over multiple years. As Florence Merriam Bailey noted on her work in New Mexico with her husband, Vernon, a "systematic survey" of the state "by Mr. Bailey and a corps of assistants" began in 1903 but extended over many field seasons.[10]

Geological surveys were one obvious and widely prevalent subtype. The US Geological Survey—established in 1879—and its predecessors sent survey parties all over the American West, including special divisions—not only Hague's Yellowstone division but also counterparts such as the earlier Rocky Mountain division based in Denver.[11] Likewise, by the early twentieth century, many state geological surveys in the Plains and Rockies were sending parties out in the field, though some of them started out as unfunded, honorific "state geologist" offices.[12] But it was not just geologists. *Survey* was a popular word for organized scientific endeavors of all kinds in the nineteenth century: topographical surveys, natural history surveys, soil surveys, ethnographic surveys—all of them accompanied the expansion of capitalism and European populations around the globe. From far-flung imperial acquisitions to domestic quests for natural resources (valuable minerals, agricultural and grazing land) and social order (for the effective control of American Indians), the survey was one of the preeminent knowledge-making strategies of science in the era of industrial capitalism, railroads, and telegraphs.[13]

Surveys were clearly associated with capitalist economic development led by an expansionist state. Yet this rubric of resource development and social control proved to be capacious. At the US Biological Survey, for example, "reports on life and crop zones" by Vernon Bailey were joined by much broader inquiries into bird and mammal life by himself and his wife, Florence Merriam Bailey, which were equally, if not more, intensive and systematic.[14] Moreover, as will become clear in this chapter, these global processes did not simply overwhelm a tabula rasa. Instead, the particular environmental conditions and work organization of surveys in the Great Plains and Rocky Mountains shaped field work in the region and gave it distinctive experiential contours. We must be attentive to these practical experiences on the ground of field scientists and their entourages. A central dynamic of the encounter of survey researchers with the environment of the West was coping with western places that were far different from the eastern metropoles and rural villages from which many of them came. To be sure, some survey team members—especially lower-status workers like cooks and packers—came from the region itself. But the seasonal pattern of survey work ensured that the seasonal alternation of eastern and western environments predominated in the experiences of survey scientists themselves.

Development of the Survey

In the American West of the late nineteenth and early twentieth centuries, the survey was one of the most important modes of scientific knowledge production, first through the US government and later through state governments and other patrons. But the survey did not emerge in full-fledged form and persist unchanged. Surveys existed before the railroad era, of course, including not only property boundary surveys but also surveys of the natural environment itself, such as nautical, topographic, and even some geological surveys. Like many other imperial settings, the advent of survey science in the American West in the early nineteenth century was closely tied to the territorial and military expansion of a capitalist nation-state—in this case, the United States. Based on the Atlantic coast of North America, the United States deployed military surveyors to produce basic geographical knowledge about the western half of the continent, occupied by American Indians and coveted or claimed, at least in part, by several other sovereign states, including Spain (then Mexico), Britain, and Russia. Later survey work in the interior West followed from and built upon the efforts of the US Army Corps of Topographical Engineers.[15] As the railroads were extended into the American West from the late 1860s onward, the region was opened up more decisively to survey science.

But with the coming of the railroad, the scale and scope of survey endeavor expanded dramatically.[16] This is not to say that survey work was itself done solely, or even primarily, by using the railroad on an everyday basis, but simply that the railroad allowed field scientists and their equipment and supplies to move much more quickly in and out of the field at the point of embarkation. Once away from the railway station, survey work was accomplished mainly on foot and with the assistance of draft animals.[17] Moreover, when survey scientists could travel up and down navigable rivers by boat, such as along the Missouri River in the northern Great Plains, survey scientists often did so, even well into the twentieth century, but this opportunity was not available in most parts of the mostly high and dry Great Plains and Rocky Mountains region.[18] Thus the usual transport geography of survey work in these regions was primarily land-based.

The transformative power of survey geography was recognized by contemporary observers. In a November 1888 paper titled "Surveys, Their Kinds and Purposes," presented in Washington, DC, Marcus Baker portrays surveys as developing organically from expeditions as part of progressive "civilization" and "mastery over nature." Notably, Baker also portrays surveys as the next stage beyond "research by single individuals, in private laboratories," which "led to discoveries of interesting and useful facts and principles . . . but which could only be tested by the study of many and widely separated localities." This epistemic need naturally proceeded, in Baker's view, to organized action by the state in the interests of the community as a whole, and these increasingly systematized expeditions that "led to the establishment of some, and rejection of other, generalizations, based upon a knowledge of only a limited area."[19] Such travel-based scientific practice developed from "exploring expeditions" into "great government surveys and international surveys,—surveys organized and conducted systematically to study a great area, and to collect and diffuse information for the benefit of the people."[20]

Beginning in the late 1860s and through the 1870s, long-distance expeditions were transformed and systematized as railroad-based surveys, in conjunction with a shift from military to civilian control over survey science. It was thus largely during the so-called great surveys of Ferdinand V. Hayden, Clarence King, John Wesley Powell, and George M. Wheeler—with varying mixtures of military and civilian control—that the shift from exploratory expedition to survey science took place.[21] The word *reconnaissance* was later used to distinguish these early surveys and, in some cases, to apologize for their deficiencies. Even while the "great surveys" were still taking place, field scientists had already begun to make the distinction. The California field geologist J. D. Whitney argued in 1875 that topographical maps resulting from true surveys "have as their essential features absolute accuracy and minuteness of detail," which was quite different from "the work of preliminary exploration, or of reconnoissance [*sic*], as this sort of surveying is commonly called."[22] This distinction may have been common knowledge, as Whitney asserted, but to some extent the boundary line between the practices of reconnaissance and the practices of survey shifted over time as standards of rigor increased.

The passage of time reveals such changes in how survey practices were regarded. In 1903, for instance, geologist S. F. Emmons, who participated in the King Survey of the 40th Parallel, recalled his summer 1869 experiences in order to contrast the scientific practices of the two eras. "Three weeks were all that could be allotted [in 1869] for the topographic and geologic reconnaissance of the whole Wasatch range south of the latitude of Salt Lake City," recalled Emmons. "It can readily be conceived that under such relations of time to area many phenomena would necessarily be but imperfectly observed." To be sure, Emmons and his colleagues were able to take advantage of the newly completed transcontinental railroad to reach the field—after all, the King Survey had been designated to follow the 40th Parallel in large part to explore the resources in a wide corridor along that pioneering rail route. Nevertheless, the work had been largely conducted under rudimentary field survey conditions that had only just begun to change from the prerailroad era. Notably, settlement infrastructure had not yet been established over most of the route. As Emmons recalled, it had been a "weary march across the desert" until, at the end of the field season in November, they reached "the smiling fields and orchards of the Mormon farmers" in one of the few settled areas of Utah.[23] Such an example reminds us that the railroad did not instantaneously transform the conditions of field practice, but rather set in motion a larger and more complex series of events.

In the immediate wake of the completion of the transcontinental railroad, others also witnessed the transition from exploratory expeditions to more systematic survey practice. The paleontologist O. C. Marsh, for example, while exploring the Uinta Mountains of southwest Wyoming and northwest Utah as the leader of the Yale scientific expedition in 1871, pursued some geological field observations in the survey mode, even while prospecting for rich fossil quarries that might be exploited more intensively. However, Marsh lamented that, much like the federally funded surveys, his "geological observations were necessarily confined to the immediate route pursued, and even then, in the limited time at our command, could not be systematically conducted." The Yale party had "passed along the base of the Eastern Uinta Mountains, as well as over portions of the range" in the course of prospecting for vertebrate fossils, and "as nearly all this region was entirely unexplored," Marsh felt it might be "desirable to note its more important geological features."[24] Much like the field work of Emmons's party two years before, this work was essentially in what would later be described as "reconnaissance" mode, and it was circumscribed by the limitations of time, accessibility, and multiple purposes. As these examples of early railroad-era survey work suggest, travels were often restricted to single pathways through new environmental regions, and the shift to more systematic survey field work would be gradual, changing as the expansion of the infrastructure of settlement and means permitted.

To be sure, the boundary line between long-distance exploratory expeditions and more systematic surveys was not always entirely clear. In fact, the survey was really a specialized type of expedition, one more closely approaching the ideal of comprehensive knowledge of a region. And many of the experiential and practical features of the survey could be found in other types of expeditions as well, such as coping with environmental challenges, navigating networks of transportation and communication, and managing team members. If exploratory expeditions always retained a whiff of selective, opportunistic travel, and sometimes even serendipity, however, the more systematic surveys that emerged in the West by the late nineteenth century aimed for more complete coverage of geographical territory.

This transformation to more systematic survey science was further catalyzed by the establishment of the US Geological Survey and Bureau of American Ethnology in 1879.[25] Survey organization achieved notable consistency in these institutions, as well as a significant regional footprint, particularly for the Geological Survey. They were the two premier institutions for survey science in the late nineteenth-century American West before the dramatic enlargement of the US Department of Agriculture in the 1890s, which catalyzed a further expansion of survey field work and its extension to new topical domains. Two significant examples within the Department of Agriculture were the biological survey work organized by C. Hart Merriam, and the soil survey work—often in cooperation with the states—organized by Milton Whitney in the Bureau of Soils. These prominent surveys were complemented by a wide variety of other federal surveys, state surveys, and even nongovernmental surveys by diverse institutions such as universities and museums.[26] In the increasingly settled American West, with its correspondingly developed transportation and communication infrastructure, the tradition of the grand expedition was more and more overshadowed by the systematic survey.[27]

As the survey became more firmly established and well known as a rigorous mode of knowledge production in the field across a wide variety of disciplines by the turn of the twentieth century, those engaged in expeditions began to reimagine their organized scientific travels as "surveys" and, in many cases, to modify them accordingly. In college expeditions, for instance, the survey mode—certainly in name and also to some extent in substance—came to predominate. As college expeditions converted to the survey style, they adopted the language of survey in a quest for the greater prestige and rigor connoted by survey terminology, but they also often genuinely intended to improve the quality of the research undertaken—and in particular to teach the latest methods to the participating students. This was especially true for efforts led by colleges from outside the region. As early as 1898, for instance, the Auburn University student party in the La Plata and San Juan Mountains in southwest Colorado was labeled a "botanical survey."[28] Similarly, when geologist Wallace W. Atwood brought summer field school parties to the same region, first for the University of Chicago in 1911 using Ouray, Colorado, as headquarters, and then a few years later for Harvard University, he organized the work "as nearly as was practicable in conformity with the requirements of the National Survey," giving details about the division of the group into smaller parties to systematically cover the territory through mapping.[29] Ensuring that student summer field excursions and state-based programs adhered to the emerging national standards for proper survey practice sometimes meant concentrating on a small enough area that it could be systematically inventoried or mapped in a rigorous fashion.

Regional institutions also started to define their work as "survey," though with more of an emphasis on frequent, shorter trips rather than just long-distance field excursions. In doing so, they often raised the ambition of their field practices in accordance with what they perceived to be the changing standards of rigor in cosmopolitan science. The Botanical Seminar at the University of Nebraska, for example, launched a "botanical survey" in the early 1890s by layering on top of traditional plant specimen-collecting activities the additional task of systematic mapping of plant distribution. Their goal was "to make the Survey Herbarium a complete representation of the flora of the state, to indicate there, as far as possible, the distribution of each species, to determine the floras of the various regions of the state, and to provide for the botanists of the state a herbarium for the study of the local flora," along with lists, catalogs, and research studies.[30] Aven Nelson at the University of Wyoming also referred to his comprehensive studies of the state's flora as a "botanical survey" in the mid-1890s.[31] Francis Cragin's attempt to lead a biological survey in Kansas during the 1880s, discussed at greater length below, furnishes yet another example of how college-based natural history programs within the West reconceived of their work as surveys.[32] Over time, many states of the Plains and Rockies instituted surveys of varying scope. Some focused on geology but others also included natural history, such as South Dakota's more multifaceted effort in the late 1910s, led by Stephen S. Visher, which applied an explicitly biogeographical, integrated approach to encompass the plants and animals of the northern Great Plains, especially western South Dakota.[33]

Similarly, the survey research of others based outside such behemoths as the US Geological Survey and the US Department of Agriculture's Bureau of Soils often covered territory at a smaller scale. For example, when Nebraska zoologist Robert H. Wolcott undertook a faunal survey in 1905 he focused his efforts on the relatively circumscribed region of the Nebraska Sandhills forest reserves.[34] These small, county-sized areas had just been established a few years earlier and were limited in size by the fact that they were intentionally planted by people. (Not only that, but the intriguing biological problem presented by how a newly planted forest might affect animal life could not have failed to attract special interest.) Studying the fauna of the Sandhills forest reserves thus offered Wolcott the opportunity to do interesting research in the survey mode but on a small enough scale to be sufficiently rigorous and systematic for his colleagues in cosmopolitan science. Individual collectors, too, described their work in relation to the survey ideal. For example, when traveling botanist Alice Eastwood reported on a collecting trip in southeastern Utah, she remarked that her three-hundred-mile "botanical survey of the country could not be so thorough as would have been possible with more time and means."[35]

The widespread appellation of "survey" to scientific work extended to all kinds of projects. Many state-led efforts delimited their efforts by narrowing the subject matter. One Kansas biologist reported in 1909 on what he called "an extended and careful survey of the prairie-dog territory of western Kansas lying between the Union Pacific railroad on the north and the Santa Fe line on the south."[36] Oklahoma's agricultural experiment station led "farm management surveys" and a "plant disease survey" in the late 1910s.[37] Other states in the region also conducted "farm management surveys," though they were occasionally known by other names, especially before this terminology had become standardized nationally, such as Nebraska's "agricultural survey" ten years earlier.[38]

As surveys came to be viewed as necessitating more comprehensive, detailed, and thorough field work by the early twentieth century, the term *reconnaissance* (sometimes spelled "reconnoissance") was not only used to retrospectively distinguish earlier, less systematic practices as noted above but also often used to describe contemporary field practices that involved the preliminary coverage of a large area. In the mid-1890s, when the University Geological Survey of Kansas initially deputized students and other geologists from across the state as official geological survey personnel, before receiving greater funding, they designated their initial report as a "reconnaissance report," perhaps in recognition of its provisional results.[39] And the Oklahoma Geological Survey's pioneering eight-week field trip across fourteen counties in the summer of 1900, for example, was described as "largely of a reconnaissance nature."[40] In other cases, these expeditions involved rapid movement by railroad across large swaths of territory with brief stops, in a single path across a large, environmentally diverse area. Such practices were typically thought of as preliminary to more detailed and systematic work to follow but sometimes led directly to scientific publications and reports in their own right. In 1909, for example, entomologists James A. G. Rehn and Morgan Hebard presented an "Orthopterical Reconnoissance of the Southwestern United States" to the Academy of Natural Sciences in Philadelphia in several parts, and this project was based on a large-scale, rapid crossing by railroad of the southern tier of the United States, stopping at various locations along the way.[41]

Elsewhere, such as in the US Bureau of Soils, the term *reconnoissance* was used to describe the initial field study of a land area that was considerably larger than the standard soil survey geographical unit, which was typically a single county or local irrigation district. As early as 1899, the first year of the Bureau's "field operations," Thomas H. Means conducted reconnoissances encompassing three quite large Utah counties, as well as in the Cache a la Poudre Valley of Colorado.[42] The brief reports resulting from these reconnoissances, corresponding with much less time in the field, contrasted noticeably with the more systematic and thorough "detailed soil surveys" that were routinely published by the bureau from then onward. Starting in 1908, nearly a decade later, the bureau began publishing what it now called "reconnoissance surveys" of large areas, encompassing the entire western, less humid halves of the Great Plains tier of states, from North Dakota southward.[43] These reconnoissance surveys differed from detailed soil surveys "according to the density of settlement, the uniformity of the soils of the region, and the degree of detail intended to be shown on the map." On the Great Plains, they pointed out, "where the country is sparsely settled, the soils uniform over large areas, the precipitation light and uncertain, and where intensive development may not be expected for a long time, a reconnoissance scale was more appropriate."[44]

The reason for undertaking these larger-scale, less finely detailed mapping projects was the high demand for information due to the land boom in the semiarid Plains region. As bureau head Milton Whitney put it: "The demand for general knowledge of the Great Plains region has been very pressing for several years on account of the rapid economic changes that have been going on. It was impossible to get this by detailed surveys."[45] Although these "reconnoissance surveys" made ample use of the railroad infrastructure for moving field parties around, as well as for the settlement infrastructure of towns and provisioning that those railroads made possible, much of the actual reconnoissance field work on the ground involved moving across the landscape—at that time, despite the unprecedented influx of new settlers, still much less densely inhabited than the typical soil survey terrain—using teams and wagons. In a sense, then, US Soil Survey reconnoissance surveys combined aspects of rapid movement associated with reconnoissance and the systematic and comprehensive coverage of a survey, albeit less detailed than a standard soil survey. The more rigorous surveys became, the more important it seemed to find other terms for less thorough practices.

Mapping, Collecting, Observing

Over time, then, surveys became distinct from other kinds of expeditions. Above all, the survey produced knowledge of a territory more extensively, more systematically, and more comprehensively. They were usually multiyear undertakings, organized in advance. As anthropologist Clark Wissler put it in his general introduction to the American Museum of Natural History's ethnographic survey of Plains Indians starting in 1907, they "planned a prolonged and systematic field survey of the many different Plains tribes."[46] It was, he believed, the first time that "the successive tribal organizations for an entire geographical area" had become "the object of detailed investigation."[47] Moreover, the results of a survey could be mapped, not just along the linear path of a single journey, but over a whole region.[48] While it is true that surveys did not all involve extensive mapping—geodetic surveys, for example, used text, tables, and diagrams to display the earth's size and shape—in the practice of field science in the Great Plains and Rocky Mountains during the railroad era, comprehensive mapping was a central goal of most surveys, whether of geology, soils, life zones, or human cultural and language groups.[49]

Previous field researchers in the natural history tradition, especially systematizers such as Alexander von Humboldt, might have aspired to such comprehensive mapping. But without the dense infrastructures of transportation and communication on land, along with the funding and work organization of the survey mode, it was a rare achievement for earlier natural history expeditions. Survey science, on the other hand, produced such comprehensive maps routinely. In other words, as a mode of knowledge production in the field, the survey was intended to obtain systematic and exhaustive knowledge of the environment.[50] In the case of field biology, Robert E. Kohler has emphasized how surveys—which, he shows, were at once both extensive in their coverage and intensive in their field methods—proliferated in the "inner frontiers" of the American landscape: those land areas not yet developed within larger geographical areas dominated by cities, towns, and farms.[51] Extending our historical purview to surveys of not just wild plants and animals but also of geology, soils, and native peoples, the larger history of survey science in the Great Plains and Rocky Mountains reveals how it involved a more

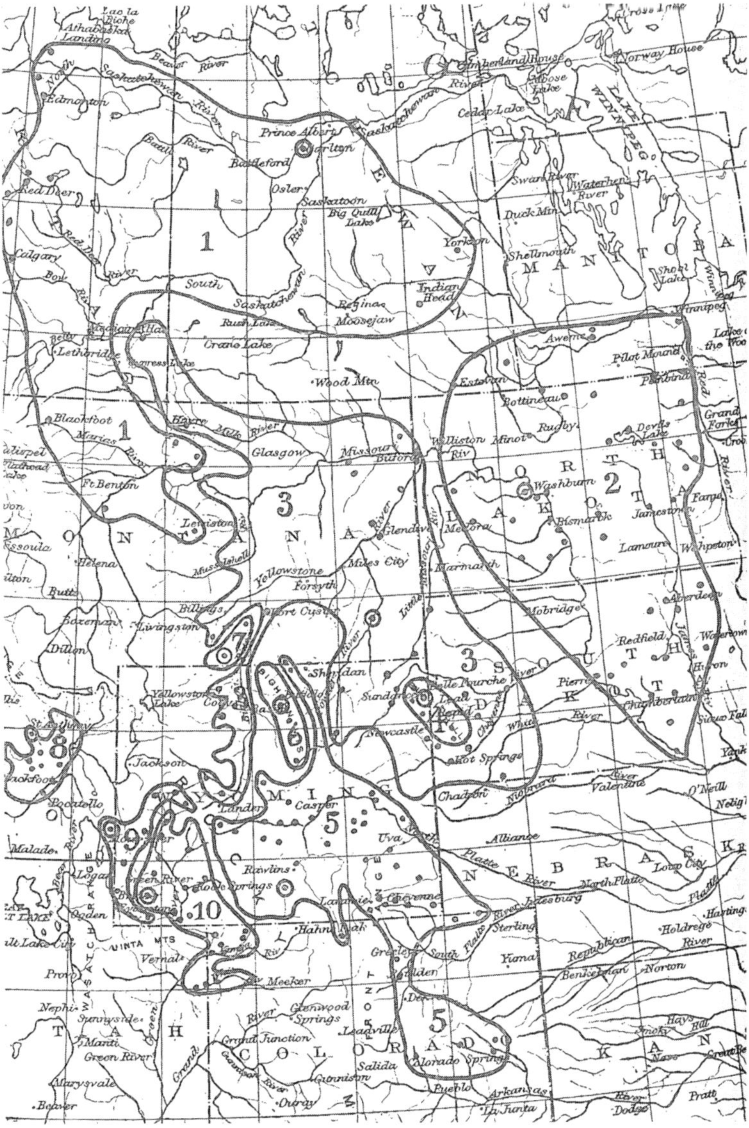

FIGURE 3.2. Drawing on Vernon Bailey's quarter century of field experience in the West, the map in his 1915 report on pocket gophers drew sweeping lines around collecting localities across the northern Great Plains and Rocky Mountains. Original caption reads: "Map showing distribution of certain species and subspecies of Thomomys, and localities at which specimens were collected." Vernon Bailey, *Revision of the Pocket Gophers of the Genus Thomomys*, US Bureau of Biological Survey, North American Fauna, no. 39 (1915), 28.

comprehensive coverage of the entire western landscape. This survey coverage came to include, especially in ethnography and soil survey, not only human settlements and farms, respectively, but also, at the other end of the spectrum, large land areas well beyond railroad lines, where field survey parties in the Plains and Rockies traveled with pack animals, in addition to the "inner frontiers" that were so important to field biology in more densely settled parts of the United States.

Though built on a foundation of both collecting and observing, mapping has thus been central to survey science. Indeed, mapping has arguably been the most important and distinctive type of activity for the survey mode of practice. The problem of what to include and what not to include has been constitutive of mapping and survey work in general, even topographical mapping. As W. M. Davis of Cambridge, Massachusetts, put it, "the truest map is only a generalization. Something is necessarily omitted, and the topographer has to choose between what he shall omit and what he shall represent."[52] Davis's claim about topographical maps applies with at least as much force to thematic maps spanning all the survey science disciplines. What ended up being represented on maps was shaped by a wide variety of considerations. In colonial settings, for example, the mapping of a boundary line between racial groups could hardly escape the social and cultural categories of colonial society itself. At the same time, such maps also simultaneously represented the dilemmas of the survey scientist trying to simplify the bewildering diversity of observations made in the complex world that constitutes the field.[53] Even when nonhuman nature was represented on maps, it combined environmental realities, such as the spatially compressed variability in mountain zones in contrast to the Plains, with the pragmatic necessity of extrapolating from patchy human collecting patterns (see figure 3.2).

Preliminary to mapping in the field, then, was the gathering of data through some combination of collecting and observing. First, consider collecting.[54] Just as manipulating physical materials in a laboratory required exacting and careful techniques, so too did collecting in the field, which also involved manual skills with tangible materials. The collection of all specimens in the field required delicacy and dexterity. The paleontologist John Bell Hatcher from Pittsburgh's Carnegie Museum made the importance of collecting skills clear when writing to the museum's director, William J. Holland, in a note comparing the work of two of his fossil-collecting colleagues who were traveling with him in Wyoming. "You have seen the two men together in the laboratory & have

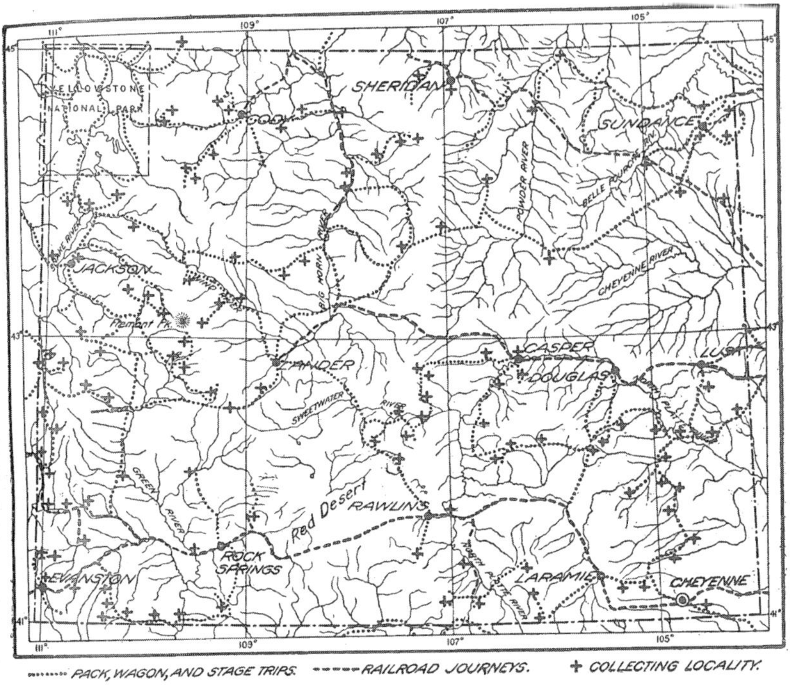

FIGURE 3.3. Typical survey routes, from US Bureau of Biological Survey field party, with stops indicated on the map, as well as the distinction between railroad and pack animal transportation. Original caption reads: "Map of Wyoming showing routes and collecting localities of Merritt Cary and other members of the Biological Survey, mainly from 1909 to 1915." From Merritt Cary, *Life Zone Investigations in Wyoming,* US Bureau of Biological Survey, North American Fauna, no. 42 (1917), 9.

remarked to me upon the difference between them," he commented, but "I can assure you that there is even a greater difference between them as collectors in the field." Hatcher recommended that the less-skilled man work under the guidance and authority of his companion, "if he expects to continue collecting."[55] But Hatcher's hopes for one collector learning from the other were not to be realized. Later, Hatcher wrote from Lusk, Wyoming, discussing the developing troubles between the two and "frankly" concluding, based on his scrutiny of Reed's collecting work, that he "[did] not want such work done for us, for in a many instances it simply means the distruction [*sic*], rather than the preservation of rare & important material & we must have the very best work possible for we are paying our collectors the highest wages of any museum in the country."[56]

During the railroad era, collecting, as a material practice, required disembarking from the railroad and moving through the field on foot or pack animal.[57] While the subjects to be studied and specimens to be collected were often specified in advance, the day-to-day movements of survey parties almost always had to be modified over time. Such a pattern involved the establishment of stopping places, which served as camping locations and temporary bases of operations (see figure 3.3). The field survey collector's path of travel might cross the entire region without regard to state boundaries and instead be based largely on a combination of available transport infrastructure and perceived environmental resources, as in Vernon Bailey's 1888 crossing as a field collector for the US Biological Survey. Or it might even track political boundary lines wherever they led, such as the Colorado Museum of Natural History's biological survey stations along the southern boundary of the state in the 1910s, though such an approach had previously attracted strong criticism on the grounds that "artificial boundaries . . . are not biological boundaries."[58] Whatever the degree of political geography that did or did not influence the routing, the collecting practices in survey mode were usually more precise and exacting than expeditions of the prerailroad era that were adjuncts of other activities.

Survey scientists didn't just collect; they observed. Most fundamentally, they observed the places and surrounding contexts where specimens were collected. But they also often observed what could not be collected. In other words, they observed not just as a valuable supplement to collecting—to know the environmental context—but also as an alternative when collecting was illegal, impractical, or physically impossible.[59] For instance, when geologists looked at landforms rather than rocks, they could not "collect" what they saw physically, so they had to be especially keen observers. Ethnographers, too, adopted a survey mode that was intended to systematically record their observations of human life and culture over vast spatial territories. In his autobiography, the American Museum ethnographer Robert H. Lowie remembered "survey work" as a style of inquiry in which "museum ethnographers were rarely permitted to stay for a long period with any one people or to revisit them for intensive study."[60]

Visual skills were often central to the observational practices of survey science. Field drawing, as historians have recognized, was essential for domains such as geology, with its "visual language."[61] During this period, survey photographers and artists aided in the process of recording observations for later use—and for the promotion of survey funding, especially during the early part of the railroad era. In fact, historians of photography have identified a distinctive style of archival, modernistic visual depiction associated with the early western surveys.[62] The famed Hayden Survey photographer William Henry Jackson described his photographic practices, in "constantly departing from the high road," as essentially similar to those of other members of the survey field party. "For every mile on the map," he commented, "we covered between two and three on the ground—up mountainside, down stream bed, across country—to gather rock specimens, to survey and map, and to paint and photograph."[63]

Even those who collected transportable specimens such as fossils, plants, animals, and human artifacts through systematic survey practices had to be skilled at observing in the field environment in order to locate their finds in the first place and to rigorously map them later on. Ultimately, with increasing emphasis on ecological context, such observational skills became even more crucial to collecting, as a necessary element of more rigorous research.[64] Even before that, however, for a systematic survey of a given region, it was important to collect a significant amount of observational data rather than just collecting objects to bring back. Lowie, for example, found himself "jumping from one reservation to another," and over time he "paid scouting visits to a good many Shoshonean groups in Idaho, Wyoming, Utah, and Nevada, sometimes not tarrying more than a few days in any one spot."[65] In combining such extensive territorial practices as mapping, collecting, and observing, field survey scientists cultivated a special combination of the knowledge of experience in many different local places with a wider knowledge of scientific categories, classifications, relationships, and distributions across space.

Vernon Bailey of the US Bureau of Biological Survey is probably the clearest example of the contextual and place-specific knowledge that long-term, extensive survey field work could foster, thereby bridging experiential and cosmopolitan knowledge. Notably, such knowledge transcended narrow disciplinary specialization. According to an obituary in *Nature Magazine*, almost certainly written by his fellow Biological Survey field naturalist Edward A. Preble, Bailey may have been "best known as a mammologist, but he was also a practiced ornithologist of the outdoor type, and had a field knowledge of the plants of the country that many a botanist might well envy."[66] An admiring profile of Bailey written by Alma Chesnut twelve years earlier had also noted this trait. "At the beginning of his field work," wrote Chesnut, "Bailey had realized that the proper study of birds and mammals required that one be familiar with the plants on which they feed." He thus became "almost as adept at identifying the different trees, shrubs, and grasses as he [was] at naming the animals."[67]

It was not just that Bailey knew a lot about plants and birds to complement his deep knowledge of mammals in the field. Bailey's holistic, contextual, and ecological knowledge of the wide range of places through which he traveled also bridged the epistemic rift between the knowledge of experience and the cosmopolitan knowledge of science. Although experiential knowledge is typically local, Bailey's unusually extensive travels had allowed him to grasp some—though, of course, hardly all—environmental features of a great many places, across whole regions. He knew enough, in fact, to write the "standard publications, telling just what plants, birds, mammals, and amphibians, and what are their habits."[68] As Preble pointed out, Bailey's "unsurpassed" knowledge of the whole American West was due to his astonishing record of travel in the field. "There have been few years, during the past five decades," Preble wrote, "that did not find him traveling, by some method, through desert, forest, mountain trail, or river valley." As a consequence, Bailey achieved an enviably high level of experiential knowledge of western environments to complement his cosmopolitan knowledge of science: "His memory of places visited was the most exact that I have ever known," proclaimed Preble. "He added to a genius akin to that of the Indian for memory of topographic details the broad geographic viewpoint of the cultivated scholar."[69] Preble's juxtaposition of American Indian topographic knowledge with the wider perspective of the "cultivated scholar" is telling, for it neatly conveys the combination of experiential and cosmopolitan knowledge—over a wide geographical area—that so distinguished exceptional field survey scientists such as Bailey.

Seasonality

As these testimonials about Bailey's unusual capabilities indicate, the mapping, collecting, and observing practices of survey science were shaped not only by culture but by the environment, too. One connection, which may seem obvious but deserves careful scrutiny and analysis, is the seasonality of survey science. The timing of many human activities is determined by seasonality, which was exceptionally strong in temperate-latitude, interior continental regions like the Great Plains and Rocky Mountains. Bitter cold winters punctuated by blizzards have swept across the Plains and Rockies for centuries, while the blazing heat and often unclouded direct sunlight of the summer have also shaped the contours of human existence in this region. The field season for survey work was most sharply defined by the summer absence of cold, snowy conditions, the accumulated remains of which could last until June at higher elevations. Arriving too early could seriously hinder field travel, even by railroad, as John Wesley Powell discovered while crossing the Wasatch Mountains of Utah in May 1873.[70] In most parts of the Plains and Rockies, the field season extended from late spring or early summer until midautumn, suspended by the first major snowstorm. On the other hand, work in some of the Plains and desert areas could be slowed or even temporarily suspended by exceptional heat, although this was only rarely a major impediment.

The US Geological Survey's Yellowstone Division from 1883 to the early twentieth century was a classic example of seasonal survey science.[71] Rather than working year-round in the field, the Yellowstone Division was based in Washington, where scientific staff lived in winter and spring. Hague and his coworkers mounted elaborate preparations each year to set up temporary field camps in the Yellowstone Park for use from July until October. Hague and his associates usually accessed Yellowstone each summer by taking the Northern Pacific Railroad through Montana, then continuing by horseback southward into the park. In the fall they would return to Washington to complete their processing of materials, analysis of findings, and writing up for publication. This seasonal pattern of return migration was typical of field research in the West, and it was widely emulated into the twentieth century on a smaller scale even by regional institutions whose headquarters were physically located within the Plains and Rockies.

In the interior West as a whole, as noted above, the field season generally lasted from the late spring or early summer into the fall. Survey scientists sometimes worked longer at lower elevations, though late fall and winter conditions of cold and snow could drive them home at any time. Those who risked the hazards of late fall to stay longer in the field might also enjoy opportunities beyond the usual seasonal boundaries of survey science. In 1919, Vernon Bailey stayed in North Dakota into late October, so that he had the rare chance (for a survey traveler) to observe the "storing season" for mice. Near the town of Cannon Ball, Bailey was delighted to find "a small cache of beans and artichokes," which he collected along with the mice themselves. "One was caught in a trap in a runway leading to the cache," he excitedly reported, "and one was taken alive in my hands as it ran out of the cavity."[72] For local residents, who experienced the place year round, such an opportunity may not have been unusual, but for Bailey it was a rare and treasured event. More broadly, when survey scientists persevered in the snow and cold, the opportunities for collecting and observing shifted with the seasons. Over three decades earlier, in October 1887, US Biological Survey chief C. Hart Merriam had warned Bailey, who was then a novice collecting in the Black Hills of South Dakota, that some mammals had probably already "'denned up' for the winter," although he also noted that "the snow will be a great help to you in trapping species which do not hibernate."[73] Thus, even snowy weather itself could bring both advantages and disadvantages in the field.

As a rule, though, the winter season with its bitter cold and wind was not a good time to be in the field in the Great Plains or Rocky Mountains.[74] Sooner or later, for the winter, field surveyors tended to move to more southern climes or return to the cities from which they came. The national Bureau of Soils formalized a seasonal pattern of north–south migration by redeploying its field surveyors working in the interior West (and other places with cold winters) to such locations as the Gulf Coast and the far Southwest in the late fall.[75] Seasonal changes also sometimes necessitated changes in survey routes. Less than a month after Bailey was tracking small mammals in the Black Hills, his federal government superiors advised him to scrap his previous plans to continue on through Idaho, instead directing him south from Utah into warmer Arizona.[76] Working in wintertime on the edge of the milder Southwest, Bailey used the small town of Kanab, Utah, as a base for side trips. Merriam urged him to visit the milder, protected areas within the canyonlands. "Undoubtedly you will find many southern birds wintering in suitable places in the cañon," he thought.[77] Even in the southern Utah canyonlands, however, the winter climate could still be quite severe. Only upon reaching the deserts of the Southwest did Bailey really find a fully suitable place to do research during the winter months.

There were partial exceptions to the seasonality of survey science in the interior American West, mostly outside of the Great Plains and Rocky Mountains. In southern Arizona, for example, winters are very mild, thus permitting survey field work by Bailey and others to take place all year round. There were few exceptions in places with more severe winters, like Colorado, however. Even those survey scientists who lived near their field sites, such as those working for state-level surveys based in the West itself, common practice still tended to follow the pattern of working primarily in the summer to early fall, then returning to the cities and towns over the winter and into the spring. Similarly, one of the US Geological Survey's earlier divisions, the Rocky Mountain Division, unlike the Yellowstone branch, was headquartered in the West itself, in Denver. But while this meant that the federal government had a year-round institutional presence in the region—precursor to what would later become perhaps the largest concentration of government bureaucracy outside of Washington, DC—it did not mean that the Rocky Mountain Division's scientists were exempt from the constraints of seasonality. They just had a shorter distance to travel between field and office.

It bears noting, however, that the seasonality of survey science was not shaped solely by the natural environment: it was also cultural. One of the most obvious cultural reasons for seasonality was the existence of vacation times at universities and colleges, from which many of the field scientists came. Not only were survey scientists freed from the obligations of teaching on campus for the summer, they also sometimes engaged in another kind of teaching off campus that offered a productive synergy between field research and instruction. These were the roving field survey parties, single-season field camps, and—as examined more closely in a later chapter—field stations, all of which typically involved teams of professors and their students, who often provided low-cost labor to the practice of field work. This academic basis for survey science seasonality was quite evident at the time. Harvard medical professor Charles Sedgwick Minot even provided a helpful chart or diagram in *Science* with data from fifteen major universities, showing that most started their fall classes in late September or early October and finished their spring classes sometime in June. In general, Minot commented, students need less vacation time but professors, who are researchers, need more, for all types of research work: "To few professors is the vacation time wholly a period of vacation. On the contrary, it is a time which he can utilize for study, for research and for the increase of all his mental equipment upon which his proficiency as a professor depends."[78]

For many survey scientists, however, especially the large number working for the federal government—and some state governments, when survey positions did not also carry teaching obligations at a state college or university—it was not the academic calendar that provided a cultural reinforcement for the seasonal pattern of survey work. Instead, the functional and cultural appeal of a balance between office work and field work made it attractive to be in the field for part of the year collecting and observing and in the office for the rest of the year, comparing the specimens collected to others in a museum or reference collection, consulting the published literature in a library, or engaging in activities that were easier to perform in the office, such as collating data and making maps. Moreover, the rich opportunities of a scientific metropolis such as New York or Washington, DC, made it possible to participate in scientific societies, community events, and family life from the late fall through the spring. Beyond these intrinsic advantages, the more mundane, yet often overriding, reason for the seasonal patterns of government-supported field work was the reality of the budgeting cycle. For field scientists employed by the US government, the fiscal year ran from July to June, and typically the field season would not begin until the appropriation was determined, a process that ideally would be completed by the first of July but would often run several weeks later, or even into August. Whenever possible, the heads of scientific agencies would try to work around such budgetary exigencies, in order to begin some field work by late spring or early summer with anticipated or conserved funds, but the fiscal realities often dictated that the government field season would run from midsummer to the fall.

The budgetary dimensions of seasonality were not restricted to the special circumstances of government budgeting, however. Another aspect of seasonality that deserves mention, because it significantly affected the finances of surveys (usually adversely), was the often significant seasonal change in the market for essential items needed for field work. Perhaps most striking was the shift in the supply and demand for pack animals, such as individual horses or teams of horses, at many western frontier railheads where field parties, along with many others, embarked on their local or regional journeys. Because the demand was higher for teams of horses during the summer field season, and because it cost money to feed and care for horses over the winter, the purchase prices in the spring or early summer were often much higher than the selling prices in the fall. In the early summer of 1905, for example, a letter to geologist Samuel Williston from one of his field collaborators, sent from the town of Rawlins, Wyoming, along the Union Pacific Railroad, complained that locals were asking "from $150 to $200 for old teams that wouldn't sell in the fall for $75."[79] If multiple years of field work were anticipated, survey parties would sometimes pay for winter subsistence for their horses rather than sell low and buy high the following year. The Hayden Survey, starting in 1873, paid local ranchers to "care for animals and equipment from one season to the next," to avoid selling low at the field season's end.[80] Seasonality was thus also produced sometimes even by local and regional market forces.

Whatever practical considerations it was based upon, seasonality had significant consequences for the knowledge produced through field science. Unlike many lay people or amateur collaborators in networks, who typically lived and worked in a particular place all year round, survey scientists only experienced the environment selectively, during certain seasons and not others. Thus, despite its spatial advantages for systematic coverage, rigorous mapping, and trained practice, the survey as a mode of field science was inherently limited by time. Much of the experiential knowledge of place *over time* could not be fully captured by itinerant survey collectors and observers. The North Dakota amateur naturalist Joel Lunell, who was so critical of power relations in networks, as noted in the last chapter, was also scathingly critical of surveys for this very reason. Even though the state had funded "a geological survey with a staff botanist," he argued in 1915 that most "rare plants will remain undiscovered" under a survey mode of practice. "The place needing weekly visits for years can not reveal its secrets on the very day when the surveyor scans it," and thus "a majority of the 700 undiscovered plants would remain hidden for 100 years to the botanist-surveyor!"[81] Unafraid to point out "vital defects of plant surveys," Lunell followed up in 1919 with a blistering review of a state survey report on Barnes County, rejoicing that surveys had recently been "abolished" in North Dakota. "There are more sensible, quicker and less expensive methods for exploring the flora of a land area," he declared. Little could be "expected from the few seconds or minutes spent on each different square yard" of the county, "perhaps never to be visited again."[82] Few, if any, others harbored the deep hatred of career survey science that Lunell did, but the problem underlying his venom was real nonetheless.

Career field scientists, including those who traveled long distances, were rarely ignorant of this concern. "Every naturalist," proclaimed J. A. Allen, then at the Museum of Comparative Zoology in Cambridge, Massachusetts, "is of course aware of the difficulties one meets with in seeking to learn something of the mammalian fauna of a locality, and how inadequate a few weeks' reconnoissance is for its satisfactory exploration." Moreover, he pointed out, some creatures have "nocturnal habits," and others are reclusive, so that "only a comparatively small proportion of the whole are readily observed or obtained, patience and strategy being requisite for the discovery and capture of the others." Allen believed that certain parts of the natural world were more easily studied in a short time than others. In particular, he thought "a few weeks of diligent collecting" could provide "a tolerable idea of the character and variety of the bird life occurring at a particular season at a given locality," but that "many months are necessary to give one an equal familiarity with its mammalian life."[83] Similarly, botanist M. A. Carleton noted that to offer "anything like a complete statement" of a place's native plants would require someone not only "to survey systematically the entire territory" but also to visit "each district at all seasons of the year and taking two or three years of time for the purpose."[84] Arguably it was Allen's and Carleton's deep study of mammals and plants, respectively, that helped them appreciate the complexity of their subjects, as would be true of any specialist, but nonetheless the phenomenon of seasonality was real to them and to other traveling field scientists.

From the earliest years of the railroad era, field scientists tried to mitigate this problem when they could. Allen himself saw a potential solution in tapping local knowledge, contending that "one can learn at second hand much more respecting mammals than birds," since "all the larger mammals are objects of special interest to the hunter and trapper, either for their furs, their flesh, or as enemies, and whose pursuit is attractive and meritorious in proportion to its dangers and difficulties." This of course raised the problem of trust, which was so central to the practice of lay networks, but he thought it was surmountable. For mammals, at least, a "traveling naturalist" could not only "consult those skilled in woodcraft for much information he has not time himself otherwise to obtain" but could approach "a certainty of results attainable in respect to scarcely any other class of animals." During a "collecting tour to the Great Plains and Rocky Mountains" in the early 1870s, "everywhere" he found "intelligent hunters, some of whom had spent many years in the vicinity of the localities [he] visited," and extracted "much valuable information" about "the occurrence and relative abundance of the larger species," though he also weighed "the accuracy of their accounts by the independent observations and reports of different observers" and his "own experience and general knowledge of the subject."[85] Nevertheless, as he reported in the case of his travels in Carbon County, Wyoming, it was only through his "constant intercourse with hunters of long experience" that he was able to "gather much information" that he "could not otherwise have obtained without a long residence" in the field.[86]

Another solution was to combine survey observations from travels at different times of the year. The ornithologist Robert Ridgway, traveling with the King Survey in 1869, boasted that he "had the good fortune to explore nearly the same ground during the breeding season, or when the summer fauna was stationary" as Allen, who had come in the fall. "Combining, therefore, the results of the two explorations," he thought, "the character of the avi-fauna of the western watershed of the Wahsatch may be pretty well shown."[87] Even then, however, Ridgway was only able to make this claim by limiting the purview of his work to listing birds and delineating their geographical distribution. If he had wanted to study their habits throughout all seasons, or to investigate the particularities of bird life in relation to the local environment in each place he visited throughout the year, he could not have accomplished this solely through a visit that lasted from mid-May to mid-August of a single year. Notably, lay networks also often filtered out experiential knowledge selectively when moving from periphery to center. But for the field survey, the time limit on collecting and observation was strikingly seasonal and was usually restricted to the summer and early fall, with only a few exceptions, and only partially mitigated by drawing on the experiential knowledge of lay people or multiple trips into the same area.

Subsistence in the Field

Beyond the key epistemic factor of seasonality, larger subsistence patterns for both horses and humans were also foundational to the environmental history of field science in the Great Plains and Rocky Mountains. As the example of Bailey's varied route through the region suggests, the interior West was by no means a homogeneous environment. The diverse natural conditions of mountains, deserts, plains, and canyons provided not only some of the environmental variety that made the region so interesting to study but also a wide range of practical circumstances to contend with. It is instructive to think of the field survey party as a group of people in search of subsistence in the natural world. To be sure, they could never have taken to the field without a wide variety not only of material technologies to procure subsistence in the field but also consumable items such as preserved foods, often made possible by canning technology. But for their own subsistence—and especially for the subsistence of their pack animals—field survey parties depended on a variety of environmental resources and at the same time were challenged by a wide range of environmental constraints.

The most determinative natural constraints were, in fact, the same as the basic environmental resources necessary to sustain the field party in its entirety, including both humans and their accompanying animals such as horses and mules. Wherever work was done before the advent of the automobile, survey work depended on the availability of these resources, especially around campsites. As Richard Irving Dodge, commander of a US Army escort for a geological surveying party in the Black Hills, expressed it, "the three essentials of a good camp are wood, water, and grass."[88] Many other expedition or survey travelers in the field, whether with the army or not, articulated essentially the same sentiment, and the tripartite mantra of "wood, water, and grass" was widely known and quoted in the late nineteenth century.

Aridity or semiaridity was perhaps the greatest environmental challenge of all, providing meager water supplies in the heat of summer, scarce tree cover, and dwindling remnants of overstressed grasslands in the late summer and early fall. Such necessities for field subsistence as wood, water, and grass might have been fairly easy to obtain just about anywhere in the eastern United States, but in the more arid West they could be quite scarce in some places. As Lawrence Bruner wrote while traveling across Wyoming and Utah as a scientific assistant with the US Geological Survey in 1881, "There is nothing which so much discourages a traveller over these western countries as the absence of good pure water. For *this* he would almost give anything at times."[89] Grass was also unevenly distributed across the interior West, though also abundant in certain places and seasons. "Knowing no other subsistence than that of the range, nor any keener relish than for bunch grass," wrote the geologist James P. Kimball, referring to his field party's pack animals, "all supreme conditions are fulfilled wherever this is to be found, even though at some distance from camp." Where bunch grass was scarce or absent, "almost any growth of grass answers in a pinch," however.[90] In the Texas Panhandle, though, a geological survey field party that often "had to subsist entirely upon grass" found that "part of the time that was very poor both in quantity and quality." As a result, their "animals arrived in Austin in a very much jaded and broken down condition."[91] Likewise, as one geological field party member noted, "movements of a party in the wilds of the Rocky Mountains are more or less governed by provision of pasturage, any other kind of forage being out of question, and indeed quite beyond the craving of range animals."[92] During the railroad era, with so much work done off the rails, the needs of work animals closely shaped the material conditions of survey field parties.

Even when they could obtain adequate wood, water, and grass, both humans and animals in field parties sometimes faced threats to their health that reflected the environmental conditions of the Great Plains and Rocky Mountains. In some places, water might be easily available but unhealthy to drink, as soil surveyor Macy Lapham discovered in 1902 when traveling along irrigation ditches near the Arkansas River on the plains of southeastern Colorado.[93] When Lapham fell ill from drinking the alkaline water in the area, he requested permission to purchase distilled water for drinking, and the head of the US Soils Bureau agreed.[94] Similarly, the availability of ample grass for animals, such as in the well-watered high mountain elevations of Yellowstone Park, did not guarantee that it would be free of disease. In the summer of 1890, Walter Weed of the Yellowstone Division wrote to Hague from Deep Creek Camp reporting that most of the animals were "in excellent condition" except for a horse named Dandy, who was suffering from an unspecified epizootic disease that had previously afflicted two other horses named Chick and Pinto. To help Dandy recover, Weed's party was letting him rest "as much as possible," letting members of the party ride him only "every other day."[95]

Given the seasonal concentration of survey work in the summer and early fall months, heat could also pose challenges to field work. Temperature is an environmental variable dividing the Great Plains from the Rocky Mountains as regions, since the high altitude of the latter generally means milder summer temperatures. On the Plains, however, as well as in the lower zones between mountain ranges of the Rockies, summer heat could strongly shape the experience of summer field survey work. In some cases, as during the prolonged suffering of unusually high temperatures during the summer of 1911, field work might be stopped or postponed entirely because of the heat, which did in fact occur for geological surveying on the plains of Nebraska. The decision was not solely due to concern for the comfort or health of survey party members but also to the interference of high temperatures with accurate usage of survey instruments. "The heat waves rendered telescopic sightings so trying to the eyes," noted the Nebraska survey's director, Erwin Hinckley Barbour, "that this party was disqualified for further work at the end of a few weeks" and was forced to abandon the field.[96] Similarly, in a report for the Hayden Survey, O. C. Marsh contended that his 1871 collecting results on the central Plains "did not equal our expectations, owing in part to the extreme heat," which caused "sunstroke and fever, weakened and discouraged guides and explorers alike."[97]

If semiaridity, disease, and summer heat posed the greatest threat to subsistence and comfort on the Great Plains, both for humans and their draft animals, the rugged topography of the Rocky Mountains was one of that region's more distinctive challenges. In a presidential address to the Geological Society of Washington in February 1899, after multiple years of survey field work in the Absaroka Mountains along the east side of Yellowstone National Park between 1885 and 1897, Arnold Hague described the "long and protracted journeys into this rugged and at the time almost unknown region." In his experience, it had required "the greater part of the summer" to reach "the wild recesses of the Absarokas."[98] Wagon roads did exist in some parts of the Rockies, but survey scientists often found themselves on rapidly ascending or descending terrain, and on mountain slopes often covered with the more dense vegetation due to the greater snowmelt and summer rainfall often produced by these precipitation islands. Moreover, the rockiness that gave these mountains their collective name, combined with frequent stream crossings, high mountain meadow wetlands, and other topographical complications, meant that horse-drawn wagons could not always be used. In such cases, survey travelers resorted to pack trains, and they often substituted mules in place of horses for these rugged journeys. One such field scientist was Marsh, whose published geological observations in the early survey mode followed up a description of his Yale field party's route with a direct comment on the need to shift transportation regimes when they crossed into a more mountainous region, moving from one river basin to another. "As the country traversed is entirely impassable for wagons," he observed, "it was only with considerable difficulty that a way for our pack-animals could be found."[99] While field survey teams aimed for a more systematic coverage, they accordingly had to adapt their pack animals to the rugged conditions of the mountain terrain.

Even when they successfully navigated the rugged topography, survey scientists in the Rocky Mountains might occasionally have their work disrupted by wildfire. In the fall of 1871, one of "great survey" leader Clarence King's field parties, the Bridger division, was camping in the eastern Uinta Mountains when the "extensive wildfires" that "raged throughout the entire Rocky Mountains" began "filling the air with such volumes of smoke as to altogether stop topographical work." Despite the smoke, they were not in any immediate danger, and they continued their geological work while delaying the more visually demanding topographical work, with its requirement for long-distance sight lines. King himself was in northern Colorado a few months earlier, where his field party "met the same unfortunate experience as regards smoke" and "were driven back on geology as the only possible labor."[100] From the available evidence, wildfire seems to have rarely, if ever, stopped field work completely, though it did often force a redirection of effort away from practices that required long-distance visibility.

In addition to larger-scale phenomena such as the availability of water and plant resources, climate, topography, and fire, survey work could also be influenced on a more local level by the fauna of the region. Game animals and fish provided a supplemental food source in many parts of the West, although expeditions and surveys usually depended at least as much, if not more, on food they brought themselves. The animals of the West, some of them common and others distinctive to the region, constituted one category of environmental resource to be harvested for both science and subsistence.[101] Even if they depended on canned and preserved foods for much of their diet, field survey scientists more fondly remembered the wild foods they enjoyed, especially the fish and land animals they and other members of their field parties had caught or hunted themselves. After retirement, the US Biological Survey chief field naturalist Vernon Bailey looked back on his most memorable meals in the field, recalling especially meals of "roast antelope ribs," salmon from the Columbia River, "fried blue grouse," brook trout, wild turkeys, buffalo tongue, and Texas catfish. He also pointed out that "the meals that stand out clearest have been served with a hunger-sauce that many people have never tasted," including one such meal that consisted of "just plain camp fare"—"soda biscuits, a kettle of beans, some fried venison and stewed tomatoes"—but came after an unexpectedly lengthy mountaintop collecting side trip without enough food to eat.[102] Even as they pursued their specialized lines of inquiry, survey scientists knew nature intimately through their stomachs as well as through their heads and hands.

In some cases there was direct overlap between scientific and culinary objects. Bailey considered the same pocket gophers he was collecting in the field to be "a palatable meal in the mountains when other food is not available," commending them for human sustenance in the same report in which he was describing their taxonomy and geographical distribution. "Two or three gophers broiled over the coals of the camp fire furnish a fairly substantial and palatable meal," he elaborated, "especially if accompanied by a stew of wild onions and bulbs of the camas." While recognizing that they were "hardly large enough" for selling their exchange value in long-distance economic markets—though, of course, he was collecting those very same creatures for long-distance science—Bailey thought that "some may be utilized for the table" in local subsistence. He further observed that "the meat is rather dark and fine-grained, tender, and in flavor not unlike that of the squirrel. Sometimes it is rather strongly flavored with wild onions, but to some this is not unpleasant."[103] Similarly, in his biological survey of Texas, Bailey could not resist fantasizing about "roast prairie dog" becoming "an epicurean dish," in order to help address "the problem of keeping them in check," and he regarded them, "properly prepared and cooked," as "a delicacy."[104]

Wild animals were not just a food source, however. They could also pose a danger to health, safety, or comfort. Survey parties complained of different animals in different places, though often of insects. One geologist traveling with a field party in Montana and in Yellowstone National Park lamented the "pest of horse-flies" attacking humans and pack animals alike, describing them as "vampires" from which "there is no escape in mid-summer except in a strong breeze or at the higher altitudes." He regarded "the horse-fly of the Rocky Mountains" as an "unmitigated nuisance and 'kill-joy,' abominated by the nerves as well as the flesh," with the "bite" of a "carnivore."[105] Creatures encountered in the field were even more dangerous when they were predatory or poisonous: occasionally the animals that scientists were collecting would bite back. In western Kansas during the 1878 field season, a rattlesnake bit Prof. Frank H. Snow while the party was trying to collect it as a specimen. The incident was reported as a particular scientific fact by another participant in the expedition: "The party was in camp about fifteen miles south of Buffalo station, upon a small stream known as Hackberry creek. . . . The bite was received upon the twenty-seventh day of June, 1878."[106] Even nonpoisonous snakes could cause problems. Another field survey scientist reported secondhand an account from 1881 about a garter snake caught stealing several chub from a riverbank, which had apparently been caught for food. "Upon opening the snakes," he reported, "one was found to contain six fishes."[107] Snakes of all kinds were a common menace to field surveys.

Snakes were not the only creatures that competed with field survey parties for food or prey. Samuel Williston reported an incident involving skunks during attempts to collect tiger beetles (*Amblychila*) in Kansas. One student in their group, John M. Walker, "while patroling his accustomed beat one morning before sunrise, discovered the fresh fragments of several half-eaten *Amblychila* scattered along his route, as if some predatory animal had just preceded him and made his breakfast upon the rarities which otherwise would have made their way into the collecting bottle." Later that evening, the same student, "while collecting in the same locality, was violently attacked by a rabid skunk, twice in immediate succession." But the next day, when another student "was similarly attacked on a neighboring clay-bank," he "had the good fortune to kill his assailant," and a subsequent "examination of the contents of this animal's stomach revealed unmistakable remains of freshly-eaten *Amblychilae*."[108] As the largest beetles on the Great Plains—and entomological prizes of considerable value—tiger beetles were highly sought after by scientific collectors. In 1877, the rare beetle commanded $12 per specimen on the market, and collectors in Kansas hoped that the removal of American Indians from western Kansas and their increasing knowledge of the creature's habits would enable them to capture many more of the beetles. During the prior year's season, the observation of its nocturnal activity by H. A. Brous and Williston had enabled the two geologists to collect about a hundred of them.[109] Still, having to share them with the region's insect-eating mammals made such "rabid skunks" the Kansas party's competitive enemies.

While the environment of the Plains and Rockies could be a barrier to the success of survey work, it was also a tremendous regional resource that obviously constituted the basis for doing field research in the first place. The minerals of the mountains, the soils of the plains, the fossils of eroding buttes, and the distinctive plants and animals that inhabited the region's diverse environments—not to mention the mobile human beings and past artifacts deposited on the landscape—were the foundation upon which all knowledge-producing activities, including surveys, rested. Insofar as the survey was a geographical, territorial, and even political mode of producing knowledge, the inventory and use of those environmental resources was all the more pronounced. Often western environmental resources also influenced survey work indirectly, through how the region's natural endowments shaped the kinds of economic activities that took place there, such as mining and agriculture. (And those activities, in turn, transformed the natural environment itself that could be studied!) These activities provided constituencies for justifying survey patronage at both the state and federal levels, and given the importance of government support for survey work this was far from trivial. Geological surveys, for example, did not just produce expectations of mineral exploitation; such economic development also in turn fed back into support for the survey work itself, thus establishing another kind of environmental influence on survey science, through the human economy.

Moreover, the bodily experience of field work for survey scientists was often one of rejuvenation and revitalization, despite the physical hardships. Field researchers looked forward to the summer field season in the West in part for the chance to live and work outdoors and out of crowded, often polluted, cities, which would have been possible even if the field work had been taking place in the eastern United States or on the more humid parts of the West Coast. But the interior West was especially attractive for many due to its dry climate, which was believed at the time to confer distinct health advantages. To be sure, members of field survey parties also faced health threats during travel in the Plains and Rockies. On the Hayden Survey in 1870s, for instance, scientists suffered at various times from "mountain fever" and other unspecified illnesses, in one case requiring "several weeks" of recovery at Fort Bridger, and in other cases travel home—and, for field assistants, this could mean an end to survey employment.[110] Without precise information on the rates of serious illness for members of the general US population with the same demographic profile, it is difficult to determine whether survey field parties were subject to greater, lesser, or similar levels of total health hazard, though certainly the specific ailments differed in their relative importance. Overall, however, the serious health concerns that haunted the surveys in practice were amply counterbalanced by a discourse emphasizing the health benefits of travel in the West.

In the most extreme examples, survey scientists suffering from health problems might deliberately seek out the opportunity to go to the West for an extended period outside of the typical seasonal work patterns. For example, in May 1883, Henry Henshaw of the Bureau of American Ethnology (and earlier a field naturalist with the Wheeler Survey) "obtained leave of absence for several months on account of ill health, and spent the time till the middle of July near Colorado Springs." (Henshaw was not content to simply rest, however, and undertook bird-collecting excursions with his scientific companion, E. W. Nelson.) For the remainder of his convalescence, Henshaw moved to northern New Mexico from July to October, and then to Nevada and California.[111] Much like the versatile British-born naturalist T. D. A. Cockerell, discussed in chapter 1, Henshaw sought out the health-promoting experiences thought to be conferred by extended residence in the high and dry land of Colorado and New Mexico.

Even more than simply living indoors, late nineteenth-century regional thinkers about the environment, health, and the body stressed the benefits of camping outdoors, which of course fit rather well with the practices of survey science. "One of the best methods of gaining the altitudes, and of obtaining the highest possible benefit from air, sunshine, exercise, and elevation is by *camping out*," declared Colorado medical climatologist Charles Denison in 1880. He especially recommended "'*roughing it*' for a month or two" in the mountains and "*sleeping out on the ground* in the pure dry air, amid the balsamic exhalations." Moreover, he described the anticipated transformation in terms that might resonate with any field survey scientists who seasonally traveled in the West: "They left the plains thin, languid, and pallid; they return, bronzed and rugged, with elastic tread and full chests, gladly owning that to the experiences of camping out they owe a new lease of life." Finally, Denison suggested living like health-seeking travelers before the railroad era, who "*must* live in the dry, open air" and "sleep under the stars, and often *do their own cooking*."[112] While survey field scientists sometimes did cook for themselves, more often, as we shall soon see, they organized survey parties to include subordinate laborers to whom such tasks could be assigned.

Organizing Survey Work

Arnold Hague's most crucial task as head of the US Geological Survey's Yellowstone Division each year in setting up the summer field work in Yellowstone was assembling the team of scientific and supporting labor.[113] Some were regulars on the Geological Survey payroll, such as assistant geologists Joseph Iddings and Walter Weed. To help these paid career researchers, Hague also invited along a volunteer assistant geologist, typically a recent science graduate of one of the east coast universities. This person often changed from year to year, although Samuel Penfield, who would later become a widely respected laboratory mineralogist, served twice. In 1886, near the beginning of his successful career in the mineralogy department at Yale, Penfield served as Hague's volunteer assistant geologist for the summer field season. In his letter offering Penfield the position, Hague implied that Penfield had already served at least one previous season, and he asked him to make a ten-week commitment from the beginning of July to be working in Yellowstone.[114] The next year, however, Hague had to find a new volunteer. He invited the participation of A. C. Gill, a young instructor at Johns Hopkins University, although not until mid-June 1887, a few weeks before field work was scheduled to begin.[115] Later volunteer assistants included Paul Dashiell, a Johns Hopkins student, in 1888 and Louis Pirsson, a Yale student, in 1889.[116]

Other field scientists besides the main geologists and their assistants also accompanied Hague's season field survey parties. A topographer and volunteer assistant topographer were part of the group in most years. Since the thermal features within Yellowstone Park were of particular interest, a physicist was sometimes part of the roster too. During the first year of field work in 1883, when a geological survey physicist, William Hallock, joined their party, Hague's Yellowstone Division did hire an assistant in physics at $50 to $75 per month, not to travel with them but, rather, to take Hallock's place back in Washington, DC.[117] From time to time other career scientific researchers joined voluntarily or contracted to work for the Yellowstone Division temporarily. (Such arrangements continued into later years of the division's work, even after large coordinated field parties were no longer taking to the field, such as when University of California botanist W. A. Setchell received $100 for travel and field expenses in 1898 to prepare a report on the algae of Yellowstone Park.)[118] In addition to the research staff, Hague employed a clerk and secretary, a teamster, and several rodmen, packers, and cooks, with overall expenditures ranging as high as $18,000 one year but typically hovering around $15,000 during most of the years in the 1880s and 1890s when the division undertook extensive field work.[119]

While the paid research staff positions were ongoing throughout the year, volunteer researchers and most support staff were selected only for the summer months in the field. Securing these seasonal workers, especially the hired laborers who served as cooks, packers, and guides, was thus one of Hague's duties in the spring and early summer months. Sometimes the selections were easy, as when someone who had worked out well before wanted to return again. It was sometimes more difficult to find reliable support workers, however, who did not have the added incentive of credit for doing science or the benefit of gaining otherwise unavailable field experience under the tutelage of seasoned scientific travelers. As late as mid-June 1887, for example, Hague was still short one cook, though he already had another new one arranged for him, who was on his way from Denver to Bozeman, Montana, for the start of the survey field season. "I hope you will hear something favorable from Cassius Carter," he wrote to Iddings.[120] Once all the necessary positions were filled, the diverse members of the field party—senior scientists, assistants, student volunteers, cooks, packers, and all—would be ready to make the steady climb from the embarkation point along the railroad at Bozeman or Gardiner, Montana, into the fabulous geological and experiential wonderland of Yellowstone National Park.

This summary of the geological surveyors of Yellowstone highlights the importance of work organization: the deployment of teams of field workers subordinate to a central officer. Often the survey party members were formally trained in some way to be part of the work structure. Typically, they were also paid employees or supervised students rather than volunteers or independent contractors. This work structure based on career scientists or student assistants made possible the tighter integration and more systematic activities of the survey as a mode of scientific practice in the field. As in the case of Hague's Yellowstone Division, a crucial feature of survey work more generally was the participation of lower-status laborers in field survey parties, including guides, who in effect offered their local experiential knowledge for hire, and the manual laborers who accompanied many survey parties into the twentieth century, principally cooks and packers. In special situations, other local workers might also be employed, such as interpreters, who communicated with American Indians in surveys of human inhabitants. While subordinate scientific staff in the field might contest the denial of full credit for participating in knowledge making—and many local amateurs also expected some form of acknowledgment—there was hardly ever any questioning, at least within the writings of field scientists, of the assumption that the manual laborers who accompanied them deserved nothing more for their contributions than monetary pay for the work they did. Hierarchical control over how survey work was organized in the field was crucial to how surveys operated.

The conspicuous failure of surveys without trained subordinates suggests how important that hierarchical work structure was to their success as a more systematized mode of field practice. In Britain, the Ethnographic Survey of the United Kingdom, whose organizers tried to depend on local amateurs, found that the uniformity and reliability of facts collected was not sufficient for more rigorous synthesis.[121] Likewise, when American surveys were organized on a strictly voluntary basis—as essentially lay networks masquerading as surveys, in other words—their results were decidedly underwhelming. In 1881, the fledgling Kansas Academy of Science called for a "State Scientific Survey" to be funded by the state government, "possessing a wider scope than ordinary geological surveys," encompassing also such natural features as plants, trees, insects, and soils.[122] While scientific boosters would eventually succeed in generating support for a geological survey based at the University of Kansas in 1889, in the meantime, Francis Cragin at Washburn College in Topeka launched the "Washburn Biological Survey of Kansas" in

1884. In a prospectus, Cragin announced "an informal biological survey of Kansas" that would last five to ten years yet would aim to be "a systematic survey of the entire field of Kansas botany and zoology." He excluded geology, which he thought to be "of too great magnitude for private enterprise," so that "its execution must be left to the State." He invited cooperation by "all intelligent and public-spirited citizens of the State" and identified certain areas of the state as lacking coverage among initial volunteers, with many of them in the western part of the state, including Medicine Lodge, extreme southwestern Kansas, the western Arkansas Valley, the western Republican Valley, and Norton County.[123]

Although Cragin prepared circulars "giving directions as to the manner of collection, preservation, and transmitting of specimens best calculated to subserve the objects of the survey"[124]—and published eight progress reports through the late 1880s—the end results were little different from the collective activities of self-organized lay networks, with many areas lacking volunteers still treated minimally or not at all. The bulk of the published reports in the *Bulletin of the Washburn Laboratory of Natural History* devolved into a series of individual "contributions," "lists," and "notes," similar to the results of many other voluntary natural history endeavors.[125] Moreover, Cragin struggled mightily to keep the survey's publications going, not only subsidizing four of the last five issues personally but also spending more than $125 for "type and printing supplies adapted especially to Bulletin work."[126] Without a more robust funding source, Cragin found it difficult even to pay publication costs, let alone to pay staff members who could be guided and coordinated to cover the territory systematically and thereby make the biological survey a reliable engine of knowledge production in the field.

Most surveys that succeeded in more systematic mapping, collecting, and observing thus relied upon a work organization of paid field agents who traveled under specific instructions and guidance, most often with state or federal government funding, or occasionally funded by a prominent urban museum. To see how these diverse surveys produced knowledge out of the complex and variable environments of the West, consider the shifting configurations of people who did the knowledge making on the ground. Early expeditions were often large, elaborate affairs, with a range of scientific specialists and accompanied by vast US Army detachments for protection. One extreme example was the summer 1874 reconnaissance of the Black Hills of Dakota under the overall command of General George Armstrong Custer, with William Ludlow as captain of engineers for topographical surveying, N. H. Winchell as geologist, and George Bird Grinnell as paleontologist, along with "ten companies of the Seventh Cavalry, one each of the Twentieth and Seventeenth Infantry, a detachment of Indian scouts, together the necessary guides, interpreters, and teamsters"—a grand total of "about one thousand men"![127]

As grand expeditions gave way to more exacting surveys, however—and as the perceived threat of conflict with American Indians lessened—the size of the parties shrank dramatically. By the end of the century most parties consisted of fewer than five people. At the US Biological Survey, for example, parties of two or three were common. US Geological Survey field parties were about the same size or slightly larger, as were the parties of the Bureau of Soils, which were "usually" supposed to include only "two men": one of them "in charge of party and a field assistant."[128] As this dramatic decline in field party size indicates, leadership of the survey party in the field shifted from an early reliance on military officers to self-reliant scientific researchers in a more thickly settled landscape.

Field Party Leaders and Scientific Assistants

Yet strong authority was still essential for the operation of a survey party, even under civilian leadership. College expedition parties, for example, were usually directed by professors who held authority over their students. In 1886, however, Princeton's William Berryman Scott was assisted by layman Frank Speir, whose leadership was crucial to making the expedition successful. "My fifth Western trip almost came to grief over the difficulties of leadership," Scott later remembered, "for it was imperative that I should be home before the end of July and Speir could not leave his work in New York any earlier than that. We finally agreed that I should take charge of the party for the first four or five weeks and then Speir should come out and relieve me."[129] Authority was also important for smaller survey groups. US Biological Survey parties, for instance, sometimes consisted of a more experienced field survey man and an assistant who was clearly meant to defer to his partner's experience. Parties without a clear leader in the field revealed the consequences when those clear lines of authority were absent. When the Colorado Museum of Natural History sent out a two-man party consisting of a mammal collector and a bird collector in the mid-1910s, for example, conflict between the two became intense and nearly brought the whole operation to a halt.[130]

Relations among the leading scientists of the US Geological Survey's Yellowstone Division proved to be more durable. Even as the various seasonal team members came and went, especially the volunteer scientific assistants, the steady constants were always Iddings and Weed. By the summer of 1890, Hague had built enough confidence in their abilities to plan for outfitting each of them with a party that season in different parts of the Yellowstone area. While Hague was paid a $4,000 annual salary in 1890, Iddings received $2,000 and Weed $1,800, thereby helping all three pursue respectable scientific careers.[131] They were also given some latitude to continue ongoing projects that interested them and were connected with the division's work, such as Weed's annual observations on changes in the hot springs areas of the park.[132] Iddings and Weed were also given a prominent level of coauthorship on the Yellowstone Division's main report (see figure 3.4). In the case of Iddings, however, the advantages of the geological survey position ultimately were not enough to keep him from accepting a position as an assistant professor at the University of Chicago in 1893.[133] By that time Iddings had nearly twelve years of experience in geological research with the US Geological Survey.[134]

Assistants to field parties might come from the region itself or they might be recruited from far away, among promising young men connected to cosmopolitan correspondence networks. Either way, the terms of employment for young assistants on field surveys during the early railroad era, from the late 1860s to the 1890s, tended to fall somewhere between pure volunteer and salaried employee. Typically, even where funds were not available for salary, assistants' field expenses were covered, and occasionally there were other benefits. For example, the naturalist Robert Ridgway, who was still a teenager when chosen to accompany Clarence King's western survey along the 40th Parallel in 1867, received $50 for his travel expenses to meet up with King in New York City before heading out, as well as money to live on while there. (Upon arriving, he found King to be "a very kind, and agreeable young man" whom he could "not help liking.")[135] But even once Ridgway succeeded in working his way onto the King Survey payroll, more improvisation was required two years later, when funding constraints "forced [King] to drop the Zoölogy and Botany from the Exploration." Unable to provide a salary, King offered Ridgway travel expenses to and from Salt Lake City, along with "subsistence in the field" and return transport for Ridgway's collections.[136]

Survey organizers also kept their eyes open for promising young locals, often natural history collectors, who might make good prospects. Vernon Bailey was himself a Minnesota farm boy whose correspondence

UNITED STATES GEOLOGICAL SURVEY

CHARLES D. WALCOTT, DIRECTOR

GEOLOGY

OF THE

YELLOWSTONE NATIONAL PARK

PART II

DESCRIPTIVE GEOLOGY, PETROGRAPHY, AND PALEONTOLOGY

BY

ARNOLD HAGUE, J. P. IDDINGS, W. H. WEED

AND

C. D. WALCOTT, G. H. GIRTY, T. W. STANTON, AND F. H. KNOWLTON

WASHINGTON

GOVERNMENT PRINTING OFFICE

1899

FIGURE 3.4. Title page of Arnold Hague et al., *Geology of the Yellowstone National Park,* US Geological Survey Monograph no. 32 (1899), Part II.

with Merriam in Washington, DC, eventually led to his appointment with the federal agency.[137] There were other examples from within the far west, although the region was far from being the amateur natural history powerhouse that the Midwest was. Such local collectors at first might be offered fairly disadvantageous terms in order to gauge their aptitudes and interest in survey work, as well as to cope with limited government funds. For instance, in the fall of 1888, Bailey recommended to Merriam a young man who lived in Ogden, Utah, whom he had met during his travels, and Merriam in turn encouraged Bailey to try him out as his assistant during his winter swing southward. While Merriam had no official funds to pay him, he did offer "to pay him $5 a month out my own pocket" in order to "defray his actual expenses," judging that "the experience would be worth to him vastly more than his time, for if he should develop into the right kind of man it is probable that we could employ him in the future." The experience would also allow Bailey to judge the young man's capabilities in the field. "What we want," Merriam quipped, "is a young man who has the right sort of stuff in him, who will profit by your experience and example, and able to make trips independently after the present season's field work" with Bailey.[138] The unnamed young man did not end up joining Bailey's survey party, but the offer suggests the terms on which many other prospects got their first taste of official government field survey work.

The advantages of serving as a volunteer survey field assistant were significant. Travel as a field surveyor, in direct contact with experienced survey scientists, provided a form of learning that could not be duplicated in a classroom or by reading books and articles. The ability to integrate the cosmopolitan knowledge of science with the knowledge of everyday experience of the western environment held the potential to bridge the epistemic rift between the two forms of knowledge but could only be learned through practice in the field. Consider once again the participants in the Yellowstone Division's summer field trips. What these volunteers received, besides food and lodging in camp, was access to an extremely rich field site for knowledge production, under Hague's guidance. In many cases this experience with the Yellowstone Division proved to be a valuable stepping-stone in building a career as a scientist, just as it was for the long-term members like Iddings and Weed. And, of course, seeing Yellowstone was itself an immensely rewarding experience, much sought after by those who could afford to see the natural wonders so enthusiastically reported on the East Coast. To see the park as a field researcher was the epitome of the middle-class vacation ideal:

making observations of Yellowstone's highly regarded scenery as a form of work.

In many cases, as we have seen, surveys relied on student labor for scientific assistance. By relying on students, survey leaders could tap into their scientific education—and, indeed, contribute to it—while at the same time avoiding the expense of paying full-time trained scientists. Having sufficient student labor for systematic survey operations, however, did require access to a regular pool of student workers with the necessary background. Such a labor pool was not easily available in most areas (if at all) until the early decades of the twentieth century, with the establishment and success of institutions of higher education inside the region. At the Colorado Geological Survey, for instance, that student labor source came from the Colorado School of Mines in Golden, just west of Denver. Beginning in 1910, state geologist R. D. George inaugurated a series of local studies of mining districts, drawing on teams of field researchers consisting of faculty and students from the geology department at the Golden campus. Advanced students earned thesis credit for their work in these district-level geological studies in various parts of Colorado.[139] This strategy had been pioneered two decades earlier, when the University of Kansas was tasked with organizing the state's geological survey in 1889, with "active operations" finally initiated under severe funding constraints in 1893, though at first just in the more accessible eastern half of the state, by sending advanced students into designated field territories for academic credit.[140] A similar work organization—deploying "students of the State University, young men well trained for such work"—was replicated a few years later when the university's geology professor, Erasmus Haworth, was charged with surveying the underground water resources of western Kansas.[141] As in laboratory work, students who acted as scientific assistants in field surveys formed the backbone of an affordable yet educated labor force.

Larger field survey parties often divided into smaller groups, for time periods ranging from one day to an entire field season. As already noted, Hague's parties in the US Geological Survey's Yellowstone Division were sometimes divided into long-term field parties under survey scientists such as Weed or Iddings, which stayed together for weeks or months at a time. Earlier surveys such as the Hayden Survey had also divided into subgroups when traveling through the Yellowstone region in the early 1870s, but not always everywhere else. William Henry Jackson, who served as photographer with the Hayden Survey, recalled these "independent excursions" with special fondness, especially on his initial

1870 foray. It was, he thought, "the spirit beneath them, that made this expedition and all the ensuing ones under Dr. Hayden so engrossing and satisfying." He described the daily practice of having "an informal conference around the campfire" before going off to do their own work "individually, or in groups of two or three." Such a "little division" for the day, he remembered "might be assigned to calculate the flow of a stream," or perhaps to "investigate the geology of the region or hunt fossils." Jackson's own group, which also included two other men, would be engaged in recording "impressions of a striking landmark, like Independence Rock or Red Buttes."[142]

When parties were unusually large or subdivided, and especially in the earlier decades of the railroad era, field survey leaders sometimes also brought along administrative personnel to help organize the expeditions. They went under various titles, such as quartermaster (by analogy with the military) or disbursing clerk and secretary (by direct extension from government office work). These officers helped to organize the survey parties by meeting those who arrived by railroad, gathering them at the "rendezvous" field camp nearby where field travel off the railroad would begin, assembling and distributing supplies and equipment and keeping track of the financial accounts and expenses in the field.[143] Alternatively, sometimes another scientist serving under the survey's field director would be charged with these organizational duties, as when Hague sent his scientific lieutenant and fellow geologist Iddings "to make the necessary preparations and outfit the party for the summer of 1888" before Hague's own arrival a few weeks later.[144] For smaller field survey parties, the organizational and administrative duties simply fell to the scientist in charge, in coordination with office staff back at the survey's administrative headquarters.

Not only did field survey parties generally become smaller over time but, as the region became more densely populated, an infrastructure for providing lodging and meals in towns partially displaced the traditional system of camping out.[145] This was particularly true for surveys in areas of agricultural settlement, such as the soil surveys undertaken cooperatively between the individual states and the US Department of Agriculture in the early twentieth century. In fact, many of the soil surveyors' wives began accompanying them on their field travels, and "often gave long hours of painstaking work in helping prepare maps and manuscript," according to the autobiographical account of soil surveyor Macy Lapham. The contributions of "these unnamed and unpaid collaborators" are difficult to discern in the historical record, and, according to Lapham, may "never be known."[146] Such unacknowledged assistance undoubtedly did help to lessen the administrative burden for field surveyors in smaller parties without paid quartermasters, clerks, or secretaries. In some cases, as Lapham suggested, they may also have been integral to the production of the scientific results.

Arnold Hague of the Yellowstone Division planned to take advantage of the expanding hotel infrastructure in the park for "the greater part of the time" during his later visit in 1902, making only "short excursions" with "saddle horses" and possibly using "tents, provisions, and blankets" that could be easily procured locally.[147] While much of the reconnaissance survey work over large areas of the Great Plains still relied on camping out, in most areas—and when reconnaissance surveys were near towns, or even homes of farmers and ranchers—rooming indoors became increasingly common. Even in the early years of surveys, of course, field scientists had taken advantage of opportunities to room and board in nearby towns or with farming and ranching families—although as geologist G. K. Gilbert's experiences in 1870s survey work in southern Utah reminds us, the facilities available might not be that much different in material reality from sleeping outdoors.[148] But with the expansion of farming settlement and town building by the early twentieth century, the overnight experience of the field survey party presented an increasingly mixed combination of camping and indoor living. Even when they were still staying in tents out in the country, survey parties in more settled areas were closer than ever to hotels, restaurants, stores, and farms from which items could be purchased or borrowed. Over time, then, as settlement became denser in the Great Plains and Rocky Mountains by the early twentieth century, survey scientists took advantage of its expanding infrastructure.

Technicians of the Field

Scientific members of the survey party, and their associated administrative personnel, were not the only participants in the process of knowledge making in the field, however. Parties typically also included lower-status laborers, especially cooks and packers, who vitally contributed to the practical work of survey science. Like the work of technicians in the laboratory, this essential labor has largely disappeared from the historical record.[149] But without those who prepared their meals, transported their material and equipment, and helped around the camp, and sometimes assisted in making scientific observations or in preparing specimens, the field work could never have been completed.[150] While the material contributions of cooks, packers, guides, and other support personnel were universally appreciated—if rarely acknowledged by name in print—their epistemic contributions were variable and often harder to credit. Moreover, survey practice involved a distinctive combination of fluidity and status consciousness, and of mobility for some coexisting with immutable social distinction for others.

Scientific survey members recognized the special skills required of the laborers who accompanied them. Packers, as their name suggests, worked closely with the pack animals (see figure 3.5). Ethnologist and ornithologist Henry Henshaw, who served as a naturalist with the Wheeler Survey in the 1870s, recalled that the packers had "to assemble and tie into bundles the camp impedimenta," which were loaded onto the animals. "As many of the packages weighed upwards of a hundred pounds apiece," he noted, "the job of lifting them into place and securing them required no mean degree of strength, skill, and patience."[151] Franklin Rhoda, who served as a topographer with the Hayden Survey in 1874, sketched out the typical start to a work day, with two packers: "One immediately starts out to hunt up the mules, while the other puts the packs and aparejos in order."[152] The geographer Alvan Southworth, also with the Hayden Survey in 1874, at first elevated the status of the "head-packer," stating that he took on "the gravity of the executive officer of a ship bringing the vessel to anchor," but then resolutely limited the packer's concerns to mundane material issues and manual labor. He was like "a seaman," who should "ignore the men, and consider the cargo and transport," as well as finding a good spot to camp. "His first anxiety is water," Southworth continued, and "must have timber for the cook, grasses for the animals, and a mountain-locked park," to keep the animals from wandering off.[153] Like Rhoda, then, Southworth conveyed a strong sense of manual laborers existing in their own realm of material life, separate from the priorities of the scientific members of the field party.

Besides the packers, though only after considering the merits of the camp mule, Henshaw also provided a vivid description of the cook. "Whatsoever he chanced to be as man," he noted, "upon his ability as a cook depended to no small degree the health and comfort of the party." By emphasizing the role of the cook's work in maintaining bodily health during field work, Henshaw's remark reminds us that the survey was especially noteworthy among modes of scientific practice for the dependence of the usually all-male field party members, when away from railway towns, on one of their own sex for this characteristically domestic

FIGURE 3.5. Subordinate field labor: packers with the George M. Wheeler's survey in the field with pack animals. Timothy O'Sullivan, "Mountain transportation. Pack mule, pack and packers." Stereo 1s00174, Library of Congress Prints and Photographs Division, Washington, DC. Part of series: US War Dept., Corps of Engineers; Geographical Explorations and Surveys West of 100th Meridian, Expedition of 1871; Lieut. Geo. M. Wheeler, commanding. http://hdl.loc.gov/loc.pnp/stereo.1s00174

task, much like the armies and sea voyages in the expeditionary tradition from which survey evolved. If the health of field scientists was intertwined with the environments through which traveling parties were passing, then the food—obtained locally or brought in from afar—was a crucial nexus for that relationship. And, of course, the camp cook was the key figure. For his part, Henshaw recognized that cooking was in part a "skill" but insisted that "patience and zeal" were foremost. "To cook a savory meal on a modern gas range, with all the conveniences at hand," he observed, "is not so difficult; but to achieve this feat in the open air, on a windy day, often under a rainy sky, with bits of brushwood for fuel, and a horde of hungry men demanding something to eat, is given only to the elect and to few of them." Tellingly, not all camp cooks were universally approved, but many were: Henshaw held "in grateful memory a number of such cooks whose triumphs over the trials and tribulations that beset their calling entitle them to golden crowns."[154] Rhoda also mentioned fuelwood and water gathering as key tasks for the cook.[155]

Social distinctions were nearly always clear on field surveys, though fluidity in practice could occasionally obscure them.[156] Henshaw commented that "polished manners and scholarly attainments were foreign to the packer's calling, and usually they lacked the gift of eloquent speech" (though when "stubborn mules needed rebuke, their outbursts of profane imaginings amounted to real eloquence").[157] Rhoda referred to the cook's "barbarous mountain dialect" in his report.[158] The forester William Pinkney Lawson—reinforcing the social distinction even as he purported to subvert it—considered "the cook . . . as the most indispensable member of the party," who "holds a position a little superior to that of the chief. He is accorded marked consideration, treated with a special and particular brand of courtesy, for upon his humour depends always the happiness, often the health, of the whole party."[159] These descriptions of lower-status survey workers clearly combine an overall tone of deep appreciation with unmistakable reminders of social distinction. On the other hand, the actual work on the ground could blur lines of distinction, especially in smaller parties. The archaeologist Neil Judd recalled being impressed as a student member of a field party led by Byron Cummings that Cummings "attended to the camp drudgery" and "did most of the fuel-gathering, water-toting, and cooking."[160]

Occasionally, however, a field scientist's description of cooks or packers would deviate from this standard script—though even then, only up to a point. Later in his account, for example, Rhoda would briefly identify one of the packers by name. In a description of the party's ascent of Mount Sneffels, he mentioned that they "were joined by Ford, one of the packers, who had concluded that he wanted to climb a peak, too, and had chosen this one, the hardest climb of the season."[161] In pointing out a deviation from typical practice, Rhoda revealed both the existence and partial permeability of a customary boundary line in the division of labor on surveys. The next year in his report, Rhoda mentioned another packer by name, Harry Yount, whom he credited for catching numerous trout and for shooting "a splendid buck deer, with enormous antlers," activities that contributed to the field party's subsistence yet were more clearly distinct from the group's scientific purposes.[162] Yount, whose name appears elsewhere in geological survey reports (thus suggesting an unusual level of recognition from his scientific bosses) had no doubt applied his abundant experiential knowledge in fishing and hunting—and on a biological or natural history survey party this might have blended more fluidly into the making of scientific knowledge. But in this case the narrator was a topographer, so it is hard to ascertain what he contributed to the content of the field party's survey science.

The contributions of guides could be even more consequential for the knowledge that surveys produced, though also reflecting how experiential differed from the cosmopolitan knowledge sought by survey scientists. When Richard Leigh, known as "Beaver Dick," helped guide the Hayden Survey to Yellowstone in 1872, his contributions appeared in the report of "Snake River division" geologist Frank H. Bradley.[163] Somewhat unusually, Bradley credited Leigh by name for his intellectual contributions to the survey field work, noting that Leigh had "long lived in this region," thereby recognizing his deep experiential knowledge of place. Leigh's observations of a stream "for many years" as well as his topographic knowledge about the extent of high cliffs, which impeded travel, were especially valuable. At the same time, Bradley declared invalid Leigh's indentification of "the next tributary, which Beaver Dick called the middle fork of the Snake," insisting that "neither by position nor by size is the stream entitled to be considered a fork of the main Snake" and renaming it "Falls River" instead.[164] This anecdote neatly illustrates the distinction between the local knowledge of experience, which corresponds well with the practical use of a place, and the cosmopolitan knowledge of science, which might require overwriting that local knowledge in order to situate it in a larger context.

While serving as guide to the Hayden Survey that year, Leigh also became well acquainted with natural history collector C. Hart Merriam, who would subsequently organize the US Biological Survey. Leigh's later correspondence with Merriam's family is preserved in the Merriam Papers, which offer us some glimpses of the larger context of Leigh's life as a guide and homesteader along the Teton River. By his own account, Leigh "moved up to the elbo of the teton river on the 25th of aprl 1876." At that location, he "bult a log cabin and fenced me a farme and rased some little fegatabls" as well as "a horse corall and a hay corall and put up 6 tons of hay in it." He served as packer for "tom Leverings and his pardner and thare camp and skins out of the mountins," presumably a fur-trapping rather than scientific party, for which he "cept a dyra [diary]."[165] But he continued to stay in touch with Merriam, sending him specimens from the field in later years. Their continuing relationship was appreciated by Merriam, whose sister Florence wrote back to Leigh in gratitude, "We all feel as though my brother Hart owes you something more than he can repay for your kindness to him, not only while he was with you, but since he has been here, in that you have sent him so many valuable skulls, bones, skins, etc. which he, by the way, prizes very highly."[166]

Into the early twentieth century, survey parties continued to employ cooks and packers. Despite the proliferation of field survey parties sponsored by such institutions as the US Geological Survey and the US Department of Agriculture, it is rare that we have access to documentary sources written by such participants in survey field work. One exception is Alcott Farrar Elwell, who was camp cook with a US Geological Survey party in Wyoming during the 1908 field season.[167] Born in 1886, Elwell decided while in his youth to devote his career "to the development of boyhood and young manhood." He joined the field party during his early adult years as a way to earn money to support his education at Harvard.[168] His widow later described Elwell's "great appreciation and enthusiasm" for the summer season in Wyoming. While the work had been "rough, tough, and challenging," Elwell had "liked it" since the Harvard student "loved the outdoors wherever he was": "He had never cooked! But like everything else, he was not afraid to try," and his widow recalled him claiming success. "He also took it on himself," she reported, "to ride to the nearest river sometimes many miles away with the men's clothing, wash it, hang it on the bushes to dry, and ride back to cook the dinner."[169]

The case of Elwell should remind us of a few important points about nonscientific labor in survey field parties. First of all, while nearly all survey team participants were men, some of them engaged in work conventionally associated with women, including not just the cooking identified in their job titles but also sometimes doing laundry (and even cutting hair on at least one occasion). Second, the subordinate labor status associated with cooks and packers did not necessarily imply a working-class background. While the cases of Beaver Dick and John Wilson, among many others, represent the majority of survey laborers who did not come from well-educated or urban middle-class backgrounds, some certainly did. As a Harvard student, Elwell participated in the federal government survey party of 1908 in much the same way as a student in the college field party tradition. Third, while scientific and nonscientific roles were clearly distinguished, the boundary lines of actual work done in the field were sometimes blurred. Thus, while Elwell's role was not to "contribute in a scientific way," he did develop some skills such as drawing and assisted with some of the mapping work.[170] Other cooks and packers likely helped with collecting or preserving specimens.

Another exceptional case was Charles A. Scott, who served as cook and teamster for a US Bureau of Forestry reconnaissance survey party in western Nebraska in 1901. Scott would later head up nursery development and planting at the innovative forest station near Halsey in the Nebraska Sandhills—thus illustrating the occasional mobility of lower-status survey labor into long-term careers—but his 1901 experiences provide another rare glimpse of survey field work from the perspective of subordinate labor.[171] Scott's observations must be interpreted carefully. Not only was he a recent college graduate at the time, but his narrative is a retrospective account based on his recollections and review of written documents many years later and thus may not fully reflect his experience of subordination at the time. Scott made a point of emphasizing the blurring of the status boundary line, revealing how his own anxiety about it was relieved—and perhaps also, indirectly, the expected way it would have operated on more typical survey teams. "Not one of the men," he reported, "ever kidded me or cast a reflection regarding my position (cook and teamster). I believe every man in the party regarded me as their equal in most lines, and readily granted that I was their superior in botanical lines."[172]

Yet even in Scott's account, subtle reminders of the status hierarchy of the field party can be detected. "When we had company, the cook ate his snack after the others had eaten," he noted with a suggestive third-person emphasis on his subordinate role.[173] Even after the season's field work was over, his status as cook might continue to be a source of commentary and amusement. Later that winter, for example, while back in Washington, he recalled, “When I would meet one of the men of our party, his greeting would be, ‘Well, Scott, let me know when you are going to serve roast beef, baked potatoes and coffee and I’ll come over.’”[174] Overcoming his lower status as cook and teamster is in many ways the subtext of Scott’s account. To show that his fellow team members respected his botanical skills, Scott first noted that the other men used his copy of Asa Gray’s *Field Botany* to identify “unusual species of trees and shrubs.” His status as the possessor of a textual authority was reinforced by two revealing anecdotes that show how Scott perceived himself as gaining the intellectual respect of other members of the party, despite his status as manual laborer. In the first, he correctly identified a wild rose stem whose thorns had been removed by the questioner as a trick. In the second, as he recalled it, the same jokester had “brought in an unusual leaf,” stumping all the others as he passed it around. They referred the leaf to Scott, who “was cooking supper at the time” and “listening to the conversation of the group,” who confirmed his authority as a botanist by identifying it.[175]

In the American West, despite occasional counterexamples like Elwell and Scott, cooks and packers were often selected from the regional resident population, both for their experience with local environments and the field survey organizers’ desire to avoid paying expensive transportation costs for people who were essentially low-skilled wage laborers. Not all local hired labor was always easy to procure, of course. Ethnographic surveys by the American Museum’s Robert H. Lowie in the early twentieth century, for example, were stymied in their work among the various Shoshonean tribes of the interior West such as the Piautes and Utes, who were “extremely reticent about divulging ethnographical information,” by the lack of suitable workers to meet their specialized needs. “Interpreters are almost everywhere of an inferior character,” he complained.[176] Several decades earlier, a US Army reconnaissance field party, with several scientists and topographers included, encountered difficulties with an unnamed guide, the “one Indian in the party who knows the country” who was becoming “proud of the power he has over us and wants to exercise it a little.” He repeatedly stalled and delayed before finally agreeing to depart, upon which “he surlily saddled up his pony and struck off on the trail, assuming from thence forward a winning smile and air of boyish oblivion.”[177]

Not only guides and interpreters but also the more prevalent cooks and packers may have varied widely in their skills, knowledge, and performance, of course, but survey scientists found them indispensable.

With their deep and intimate experience of the land in their local and regional places, they offered more than interchangeable manual labor. People familiar with local conditions were sometimes expected, with good reason, to possess experiential knowledge of the areas in which they lived and worked, which might contribute to the objectives of the survey. Indeed, it was central to the alchemy of survey science.

Local Assistance to Surveys

The experiential knowledge possessed by hired laborers who assisted field survey parties was not just appropriated from guides or other workers employed by field parties for the entire season. Other local people were employed for short periods, or simply offered to volunteer advice and guidance to survey parties—advice that was grounded in their place-based experiences. The assistance offered by local people to survey work thus extended beyond paid employment for the field season. In addition to the ongoing members of the party itself, survey science also relied on local people for support and hospitality during their travels and in the particular local places where they stopped to do their survey work. Generous assistance was the norm in the sparsely inhabited interior West. An ethic of helping out travelers, including survey scientists, prevailed throughout the region. It was so normal, in fact, that specific acts of kindness were rarely reported in letters or memoirs. However, rare violations of the frontier ethic of cooperation reveal the customary expectations. One example of denied help occurred in the South Park of central Colorado during Princeton University's 1877 travels through the area, when the field party members were "refused water for their horses." Such behavior outraged the Princeton party, whose chronicler noted that "refusing water to travellers used to be considered in the West as something like a crime." The scarcity of such a precious resource in a dry region made such a refusal perhaps somewhat understandable, given the large volume of traffic through the area, which was near a significant mountain pass. Yet it still rankled with scientific travelers accustomed to the hospitality and assistance that locals customarily provided.[178]

William Berryman Scott of Princeton was particularly diligent in recording instances of bad hospitality or refused assistance. For example, he remembered an incident when a local "man driving a light wagon" did not help near Fort Robinson in Nebraska during 1882 expedition. "In all my wanderings through the Far West," Scott declared, "that was the only time I met a man who did not gladly volunteer help to any one who needed it, not to mention the refusal to help, when asked for it." He pointed out that "in so thin settled a country, mutual help was often a matter of life and death and no man could foresee when he might stand in desperate need of it himself." According to Scott, this was not just his own opinion of the situation. "When I told of my adventure to a group of ranchmen in the trader's store at Robinson," he added, "they were extremely indignant."[179] Indeed, ranchers as a group were usually quite helpful to traveling survey parties, though in some cases they did try to charge for feeding animals and people.[180]

On the other hand, initial local hostility could turn to hospitality, as William Rush Taggart of the Hayden Survey discovered during summer field survey work in Colorado in 1873. As he wrote in a letter home to his sister, a "funny scene took place" one Sunday. After fording the South Platte River, they found "a good camping place in an enclosed ground beyond a field," but then the trouble began: "Pretty soon a little old man came tearing down so angry he could scarcely speak and wanted to know why we were camping on his ground and said he would not stand it, he would not stand it, he would shoot every mule we had, etc., evidently wanting to get up a squabble and to tell us that he owned the land." Yet instead of the situation devolving into conflict, the Hayden Survey party "all fell sympathizing with him and endeavoring to fix things to suit him," which "so beat him" that he became "the most generous man" they had encountered. They purchased "corn and butter" from him, and they received "all the milk we wanted" as a gift, as well as advice about "a good route along the mountains" to top it all off.[181] What had started out as a confrontation over the right of government field agents to camp on private land became instead a textbook case of local assistance, including not just selling and donating food supplies but also providing geographical advice. While such experiences did not happen all the time, smaller acts of hospitality and kindness from local people who knew the land more intimately are ubiquitous in the documentary record of survey work in the West.

In addition to private individuals, survey scientists also received hospitality and assistance from US Army outposts. As with other locals, the officers leading army posts were usually generous, with rare exceptions eliciting the most comment. Scott recounted only one "bad experience" at a frontier fort, which was during the 1885 field season. "That stay at Fort Bridger was unique in my experience," he recalled, noting that on "all other occasions, before and after that date, when we visited a military post, we were received with the utmost hospitality and kindness."

The problem at Fort Bridger had not been any failure of people there to provide practical help to the survey party, but rather that Scott and his companions "were ignored" socially. "Officially, everything has been done for us that I could ask," he wrote, "but no one has called on us, or asked us to their houses, or paid the slightest attention to us in any way."[182]

Generalized environmental resource provisioning and social hospitality were two aspects of the assistance provided to survey parties by local people, both of which were comparable to what was offered to travelers with aims other than scientific field work. But another facet was the direct contribution of local knowledge about the environment, thereby placing the efforts of locals closer to the heart of the survey science's distinctiveness from other kinds of travel. Especially when coming from the Midwest or East, survey researchers were on unfamiliar ground and could be quite disoriented in a novel environment. Consider the case of the US Biological Survey's Vernon Bailey. He would later become as familiar with the diverse environments of the American West as anyone during his multiple decades as the chief field naturalist, yet when he was first starting out, he felt disoriented. Upon arrival in the Sandhills region of western Nebraska, "about 40 miles south of Valentine" he found himself "in a little farm house, partly of boards & partly of sods." But to get there, he "had no idea where to go but by asking questions," riding along with "a mail carrier [who] was just starting out for his place," who was "a first rate fellow, well educated & interested in my work."[183] In this case, Bailey found a knowledgeable and helpful local guide through serendipity.

In other cases, however, people on the ground were already known in advance to the survey staff. For instance, Merriam wrote to Bailey later that same summer urging him to "hunt up Mr. Frank Bond, of the Surveyor General's office" after he arrived in Cheyenne, Wyoming. "Mr. Bond has been at Cheyenne for some time," Merriam noted, "and will be of the greatest assistance to you in various ways."[184] Likewise, Merriam also pointed Bailey to Mr. John Hague in Nephi, Utah, who came recommended by a US Geological Survey researcher (presumably on the basis of prior field experience), along with a whole series of local names in southern Utah towns whose sources were not revealed.[185] Another strategy was for national surveys to hire local naturalists who knew the area to accompany their surveys as scientific assistants, as when the US Geological Survey hired Lawrence Bruner, who would later become an entomologist at the University of Nebraska, to work "in company with one of them," in 1881, since the young Nebraskan was "better acquainted with certain of the formations" than the visiting geologists were.[186]

State surveys and expeditions, even without the far-ranging contacts shared among well-staffed federal agencies, still managed to find their own local collaborators. During their brief collecting trip to the small village of WaKeeney in western Kansas in 1878, two naturalists from the state's university relied on knowledgeable locals throughout their visit. First, they were directed nine miles out of town "to look after some bones reported to be of large size," which turned out to be from a bison. At the same location, where someone (presumably a farmer) named J. M. Davis lived, the two "found numerous fragments of stone implements lying about the premises." The local man agreed to give the university men the implements—including "stone mallets (many of them of large size), pestles, lap-stones, grinding-stones, and smoothing-stones"—as well as to save any found in the future for them. However, all was not good news, for according to Davis's neighbors, "the best and choicest hammers and other implements" had already been sent to "relic-loving friends" in the East.[187]

Foreign travelers could not always rely on the dense social networks of local contacts that federal and state survey scientists often used but they nevertheless improvised, often with the help of American field scientists whom they did know. The British naturalist Alfred Russel Wallace, while on a brief expedition with Denver botanist Alice Eastwood in the Rocky Mountains, received advice from local miners about where to look for plants. (Eastwood was in her own way a local assistant offering her general knowledge of the area to Wallace, although he did not acknowledge her by name in some accounts of the field work.) As they hiked up into the Rockies, the two collectors passed by many miners' houses, including some in the Grizzly Gulch valley area on the north side of Gray's Peak. Wallace later recounted how some miners had offered hospitality as well as local plant knowledge. When they had arrived "just below timberline," they reached "a miner's house," where they were invited to join "two miners who had come home to dinner." They "offered to show us a fine place for flowers," he noted, taking the traveling naturalists "through the wood for half a mile" to "a rocky and grassy slope with great snow-patches in the shady hollows," where "the ground which the snow had left was literally starred with flowers." On the same expedition, Wallace and Eastwood also ascended Gray's Peak after staying overnight at a miner's cabin located at the alpine elevation of 12,500 feet.[188] While local assistance sometimes failed to merit any mention by name, it was still a crucial part of the authenticity of scientific reporting. The help of locals did not always go completely unrecognized in the publications of field researchers, but the full extent of such connections is only fully clear in the correspondence and journals of survey scientists.

Geographical Relations

All survey field parties, regardless of their affiliation, faced similar environmental constraints and adopted similar forms of work organization. However, these experiences and activities were often further shaped by the structured spatial interactions between the members of survey parties who traveled in the field and their supervisors in the cities and towns where their sponsoring institutions were located, whether in distant metropoles such as Washington, DC, and New York City, or in closer regional centers, such as Lincoln, Nebraska, and Denver, Colorado. The most common work rhythm was to move from making observations and collections in the field back to a place with better libraries and other facilities for turning that field work into publishable science.

In a few rare cases, survey institutions might establish branch offices closer to the field territory being surveyed, as when the US Geological Survey formed a Rocky Mountain Division shortly after its founding in 1879, based in Denver, Colorado, and directed by S. F. Emmons. Writing to Spencer Baird at the Smithsonian in 1882, Emmons described the division's "headquarters" in Denver as possessing "the nucleus of a considerable working collection both of rocks and minerals, and the commencement of a scientific library." By converting field work into printed reports in Denver, Emmons noted, his "assistants do not go east," a practice which vitally depended on maintaining "access to all the current periodical scientific literature of the day," which included receiving "most of the foreign geological and chemical journals."[189] By indicating the working requirements for geological collections and an extensive library of scientific publications that were required for field scientists there to write up their research effectively, Emmons succinctly conveyed the necessary features of survey branch headquarters for situating field work in the larger context of cosmopolitan science. However, given how expensive and cumbersome it was to set up such branch offices in the field, especially when survey scientists were already operating over vast geographical spaces far beyond any single location that might be chosen, the Rocky Mountain Division's headquarters in Denver was an exception to the general pattern.

The geography of survey work in the West was further complicated by the overlapping jurisdictions of national and regional institutions. This produced a recurring tension inherent in survey work between field researchers (usually sponsored by state governments) who were identified with the region itself and their counterparts (often sponsored by the federal government) who were identified with the metropoles of the East Coast, especially those in Washington, DC. In 1888, for example, the newly established journal *American Geologist* gave voice to those who, in the wake of the national survey's geographical growth, were "fearing that the concentration of all authority and control at the national capital, and by the extensive accumulation at one centre of all the material illustrating the state and local geology of the country, the local interest and effort may die out, and that ultimately the weight of public sentiment favorable to geologic investigation may suffer diminution."[190] While these tensions were usually worked out so that relations could be cooperative, conflicts did frequently emerge, and they were a constitutive element of the structural organization of survey field work.[191]

As survey work in the region proliferated by the late nineteenth century, the locus of directorship often shifted from the field itself to a metropolitan center, either in the East or in a provincial capital within the region. In other words, directors of surveys ceased to accompany their field parties and had to guide their operations through advance planning and frequent communication along the way. Some authority was delegated to the field party's leader, although distant directors retained some control. The advantages of formal authority in the metropole were often counterposed against the practical reality of decision making on the ground in response to local environmental conditions and the contingencies of actual experiences in the field. From the perspective of the survey director situated in Washington, DC—or even in regional centers like Denver, Colorado—the most crucial thing was to keep in regular contact through frequent correspondence. Most commonly, communication was by letters mailed through the postal service (thus also providing a rich archival record to later historians) with occasional urgent messages sent over the telegraph lines. The expansion of a railroad-dependent postal and telegraph infrastructure was therefore essential to the operation of field survey work, as much as it was for the lay networks.

Metropolitan directors often reiterated the necessity of regular communication. As Merriam wrote Bailey near the beginning of his field travels in 1888: "Please write often, as I want to be in frequent communication with you from this time on. Always notify me in advance of change in your address."[192] Compared to other regions of the United States, travel in the interior West presented greater challenges for maintaining the constant correspondence that survey directors coveted. Some of the places with the most attractive environmental resources for scientific study were also the most remote from the communication lines of American settler society. For example, in the thinly inhabited and rugged canyonlands of southern Utah, Merriam feared losing contact with Bailey. While the region's animals were "practically unknown" and thus extremely valuable to science, it would be difficult to stay in touch. "Hereafter it will be necessary to notify us of your post office address a number of weeks in advance," Merriam warned, "as the mails are slow in penetrating the region you are about to enter and post offices are few and far between."[193] J. D. Figgins of the Colorado Museum in Denver experienced similar anxiety keeping up with his field surveyors in the remote southwestern part of the state with its poorly developed communication lines and widely spaced post offices. One of his field men wrote in May 1913 explaining the situation: "I presume you do not understand just how we are situated in regard to mail. We are 22 miles from the nearest P.O. and of course can not get to it except at rare intervals."[194]

Acrimony between survey scientists and their city-based bosses often revolved around disagreements regarding the pace of movement and the choice of stopping places. Merriam's surrogate while he was away in August 1888, A. K. Fisher, expressed disappointment with Bailey's failure to await instructions in Wyoming, stating that he "regretted that [Bailey] did not wait for orders at Fort Fetterman."[195] Merriam himself wrote with even greater dissatisfaction to Bailey in southern Utah later that winter. "Your letters of Dec. 24 & 25 from Kanab gave me the bitterest disappointment I've had in many a day," he lamented, adding that Bailey "must have forgotten your instructions from the Commissioner . . . or you would not have hurried through this Kanab country where we *particularly* wanted you to stay long enough to secure all of the evidence obtainable of the northward extension of the southern fauna there."[196] On the one hand, here was a lonely survey scientist in the field, struggling through winter conditions on Christmas Day, trying to make the best decisions he could with only sporadic advice sent by mail. On the other hand, there was a director frustrated by his inability to communicate regularly with his agent in the field—frustrated, that is, by how difficult it was to exercise formal control over a distant and rugged field region, with only sporadic and very limited mail outposts.

In the early years of the railroad era, research programs within the region received little or no funding, so any work had to depend on eastern patronage. Some Kansas field collectors made a creative attempt in the 1870s to build a program of state geological survey work out of their subordinate positions in Marsh's fossil-collecting network based at Yale. One year, Marsh sent Kansas University professor of chemistry and physics G. E. Patrick, along with four assistants, to collect ("in the pay and interest of Yale College") from within the state's Niobrara cretaceous deposits. But as they collected fossils for Marsh, the Kansas collaborators also made their own geological observations. Thus, as Patrick commented, "while Yale added to its museum, we at the same time added to our knowledge of geology much that is important and valuable to us as a State."[197] Moreover, some members of the Kansas party collected nonfossil specimens. Samuel Williston brought back soil and chalk samples from Wallace County, which Patrick later analyzed in Lawrence.[198] Another party member, H. A. Brous, also made observations on the habits of distinctive Great Plains animals (prairie dogs and tiger beetles) he had seen in western Kansas.[199] In order to accumulate specimens entirely for their state's institution, by the end of the decade Kansas University parties were collecting on annual expeditions to western Kansas and Colorado, led by Professors Frank H. Snow and Benjamin Mudge.[200]

Over time, these furtive attempts to produce survey knowledge at the state level became more firmly established, and then the question of geographical relations shifted to the linkages between those based within the West itself and those headquartered elsewhere—typically the former working for the state government and the latter for the federal government in Washington, DC, or at a large urban museum back East. One way to manage the tensions between national and state surveyors was to posit a functional division of labor between the two levels. In geological surveying, for instance, this sometimes reinforced a distinction between the national US Geological Survey's focus on large-scale scientific geology and the regional- or state-level focus on more fine-grained, practical economic concerns. As one survey booster, A. H. Thompson of Topeka, argued based on his conversations with US Geological Survey director John Wesley Powell, the work of national surveys was properly "supplemented in all the States in which they work, by State surveys, to carry forward this work in greater detail." While the federal government "furnishes the exact scientific basis," in his view "the States and individuals can elaborate the minutiae." In particular, states should focus on the "economic part of geological surveying," such as the location and description of mineral beds, the economic value of deposits, and the methods for their development.[201] The expansion of US Geological Survey research in "economic geology" by the end of the nineteenth century belied any reservation of such work to the states, however, and some state geologists and geological survey leaders continued to grumble into the early twentieth century about the overlap.[202]

To some extent, the division of labor between state and federal geology transmuted into a distinction between larger-scale mapping and more detailed local studies. But even that boundary was hopelessly blurred by the US Geological Survey's vast structure and its tendency to do all kinds of work. One strategy, both to accomplish more work and to promote collegial relations between state and federal geologists, was for state geological surveys and the national Geological Survey to formally cooperate with one another in the same field parties, dividing the expenses and credit. This approach was widely attempted, though initially it often encountered serious funding obstacles in many states of the Great Plains and Rocky Mountains. The Nebraska Geological Survey, for example, was at first unable to arrange much cooperative work with the US Geological Survey at the beginning of the twentieth century, because they could not obtain their share of funds from the state legislature, although some work was performed cooperatively through funding from the federal government, "in consideration of the youthfulness of the state and its lack of means at that time."[203] Finally, in 1911, the Nebraskans secured up to $3,000 in state funds for the next two years as matching funds for cooperation with the US Geological Survey. Field parties were launched as soon as "the exceptional heat of the summer" dissipated. In their cooperative work, they adopted a division of labor in which the US Geological Survey surveyed the selected counties topographically while the Nebraska Geological Survey surveyed them geologically.[204]

Behind the scenes, however, any formal division of labor based on geographical scale or subject matter tended to break down. For one thing, if state survey scientists wanted their categories and standards to be consistent with cosmopolitan forms of knowledge outside their state boundaries, they had to consult with and defer to the comparative knowledge of national (or even international) scientific bodies. Even more strikingly, federal survey scientists recognized that their state counterparts often possessed distinctive knowledge that integrated scientific perception with the knowledge of local experience that could not be captured in written reports. In North Dakota, for example, the state's geological survey at the University of North Dakota in Grand Forks eventually cooperated extensively with the US Geological Survey, including a multiyear lignite coal survey in the mid-1910s, which opened the way for informal correspondence.[205] In January 1918, the national survey's A. J. Collier wrote to North Dakota's A. G. Leonard for his advice about classifying coal land in the state. After proposing a provisional boundary line, Collier confessed that he did not "know the State of North Dakota anything like as well as you do."[206]

Like the US Geological Survey, many subdivisions of the US Department of Agriculture also sought cooperation with the states. One prominent agency that worked widely in the Plains and Rockies regions, among other places, was the Bureau of Soils. However, the bureau experienced more regional opposition due to its focus on small areas often thought by state scientists to be the province of their own authority. States in the West were among the most recalcitrant, cooperating less frequently than other regions of the country such as the Midwest, South, and Pacific Coast. To be sure, some Plains states, such as Nebraska and North Dakota, cooperated regularly, with occasional participation by others such as Kansas. Yet other states in the region, such as Colorado and Utah, were largely unwilling or unable to engage in cooperative soil survey work. Nevertheless, Whitney did attempt to inform state officials of his plans. For example, he wrote to Colorado agricultural experiment station director L. G. Carpenter in February 1901 and 1902 outlining his plans to survey the San Luis and Arkansas valleys, two emerging agricultural areas of the state, during those two summer field seasons. In the first case, he requested Carpenter's assistance in finding a good base map of the region. In the second case, he referenced a political motivation for initiating Bureau of Soils work: "We have had requests from the Colorado delegation that this work should be undertaken," he noted.[207] After the work was done, he wrote again to Carpenter offering that the Bureau of Soils report on Arkansas Valley could be reprinted in the state for local distribution.[208] Even when formal state cooperation could not be secured, Whitney sought informal help and offered state scientists some benefits from his agency's field work.

Another branch of the Department of Agriculture, the US Forest Service, also undertook cooperative field survey work in the western states. For example, in the summer of 1901 a series of tree surveys was undertaken across several counties in Nebraska, including many in the Platte Valley, Sandhills, and High Plains areas. Like the Bureau of Soils reports, the results of these tree surveys were given for different counties by individual small groups of field surveyors, usually just one or two men.[209] Likewise, in July 1902, the forest service sent its chief grazing officer, Albert F. Potter, to conduct a series of surveys across the state for five months. In the words of one historian, this particular forest survey carried Potter "on a crisscross exploration through the Wasatch Mountains and Colorado Plateau from the Idaho border on the north to Escalante on the south," which "led directly to the establishment of a series of forestry reserves in Utah." To promoters of state economic development, Potter's emphasis on practical utility contrasted favorably with the "grand theory" of Powell's despised irrigation survey over a decade earlier, which had upset many boosters and developers in the West with its emphasis on broad regional climatic limitations.[210]

Cooperation with state scientists and proponents of economic development were thus central to the operation of national efforts to survey geology, soils, and forest resources. But not all attempts at cooperation yielded valuable results for national researchers. Problems were more likely when federal agencies did not send their own survey scientists to the region to work directly with state researchers. For instance, the efforts of the US Bureau of Soils to cooperate with Nebraska in 1912 in studying the saline lakes of the sparsely inhabited Sandhills turned out to be a waste of time and resources. An agent of the bureau sent twenty special "16 oz. glass bottles in pasteboard mailing cases" for University of Nebraska researcher Raymond J. Pool to use for collecting and sending samples from the lakes. The bureau instructed Pool to take a sample and give a verbal description of the water in each lake. "When samples are to be sent through small post-offices," the letter cautioned, "the Post-master should be cautioned to have them handled carefully on stages and not to put them in sacks which will be caught by hooks on moving trains." The same letter was sent to Pool's in-state colleagues who were nearer to the field sites in the Sandhills.[211] However, when the US Department of Agriculture analyst completed his study of the samples sent from Cherry County, the waters turned out to be "very much more dilute than [he] expected. So much so, indeed, that they have only a negative interest for our investigations. We shall probably not make any further examination of the samples unless you desire it." If the bureau had had someone in the field to examine the lakes, their unsuitability for the agency's studies would probably have been quickly evident. But for regional scientists not highly trained in specialized areas of study, it was not apparent that they would be so "uninteresting" to the federal government scientists.[212]

Even when economic development issues or gaps in local or regional awareness of specialized cosmopolitan standards and knowledge were not directly involved, there was often still tension between federal and state survey researchers. In such cases, the conflict might be properly attributed almost entirely to the scientific economy of credit circulation. Take Colorado, for example, the state of the interior West that was most advanced in its metropolitan development by the early twentieth century. Scientists at institutions in the densely settled Piedmont region including developing centers along the Front Range such as Boulder, Fort Collins, Colorado Springs, and Denver engaged in their own survey research. They often defended their turf against the domination of federal government scientists, whom they viewed as too hasty in their travels through the state and too stingy in their acknowledgment of regional achievements. The University of Colorado in Boulder and the Colorado Museum in Denver both undertook substantial survey efforts in the early twentieth century, claiming the state's territory for themselves rather than for national researchers headquartered in Washington, DC.

One example of their indignation at federal presumption was the blistering review offered by Francis Ramaley of the University of Colorado to the US Biological Survey publication titled "A Biological Survey of Colorado," authored by Merritt Cary. Although Cary himself was a child of nearby Nebraska, Ramaley was still upset by his unfamiliarity with Colorado. "It is little short of marvelous," he sarcastically commented, "that a young man with so little knowledge of the state, of its climate, its flora and its fauna has been able to prepare so good a report." In Ramaley's view, Cary had not sufficiently credited regional publications, such as the mammal survey published by Colorado Springs collector E. R. Warren in 1910, thus resulting in a report "conspicuous for that perfect independence of thought characteristic of those who know nothing of the writings and opinions of others." Moreover, Cary had failed to consult local collections, so that despite starting from Boulder he "apparently did not take the trouble to visit the university there nor the museums in Denver, only 30 miles away."[213] Unlike many of the federal researchers who graciously sought the collaboration of state-based scientists, Cary had exposed the latent tension between knowledge makers in the West and those in the East.

The tension between regional and national institutions in the exploitation of the environmental resources of the West through survey field work played out in a variety of ways. Aspiring regional upstarts like the Colorado Museum challenged the domination of distant metropolitan centers over scientific knowledge production, but they also felt obligated to play by the standards of metropolitan science, as they perceived them. Interactions between the federal and the state governments were more regularized and sometimes even formalized in cooperative survey field work, though this did not mean that structural tensions were absent.

Both Arnold Hague's Yellowstone Division and Vernon Bailey's extensive field work for the US Biological Survey across the vast territory of the West exemplify the distinctive symbiosis of cosmopolitan and experiential perspectives on knowledge making in the field in the survey mode. In Hague's case, this deep knowledge led directly to public activism. After nearly a decade of leading Yellowstone Division field survey parties, in 1891 Hague was lauded for his "intimate knowledge of the Park [that] extends back over a number of years" by the magazine *Forest and Stream*, which cheered that he had "long been a consistent worker" for Yellowstone, "with an eye single to the public good."[214] More broadly, Hague's and Bailey's surveys are indicative of a much larger number of national and regional projects by aspiring career researchers to foster the development of science through survey field work. By inhabiting the Great Plains and Rocky Mountains seasonally through long-term travel, subsistence, and material experience, field survey scientists intimately learned the particularity of nature in distinct places while mapping those places more fully on a regional scale and situating both localities and regions in the larger context of cosmopolitan science.

At the same time, however, their perspective of nature was inherently limited and partial, just like any other mode of field practice. Most obviously circumscribing the fullness of their knowledge was the transitory status of their residency in the Plains and Rockies, especially the marked seasonality that has often attended field surveys everywhere, but especially in environmental regions with such strong patterns of seasonal variation. Moreover, the bifurcated envirotechnical system on which survey science depended, combining selective railroad access corridors with pack animals to enable their movement off the rails and into the field, meant that travel patterns and perceptions of nature were mediated by the two travel regimes. Finally, the geographical relations that connected field survey parties with their headquarters relied on selective lines of communication, mostly postal service routes with occasional use of the telegraph, which channeled the interaction of ground-level experiential encounters with the cosmopolitan concerns of distant survey directors.

In narrating and analyzing the contours of field life on survey parties, I have also emphasized not only the larger political economy of survey science but the everyday experiences of the environment and work organization of survey field work on the ground as well. If research in the Plains and Rockies was governed in part by larger historical processes such as state expansion and capitalist economic development, it was also subject to the particular regional conditions that prevailed in the interior West. The focus on survey science in the West neatly combined the interest of regional political and economic decision makers in securing the benefits of scientific expertise for economic development with the desires of career-making researchers in the same region to gain epistemic authority over the natural resources that were closest at hand.

Surveys enabled field scientists to study problems of interest with more or less trained and supervised field personnel, following selected routes rather than depending on the serendipity of finding collaborators who lived in particular places. But such training only partly released them from their dependence on the people and nature of the West. Surveys faced the practical constraints of transportation and communication infrastructure, cooperation and assistance from local people, and the tension between the supervising scientists in the metropole and subordinate collectors or observers in the field. At the same time, researchers were usually venturing out of their familiar home bases—often eastern and urban—into the very different environment of the American West. Limitations such as aridity, strong seasonality, and rugged topography would present special challenges to survey scientists. Strikingly similar constraints were faced by those scientists who wished to work more intensively in locations with especially rich resources—quarries—that could be extracted and processed in distant universities, museums, and laboratories, when they were engaged in the prospecting stage of finding promising sites, but a whole new set of limitations and opportunities would attend the on-site extractive practices of those quarries once they were identified.

Chapter 4

Quarries

"Big science" in the United States is usually dated to the middle of the twentieth century, with its particle accelerators and massive nuclear projects, but certain physically large and imposing field sites in the late nineteenth and early twentieth centuries were a kind of big science of their own time. The opening and ongoing occupation of an enormous dinosaur fossil quarry in the lower part of the Uinta Mountains in Utah near the Colorado border during the 1910s was one prominent example. This Carnegie Museum site, supervised and operated for over a decade by field man Earl Douglass, would later be protected as Dinosaur National Monument (now Dinosaur National Park).[1] Upon visiting in 1922, William Diller Matthew of the American Museum enjoyed a guided tour and proclaimed it "a wonderful quarry," having "yielded more and better Jurassic dinosaur skeletons than all those previously known put together."[2] Today, the site is still very much recognizable as a scientific quarry, and in fact, if you visit this remote corner of Utah you can see a fossil display wall in situ that preserves a portion of it, though it has been a full century since the first railcar loads of fossils were extracted from gashes dug or blasted deep into the earth and transported to museums in Pittsburgh and elsewhere.

Your chances of seeing physical quarry remains on site are even better at another place in the Rocky Mountain region, also preserved by the US National Park Service. At Mesa Verde National Park, named for a sprawling, dissected plateau just south of the San Juan Mountains of southwestern Colorado, not only are numerous excavated artifacts on display at the on-site museum (though many were taken elsewhere) but a myriad of named structures have been restored for tourist visitation, such as the Cliff Palace, Balcony House, Spruce Tree House, Step House, and Long House, among others. Several of the most prominent quarries on Mesa Verde were discovered in the late nineteenth century by a local ranching family, the Wetherills, who guided visiting travelers such as Frederick Chapin and Gustaf Nordenskiöld. These visitors soon published their findings in books and scientific journals to much public acclaim. As excavations became more intensive and systematic, organized scientific extraction replaced the seemingly haphazard artifact collecting by amateurs and lay people. After its establishment as a national park in 1906—thanks in part to the lobbying efforts of the Colorado Cliff Dwellings Association and its outspoken leaders Virginia McClurg and Lucy Peabody—Mesa Verde's quarries were further studied by government-authorized archaeologists, especially Jesse Walter Fewkes of the Smithsonian's Bureau of American Ethnology starting in 1908.[3]

The Dinosaur National Park and Mesa Verde quarries were not the only scientific quarry field sites opened by the early twentieth century. Nor were they the only ones that would be protected and managed by the National Park Service—others in the Plains and Rockies have included Agate Fossil Beds in Nebraska, Florissant in Colorado, Hovenweep along the Colorado-Utah border, and Fossil Butte in Wyoming, all of which will make their appearances below. But Dinosaur and Mesa Verde were two of the most prominent, and their juxtaposition helps us to see some of the common features that united scientific quarries in different disciplines. Within the much more extensive territories covered by both surveys and lay networks, some specific places had such rich and highly concentrated scientific resources that field scientists developed a more intensive mode of field practice. At these sites, valuable resources such as fossils, minerals, artifacts, and even biological specimens could be removed from their natural settings and transported to distant locations for processing. We thus might usefully call these sites quarries in recognition of their similarity to rock- and mineral-extracting enterprises.

Indeed, observers of quarry field work at the time explicitly likened such efforts to mining. "As in mining for precious metals," one writer on fossil digging stated, "the earth seems to withhold her secrets jealously from a hundred men that she may bestow them freely on the one man following after"—encapsulating the key element of serendipity sometimes involved in initially finding quarry sites.[4] Sometimes field scientists even explicitly identified such places as quarries in their writing, suggesting that they functioned as such for both economic and scientific purposes, as when New Jersey mineralogist E. H. Wilson described a Colorado site he visited to collect zeolite. "Specimens may most readily be secured at a quarry close to the summit of North Table Mountain," he explained, that could be "reached by a steep climb of several hundred feet from the road along Clear Creek, leading into Golden, where runs a trolley line from Denver."[5] Moreover, at least one archaeological field site in eastern Wyoming was determined to be a "vast aboriginal quarry" and visited by field parties from several museums and universities large and small, but all from out of state.[6]

As this example suggests, while stations—as we shall see in the next chapter—could sometimes be organized by regional institutions and researchers to concentrate the value added in knowledge production within the region itself, as well as to address research to pressing regional problems and issues, quarries were usually dominated by more distant institutions and researchers based outside the region. By removing natural objects from their regional settings and moving them elsewhere for scientific investigation, distant researchers gained the largest share of the payoff in credibility from producing knowledge about them. In this period, quarries therefore usually operated according to a model of scientific practice focusing on collecting physical specimens more often than observations. To be sure, other modes of field production such as surveys and networks also frequently extracted specimens, but not so intensively or over such a lengthy time period as in the case of quarries.

During the railroad era, scientists involved in intensive quarry extraction tended to eschew the detailed investigation of the field site itself—though this practice would later change—in favor of simply getting as much material out as possible. Through quarries, field scientists based elsewhere attempted to concentrate scientific materials in their own repositories. At the quarry, the control of the site itself and the ongoing extraction of high-value materials over a significant period of time—an entire field season or in some cases many years or even multiple decades—was paramount. Because the quarry often represented an intensification of natural history collecting, it predictably flourished in the museum setting. In understanding quarries as historical phenomena, therefore, the museum context is crucial. It was in the large urban museums—especially the American Museum in New York, the Field Museum in Chicago, the Carnegie Museum in Pittsburgh, and the US National Museum (Smithsonian) in Washington, DC—that changes occurred by the late 1890s driving a new expansion of quarry-based field work in the interior West.[7]

This expansion of quarry activity occurred first and most decisively in vertebrate paleontology, but it also included significant work in other object-based fields such as archaeology, zoology, and ethnography. The development of innovations in public display such as life groups and habitat dioramas, as well as the increasing popularity of dinosaur and extinct large mammal displays, made the collection of particular high-value objects crucial to museum patronage. In fact, following the separation of display and research collections in the late nineteenth century, both purposes flourished in the wake of vastly expanded funding for museums from urban capitalists made wealthy by the rise of big business.[8] What made quarries distinctive was the enormous demand for actual physical specimens in both display and research contexts. The control of such finite resources in the vast and thinly inhabited interior American West thus became central to the practice of science in the turn-of-the-century museum.

The competition to expand the collection of physical materials in the West predictably raised the possibility of resentment by much poorer regional institutions loath to see their own states' precious resources shipped eastward. Emerging East-West tensions in fact were a central feature of quarry science in the late nineteenth and early twentieth centuries. During this time period, "East vs. West" was effectively another way of describing the hierarchical rivalry between two different regions. As the metropolitan eastern United States improved its own position relative to European centers—shifting from semiperiphery to incipient core status, in part based on leveraging its institutions' privileged access to research materials in the western United States—the interior West became an arena of regional conflict within American science.[9] (Very little quarry research was done by scientists from foreign countries, as opposed to metropolitan easterners merely viewed as "foreign" by the residents of western states.) As eastern museums engaged in fierce competition with one another—none more than the American Museum and the Carnegie Museum in vertebrate paleontology—upstart western institutions struggled to get into the game.

The tension between East and West was also reflected in the relationships of individuals. Eastern museums recruited labor, both scientific assistants and manual workers, from the people who already lived in the West. Even if control in quarry science was decidedly metropolitan, collaboration with locals was a crucial aspect of this mode of field work, just as it was for lay networks, surveys, and stations. Collaborators included both American Indians and white settler residents such as ranchers. In addition to their sweat and toil, residents of the West often had first-hand experiential knowledge of the land: how to get around, where to find particular resources, how to cope with its climatic extremes, and so forth. Moreover, locals sometimes possessed important means of leverage in the form of legal control over land.[10] Other westerners participated more extensively in scientific quarries, ranging from assistants on field teams to coordinators of quarry work, and occasionally even to formal positions in metropolitan museums. Nevertheless, despite career mobility for a few, most did not have such opportunities. And even when they did, some social tension often remained. Finally, many locals, both those who became employed in scientific quarry work and those who simply took an avocational interest, articulated their own meanings for the quarries and what was excavated from them. From the prospecting phase to the eventual display and interpretation phase, local people living and working around quarries were closely involved in them.

The Shift to Quarry Science: Going Vertical

In principle, anything that can be extracted for scientific purposes over a long period of time might be considered an object for field work in the quarry mode. Most important in this region and period were fossil quarries, for which the interior American West was the epicenter of the tremendous expansion in the late nineteenth century. Other kinds of quarries were also exploited, however, including more obvious analogs such as artifact quarries in archaeology at Mesa Verde and elsewhere, as well as a few other examples in zoology—regularly visiting sites to supply museum habitat dioramas, for example—and ethnography, where extended field work in a particular place often entailed not only the ongoing collection of large quantities of material artifacts but also the mining of stories, rituals, languages, and other cultural phenomena, often with the close assistance of long-term American Indian informants.[11] Examples of deepened, intensified, and often technologically enhanced study of human culture in the Plains and Rockies during the railroad era, to be explored further below, included James Mooney's work with relocated tribes in Oklahoma and elsewhere, Alice Fletcher's comprehensive study of the Pawnee Hako ceremony in Nebraska, and Frances Densmore's investigations of American Indian music among the Northern Utes, Teton Sioux, and others. While intensive, multiyear extraction of cultural knowledge had previously been undertaken by long-term missionaries, both before and after the building of the transcontinental railroad—such as by Stephen Return Riggs, who compiled a 665-page dictionary, along with extensive collections of grammar, texts, and ethnography while residing with the Dakota—later efforts were increasingly targeted to specialized academic topics that were every bit as exacting and intensive as work in paleontology or archaeology.[12]

One might even consider some examples of intensive observation, which required repeated occupation and visiting below the earth's surface—progressively deepening—over a period of many years, to display many of the key features of science in the quarry mode, such as when the US Geological Survey under S. F. Emmons mapped the complex underground geology of Leadville, Colorado, from the 1880s to the early 1900s, in the same expanding network of mining shafts and tunnels that was also the scene of considerable quarry extraction in the conventional economic sense as well. The Leadville geology example illustrates how, in most cases, a distinguishing feature of quarry science was its verticality: probing underground required a more intensive, long-term form of work organization than skimming from the surface.[13] Similarities in intensity and deepening—that is, the verticality—of field practices permit a cross-disciplinary analysis combining evidence from paleontology, archaeology, ethnography, and underground geology. Indeed, the analogies between different types of quarries for scientific extraction were evident even to railroad-era field scientists themselves, such as paleontologist and museum director Henry Fairfield Osborn. "Very much after the manner of the archaeologist," he wrote, "the paleontologist digs and excavates."[14]

Even within the great fossil-hunting domain of the interior American West, in which vertebrate paleontology clearly reigned supreme, many other types of point-source resources existed in the region, such as the globally significant fossil insect quarries repeatedly exploited near Florissant, Colorado, the striking fossil plant quarries found not only at Florissant but also in the Black Hills of South Dakota, and the famed fossil fish quarries of southwestern Wyoming. The last of these, located near Green River, Wyoming, were regarded in 1880 as "one of the finest fish-deposits in the world," and the most famous specific fossil fish quarry site was literally carved down into the earth by the building of the transcontinental railroad "some three miles out of the city," revealing not only "fragments of fish" but also "water grasses" and "leaves" preserved in rock. "Travelers over the Union Pacific Railroad are familiar with them," declared John B. M'Master of Princeton, reporting that they were "exposed for sale at almost every station from Cheyenne to Salt Lake," although their "distribution in nature is by no means so diffuse." The two main localities "where they occur in great numbers" were the quarry site cut by the Union Pacific and another quarry "at Morgan's meadows, some 25 miles south."[15]

On the ground, then, quarry geography was even more complex, with similar materials often found in nearby, if distinct, quarry sites.

This was true also of the fossil cycads of the Black Hills Rim complex, as it was labeled by G. R. Wieland in 1916, which encompassed "a whole group of larger and lesser localities, easily the most remarkable known, forming a complete cycad girdle of the hills, along which new and striking discoveries of trunks, either singly or in groups, may be expected at any time." While this area had been known for only a short time, Wieland observed that "the 'rim' [had] already yielded far more and far handsomer trunks than all the rest of the world beside." Specific quarry sites were not randomly distributed, even within the Black Hills area, and Wieland pointed out their heavily localized concentration, so that "by far the greater number of specimens [came] from the Minnekahta neighborhood on the south side of the Hills." It was not simply a matter of quantity but also of quality: "Here the trunks are not only more numerous than elsewhere," he noted, "but often exhibit the finest types of conservation, while some of the specimens reach the maximum size."[16] This example from paleobotany illustrates the typical microgeography of quarry distribution.

As noted already, however, it was in the realm of vertebrate paleontology—dinosaurs, mammals, and great sea creatures—that the American West was one of the greatest fossil-hunting regions in the history of the world. The underlying environmental context of this quarrying material was based on the distribution of eroded strata across a vast, semiarid land area. Fossils were one of few visible resources on this treeless and sparsely inhabited landscape that provided only limited ways to make a living. While prospecting near Edgemont just outside the Black Hills of South Dakota, the American Museum's William Diller Matthew concisely located the "fossil region" in "the high plains and foothill country."[17] By emphasizing the abundance of fossils on the High Plains and in foothills near larger mountain islands, Matthew drew attention to how outcroppings of fossil deposits were often concentrated in the transition zone between the Rocky Mountains and the prairies of the Midwest. Other fossil quarries, such as the fabulous Dinosaur quarry site in the Uinta Mountains of northeastern Utah, opened up by Earl Douglass and the Carnegie Museum, were located instead on the western side of the Rocky Mountains, in the foothills where they transitioned into the arid lands of the cordilleran interior.

Not all collecting work merits separate analysis as a distinctly "quarry" mode of field production, however. I use the term not to analyze a particular kind of subject matter, such as fossils, but to highlight how work was organized in relation to the environment in order to produce knowledge. Thus, a key feature of the quarry mode was prolonged work operations over a period of many months, perhaps even years, at a remote site distant from scientific centers. At the famed Bone Cabin Quarry in Wyoming, for example, American Museum field men took "over six years working with the plow and shovel and spade" to reveal "parts of seventy-three different animals," including forty-four dinosaurs (see figures 4.1 and 4.2).[18] Likewise, anthropologist Alice Fletcher's intensive study of the Hako ceremony "in its entirety" required "four years of work," involving "close friendly relations" and "much patience," so that she could "overcome the scruples" of her chief informant, a Pawnee elder named Tahirussawichi, and thereby extract the closely held ritual knowledge on "very sacred subjects" from this "man of truth."[19] Unlike a temporary visit during an expedition or survey, which depended on the materials to be collected lying on or near the surface, or information that could be obtained in a brief interview, the quarry became the site of prolonged physical or cognitive excavation that justified keeping people digging (literally or figuratively) at great expense for longer than a few weeks. The Dinosaur quarry in the northeastern corner of Utah, near the Colorado state line, was another prime example. By establishing semipermanent camps and often employing a skeleton staff to supervise the site all year round, research teams remade a mere fossil collecting locale into a full-fledged quarry operation. Team members would typically return to the same place in subsequent years to continue working, progressively deepening extraction over a long period of time.

Such quarries were most necessary when the fossil materials were unusually large, such as dinosaurs or big mammals. But a few exceptionally rich sites for smaller types of fossilized animals, such as the Florissant site in south-central Colorado, became long-term quarries for fossil insects or other diminutive creatures. Likewise, at Florissant and elsewhere, unusually valuable fossil plant materials of various sizes could be subject to quarry-style extraction when present at a few favored localities in sufficient quantity and depth. In the Black Hills of South Dakota, for instance, which were so rich in fossil cycads, as noted above, paleobotanist G. R. Wieland of Yale had begun by 1916 "to insist on the paramount importance of field-work and the necessity of developing quarries on a large scale." Comparing the new American sites favorably to already opened "quarries of the Isle of Portland" off the coast of Britain, Wieland argued that previous collection and observation work in the Black Hills "showed conclusively the relation, abundance, and position of material in the terranes," thus justifying more intensive opera-

FIGURE 4.1. "Working in the Bone-Cabin Quarry during the first year, 1898." Henry Fairfield Osborn, "Fossil Wonders of the West: The Dinosaurs of the Bone Cabin Quarry," *Century Magazine* 68 (1904): 680–94, on p. 683.

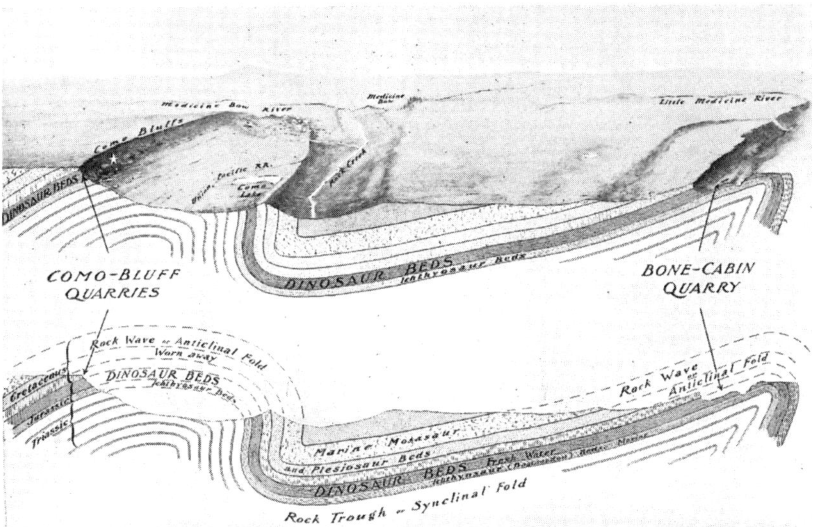

FIGURE 4.2. "Diagram key to the geology of the region." Henry Fairfield Osborn, "Fossil Wonders of the West: The Dinosaurs of the Bone Cabin Quarry," *Century Magazine* 68 (1904): 680–94, on p. 684.

tions. "At Minnekahta," he predicted, "it would be practicable, by means of a shallow quarry of some width, to secure an extraordinary wealth of material illustrating the extent and appearance of the cycad forest with a completeness such as no series of eroded out and often surface-work trunks could possibly yield." While he admitted that "the cost of such operations would be large"—though, he hastened to add, "scarcely a tithe of that devoted to quarry for Dinosaurs in recent years"—he proclaimed that "the results must unfailingly prove both instructive and striking."[20]

As Wieland's plea implied, opening (or reopening) quarries required a much greater commitment on the part of scientific institutions than did collecting in the survey mode. Commenting on the disappointment felt by members of the sprawling railroad-sponsored Fossil Fields Expedition of 1899 in Wyoming, Charles Schuchert of the US National Museum perceptively noted that a traditional seasonal expedition would be insufficient. "Many of the members also expected to secure a Dinosaur each," he noted, "but the magnitude of the work soon changed enthusiasm into regret." He also identified some of the key features of the quarry mode of scientific field work in the case of dinosaur paleontology, including its long duration, its labor intensiveness, and its high cost. "In no one place are complete Dinosaur skeletons found," he noted. "Sometimes a 'quarry' will yield a lot of vertebrae, or a number of either hind or fore limbs, or there is a general mixture of parts of animals of different genera." Schuchert thus concluded that an "adequate collection of Jurassic Dinosaurs . . . requires several successful field seasons." Further costs back in the metropolitan laboratory would include having "the bones . . . cleaned, hardened and restored . . . for study and exhibition." Due to all these factors, and since "Dinosaur skeletons are very large," he judged quarry work to be "extremely expensive."[21] The great investment required to become a major player in the competitive game of quarry science would have been especially evident to a government museum scientist like Schuchert, who undoubtedly observed the dramatic expansion in funding for museum expeditions by the large urban museums of the day.

A few liminal examples from the 1870s, earlier in the railroad era, further illustrate the shift from expeditionary to quarry science. First, consider intensive collection from American Indian burial sites, which in a sense functioned as highly valued quarries for human remains. As H. C. Yarrow noted in reporting on an 1872 visit to a high-mountain burial site in the Wasatch Mountains of Utah—the location of which was treated as a valuable secret, not even revealed in a later article—they were successful in their quest for "a rich harvest of crania and archaeological material for the National Museum." However, as he explained, they had to work secretly and quickly, in order to avoid detection by those "who might object."[22] Human quarries, though sharing features with more intensively developed scientific quarries, were exhumed rapidly, without a long occupancy. As extraordinarily valuable as the human remains at the burial site may have been, the site was exploited in the mode of short-term collecting by expedition.

For a second and more conventional example, consider paleontology in western Kansas around the same time. When they visited a site along Hackberry Creek in Gove County, Kansas, a party from Kansas University under Benjamin Mudge, the erstwhile Marsh collector, "hardly expected that the gleanings from a field so thoroughly worked would contain anything of unusual value." Several years of surface or near-surface collecting by various field parties should have cleaned out the site. Yet, as Frank H. Snow pointed out in his report of the expedition, "the frosts and rains of a single year had exposed to view many remains of fishes and saurians, and our labors were unexpectedly and abundantly rewarded." As a result, Mudge's party was "in less than three weeks" able to collect hundreds of fossil specimens, "many of them in almost perfect condition, and all of them of sufficient value to justify their transportation for a distance of 300 miles, to take their places in the cabinets of the University."[23] Spending just a few weeks on site, collecting what could be obtained near the surface, the efforts of the Kansas party were, in some respects, more typical of expedition or survey work. This is not to say that the work was easy. Indeed, one especially attractive saurian was "imbedded in the solid rock," thus requiring the removal of fully "eight feet of superincumbent material" to collect the specimen properly. That arduous task took "three days and a half of hard labor" by three members of the party.[24] It was on the cusp of a quarry mode of practice and shared many of its features.

Even further along the continuum toward full-blown quarry science was the work of independent fossil prospector Charles H. Sternberg with his sons in the fossil-rich region where Wyoming, Nebraska, and South Dakota meet. In 1910 along the Cheyenne River, they excavated a dinosaur together at a place he did not hesitate to label as a "quarry" site. "It was a big undertaking," Sternberg recalled, "to get that dinosaur out of the quarry and haul it to the railway at Edgemont, South Dakota, 75 miles away." According to Sternberg, it took them no less than "two months and a half of tireless effort" to accomplish the work.[25] A few years later, at "Gidley's quarry" for horse fossils in Texas, named after a rancher named Gidley Mayfield, it took four days of "stripping the upper layers" before "the spading" could even begin, and eventually a "summer's work" removing "approximately fifteen hundred tons" of earth over a seven-week period, covering "an area of over three hundred square yards."[26]

However, even a continuous effort extending over a few days, weeks, or even months can hardly be compared with the much larger operations that occurred elsewhere, which necessitated a distinctive scale and employment of labor. Another characteristic of the longest-lasting quarry field work was the tendency for it to become more self-consciously systematic and carefully documented. Meticulously prepared site maps, often showing the precise locations of specimens that had been unearthed, and detailed descriptive notes were not yet universal features of quarries during this period, but they were becoming increasingly common. As Melinda Elliott points out, the field practices at Mesa Verde for "unearthing artifacts" changed dramatically in one year with the arrival of Swedish traveler Gustaf Nordenskiöld in 1891. Assisted by the Wetherills, Nordenskiöld implemented "scientific note-taking," along with "meticulous drawings to serve as precise records of architecture, artifacts, and burials, and their locations at each site." Such practices may have been "slower," but they were also "more systematic than the rough pick-and-shovel approach to which the Wetherills were accustomed." Nordenskiöld taught them to use smaller and more precise tools in their work assisting him.[27] Adopting more systematic practices could be expensive and time consuming, of course, but Nordenskiöld had sufficient resources.

Just as the shift from lay network or survey collecting at the surface to more intensive quarry operations typically involved a greater investment of both funds and time in order to dig deeper into the earth—and let it not escape notice that a similar pattern has often occurred in the history of mining more generally—a second transition sometimes occurred at some of the richest sites to an even more intensive level of quarrying. During a second era of exploitation, around the turn of the twentieth century, many former sites were reopened for deeper digging, even after years of dormancy. Of course, many such visits to famous quarry sites seemed more like pilgrimages than renewed quarry operations, such as when Wesleyan University's S. Ward Loper traveled in 1898 to already established quarries such as Fossil Butte in Wyoming and Florissant in

Colorado, or when Samuel W. Williston revisited the Cañon City, Colorado, quarry on one of his last field trips in the summer of 1912.[28]

Other visits to previously exploited quarries were bona fide attempts to reopen and intensify quarry work, and such efforts often resulted in renewed quarry activities that further deepened the extraction of scientific materials from the earth. In the spring of 1897, an American Museum field party reopened the Como Bluff quarry, and in November 1900 the Carnegie Museum reopened the Cañon City quarry.[29] In 1906 and 1907, T. D. A. Cockerell returned to Florissant, which he regarded as a "Miocene Pompeii," along with his wife and colleagues from the University of Colorado, the American Museum of Natural History, and elsewhere, to mine fossil insects from the rich quarries there, where a "large collection" had been extracted by Samuel H. Scudder of Harvard in the summer of 1877, and others since then.[30] Some quarries might also have been revisited for the purpose of studying their geological contexts and collecting other materials, as when Willis T. Lee of the US Geological Survey visited the Morrison formation west of Denver, Colorado, in the summer of 1916, accompanied by someone who had worked there previously, who pointed out "the old quarries where the dinosaurs were found."[31] Lee collected fossil plants there to reveal aspects of the prehistoric environment besides the spectacular dinosaurs that had first attracted attention to the site in 1877.[32] Across the region, then, many of the famous quarries of the 1870s and 1880s were reopened during this later period, in order to exploit the scientific riches they still were thought to contain.

But this period also witnessed an unprecedented explosion in the number of quarry sites opened up, in part because so much more of the terrain was being explored by the burgeoning ranks of urban museum field parties, as well as by curious local people. When discoveries were made, if they were promising enough, a multiyear investment in the quarry mode of operations could seem well justified. In 1906, for example, William Diller Matthew suggested making Porcupine Creek, South Dakota, a quarry site, putting Albert Thomson in charge of the operation for several years. Matthew's comments make clear the logic of moving from extensive collecting over a wide area to a more intensive and sedentary quarry mode. He began by observing that "if Thomson comes out and takes charge every year, he will become thoroughly expert in the work." He estimated that it would "take two or three seasons" for Thomson "to round up this South Dakota country alone. But carried through in this systematic way we should have in ten or twelve years an invaluable collection from the Plains Tertiary, with a minimum of time wasted in securing an outfit or money in profitless travelling."[33]

Prospecting was also a vital component of quarry science, but it undoubtedly consumed time and resources. Indeed, prospecting often entailed reliance on the experiential knowledge of locals, whose identification and assessment itself could take considerable time. But Matthew recognized the indispensable need for collaborating with locals to combine their experiential knowledge with some cosmopolitan scientific instruction for the most promising local collectors, adding that Thomson could "train in local men to collect for us after he leaves and sell any specimens that we want to us, while we can dispose of other good specimens for them to other museums." Matthew thus concluded that if Thomson could be funded at $1,500 annually, the American Museum could "be confident of its being well spent."[34] This plan was apparently persuasive, because it met with the general approval of director Henry Fairfield Osborn.[35] This example suggests that museum researchers were learning the value of committing to more intensive operations at particularly rich sites, both for the payoff from extracting scientific objects at already identified sites and for developing local people's skills in prospecting throughout the surrounding region.

Prospecting

The first step in opening up a quarry for scientific research was finding it. Even on the wide open, treeless landscape that pervaded much of the West, locating places rich with fossil, artifact, and biological materials was no trivial task. For one thing, the spatial expansiveness of the territory made it time consuming to search out promising quarries. A typical strategy in field prospecting is revealed in Barnum Brown's explanation of the more common badlands formations of the interior West in comparison with his later water-based exploration of the Red Deer River region in Canada. "Usually fossils are found in 'Bad Lands,'" he explained, "a name applied by the Jesuit missionaries to desolate regions denuded of grass and eroded into picturesque hills and ravines. In such places camp is located near some spring or stream, and the collectors ride or walk over the exposures till the region is thoroughly explored."[36] Similarly, when Amherst College's Frederick B. Loomis brought two students to a prospecting camp along the Grey Bull River in Wyoming in the summer of 1903, they made "daily excursions of from five to ten miles into the bad lands, seeking fossil bones," and he described their experiences in an evocative third-person narrative. "Hour by hour the men prowled slowly along the base and on the sides of the ridges," he reported, "looking for tell-tale fragments which might be weathered out."[37]

The challenge of navigating a vast land area was compounded by the fact that so many of the vertebrate paleontologists, archaeologists, ethnographers, and biologists who prospected for rich scientific materials were outsiders to the region. Often affiliated with eastern museums and universities, these field researchers found themselves on unfamiliar terrain, so different in both its environmental and social contours from cities like New York, Pittsburgh, and Washington, DC, from which they came. The intimate experiential knowledge of the terrain—and especially knowledge about the location of promising deposits—that was possessed by lay people living in the area was thus of special significance during the prospecting phase. This included amateur naturalists, to be sure, but also the ranchers, farmers, miners, and others working the landscape, whose everyday experiences often led to deep, detailed knowledge of place. Consider the Wetherill ranching family near Mesa Verde. When their "range became limited at the ranch," they ran their cattle "on the high plateaus and cañons of the Mesa Verde" for a full decade, as one of them later recalled. "The cattle became like wild animals and it was necessary to hunt constantly and everywhere for strays so as not to overlook the hiding places of a wandering bunch of the piñon splitters." As he and his brothers grew up, they spent much time "scouting around the mesas and cañons of the Mancos River."[38] Such experiential knowledge, developed over many years, almost perfectly complemented what visiting cosmopolitan scientists knew. Indeed, often it was the prospecting phase of quarry science that most clearly revealed the role of local collaboration, not only in the hard labor of digging out highly desired materials, as discussed later, but also in the provision of experiential knowledge to scientific prospectors.[39]

Even when prospecting did not directly involve local people brought along for their deep experiential knowledge of place, it was still often most successful when it combined elements of practical topographical use knowledge with the cosmopolitan insights of science. Perhaps the most vivid example of this comes from the discovery of the Dinosaur quarry in northeastern Utah by Carnegie Museum field man Earl Douglass in August 1909, after he had traveled with museum director William J. Holland through the area in the fall of 1908. In Holland's recollection, one evening under a beautiful sunset with Douglass he had "remarked to him that a study of Hayden's Survey disclosed the fact that in those mountains there were extensive exposures of Jurassic strata," making dinosaur remains likely. "We started shortly after dawn," recounted Holland, "and spent a long day on the cactus-covered ridge of Dead Man's Bench and in making our way through the gullies and ravines to the north." That night, they descended "a wild declivity" to "the edge of a small brooklet," sleeping beneath blankets "while the mules grazed upon the scanty pasturage which was found at that spot." The experiential particularity of the terrain, as well as of human and animal subsistence, comes through clearly in Holland's account.

Situating the terrain in its cosmopolitan context was important too, however. On the following day, they explored "the broken foot-hills which lie east and south of the great gorge through which the Green River emerges from the Uinta Mountains," and as they "slowly made our way through stunted groves of pine we realized that we were upon Jurassic beds." They split up, agreeing to fire rifle shots upon discovering anything, though it was the experienced field man Douglass—who had already spent many years traversing western landscapes, looking for fossils—who found a "weathered-out femur of a Diplodocus, lying at the bottom of a very narrow ravine into which it was difficult to descend." The next year, Douglass returned without Holland and found nearby the Dinosaur quarry, which would become the object of so many years of the Carnegie Museum's field work.[40] In Holland's telling, at least, the discovery had involved some scientific knowledge of paleontology and geological formations, which arguably they both possessed, but it was the experienced field man Douglass, whose mind was deeply imprinted with the topographical countenance of western landscapes, who found the initial dinosaur bones near the quarry and, eventually, the quarry site itself.[41]

During the prospecting stage, producing knowledge through quarries closely resembled survey science in many respects, because of the need to cover a geographical territory through systematic travel by wagon, pack animal, and tent camping. Accordingly, although the focus of work effort in quarry science would ultimately be much more concentrated at especially rich sites, the practical experience of the prospecting phase overlapped with survey work. Quarry prospectors faced the same constraints of climate and utilized similar kinds of work organization. In rare cases, museum paleontologists even joined survey parties in order to prospect more systematically over large land areas. For example, the Carnegie Museum's John Bell Hatcher joined US Geological Survey parties examining the White River deposits of South Dakota for three weeks in June 1901 and then again in a "geological reconnoissance" led by survey geologist Dr. T. W. Stanton across Montana and parts of Canada in June and July 1903.[42] Yet quarry prospecting work was usually distinguishable in its aims from survey work. Above all, field researchers sought to locate promising places for removing large quantities of physical materials that would be taken elsewhere for processing. Rather than mapping for extensive coverage, therefore, quarry prospectors hunted for places that promised intensive extraction.

As for survey researchers, the patterns of movement prospectors followed rested on the foundation of the railroad system. As the primary conduits for bringing people and supplies in and out of the field, the railroads defined the baseline geography of quarry prospecting from the mid-nineteenth to the early twentieth centuries. Some notable quarries emerged right beside railway stations, such as Como Bluff along the Union Pacific line in southeastern Wyoming, a site initially discovered by the station agent and his partner, and the well-known "fish cut" near Green River, Wyoming, which had been created by the construction of the Union Pacific.[43] But most sites were not so fortuitously located, and prospectors struck out across the open landscape by wagon, by pack animal, and on foot. Even in such cases, however, the railroad infrastructure might offer shelter during the winter, as fossil prospector H. W. Menke found in 1897–1898, when "an old-abandoned section-house" near Aurora, Wyoming, provided "winter quarters" once "cold weather and snow rendered field work impracticable as well as very disagreeable."[44]

As field scientists moved outward from the railways, their patterns of quarry prospecting were dominated by a ranching geography. In such a sparsely inhabited and resource-poor region, scattered ranches constituted just about the only human points of reference. Prior to the spread of ranching into some areas, prospecting was even more challenging. On his "first trip into the Big Horn Basin in 1880," the paleontologist Jacob L. Wortman recalled, "the country was a wild, uninhabited region," with perceived threats from "roving bands of hostile Indians" that posed "no small amount of risk to one's personal safety." The army officer in charge at Fort Washakie refused to guarantee the prospecting party's safety, so that "collecting was not as thoroughly done as it probably would have been under less trying circumstances."[45] Ranches, on the other hand, provided not just human interaction and assistance but also an effective map to the vital resources needed to maintain prospecting expeditions in the field. A typical ranch for the region was described by one field scientist as having "buildings and barns and corrals scattered around higgledy-piggledy over two or three acres of ground, with a very pretty little stream which carries water all summer. . . . No garden and no pretense at neatness or order."[46]

Disordered though they might appear, the buildings of ranchers provided bulwarks against the elements when late spring blizzards and summer thunderstorms attacked. For example, when an American Museum paleontology party sought refuge at a ranch in South Dakota from a "pretty fierce" storm with "more thunder and lightning" than they had seen in years, they had "just got [their] horses in when [the rain] came down in bucketsful."[47] Western ranches also tended to be located at the best sources of the field's holy trinity of water, wood, and grass. In addition, ranches provided prospectors with a convenient way to order the landscape in their reports and letters, anchoring their movements on such a vast terrain with few landmarks. O. A. Peterson of the Carnegie Museum remarked in a typical reference to ranch geography: "We are camped a while this side of Bob Harris ranch 5 miles from here and expect this will be our address for some time."[48] As clusters of animals, buildings, and humans located along the rare oases of water and vegetation on a resource-poor landscape, ranches were central to the geography of quarry prospecting—as they often would be for intensive quarry operations as well.

Another essential aspect of ranching geography was the contribution of ranchers and their employees not just to the material basis of quarry prospecting but also to its cognitive side. As people who worked outdoors and traversed the terrain in their daily work patterns, ranchers and ranch hands knew the territory and its resources. Often this included ideas about where fossils, artifacts, and other quarry materials could be found. Ranch people frequently acted as guides to prospectors, offering their ample local knowledge of the terrain. Despite the fact that few of them had extensive formal education, their firsthand experience provided a different sort of knowledge that could be equally valuable, especially at the prospecting stage.

The pivotal role of the Wetherill ranching family of southwestern Colorado, for example, has long been acknowledged as crucial to the guiding of archaeological field parties during this period, not just at Mesa Verde but across the Four Corners region, though their relationship with federal government science agencies has been much debated.[49] Despite some brief earlier observations on ruins in the same area by members of the Hayden Survey in the 1870s and others, it was the

Wetherills who explored the area most thoroughly and intimately, after settling in the Mancos River Valley nearby. When Gustaf Nordenskiöld arrived in 1891, he immediately recognized the value of the Wetherills' experiential knowledge of the land around Mancos and Mesa Verde. "I must remain in the good graces of the Wetherills," he wrote home to his father, since they were "the only ones who know their way around this labyrinthine area."[50]

But the first outside explorer to document the ruins in the wake of the Wetherills' explorations was Frederick H. Chapin, who had visited in 1889 and 1890 with assistance from the ranchers, especially the two sons Richard and Al.[51] In *Land of the Cliff Dwellers*, Chapin gratefully acknowledged his dependence on their knowledge for prospecting. He called meeting the Wetherills a "good fortune" and acknowledged their vast store of experiential knowledge from following their "large herds of cattle, which roam through the valleys and over the mesas," especially their "long rides" southward to chase after stray cattle. He fully credited the Wetherills with the prior discovery of "many cliff-houses of great interest" in the multiple branching side canyons of the region, to which they directly guided Chapin. Together, they also explored new, "hitherto unknown ruins," but Chapin nonetheless credited the Wetherills for their superior local knowledge in identifying sites. In passing, Chapin also revealed that they were often "journeying by Indian trails," thereby revealing that at least some of the topographical knowledge for their prospecting excursions was transmitted materially from the region's former American Indian inhabitants, whose knowledge was encoded in those trails.[52]

Chapin's praise of the Wetherills was echoed by another learned traveler the next year, W. R. Birdsall, who boasted to the American Geographical Society of his "good fortune" to experience Mesa Verde "under the guidance of Richard, Al, and John Wetherill," who, "although not professed archaeologists," were "in possession of a vast number of observations and facts concerning them." In his view, "no one knows this part of the Mesa Verde as they do." Yet Birdsall's appreciation of the Wetherills by name was also mixed with anonymous criticism of the practices they and other local people employed. Later in his presentation—carefully distanced from his lavish praise of the Wetherills—Birdsall lamented that no field work had yet been "carried out according to modern scientific methods for ethnological and archaeological research." While he recognized that "tourists and prospectors" who "made a business of forming collections for exhibitions or sale," had, "in consequence of the time devoted to the subject," found themselves able to "come into possession of a much greater number of facts than other observers," it was nevertheless "unfortunate that their work was not done under competent archaeological advice" and "upon a definite plan of investigation" to improve the quality of the scientific knowledge produced. In Birdsall's view, mere "possession of a much greater number of facts than other observers," while commendable, was not enough.[53]

Still, these published accounts make it clear that even criticism of inadequately scientific practice had to be matched by a robust appreciation for the value of the experiential prospecting knowledge that the Wetherills and other local people possessed so abundantly. Moreover, as Chapin and Birdsall—and, to some extent, even the Wetherills'—reliance on prior topographical knowledge indicates, local knowledge need not have been specific to the materials that were being collected but could simply be of the general use, geographical sort. Another example of this comes from archaeologist Earl Morris, who in 1913 "hired as guide a Mormon cattleman familiar with the region" when exploring the La Plata country of southwest Colorado.[54] To be sure, prospecting researchers also made use of their own expertise to follow geological formations, cultural areas, and habitats to find likely places where valuable materials might be found. (And not all attempts to use local knowledge to find valuable fossil quarries paid off. When Hatcher heard from a "stockman" in Wyoming about "some bones that [he] had found," Hatcher rode 110 miles only to find "a weathered out & worthless skull of Triceratops!"[55]) But it is really quite remarkable how heavily quarry prospecting relied on local, experiential knowledge in its earliest stages. Indeed, as historian Paul Brinkman has pointed out, it was "common practice" at large urban museums, especially at the American Museum under the wealthy Henry Fairfield Osborn, to pay locals for identifying fossil sites.[56]

American Indians, and not just those who engaged in ranching, occasionally played a role in providing local knowledge to the quarry prospecting enterprise, such as along White River on the Pine Ridge Reservation, where one, Yellow Bird, is even mentioned by name in William Diller Matthew's correspondence with his wife.[57] Along Porcupine Creek, "various Indians" indicated the presence of fossils all through that area, which would become an important quarry site for the American Museum. "We are on friendly terms with the Indians," wrote Matthew, "who are very willing to tell us all they can about likely places to find fossils."[58] Sometimes the local intelligence was less useful. For example, at the Indian School two miles out of Rapid City, the American Museum party learned of fossils discovered in a ditch they were digging, although it turned out to be merely Pleistocene rather than an older and more valuable Miocene bed.[59] Occasionally, however, the advice might be completely worthless, as when "an old Indian," who was "pretty deaf" and unable to speak English—but also "a very decent fellow, [who] keeps himself neat and clean"—claimed to know the locations of many bone deposits but was apparently just seeking food handouts.[60] Within a few days, their "Indian friend, whose name is George Red-Rock," was "proving to be something of a grafter," and soon they were attempting to fool their "Star Boarder" by adjusting meal times.[61] Finally Red-Rock left ("*mirabile dictu*," wrote Matthew) to work elsewhere.[62]

As this anecdote suggests, the practical, everyday problems faced during the prospecting phase of quarry operations could be quite similar to those encountered in survey practice. For another example of the similar difficulties encountered in the management of field parties, consider the experiences of Carnegie Museum field paleontologist O. A. Peterson in 1912 while working near Chipeta Wells in Uinta County, Utah. For Peterson, his chief problems were "hot weather and camp cooking," which he tersely summarized to museum director Holland: "I was disappointed in my cook who left me after we got started from Vernal. On Green River I got hold of a boy who said he would cook and attend to camp. I found, however, that the boy did not fulfill the requirements of a cook and shipped him a day before yesterday. I am making other arrangements for a camp cook, and will try that a while."[63] In the interior West, much of it the sparsely settled domain of isolated ranches, hired laborers could be hard to find, during the prospecting phase or, as we shall see, during intensive quarrying operations. As for field surveys, then, the practice of prospecting for quarries meant managing an uneven traveling hired labor force in an area with very few local residents, and the vast domains prospected for quarries tended to be among the more remote areas within the region's geography, even in comparison to the varied sites traversed by survey parties.

To be sure, some prospecting did take place in nonranching landscapes, such as farms along the eastern edge of the region, where rainfall was just marginally sufficient, or near river valleys. In more thickly settled areas, however, any potential fossil quarries might already have been ransacked by settlers, as when a party from Kansas University visited Elam Bartholomew's farm in July 1885.[64] A more successful example of a scientific quarry in this farming area was the Sternberg, or

Long Island, quarry, in north-central Kansas. Likewise, in other forms of quarry science, such as archaeology, farming areas were especially prominent given the likelihood of past residential settlement by American Indians in areas suitable for agriculture. Occasionally, prospecting might even occur in areas covered with irrigated farm lands. Yet, like the rainfall farm belt, such areas contained few workable quarry sites. Moreover, such experiences brought their own environmental challenges, such as the mosquitoes that plagued an American Museum party in 1902, whose members "had to build smudges of sagebrush about every other night" and take "refuge . . . up on the dry plains, behind a smudge; or in bed with the 'tarp' pulled over your head and carefully made 'skeeter tight."[65]

The difficult seasonal climate with its harsh weather extremes often challenged prospecting in the interior West. Arriving at Laramie, Wyoming, in early April 1900, paleontologist John Bell Hatcher was impeded by a "blinding snow storm" that raged in the town when he arrived, with continuing snow for several more days. "So much snow is on the ground," he noted, "that it has been useless for me to take the trip to Ogden [Utah] or Cañon City [Colorado], since I could see little."[66] The weather "moderated" toward the end of the month, but by the beginning of May, Hatcher's party was again "held up" in the nearby town of Medicine Bow "by a blizzard, it having snowed continuously from Saturday night until Monday (last) night." (Without so much as a hint of noticing the irony, Hatcher was at that very time negotiating with Holland for an Antarctic expedition.)[67] These Pittsburgh researchers were finding that the unusually high altitude of southeastern Wyoming—much of it over seven thousand feet above sea level—made for unpredictable and sometimes wintry weather long after the onset of springtime in the East.

In addition to enduring challenging environmental conditions, fossil prospectors had to compete with one another for access to the best prospecting sites.[68] Military metaphors often infused the planning campaigns of the museums. In the spring of 1900, Holland wrote to Hatcher calling for "a council of war . . . [to] form our plans for the coming collecting season."[69] Hatcher's ambitious plan, endorsed by Holland, involved sending out three separate field parties, all to be prospecting in various parts of Wyoming.[70] There were shortcuts, or at least accusations of shortcuts. Eight years earlier, the geologist and federal science administrator John Wesley Powell had indicted Osborn at the American Museum for trying to "gather new fossils without discovering them himself," by poaching O. C. Marsh's experienced field assistants from the

US Geological Survey, who already knew where to look.[71] Whether this alleged motivation was strictly true or not—and quite possibly it was—experiential knowledge of local place might be as valuable an outcome of quarry prospecting as the fossils themselves.

In their reports back to their metropolitan supervisors, field prospectors often reported on the movements of other parties from competing institutions, when they could gather such intelligence. This could be true even when the other parties were not looking for the same thing. For example, O. A. Peterson reported from Harrison, Nebraska, that "Prof. Bruner and party of 4 from Neb. State Univ. is camped up in the canyon of Warbonnet cr[eek] hunting bugs."[72] Such instances could, of course, be interpreted simply as friendly curiosity about local scientific activities, especially in a wide open and sparsely populated area like the northwest corner of Nebraska. More often, however, reports on other parties pointed to the intense competition between institutions for access to rich quarry sites.

Mentions of competitors' field campaigns centered not just on their movements but also sometimes on comparative judgments of how well they were succeeding at bridging their scientific knowledge with their experiences of the local terrain. For instance, in May of 1901, Holland replied to Hatcher assessing his chances at success upon reopening a quarry site at Cañon City, Colorado. "When it comes to a scientific proposition," Holland bragged, "you will permit me to say that I will bet on you every time as against Osborn. However, this is of course said confidentially."[73] Thus, despite some setbacks in the Carnegie Museum's attempts to implement Holland's "war" plan in 1900, they tried a three-team approach again in 1901 with some success, but now spreading the parties across a greater geographical (and stratigraphical) area, including the Tertiary of northwestern Nebraska, the Jurassic of Wyoming, as well as the dinosaur beds of Cañon City.[74] The next year, after some disappointing results in Colorado, the third party was shifted to the Tertiary formations of western Montana.[75]

What might such prospecting expeditions cost? For the summer field season scheme Holland outlined for the Carnegie Museum in 1900, for instance, a grand total of $2,414 would be divided among three field parties, estimating between $735 and $857 for each of them. The three major expense categories for each party were projected to be personnel, supplies, and field transportation. Paying and provisioning field staff members were unavoidable expenses, of course. General travel expenses —presumably by railroad—constituted between $10 and nearly $100 for each party, but a much larger transport category, "pack animals/wagon," ranged from $150 to $260 for each party. This final category might involve the purchase of horses and wagons at a railhead disembarkation point in the field.[76]

Before the 1910s, almost all prospecting, like field surveying, was done with pack animals. At the beginning of the 1905 field season, Chicago-based paleontologist Samuel Williston's field collaborator reported discouraging news from Rawlins, Wyoming, the Union Pacific town where they would begin that year's prospecting: "Horses are up in the air. The owner of Holly asked me a hundred dollars for her. I paid sixty for her last year. They asked me from $150 to $200 for old teams that wouldn't sell in the fall for $75. I finally bought an outfit that wasnt quite so high. The team is as good as the one we had last year but slightly older."[77] Buying teams in the spring and selling them in the fall—and facing the transaction costs involved in annual exchanges—left prospecting parties at the mercy of adverse frontier market forces and inflation. On the other hand, paleontologists did not relish finding someone to care for horses over the winter or paying to feed the animals in a place where winters were often harsh with little forage available.

For the Carnegie Museum, field prospecting expenditures in 1900 and subsequent years were covered by a Fossil Fund sponsored by Andrew Carnegie himself, which supplemented the Museum's regular budget.[78] Carnegie chose to fund vertebrate paleontology in particular because of his strong desire to have impressive dinosaur skeletons on display, as well as to exhibit casts of the Carnegie Museum specimens as he attempted to gain status as a social equal with European royalty.[79] Even the Fossil Fund had to be wisely husbanded, as Holland implied when he urged against spending too much on a team of horses that might have to be sold later at a loss—"getting the hot end of the poker," in his characteristically memorable words.[80] When the earmarked fund dried up completely, field expenditures had to be curtailed accordingly. By 1915, for example, Earl Douglass at the Dinosaur quarry had to be satisfied with $300 per month including his own salary, even though he tried to request more.[81] The American Museum also enjoyed a special dedicated fund, initiated in 1897 by trustee William C. Whitney's $15,000 donation at Osborn's behest in order to study fossil horses (Osborn, Whitney, and other trustees were "avid horse owners and fanciers") through field expeditions, research, and museum display. Another wealthy New York capitalist, E. H. Harriman, sponsored a dinosaur expedition in 1899.[82] Similarly, archaeological sites in the region, such as the spectacular Mesa Verde in southwestern Colorado, benefited from the widespread public interest in such material remains, especially among wealthy potential patrons.

Land Control

Once a promising quarry was found, the next step was to ensure control of the site. While field surveys or traveling collectors on expeditions, with their short duration of residency and characteristic mobility, could usually operate without special permission—or with simple permission to camp for a few nights—the extended occupancy and extractive nature of quarry field work meant that it was crucial to secure control over the land.

Some quarries were on private land. And often the local owners were happy to allow scientific collecting on their lands, through courtesy. In contrast to economically valuable minerals and other materials, the objects found in scientific quarries were often viewed as curiosities free for the taking. Moreover, if specimens were on or near the surface it was nearly impossible to stop people from simply carting them away. Occasionally disagreements did erupt, however, and the more that a collecting site became a full-fledged quarry with long-term occupation, the more likely it was that the local owner might try to assert control, sometimes by requiring a cash payment for the right to quarry materials from private land. For instance, in 1900 at Cañon City, Colorado, Carnegie Museum paleontologist John Bell Hatcher indicated that he was quarrying for fossils "as rapidly as possible" since he was "paying $25 per month for the privilege," which would, however, be waived "if there are no bones."[83] By the early 1920s, such arrangements had become increasingly formalized. In one typed agreement, George A. Henderson of Sterling, Colorado, granted the Colorado Museum of Natural History "the sole possession of and the exclusive right and privilege of excavating and removing from his property, fossil skeletons of prehistoric animals," in return for monetary payment and a pledge "to avoid damaging crops, removing or damaging fencing," and to "fill up and grade all excavations."[84]

Cash or courtesy eased relations with many private landowners, but not always. One place where conflict did emerge was at the Long Island—sometimes known as the Sternberg—quarry in Kansas, where the land owner was a man named Mr. Overton. The quarry was first discovered by freelance collector Charles Sternberg in 1882. When Sternberg persisted in collecting after an initial warning, Overton objected and tried unsuccessfully to bring in someone to arrest him.[85] In 1884, Sternberg was again working at the quarry, this time for the federally funded fossil warrior O. C. Marsh, and joined by John Bell Hatcher, who was at that point a recent Yale graduate. However, they wisely secured Overton's consent by hiring him or his son to assist the party in excavation.[86] Later, Overton's son became Samuel Williston's paleontological assistant at the Kansas University Museum in the 1890s, helping the geology professor acquire and restore another skeleton from the "bone-quarry" on his father's land.[87]

Despite a few cases of fossil digging on private farms, most of the greatest quarries were public or quasi-public lands, which might be used by ranchers but were ultimately controlled by the government in some way. (That is, unless some intrepid dry-farming homesteader came along and tried to make a living on such marginal lands.) Understanding the dilemmas of eastern museums and other institutions seeking to maintain control over resources in the West thus requires an awareness of the land laws of the time, and the dangers posed by the claiming of homesteads on public lands. The most basic law was the Homestead Act of 1862, which permitted settlers to claim 160 acres of land for a nominal fee and evidence of improvement within five years. The Mining Law (1872), the Timber Culture Act (1873), and the Desert Land Act (1877) presented further opportunities to gain control over land in return for the discovery of minerals, the planting of trees, or the irrigation of the land, respectively.[88] In 1904, another trial effort permitting homesteads of 640 acres was the Kinkaid Act, which applied only in western Nebraska.[89] This legal expansion of homestead size was extended beyond Nebraska with the Enlarged Homestead Act (1909) for 320 acres and the Stock-Raising Homestead Act (1916) for 640 acres. In addition, some areas were American Indian lands, regulated through the Bureau of Indian Affairs. The landscape of the West was covered by the whole tapestry of federal land law, and quarry scientists could exploit it as much as anyone else.

Furthermore, quarry scientists, who were often based at or near the same location for many years in succession, might seek their own personal homesteads in the field. If homesteads were located before large-scale quarry operations, the geographical distance could prove to be a challenge. In 1902, for example, the recently hired Carnegie Museum paleontologist O. A. Peterson wrote from the town of Adelia—in the northwest corner of Nebraska near promising quarry sites, such as Agate Fossil Beds, which he would find two years later—that he "expect[ed] to make final proof on my Homestead June the 11th and shall then be foot loose to do or go where I choose."[90] Other letters around the same time suggest that the site was located near Long Pine, Nebraska, a few hundred miles to the east, and it was therefore important that he prove up on his homestead before basing himself semipermanently in the more promising region for quarry field work. Another homesteading field researcher was Earl Douglass, the Carnegie field paleontologist who managed the Dinosaur quarry. Douglass lasted longer than any of the others as a homesteader—in large part due to his many continuous years of work nearby managing the fossil quarry—though he never succeeded in his irrigation schemes intended to overcome the challenging arid climate of northeastern Utah. Moreover, his scientific work took so much of his time that the homestead suffered and eventually reverted to sagebrush.

Other scientists interested in claiming land took advantage of more permissive laws such as the notorious Timber Culture Act before its repeal in the 1890s. University of Nebraska entomologist Lawrence Bruner entered a timber claim in Holt County—located in the state's Sandhills region—in 1883–1884.[91] On the other hand, biological field surveyor Vernon Bailey was tempted by a timber land claim in Nebraska in 1888 but decided against it.[92] Unlike Douglass or Bruner, who conducted their research in the West all year round, Bailey almost always returned to Washington, DC, in the winter. Thus, deciding whether or not to take up a homestead claim on the side effectively meant deciding whether or not to live in the West. Choosing to maintain a homestead involved far too much work to avoid committing oneself to settling in the region.

While homesteading was sometimes pursued separately from scientific field work, the homestead laws could also be used directly to protect access to a favored quarry site. In other words, land laws could be a potential weapon in the arsenals of competing museum teams. When the Carnegie Museum reopened the quarry at Cañon City in 1900–1901, they ran into trouble with their local collaborator. "Mr. Felch was trying to play us false," noted Hatcher, "but fortunately I was on the ground in time & have nipped the whole thing in the bud so that I feel sure all will be well now." What had Hatcher decided to do? As he put it, he would "go into the real estate business," filing four land claims under Holland's name and his own, thus feeling "sure we have here the key to the development of the *Sauropoda*" and proposing "to corral everything in site."[93] Hatcher had apparently done the same thing the previous year, filing a claim (of unspecified type) for the Diplodocus quarry at Sheep Creek in Wyoming.[94]

Another option besides the homestead and other land laws designed for agriculture was the filing of mineral claims. The Carnegie Museum attempted this strategy to take firm control of the fabulous Dinosaur quarry in the Uinta Mountains of northeastern Utah. The logic of such a filing was straightforward: fossils were just like other mineral resources that could be extracted by their discoverers. "Being concerned for the continuance of the Museum, and apprehensive that some wandering prospector for the purpose of levying tribute upon the Museum, might enter to the tract upon which we were working," as the Museum's *Annals* later put it, the institution's field paleontologist Douglass was directed "to take steps to file a claim to eighty acres covering the site of the quarry, under the mineral laws of the United States, we having been advised by eminent legal authorities that such a procedure was proper."[95] But land office officials in Washington, DC, did not agree with this reasoning, and the claim was denied on the grounds "that fossil bones cannot be regarded as 'minerals'" despite their apparent similarities.[96] Holland vehemently disagreed with the negative decision, disputing the argument upon which the denial was based as "too preposterous to be considered for one moment" and "a most outrageous piece of quibbling" for good measure. "We are engaged in quarrying fossil bones," he pointed out, "and the opening we have made is a *quarry* and it is quarry land, and we take it under the law made and provided, as has been done in scores of cases before."[97] Despite Holland's protests, the mineral land claim strategy failed to secure long-term control of the quarry land.

Some desirable quarry sites were beyond the reach of public land laws. Among these were the American Indian reservation lands, designed to protect against incursions by white settlers. One might think that field scientists would be barred as well, but instead the government gave them special access for noncommercial purposes. Such access was granted not just to official government collectors but sometimes to the private urban museums as well. When Marsh attempted in 1893 to use his position as federal paleontologist to keep Osborn and the American Museum from digging fossils on the Uncompahgre Ute Indian Reservation in northeastern Utah, for example, Osborn used his connections—including a former secretary of the interior who was on the museum's board of trustees—to gain access. (Marsh had also tried and failed to use the same tactic on regular public lands in Wyoming a few years earlier.)[98] While American Indian lands were not open to quarry field work without special permission—and certainly many aspiring collectors would have lacked Osborn's high political connections—such access could be obtained by all the major institutional parties in the competition for western quarry sites. In March 1908, for example, with the support of US Senator and Rules Committee chairman P. C. Knox, the Carnegie Museum instantly received a permit "to search for the bones of extinct animals" on the Uinta and Uncompahgre Indian Reservations, which precipitated the discovery of the Dinosaur quarry by field man Earl Douglass during the next year's field prospecting season.[99]

Such approval was not always automatic, however, and the agreement of local American Indian agents sometimes had to be secured before quarry prospecting and digging could begin. In 1904, the Carnegie Museum sought permission to send the fossil-collecting party under Peterson, "with two assistants and four horses," to White River badlands on the Pine Ridge Indian Reservation.[100] Because the incursion of white Americans on American Indian lands was often "displeasing to the Indians and sometimes leads to trouble," the response from the acting commissioner of Indian Affairs was lukewarm. But here the commissioner drew a distinction between quarrying for fossils and quarrying for artifacts. If they had been looking for "pre-historic remains," they would have been refused a permit, but it was less likely "that the Indians [would] seriously object" to a search for fossils. So a permit would be allowed, with the consent of the American Indian agent at Pine Ridge, John R. Brennan, who would speak for the opinions of local American Indians on the matter.[101] Despite receiving approval, the Peterson party never really reached the Pine Ridge reservation, delayed as they were in their search for fossils in northwest Nebraska by the riches of the spectacular Agate quarry site.

At around the same time, the American Museum did end up actually opening quarries on the Pine Ridge Reservation, but their field researchers found that the permission of government agents did not always meet with approval from local American Indians themselves. At first the American Museum looked to open quarries on the Rosebud Indian Reservation in the summer of 1903. Here they had their first experience of confronting Sioux who vigorously guarded their reservation lands against the incursion of white men who did not have government-issued passes. American Indian policemen would "come over occasionally and inquire what we are doing and whether we have authority to be on the reservation," he noted.[102] Upon returning to the nearby Pine Ridge Reservation to open the Porcupine Creek quarries in June 1906, they again encountered suspicion from American Indian residents. An American Indian landowner "came 'round with his daughter for interpreter," and, Matthew believed, "the design of getting a dollar or two out of us for camping on his land." Translating her father's question, the daughter reportedly asked them, "Who tol' you sit here?" Moreover, the man "wanted to know how long we were going to stay and what we wanted and whether our people (families) were coming." However, when the American Museum party claimed to have the Indian agent's permission "to come and camp anywhere we liked," the father-daughter duo gave up and left. "Most of these Indians are quite good-natured about camping," judged Matthew, "but now and then one comes across these surly ill-conditioned fellows who want to make trouble."[103] Despite this occasional (and often understandable) hostility, the American Museum did manage to make use of permitted access to fossil resources on American Indian lands.

Besides fossils, field scientists also sought other scientific materials through the quarry mode of practice on American Indian lands, including especially archaeology and ethnography. Unlike fossils, human artifacts were generally more contentious as objects of scientific interest. At Mesa Verde, for example, the excavation of the ruins and remains was vigorously disputed by local Utes, and protracted boundary disputes resulted even after much of the area was established as a national park.[104] On the other hand, quarrying for cultural information through ethnography, such as stories—intangible, yet recorded materially on paper—was more common, and in fact it was very difficult to obtain such information without going onto American Indian land. Indian agents on reservations typically had to give permission for intensive long-term ethnographic work, which might extend beyond mere approval to include practical assistance. For example, when one of the principal ethnographers of the Great Plains Indians during this period, James Mooney, who worked for the Bureau of American Ethnology, traveled to the Wind River Agency in Wyoming in the early 1890s, he was aided in his efforts to meet with the Arapahos who lived just north of the agency. In this case, Mooney was allowed to use "a tent and cooking utensils, and furnished . . . with an interpreter and the agency clerk to act as guide."[105] Later, in 1901 and 1902, when Mooney was engaged in intensive ethnographic work with the "Kiowas and nearby tribes," the local Indian agency let him use "a government house" that had "originally been built for a government physician but had remained unoccupied for years."[106]

Once control over land was secured, it was crucial to protect and defend the quarry site against rivals, thieves, vandals, and well-meaning members of the public. Sites with materials on the surface were often visited by lay or amateur collectors before any publishing scientific researcher arrived. Local discoverers of quarry sites were often the first to discover such plunder, retroactively labeling it as theft or vandalism. Arthur Lakes of Colorado, for instance, brought a field party to geologically examine "some properties in the Florissant basin" and found "the remains of stumps of gigantic fossil trees" that were "8 to 10 feet in diameter." Previously, he surmised, they had "stood up several feet above the level meadow," yet had been "mostly razed to the ground by visitors and vandals." Their most remarkable find was a "monster" fossil tree still standing, with "the stone saws that attempted to cut it in sections" stuck in it.[107] Similarly, at the cliff dwellings on Mesa Verde, in southwestern Colorado, T. Mitchell Prudden complained that many local settlers would picnic around the ruins on Sundays and holidays and collect items.[108] The creation of local meaning through the collection and interpretation of surface objects was a hazard at many quarry sites, even before they were opened up and deepened, since such use competed with scientific research.

The danger of competition increased after the labor-intensive digging of the quarry, which revealed more material. Eastern museums often saw the advantage of keeping someone around even during winter to stand guard and continue the work of digging whenever possible. Institutions in the interior West, however, rarely had the funds for such ongoing operations. Consider the experiences of the Colorado Museum of Natural History in the late 1910s. With the permission of rancher H. D. Boyes near Wray in east-central Colorado, a fossil quarrying team from the Colorado Museum spent two years working the "extraordinarily rich pocket" of fossil bones at the quarry. However, lacking the staff to guard the site year round, the museum was unable to protect its treasures. "Unfortunately, vandals had already destroyed many specimens of priceless scientific value," museum paleontologist Harold J. Cook sadly reported, "and even after the expedition was at work collecting here, people slipped in, broke up, and destroyed and stole various specimens, including parts of the specimens herein described and figured." In explanation, Cook decried "the unthinking curiosity of people in general who find fossil remains," which "commonly results in the total or at least partial destruction of specimens of the greatest historic importance." Nevertheless, he held out the seemingly universal hope of all field scientists: "If the general public could only be made to realize the importance of reporting to some responsible institution or person, when they find fossils, instead of trying to collect or examine them unassisted, our general knowledge of pre-historic life would be greatly increased, and many specimens, the duplicate of which man may never find again, would be saved to science."[109] The Colorado Museum's experience suggests that while most of the perceived competition for quarry material came from other research institutions, the takings of amateur collectors—especially damage caused by those labeled as "vandals"—could also be significant.

The control of professional scientists over quarry sites was even more of a pressing dilemma in archaeology. Archaeologists grappled with the problem of pervasive "relic hunting" by local people or visiting tourists as a veritable crisis. Part of the reason that so much Mesa Verde field work was focused on restoring structural ruins was that so many of the artifacts lying on the surface had already been taken. The disdainful attitude of professional archaeologists toward amateur collecting was starkly revealed in a 1924 report written by Neil Judd, with fellow committee members Alfred Vincent Kidder and Earl Morris, for the American Anthropological Association. For them, the "exposed and unprotected condition" of artifact quarries constituted "a constant invitation to the passerby to excavate for chance curios whenever the opportunity admits." Like the Colorado Museum's fossil collectors, their comments also reveal a normative rift between institution-based professionals and local amateurs. "Most residents of the Southwest regard it as their inalienable right to dig for 'relics' in any ruin that tempts their enthusiasm. This spirit has rarely been contested," they remarked. Such locals also had easy access by virtue of their nearby residence. "Except in the more inaccessible districts," Judd, Kidder, and Morris lamented, "it is now extremely difficult to find a prominent ruin that has not been measurably mutilated."[110] The archaeological committee also complained of problems with traders on reservations, who encouraged American Indians to dig by buying their specimens. The committee leaders continued to press the government to more strongly enforce government regulations, which had been mainly intended and used to protect archaeological resources.

To justify the superior claims of institutionalized professionals over their amateur counterparts, as well as those who simply collected objects for sale, field scientists typically invoked not only the obvious preservation of the quarried objects in organized public collections but also the loss of environmental context and other related facts in the field. The archaeologists Judd, Kidder, and Morris, for example, contended "that all data connected with objects of antiquity recovered during illegal operations is wholly ignored by the commercial or amateur collector." And even "the objects themselves," while "perhaps temporarily cherished by the collector, are invariably soon broken and discarded or sold to any person offering a satisfactory price. In either case the specimens are lost to science and the unwritten story held within the ruins from which they came is rendered fragmentary for future investigators."[111] Such an argument would become increasingly common across the field science disciplines engaging in the quarry mode of research, as intensively site-based science itself transformed into a more elaborate, context-oriented mode of practice.

Archaeologists' efforts to use government authority to control access to ruins and relics can be traced back at least to the early twentieth century, when prominent archaeologists such as Edgar Hewett urged the adoption of regulatory measures to protect "prehistoric ruins" all around the Southwest.[112] Notably, such efforts brought together experts and professionals with supportive lay people and amateurs, so that historians cannot place all of the latter in normative opposition to institutionalized, regulated field science. The Antiquities Act of 1906, adopted in response to agitation by archaeologists and reformers (especially women's clubs), ultimately brought many quarry sites into the national park system.[113] Sites all around the Southwest were incorporated into the system, but not every effort to protect a vulnerable site succeeded. At the Big Horn Medicine Wheel, a prehistoric American Indian site in north-central Wyoming, a reformer named Mrs. W. H. Walling from the nearby town of Sheridan, Wyoming—armed with a resolution from the State Federation of Women's Clubs—wrote in 1915 to her US House representative, F. W. Mondell, asking for a fence to "protect the Medicine Wheel from 'curio hunters.'"[114] Mondell, in turn, wrote to the secretary of the interior, suggesting that the site be considered for national monument status.[115] Yet despite some support from government scientists, the effort was unsuccessful.[116] Other regional sites were designated as national monuments, including Hovenweep and Yucca House near Mesa Verde (which itself was already protected as a national park). Even the largest structures at such sites faced the dangers of what archaeologists regarded as intolerable theft and vandalism. In 1918, for example, Jesse Walter Fewkes worried that some large structures at the proposed Hovenweep National Monument might face vandalism, not from relic hunters but from "dry farmers [who] might take the opportunity to transport stones from these beautiful well-constructed buildings for their houses."[117]

Field scientists and their reformist supporters also sought some protection for quarry sites at the state level. Sites identified through collaboration in lay networks might even form the basis for identifying and controlling land containing archaeological remains and human artifacts, which could later be studied more intensively as scientific quarries.[118] Other protection efforts were more strictly the work of institutions and their professional staff members. One innovative use of the Antiquities Act was when the Carnegie Museum reluctantly agreed to support national monument status for the Dinosaur quarry site, for which they had been unable to secure a mineral land claim. Holland had "heartily opposed" such a "scheme" when it was first proposed in 1912, perhaps fearing that it would give federal government officials—including agents of the rival US National Museum—too much power over the site.[119] But seeing no other workable options to protect the land against land law claims, he eventually relented in 1915, spending some tense weeks in December of that year waiting for approval from the federal authorities. Holland did not relish subjecting his museum's operations to government decision making and the self-interest of Smithsonian scientists, descending to racial epithet to underscore his disgust, then speculating whether their work had been sabotaged "or whether the delay is simply due to the fact that we are dealing with large bodies, which traditionally move very slowly."[120] Nevertheless, in January 1916 the Carnegie Museum received permission to continue excavating at the quarry site under its new protected status.[121] Much like the more collaborative efforts of field scientists and their local allies, the Carnegie Museum's campaign to control a quarry field site entailed a significant element of uncertainty and improvisation in its outcome.

Working the Quarry

After a quarry site had been more or less effectively brought under the legal and social control of those who wished to exploit it for scientific purposes, the arduous physical labor could really begin. Whether digging out fossils or artifacts, quarry work had a certain rhythm to it. Removing specimens from a quarry, however, was often very hard, backbreaking work. Probing vertically into the earth and removing material—many of the objects being quite large—generally required more manual labor than any other kind of field work. A popular account in

1900 explained that developing a large quarry often involved "stripping away the bank to the bone level" by using "horses and scraper, and sometimes dynamite." The bones could then be carefully removed with "large awls, or hammer and chisel point," after which they were "thoroughly saturated with a solution of gum arabic, which upon drying gives to it greater strength and hardness." Finally, the bones were covered with tissue paper and "strong bandages of burlap saturated with ready-mixed plaster of Paris."[122]

At some of the largest quarries, such as the Dinosaur quarry in Utah, the excavation stage where the earth was stripped away could take a very long time. Even after the quarry had already been opened to a considerable vertical depth by the Carnegie Museum, when the Smithsonian Museum's Division of Vertebrate Paleontology came afterward to get a *Diplodocus* skeleton in the 1920s, the "work of collecting . . . occupied the energies of 5 men for 4 months to quarry out, transport, box, and ship the 26 tons of rock-enclosed bones. From first to last, 2,545 working days were spent on this specimen, which translated into Government time, means 1 man working steadily for nearly 9 years." In the estimation of a Smithsonian paleontologist, it was "the most ambitious single piece of exploration work ever attempted by this division."[123] And yet, considering that Earl Douglass of the Carnegie Museum spent over a decade with a team of laborers at the same site beforehand to unearth a variety of fossil specimens, many of them quite large, the Smithsonian's effort was not even the most intensive example of quarry science.

Paleontologists themselves engaged in considerable hard labor, so much that they might lose all traces of urban refinement in the field. Discarding the "effeteness" of wearing a coat and collar, Matthew described his field self to his wife in self-mocking terms. "The bone digger is about as disreputable looking a specimen as you'll often see," he observed," with his clothes all stained with flour paste and miscellaneous camping stains, an old felt hat and a battered looking pair of shoes." He promised to "get a couple of camp pictures for you while we are out, so that you can see what a tough-looking specimen you are really tied to," indicating to her the very material and bodily transfiguration effected by participation in intensive quarry labor.[124] For urban easterners, in particular, becoming a fossil quarry worker, even as leader of a field research team, entailed a marked transformation in identity and appearance. At the same time, the highest-status people in a museum, such as curators and directors, made only brief and rare visits to the quarries. They might even remain attired as easterners, as did Henry Fairfield

Osborn, whose brief field reconnaissance visits were made "often with his wife and one or two children in tow."[125] Such visits were important, nonetheless, for they might be used to smooth out tensions with local collaborators (e.g., when the Carnegie Museum's director Holland visited the Agate Spring Quarry in the summer of 1908) or inspect ongoing operations (e.g., when Holland visited the Dinosaur quarry in the spring of 1910).[126]

Most of the physical labor on site was done by museum subordinates and hired locals. The key distinction was between skilled scientific assistants who worked for the museum and unskilled labor for hard digging that usually came from the local area. Yet there was some blurring of the boundary line; some quarry field assistants even in the former category hailed from western backgrounds themselves. For example, Albert Thomson came from a family with a ranch in the southwest corner of South Dakota. Starting at the age of twenty, "as a cook, teamster, photographer, and general handy man" in the 1894 summer field team led by Jacob L. Wortman of the American Museum, Thomson gradually learned more about paleontology through the five following seasons, three of them at the Bone Cabin Quarry in Wyoming.[127] Eventually he was hired to work for the American Museum, although he still spent most of his time in the field.[128] Likewise, Hatcher corresponded with a potential field assistant in Denver, Colorado, named William Patton, who had sent Hatcher a handwritten letter of inquiry. Other paleontologists, however, including many scientific field assistants, came from the East. During the summer of 1904, for example, Theodore Olcott of Staten Island, New York, was invited to join Peterson's field party for the Carnegie Museum in Nebraska for the typical field season of mid-May to mid-October, at $40 a month.[129]

Hired laborers, by contrast, almost always came from the West, since museums were not willing to pay for the transportation of unskilled and uneducated labor from the East. Such local labor was often described with the same anonymity—and often in the same sentence or phrase—as draft animals.[130] Quarry workers were recruited from nearby railroad towns and ranches, or from frontier colleges and farms. Gustaf Nordenskiöld, excavating at Mesa Verde in 1891, got by with "John Wetherill as foreman and two Mexican labourers," even for his "excavations on a larger scale."[131] Nearly two decades later, however, at Mesa Verde's Cliff Palace in 1909, government archaeologist Jesse Walter Fewkes worked the site more intensely, with a "force of laborers, numbering on an average 15 workmen," from the nearby town of Mancos.[132]

When laborers with quarrying experience and proven work habits were available, they were especially prized by field research leaders. For instance, when John Bell Hatcher of the Carnegie Museum was approached in Harrison, Nebraska, by a man who had labored with him for three years during his earlier fossil work with O. C. Marsh, he was happy to have an "experienced man" help him. Offering the man $50 a month ("$25 less than he received from Marsh"), Hatcher described him as "a good workman, industrious, quiet & unassuming," even suggesting to the director that he be brought to the museum after the field season to help out there.[133] Field scientists excavating quarries thus prized in their hired men not only a strong work ethic but also a willingness to cheerfully subordinate themselves to authority without complaining or trying to claim higher status or prestige. Hired labor was just that; workmen should know their place.

Douglass's operation of the Dinosaur quarry was a noteworthy example of full-fledged labor management at a quarry site. In many ways, the hiring of labor at Dinosaur closely resembled the labor practices of stations, which will be examined in the next chapter, since workers were hired with the intention of employing them for several years at one place, and often seasonally. As a result, quarry laborers could have nearby homesteads with their families and sometimes viewed their employment as part of making a living in the community. The ample labor records kept by Douglass reveal how much he acted like an industrial or mining site manager, detailing the work hours of all his employees. He referred to most of them by their first names and tended to evaluate them on the basis of effort and obedience. In 1914, for instance, Douglass notes how a "good worker" named Clarence, who was his first employee back in September 1909, took up a homestead on the other side of the Green River. When Clarence "wished to work at the quarry," Douglass hired him again.[134] Each month, Douglass summarized the hours worked and pay earned for each laborer, as in August 1917 when Clarence Neilsen received $73.50, "Mr. Kiser 16 days at $2.73 = $43.70 with board 17 days $.645=$11.70, and Mr. Wm Schaefermeyer 13 days at $2.777 = $36.10."[135] By the early 1920s the quarry operation at Dinosaur was so much like a mining operation that it found itself obligated to comply with various Progressive Era labor laws. In 1921, the Carnegie Museum was compelled by Utah law "to purchase insurance for every employee in the quarry" and to post a notice specifying "that accidents must be reported within 48 hours to be covered."[136] This odd circumstance came about because, according to the Industrial Commission of

Utah, workers' compensation insurance was required for any "operating quarry with three or more employees."[137]

Few of the workers at Dinosaur had any opportunity to enter scientific careers from their labor at the quarry site, although in the case of J. LeRoy Kay, Douglass did support the chance for him to go to Pittsburgh to try out museum support work (he apparently failed to flourish, and gave up museum life).[138] To be sure, as examples like not only Kay but also Thomson, the frontiersman turned researcher, suggest, class and status lines were blurred or crossed in the field. Yet that did not mean that class tension was absent. Museum paleontologists accustomed to the urban culture of the East sometimes found the experience of working for long periods of time with less-educated westerners a frustrating (if tolerable) experience. "It would be more pleasant if the rest of the party were a little better companions," William Diller Matthew commented while leading a prospecting party near Sterling, Colorado, in June 1898, "but of course one can't expect everything, and though their language and ideas are not exactly Emerson's we get along very well."[139]

Despite their acceptable company, uneducated workmen were rarely trusted to do quarry work alone without the supervision of a trained paleontologist. Leaving "matters in the hands of the ordinary workmen" was simply not an option in the minds of museum directors.[140] In this regard, workers like Kay were exceptions to the rule, for Holland's general impression of the Douglass's workers was that "aside from K[ay], . . . they seem to be better fitted for growing alfalfa than extracting fossils for the Museum; no doubt good enough for the rough work, but when it comes to close work with the fossils themselves, doubtfully competent, unless constantly directed and guided as to their duties."[141]

Social tensions also existed further up the hierarchy, between field men and museum directors. At the American Museum, Osborn permitted career advancement through publication only for field leaders whom he viewed as sufficiently well educated. For instance, while he let Matthew publish his own research, he did not give Peterson that chance, contending that he "lacked a facility for writing," in Ronald Rainger's words.[142] The saga of conflict between Carnegie paleontologists and the frontier fossil prospector William H. Reed provides another example of social tensions that were based simultaneously on class and geography. After Hatcher forced Reed out following conflict over his inability to locate promising fossil caches and his unwillingness to work under the direction of Peterson, he justified the move to Director Holland in Pittsburgh by describing Reed as "without the necessary qualifications for the work" and therefore unworthy of the salary they had proposed to pay him.[143]

Holland's evocative reply even more starkly limned the class and geographical rift in scientific labor. "Reed is afflicted," he proclaimed, "as unfortunately so many partially educated and so-called 'self-made' men are apt to be, with an exaggerated idea of their importance and the value of their attainments." Far from being a unique case, Reed was merely "one of the long list" of such men victimized by what Holland called a "megacephalic disease," and he was "relieved to know that he is no longer on our force." The director then warned Hatcher "to be very careful in the selection of men," preferring "the man who has had the benefit of a liberal course of training in a literary institution," who was, in his view, "far less apt to be conceited and troublesome than those who have not had such advantages." Not only "attainments" but also a class-based social character mattered. Thus he also urged Hatcher to "look well into the spirit of these men," seeking out as personality features "docility, willingness, [and] the disposition of the soldier, who obeys orders." He explained, "Men who imagine that the holding of a minor position in an institution of this sort entitles them to assume the airs and to talk in the tone of men who have attained to scientific distinction and are leaders in the world of thought are to be avoided." Again, he emphasized that Reed was not alone in failing this test of attainment and character, referring to one or two other "unfortunate experiences" in which men "who, having published a page or two of their exceedingly insignificant observations, have suddenly blossomed out in their own estimation as full-fledged scientists, capable of criticising the opinions of everybody from a Kelvin to a Cope." Rather than such "bumptious, verdant youths," then, Holland instead cried out for "intelligent, capable, willing men possessed of good common sense."[144] Holland's lengthy tirade reveals not only his contemptuous attitude toward uppity, uneducated frontier collectors but also his desire to reinforce a strong hierarchy within the museum research staff, firmly centered on the metropole of Pittsburgh.

Another important category of workers was student labor, which came from both eastern institutions and universities within the region itself, who were closer at hand and often already accustomed to frontier conditions. The Carnegie Museum, for example, hired University of Wyoming mining engineering student Charles W. Gilmore in the summer of 1900. Judging that Gilmore possessed "a natural talent" for paleontology, which he had developed over four years in the region, Hatcher proposed hiring him for the field season from June through October, with the promise of hiring him permanently after his graduation the following year if the museum was "totally satisfied" with his work.[145] Thus, experience with students could also reveal if some among them had prospects for longer-term employment in the field sciences. (Gilmore would eventually have a long career in the US National Museum's Division of Vertebrate Paleontology.) In such cases, students occupied a hybrid position within the work organization of the quarry, combining the activities and subordinate status of hired labor with the potential for mobility into a scientific career.

At least as much as in paleontology, the centrality of student labor became legendary in archaeology, with Edgar Hewett's first archaeological field school in the Four Corners region in 1907 achieving nearly mythical status. The three student workers from that initial year—Alfred Vincent Kidder, John Gould Fletcher, and Sylvanus Morley—all had outstanding careers, though in Fletcher's case not in archaeology but in literary work. "Their adventures and misadventures are one of the primal legends of Southwestern archaeology," writes one historian.[146] Of the three, only Kidder returned for Hewett's second "field school" in 1908, but in that and subsequent years, many other students learned the practice of archaeology while serving as the labor force for excavation and restoration efforts throughout the region. The first "official" field school at Hewett's newly formed School of American Research in New Mexico was held in 1910. In one history of the school, Hewett is said to have "believed in the 'sink or swim' method of initiation to the hardships of the field," leaving his first group of students "with minimal supplies and an immense survey task."[147] The student field school model was widely replicated, with students accompanying the expeditions of Jesse Walter Fewkes, Byron Cummings, and others.[148]

Working as a student laborer was not necessarily a glamorous experience, however. Unlike the earlier college expeditions, whose participants enjoyed varied itineraries resembling vacation tours at a time when such opportunities were truly rare, students at quarry sites usually had to stay in one place and work systematically at exhausting or mind-numbing tasks. Fletcher, for example, during the legendary 1907 field school, found the work "not difficult, but somehow . . . monotonous. After a week or two of it I grew weary holding the tape measure while Kidder drew plans of some ruined building and Morley checked off the measurements." Quarry work could be rather tedious and unpleasant, he was finding out. "Slowly during these weeks there was growing up in me an increasing resentment against the toil we were undertaking," Fletcher remembered, "and against Dr. Hewett in particular for making us do it." He wondered: "Why did he keep us, day after day, working over the same old dreary piles of rock, finding nothing of any interest, when all about us, as our eyes could see, lay a magnificent, a marvelous country?"[149] Given the lack of appeal for such work, perhaps it is not surprising that most quarry labor ended up being done either by students aspiring to scientific careers or simply hired workmen who were paid market wages.

Only rarely did scientific quarry operators in the Great Plains and Rocky Mountains rely on American Indian labor, despite its importance in many other parts of the West. Over a wide swath of the Plains and Rockies, this was in large part because most American Indians had been removed to reservations on the edges of the region by the late nineteenth century. A few quarry sites were close to American Indian populations, including Mesa Verde in southwestern Colorado, which had been partially carved out of the Ute Indian Reservation. However, even at Mesa Verde, officials did not initially view American Indian labor as a viable option. Mesa Verde National Park director Thomas Rickner held deeply racist views in common with many Colorado settlers of that time, which he expressed in 1914 when he formally opposed the idea of leasing park grazing lands to the Utes: "The Indian has no idea of cleanliness or sanitation," he complained, "and no power on earth can control the impulses of the present-day Indian." To employ them, in his view, "would require a much greater amount of policing" in the park, "and then conditions would not meet the approval of the traveling public." While he acknowledged that "a very few" were "educated," he saw the typical American Indian as "the same uncivilized being that he always has been," lacking understanding and "regard" for "the white-man's laws," especially rules against taking the park's resources. "They are a remnant of the past, left over," he concluded, "and do not, as yet, fit into present conditions."[150] For a park administrator holding such derogatory views, it is predictable that he discouraged archaeological field work teams from hiring American Indian laborers.

Nevertheless, the labor shortage during World War I eventually compelled even Rickner to see the merit of opening up the lowest-status and least-skilled jobs to American Indians. After noting that "the pay of all workmen has advanced here as well as elsewhere," he told Fewkes that he anticipated employing "a good deal of Indian labor on the road work." He did not, however, endorse their suitability for excavations. "The Indians are good workers," he admitted, but only "in such labor as you can trust them to do."[151] A later park superintendent, the archaeologist Jesse L. Nusbaum, apparently did begin to employ American Indian labor more widely in the park. When local workers were not available in 1924, he brought in up to 148 Navajos.[152] In the aggregate, beyond just archaeological quarry labor, Navajos made up "more than ninety percent of the park's unskilled labor force during the 1920s."[153] Still, through the 1910s, Indian labor at quarry sites was rare, even in those few cases when their reservations were nearby.

Whatever the source of the hired labor they were supervising, an important tactical dilemma facing field scientists who operated quarries, with their typically multiyear operations, was whether they would remain in the field during the winter. If field researchers and labor supervisors did return to urban centers, sometimes locals such as ranchers would help them out by keeping equipment during the winter. In 1900, for instance, John Bell Hatcher judged from his camp on Warbonnet Creek, twelve miles north of Harrison, Nebraska, that it would be "wise to continue both parties in the field as long as the weather holds good." Work organization was thus closely intertwined with environmental conditions. "We are having fine weather at present & are doing very effective work here & at Med. Bow," he proclaimed, noting that they were "getting a fine series of White River mammals" from the quarry. He therefore "decided therefore to work on" at the site through early October if the director concurred, "then leave Mr. Utterback to go on with the work here until about Nov. 1st or such a time as the weather suggests his closing up, when he will close up the work & load the boxes in the car ready for shipment, leave our outfit here with a ranchman to winter & proceed by rail to Cañon City to take up this work of stripping there."[154] By staging the autumnal departure in descending order of the field hierarchy, as many quarry parties did, he intended to rationalize the organization of work according to anticipated outdoor conditions, concluding with the disposition of field equipment, especially horses, to the supervision of locals. But such arrangements were not always free of conflict. When O. A. Peterson of the same institution came to Harrison—another storage and embarkation point for seasonal field work in the region—in early March 1901 he "found [the] outfit all ok," although he "had to send to Omaha for one." Yet "Mr. Keel is kicking because he expected us to take out the horses Apr. 1st. He seems to think we have rubd it in to him for about one months pasture bill."[155]

Back in the metropole, where the rich materials from western quarries were typically brought for further processing, study, and display, the work process initiated in the field continued indoors. This typically meant considerable further material transformation in the preparation laboratory of a museum. Although such laboratory work is not the focus of this book, the interconnection of lab and field practices as sequential parts of the same work process—not to mention that many skilled field men were also lab preparators as well—makes it a topic worthy of at least brief consideration. For instance, Peterson not only was a field collector of fossils for the Carnegie Museum but was viewed by Hatcher, the lead paleontologist, as "a rapid and skillful workman in the laboratory." Since he "need[ed] Peterson very badly in the Museum," Hatcher also recommended hiring another field assistant named Mr. Utterback to help with fossil quarry work at Cañon City.[156]

The material technology of the fossil preparation laboratory is a complex subject, and here there is space only for a few examples, which perhaps can also offer something of an instructive contrast with field practices, for the environmental setting of an urban museum laboratory was obviously very different from the distant western field.[157] Yet despite the great shift in environmental context as scientific objects crossed the threshold from the field to the lab, in both cases the encounter of skilled hand and head workers with natural materials, mediated by instruments of various kinds, was central. Just as new technologies transformed field work during this period, so too did new instruments of the museum lab. Consider the innovation of pneumatic tools, as described by geologist Elmer S. Riggs at the Field Museum in Chicago. "The tedious work of removing fossils from their matrix by means of the hammer, chisel and awl," he reported, "has led to various experimentation with machine tools in the hope of devising some more rapid method. The dental engine and the electric mallet have been in use in some laboratories for a number of years, and have proved very efficient in such work as the removal of hard matrix from small skulls. However, their efficiency has so far been limited to light work."[158]

In touting the instrumental developments within his museum, Riggs was implicitly acknowledging the physicality of fossil preparation work in the indoor setting of the laboratory. He included a diagram and description of the tool, observing that the hammer's "advantages . . . over the old-fashioned hammer and chisel are its much greater cutting capacity and its freedom from the jar which causes so much breakage in specimens encased in hard matrix."[159] Although he was clearly quite familiar with the material technologies, the experiential knowledge of their use would have been developed by the fossil preparators under his charge, whom Riggs referenced only obliquely, noting that "skill in the use of these tools is readily acquired. By adapting the size of the chisel to the work in hand and gauging the amount of air admitted to the tool by means of a push-button throttle valve, the stroke can be reduced so that a scale may be removed from the most delicate surface."[160] In moving between field and lab, the practitioners of quarry science knew and transformed nature through their very physical labor in both settings.

Not all museum personnel moved from field to lab, however, since the richest quarry sites often stimulated year-round work. Many vertebrate fossil quarry sites, for example, were in foothills and hogback ridges beside towering mountain ranges—and thus at relatively low elevations compared to the surrounding terrain—so that work could be continued intermittently through the winter.[161] Probably the most sustained labor at any quarry site was that performed at the Carnegie Museum's Dinosaur quarry, where immense dinosaur fossils were buried in a stratum inclined at a steep "angle of 60 degrees to horizon." In a retrospective account written by museum director Holland, the quarry takes on all the attributes of labor and capital—and environmental transformation—of a conventional mining site. For the museum's agents "to remove the overlying shales and rock" in order to "expose the bone-bearing strata" necessitated "much labor and consequent expense." In addition to hiring "men from the neighborhood," they built "a track of rails, such as are used in coal-mines," along with "a dumping-car."[162]

In describing the actual labor process, the intensity of quarry operations became even more palpable, as Holland's synopsis makes clear: "We tore away the rock covering the remains imbedded in the lower layer of sandstone. We went down from month to month and year to year, and finally, at the time we ceased operations, we had made an excavation nearly six hundred feet long and eighty feet deep from the point where we had begun the process of uncovering." Besides Holland's deployment of a typical, if fanciful, managerial-industrial metonymy that implicated himself in the agency of physical labor, we can also see in this set of operations a tangible manifestation of the fully developed and intensive variant of the quarry mode of field science. In Holland's assessment, it had most likely been "the largest undertaking of its kind in the history of paleontology."[163] While many other quarry sites were much smaller, and thus with correspondingly more modest concatenations of human labor and transport technology, the Carnegie Museum's year-round work program reveals in magnified form the ground-level (and below-ground) realities of quarry science.

The experience of Earl Douglass and the Carnegie Museum also demonstrates that an intensive, long-term quarry excavated deep in the ground was made possible not only by the deployment of copious human labor but also through novel technological interventions to transform the environment, such as explosives. "The man who uses the giant powder can do as much as 4 or 5 men with pick and shovel," Douglass commented shortly after beginning excavation work at the Dinosaur quarry.[164] By that time, he had already "engaged a man who thoroughly knows the art of using powder," using this more powerful technology to deepen the quarry and increase the pace of extraction. The technological system he devised integrated explosives with animal technologies and human labor: "We can use horses, plow, scraper and dynamite to advantage without damage to the specimen," he declared, "and remove dirt quite rapidly."[165] Even during "extremely cold weather" during the winter, when other quarry work was difficult or impossible, he found it possible to continue working with "giant powder," since "the rock is so hard in places that it is useless to try to do much without it."[166]

Likewise, even field work that might be subsumed under the quarry mode of science despite its not involving actual physical and vertical penetration into the earth, such as ethnography, could involve special tools to facilitate long-term intensive extraction. Such was the case for James Mooney's ethnographic work with the Caddos and Wichitas served by the Indian agency at Anadarko, Oklahoma. Over a period of several years in the early 1890s, Mooney not only recorded their songs on paper but also began making use of a novel scientific instrument: the "Edison Graphaphone," or phonograph.[167] So did Alice Fletcher while recording the Hako ceremony, with the help of a man named Edwin S. Tracey, who possessed the technical capability to transcribe the music "from the cylinders" for further verification and study.[168] The ethnographer who first published on using the phonograph, in 1890—none other than Jesse Walter Fewkes, the Mesa Verdean archaeologist—analogized the use of phonograph cylinders in the study of language to the specimen collected for the naturalist.[169]

An even more elaborate sound-recording technology deployed at ethnographic quarries—usually at particularly rich local sites where cooperation could be obtained—was introduced by the ethnomusicologist Frances Densmore, whose pathbreaking studies of American Indian music in the early twentieth century included intensive work on several tribes within the Great Plains and Rocky Mountains regions. Densmore's multiyear project on the Teton Sioux from 1911 to 1914 recorded music mainly on the Standing Rock Indian Reservation in South Dakota.[170] In subsequent quarrylike field data collection, "recorded among the White River, Uinta, and Uncompahgre bands of the Ute in 1914 and 1916" and "conducted chiefly at Whiterocks, Utah," Densmore tried out a new device called the "phonodeik" that enabled the production of "tone photographs," which were then analyzed by a Cleveland physicist.[171] If handwritten paper notation technology was the equivalent of digging in the ground with human power, then the graphaphone and the phonodeik were its more powerful, earth-blasting successors.

Notably, Densmore herself made an explicit connection between her own metaphorical cultural excavation and the physical vertical excavation of other quarry sciences. She later recalled thinking of her work as an "archeology of the mind," in which she "was digging down into the *minds* of old Indians, going down through one layer after another until I got down to what they remembered of the oldest traditions." In Densmore's analogy, the "old Indians" she recorded "had something worth while at the bottom of their minds."[172] And just as explosives and other technologies of intense physical extraction remade fossil quarry science, so too did recording devices in the realm of cultural extraction. The two examples may seem rather different, but the central point is the same: deepening exploitation of any resource for scientific use, whether literally in the ground itself or in the figurative fertile ground of human culture, meant developing or adapting new technologies for relating scientific work organization with its material or experiential substrate. Going vertical often required technological innovation.

We can clearly anticipate and observe the reactions of human informants who heard their own voices singing back to them, but we might also consider the response of the rest of nature to the deployment of increasingly intense technologies of extraction. Given the magnitude of environmental transformation wrought by some quarrying operations, it should not be too surprising that occasionally the earth talked back. In the spring of 1917, for example, the Dinosaur quarry in Utah experienced a "serious interruption" due to the "sliding down of a very large mass of crumbling sandstone and rock . . . which, loosened by rain, obeyed the law of gravity and fell."[173] Just as in the coal mines, the instability of the earth being carved apart was an ever-present danger at the Dinosaur quarry (although in this case, no one apparently was hurt). When one gazes at the reconstructed dinosaur skeletons on display in large urban museums such as the Carnegie Museum in Pittsburgh, one should not forget that the earliest work that put them there had as much in common with a risky coal mine as a laboratory.

Quarry work was also undertaken in the vicinity of other sudden hazards besides the movement of the earth itself. Among the biota of the High Plains and foothills, rattlesnakes were perhaps the most troublesome creature. On one "very hot and sultry" day in July 1906, paleontologist William Diller Matthew of the American Museum killed and skinned "another rattler, and the smell of the varmint almost turned me sick again."[174] Snake bites were not uncommon, and the hazards of dealing with the omnipresent rattlesnake were important to quarry workers, who were, after all, tearing apart the very earth in which those creatures resided. Less venomous, but nonetheless irritating, creatures also came up from underground, including the field mice that tormented Arthur Lakes and his field quarrying team at Como Bluff, Wyoming, in 1879. They swarmed Lakes's tent, and to build nests they "tore up [his] volume of Hayden's surveys" and pilfered his packing cotton.[175] Other unpredictable environmental hazards whose appearances punctuated the steady, hard labor of quarry excavation included thunderstorms, which were dangerous, especially when they involved hail or tornadoes. (At the same time, such sudden, ferocious acts of nature were also essential for recharging dry streams that quarry workers depended on as a source of water.)[176] Local people might also disrupt operations through their own exertions of will, regardless of whether or not they were the intended object of study. Working a quarry involved constant immersion in local contingency and chance.

The environment talked back not just in sudden, violent interruptions but also in the ongoing natural and cultural rhythms that surrounded quarry field work. The extreme seasonality of the interior West meant that environmental influences palpably shaped the way of life of quarry researchers and their employees. Heat and sun were omnipresent burdens in the working lives of quarry scientists, their assistants, and hired laborers during the summer season. Almost always, members of the party needed to carry lots of water. Many quarry sites were located inconveniently far from water sources, such as the Mug House on Mesa Verde, where Nordenskiöld worked in 1891.[177] The experience of excessive heat was a source of frustration for field parties. "It is hot and hotter each day and promises to be another buster today," wrote Matthew, who noticed a "queer smokiness about the air" and recorded temperatures of 100 degrees in the shade, 116 degrees in the sun, and "considerably hotter" in the badlands fossil quarry work area.[178] Wind was also a constant presence on the treeless plains and eroded formations. "This afternoon it was a very furious blow," reported Matthew in July 1906, adding that such conditions make "it unpleasant to work, for the wind nearly blows you off the badland ridges and makes it hard to concentrate your attention on fossil hunting. . . . It blows the dust and scraps of rock into your eyes and tries to pull your hat off and whips away your tools and generally is as obnoxious as it knows how."[179]

The toll of the heat and the strain of work in such a challenging environment could have negative impacts on the bodies and health of field researchers. Scientists often fell ill in the midst of the intense summer field season. Hatcher, for example, was so concerned with his health in July 1901 that he wrote to Holland for advice. In reply, Holland offered a folksy nugget of wisdom derived from his "long ago" observation of "an Irishman working in a railroad cut": "He swung his pick slowly, but in the end, though apparently working slowly, he accomplished more than the man who rushed in and soon exhausted himself by his exertions." Drawing on this example, Holland urged that Hatcher "not take any undue risks, and, get well." It might not be an exact analogy, but the underlying principle was the same. "If you are feeling badly," Holland advised, "go to Hot Springs and get proper medical attention, and let the fossils go."[180] A few weeks later, as the "terribly hot weather" continued, Hatcher finally bailed out, asking for a railroad pass so that he could return to Pittsburgh.[181] Likewise, during the same summer (1904) when he found the Cooks of Nebraska and their amazing fossil beds, Peterson was suffering from health problems due to his exertions in the field (exacerbated by "greasy camp cooking"), and he desperately wanted to return despite Holland's entreaties that he stay in the field.[182] For scientists engaged in quarry work, field labor was at once both mental and physical. If men like Hatcher were educated producers of scientific knowledge, they were simultaneously manual workers in a harsh environment.

Despite the shared material experiences of workers at different status levels of the same quarry environment of the interior West, social tension and hierarchy were still ever-present realities. In some cases, conflicts crossed the boundary line between museum and field. Running competitive and ambitious operations often involved strong authority from the metropole that grated on field men of even the highest status. Holland, the Carnegie Museum's director, was an especially difficult boss. Hatcher once complained bitterly to him of the "personal abuse visited upon [him]" by Holland during a meeting in Pittsburgh, during which Holland apparently "called [Hatcher] a jack-ass & a d—d fool." In Hatcher's judgment, such strong words "cannot but tend to destroy that harmony, enthusiasm & interest so essential to the welfare of the institution."[183] These examples underscore the claims of historian Ronald Rainger, who has highlighted the frustrated ambitions of field paleontologists like Hatcher and the American Museum's Jacob L. Wortman, emphasizing the hierarchical structure of the discipline in big museums with strong directors.[184] Such evidence thus prompts us to consider not just the organization of field work on the quarry site but its relationship to a larger, long-distance geography as well.

The Geography of Quarry Knowledge

Status tensions between museum and field overlapped with the geography of quarry knowledge. That is, fault lines such as director vs. field paleontologist, and scientific vs. hired labor, were often correlated with East vs. West identification. What made quarries significant was not just how labor was organized in the field; it was also how such an extractive mode of producing knowledge enabled the metropolitan institutions of the East (and occasionally overseas) to exploit the natural resources of the West. Thus large urban museums removed the largest, most scientifically interesting, and best-preserved specimens to their own exhibit halls and research chambers for value-added processing into credible knowledge rather than circulating those specimens within the region to build institutions there. In 1920, Freeman Ward, the state geologist of South Dakota, lamented the "distressing fact" that "the institutions of South Dakota do not have fine collections of their own material," noting the high value of fossils extracted from South Dakota and estimating the worth of South Dakota fossils in larger museums outside the state. "The collection of the State Survey is decidedly ordinary not to say poor," he worried.[185] Examining geographical patterns of specimen movement is crucial to fully understanding the practices and consequences of quarry science.

Whether quarried materials would stay in the region or leave it proved to be a matter of significant debate. One of the more divisive and well-known conflicts over objects removed from a scientific quarry was between the Swedish traveler Nordenskiöld and local residents in Durango, Colorado, who resented having artifacts from quarries on Mesa Verde taken out of the country.[186] "Some slight difficulties which had previously been thrown in my way on the part of the authorities, but which I had imagined were already obviated," Nordenskiöld obliquely wrote, “were now renewed in a more serious form.” Although he was forced back to Durango, the charges were “formally abandoned.”[187] Despite some initial efforts to use the police to stop him, it turned out that there were no definite laws against expatriating the materials. In the meantime, however, he wrote home to his father to ask for emergency funds, since locals had “begun to oppose my excavation and work in a way that makes it desirable for me to soon leave this area.” As he succinctly summarized the sentiment prevailing in Durango: “Americans would rather that cowboys, miners, etc., dig amongst their antiquities, than foreigners.”[188]

Controversies over the export of scientific materials out of the region erupted periodically in state legislatures, and they heightened the demands of regional scientists for better patronage from their characteristically stingy and poorly funded state and local governments. For instance, Erwin Hinckley Barbour of Nebraska declared the need to keep fossil resources in the state in his 1912 appeal for a new state museum building. The existing museum building was too crowded already, he explained. Specifically, he cited collections that had to be “turned down and instead went to other museums,” such as the state’s “largest collection of minerals and crystals,” which went to Oberlin College in Ohio. Meanwhile, he observed, collectors from Harvard were scouring ancient sites around Omaha, so that “Nebraska material” was going “to enrich those of other states having readier means and more ample buildings.” He was only just getting started. “Now we are put in the ridiculous position,” he complained, “of sending to New York for two old and weather-beaten native Nebraska bison, the only representative of the most remarkable herds that ever roamed the plains of Nebraska. Our state has not saved even a single representative of its native animals, the elk, deer, antelope, or bear.” He followed with a section heading dramatically headlined, “World-Famous Nebraska Collecting Grounds Stripped By Others,” and sounded an ominous warning about eastern museums posting “one or more collectors permanently in Nebraska with the result that more choice collections go out of the State each year than go to our own cabinets each decade.”[189]

Such tensions—and the heated rhetoric they inspired—often strained relations between western and eastern institutions. One way in which the largest urban museums attempted to mitigate resentment of smaller players was to form cooperative relationships with them. Thus, once the American Museum or Carnegie Museum had opened up quarries and taken out the material they wanted, others might be permitted to follow up in the same locations, effectively as gleaners after the main harvest. And like those gleaners in a wheat field, less wealthy institutions sometimes anticipated useful treasures that were quite desirable to them. Frederick J. Pack of the University of Utah hoped "to extract one or possibly two fossils" from the Dinosaur quarry after the departure of the Carnegie Museum in the early 1920s.[190] At the Agate Springs quarry in Nebraska, the American Museum cooperated with the University of Nebraska to make sure that, while many of the best fossil skeletons would still be taken to New York, some materials could stay in the state itself, including one of the site's fabulous *Moropus* skeletons.[191]

Despite such cooperation, however, protectionist forces within the State of Nebraska, led by the State Federation of Women's Clubs, conspired in the late 1920s to bar fossil collectors from out of state. Harold Cook wrote his friend William Diller Matthew, by then at the University of California, informing him of the legislation.[192] In fact, some states had already enacted similar laws a few years earlier to keep archaeological materials from leaving their boundaries. Researchers associated with national institutions predictably opposed such measures in the strongest terms. Bureau of American Ethnology archaeologist Neil Judd wrote to another federal government scientist, "It is my personal conviction that the future of archeologic research within the United States will be seriously jeopardized unless a more generous, or national, attitude is fostered within the respective states." In this instance, he was careful to avoid attributing such misguided rhetoric directly to professional scientists. "State officials and local enthusiasts," as he put it, "ignoring the fact that prehistoric man was not governed in his migrations by our present state boundaries, have shown an increasing disposition to limit archeologic explorations to residents of their respective commonwealths." He blamed "Alabama and Ohio," in particular, for having already "taken the most far-reaching steps in this direction," worrying that "other states have attempted similar prohibitions," while also noting that some "have passed protective laws which are not objectionable." Even if the possibility of further efforts "in several localities to introduce laws that deny non-residents the right to examine prehistoric sites or purchase archeologic objects" might be due to the well-meant "plain desire of those interested" in stopping "irresponsible digging," he foresaw that "the laws favored would also deny to non-resident investigators, however competent, the privilege of pursuing meritorious researches." Judd further argued that most other countries claimed archaeological sites and artifacts as national patrimony, "even when found on private property." In his view, "a nation's prehistory belongs to the whole nation," not to the individual states that might be tempted to enact protectionist laws.[193]

In paleontology, likewise, the response from institutions undertaking field work on a national (and international) scale was similarly hostile. Matthew's reply to Cook's warning letter, which was noted above, reveals the rhetorical tactics that scientists used to defend themselves against such maneuvers. Matthew was "very sorry" to see restrictions advocated "by so public-spirited an organization as the state federation of women's clubs," who, he judged, "do not understand what it really means. You and I do." He grounded his disdain for such laws in the widespread early twentieth-century elite discourse of race and civilization. Restrictive expatriation policy, he noted, had "long between adopted by inferior, petty and semi-civilized nations," such as "China and Russia, Greece and Turkey and Egypt, [and] most of the Central and South American countries," whereas "the higher races, the United States and Canada, England, France, Germany, the Scandinavian countries, Switzerland and others, have hitherto been broad-minded enough to avoid such practices." Such a descent into barbarism was not only a slide down the hierarchy of civilization but also inexpedient. "The prospect of the State of Nebraska taking care of all its fossils is absurdly impossible," he declared. "The practical result of such laws has always been to hinder scientific exploration by anyone, and the fossils, instead of being of use to science and adding to the reputation of the state, are destroyed by rain and frost and erosion and are valuable documents of science lost forever through selfish and ignorant laws devised by petty politicians, and backed all too often by those who ought to know better."[194] This last line may have been a veiled reproach against Barbour (who was both a colleague and Cook's father-in-law). Although Cook replied with reassurances that Barbour had not supported the idea, Matthew can perhaps be forgiven for wondering, given the overt appeals to state pride and protective sentiment in many of the Nebraska geologist's writings.

Individuals faced the same geographical dilemmas as institutions. The danger for people working in the field who desired scientific status was that so much of the material was physically taken out of the region before being processed in any way to produce publishable knowledge. Such a knowledge production sequence—similar to the "commodity chains" analyzed by world historians in which raw materials are extracted from one country to be rendered into more valuable products elsewhere—defined a geography of circulation that generated tension and resentment. Some field scientists did manage to move quickly and effectively from field to publication, such as John Bell Hatcher, who, according to fellow paleontologist Wilbur C. Knight, had "surpassed all records of Paleontology works in the U.S. for getting material from the field into press."[195] Many field workers, however, might indeed wonder if they were being treated fairly when scientists in the metropole took the specimens they had found, excavated, and removed, and turned them into publications without the proper credit going back into the field. Over the years, hints of the Carnegie Museum field paleontologist Earl Douglass's resentment trickled back to his counterparts in Pittsburgh.[196] In 1917, Peterson—himself an accomplished field man who had made the transition to the metropole—felt obliged to bring up the delicate issue, writing Douglass directly. "Have you ever intimated to any one," he asked, "that you are displeased with the idea that I have taken up the work of the description of the Uinta material which you have collected for the Carnegie Museum? If you have I shall regard it as a guide and try to act accordingly. If not I shall feel free to go on with one or more papers which I expect to present before the paleontological section of the A.A.A.S."[197] While scientists might appropriate the labors of their hired field workers with relative impunity, viewing it as wholly absent of any cognitive component, they occasionally might have to tread more delicately with a trained colleague who was doing similar work.

One thing that distinguished the local commitment of quarry researchers as opposed to their survey and network-building counterparts was their propensity to build semipermanent structures. Field scientists at quarries often located themselves on site for several years, even enduring the harsh winters to continue working—and perhaps more importantly, protecting valuable resources against theft or appropriation by others during the off season. As a result, tangible buildings similar to those that would have been constructed by miners or ranchers sprang up around the quarry landscape. In some cases these structures were built for some other purpose and merely reimagined as useful in quarry work, as when an American Museum field group sought refuge from hail and rainstorm "over night in a little shack that Albert Thomson has built on a quarter-section of the land that he owns on the bank of the Cheyenne."[198] Another example was the homestead cabin built by Harold Cook at the Agate site in Nebraska, which was later used to house paleontological parties.[199]

In other cases, however, these buildings were intentionally constructed for ongoing work at a quarry site. The American Museum paid to have a field shack built near Cook's homestead cabin in 1911. Another example was the fossil cabin built by freelance fossil fish quarryman Lee Craig at Fossil Butte in Wyoming.[200] And as early as 1899, the University of Wyoming had already built a "cabin" amid the dinosaur fossil quarries of the Freezeout Hills, which was also used by Lester F. Ward when he came to study and collect fossil cycads at a quarry site that was "only half a mile" away.[201] From one point of view, these buildings were merely shacks put up for convenience to protect field workers from the elements. But in many cases they were also attempts by locals to concentrate specimen processing on site or in state. Cook's idea for the buildings around Agate, for example, was to transfer the processing of materials from New York to Nebraska, so that only prepared specimens (and not all of them) need be transported East.[202] Despite the rugged and rough character of quarry labor, the significant length of time on site often made it worthwhile for field men and their associates to erect buildings, which might in some cases replace the tent camping that prevailed on many surveys and quarry prospecting trips.

Local Meanings for Scientific Objects

As these efforts to establish local permanence suggest, many field scientists through their long residence in the field came to identify as much or more with the western locale rather than with the urban museums that paid their salaries. Earl Douglass settled on a farm site near the Dinosaur quarry shortly after he began leading the Carnegie Museum's operations there, as noted earlier. Upon purchasing some land along the Green River in 1910, he built a house there and brought his family to live with him while the quarry work was going on. "He labored the year long," wrote Holland, "even in the cold of winter, when it was possible to work in the quarry." For fully twelve years, Douglass worked on the research staff of the Carnegie Museum while residing near Dinosaur, becoming so attached to the area that he resisted moving back to Pittsburgh when Holland decided to close up operations. Despite his family's failure to irrigate the farm site—mostly due to the insufficiency of snowmelt to supply enough water to the reservoir they had constructed—Douglass came to identify more strongly with the place of his field research rather than the metropolitan museum that employed him. As an uncomprehending Holland put it: "He continually refused the urgent entreaties of the writer of these lines to return to the Museum in the winter months, preferring as he expressed it, the life of the wilderness, with which he had become entranced."[203]

After living so long in Utah and identifying with the local point of view, Douglass sympathized with state inhabitants' "widespread desire to get a skeleton in one of the museums of Utah." So many years of expatriating material to the distant city of Pittsburgh had made Utahns sensitive to the issue. Whenever Douglass traveled around the state giving talks about the Dinosaur quarry, he invariably encountered requests for preserving material within the state. He communicated such concerns to Carnegie officials, telling one in September 1920: "I told them that when the proper time came I would do what I could to help them to get a specimen . . . I gave them a rough estimate of the cost of getting out and mounting a big specimen provided such could be found."[204] And there was the nub of the problem: as much as paleontologists at the Carnegie Museum viewed themselves as perennially short of funds, they still had far more resources at their disposal than any institution in Utah. The University of Utah might want to have a dinosaur to display in Salt Lake City, but only by securing the agreement of the Carnegie Museum to take one of the specimens already partially uncovered could the operation be remotely feasible. Given the protected status of Dinosaur National Monument, such an agreement also required the cooperation of government scientists and administrators in Washington, DC. Early attempts stalled out, but University of Utah scientist F. J. Pack renewed his institution's dinosaur quest in 1922, when he sought and received guidance from Douglass in this endeavor. Douglass wisely warned him about the delicacy of negotiating with the federal government given that powerful institutions like the American Museum and the US National Museum might have designs on remaining dinosaurs.[205]

These efforts to preserve something to see in Utah were not just at the state level but at the very local level, too, and are reflective of local meanings given to scientific objects and the quarries from which they came. It later became Douglass's dream to establish an on-site museum at the Dinosaur quarry to display specimens for the public near the place where they were actually taken out of the earth. However, in a May 1922 letter to the nearby Vernal Commercial Club, which had written about the "national importance" of the "wonderland" visited by "thousands," the acting director of the National Park Service had roundly dismissed the Club's modest proposal "for guards for the exhibits at the Dinosaur quarry." While he acknowledged that locals "have been in the habit of making Sunday trips to the quarry ever since the operations began, and have been interested in the glimpses they have gotten of the fossil remains as they have been taken out," he contended that now the site was "simply a hole in the ground from which blocks of stone have been taken, which promise to be of interest to science." It had become "in truth nothing except a gash in the rocks on the mountainside," he declared.[206]

This reply echoed, and in some cases directly borrowed, the dismissive language of Carnegie Museum director William J. Holland, who had written to the National Park Service in a letter six months earlier, derisively rejecting Douglass's idea "that the scene of his immortal labors ought to be marked by the erection on the ground of a stately edifice in which there should be assembled plaster-casts of the dinosaurs which we have extracted from the spot." Realizing Douglass's vision, Holland judged, would require "a very formidable sum of money" for "a structure like the famous 'Walhalla' not far from Munich, which cost the Bavarian government nearly one half a million to erect." While it might provide some local employment and "enhance the value of certain acres at present covered with sage-brush in the vicinity," Holland mockingly dismissed the idea as a "wild scheme" unworthy of federal government funding.[207] For his part, Douglass participated in an appeal to Congress to fund a "Natural Museum" at Dinosaur National Monument, but that effort was not successful during his lifetime.[208] It would be many years before the tables would finally turn.[209]

The local people living near other scientific quarries also tried to set up their own on-site museums. Harold Cook sought early on to set up a museum to display the fossils at his family's Agate Ranch in Nebraska. This idea was in accordance with his rancher father's wishes, which he expressed to William Diller Matthew at the American Museum of Natural History.[210] At least one prominent state resident, the historian Addison Sheldon, still agreed with the Cooks' idea in the early 1930s, declaring that "no other spot in Nebraska is so widely known in scientific circles around the world," and that an on-site museum would make it "a permanent point of pilgrimage for Nebraska tourists."[211] And the Wetherills set up their own modest museum of Mesa Verde artifacts for several years at their ranch near Mancos.[212] Not surprisingly, these homegrown efforts received little initial support from metropolitan institutions.

Gradually, however, support did grow, both within and outside the region, for displaying quarry materials on site for tourists and educational purposes. Using Mesa Verde as a prime example, a spokesperson for the Committee on American Archaeology argued in 1908 that "cultural material is nowhere else so instructive both to students and the public as in conjunction with the buildings and in the environment where it was produced."[213] Efforts to set up a museum to display artifacts inside Mesa Verde National Park were more successful after the park's official establishment, due to the control of the site by a tourism-minded National Park Service. Established in 1918 in a two-year-old cabin that had previously functioned as "a combination ranger station and equipment storage shed," the Museum at Mesa Verde was one of the first museums in the park system.[214] Even there, however, artifacts first had to go to the US National Museum in Washington, DC, and only when "duplicate specimens [were] obtained" did they "permit of their being retained within the metes and bounds of the park."[215]

At the Dinosaur quarry in Utah, as mentioned in the introduction to this chapter, the National Park Service eventually would set up a museum at the quarry site itself, which includes an uncovered and enclosed slab of the fossil strata that can be directly viewed by visitors in situ, belatedly fulfilling Douglass's dream. (By then, locals in the nearby town of Vernal, and their supporters at the state level, had long since already built their own Utah Field House of Natural History, which opened its doors in 1948.)[216] Thus, at several of the quarry sites, on-site museums were eventually established, but only after metropolitan institutions—whose funds, of course, had opened the quarries in the first place—had finished getting out the scientific materials they wanted.

Complementary efforts at on-site display other than the museum began even earlier at Mesa Verde, where restoring ruins for tourist development was a high priority upon the establishment of the national park in 1906. A preliminary report to the director of the US Bureau of American Ethnology by New Mexico archaeologist Edgar L. Hewett that same summer involved "an exploration of the mesa, preparation of photographs and plans of the ruins with notes on their state of preservation," and recommendations about "what ruins were of sufficient importance" to help determine the park's boundary line.[217] (Errors in drawing the border would later force them to reopen negotiations with the neighboring Utes.) Jesse Walter Fewkes's initial stated goal when he arrived at Mesa Verde's fabulous Cliff Palace in 1909 as the bureau's lead on-site archaeologist was "to repair completely this great ruin and to leave it in such condition that tourists and students visiting it may learn much more about cliff dwellings than was possible before the work was undertaken." He sought, in particular, "to increase the educational value of Cliff Palace by attracting tourists and students of archeology."[218] From their initial state as plundered quarries (see figures 4.3 and 4.4), the Cliff Palace ruins would be not only restored but also mapped

FIGURE 4.3. "Central part, before repairing." Jesse Walter Fewkes, *Antiquities of the Mesa Verde National Park: Cliff Palace*, Bulletin of the Bureau of American Ethnology, no. 51 (1911), between pp. 8–9.

FIGURE 4.4. "Cliff Palace, from the Speaker-Chief's House to the southern end." Jesse Walter Fewkes, *Antiquities of the Mesa Verde National Park: Cliff Palace*, Bulletin of the Bureau of American Ethnology, no. 51 (1911), between pp. 14–15.

and studied for improving their interpretation (see figure 4.5). Similar plans would be adopted at other quarry sites spread across Mesa Verde.

In addition to "scientific data revealed by [the] excavation" at Cliff Palace, however, Fewkes also reported caustically that "the vandalism wrought by those who had dug into it had [already] destroyed much

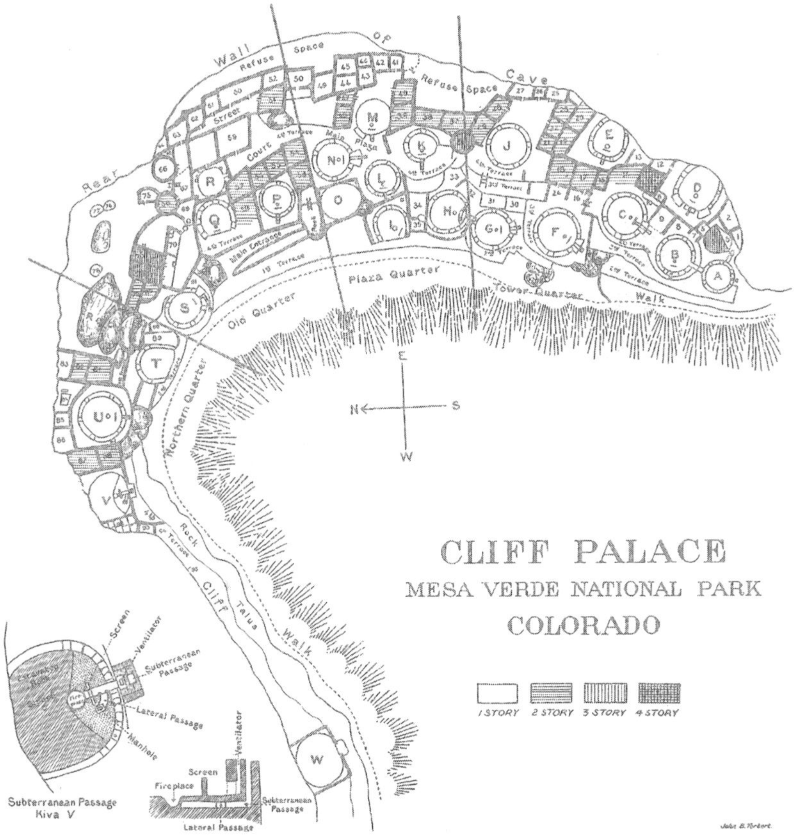

FIGURE 4.5. "Ground plan." Jesse Walter Fewkes, *Antiquities of the Mesa Verde National Park: Cliff Palace*, Bulletin of the Bureau of American Ethnology, no. 51 (1911), between pp. 22–23.

data," and that the Cliff Palace "had been almost completely rifled of its contents, the specimens removed, and its walls left in a very dilapidated condition." As they dug into the ground, they realized that they were excavating "earth . . . that had already been examined," devoid of many artifacts already taken out. "No ruin in the Mesa Verde National Park had suffered more from the ravages of 'pot hunters,'" he declared, and "it had been much more mutilated than the other ruins in the park." He blamed local people, or at least "parties of workmen" who had resided nearby for the whole winter season, "and many specimens had been taken from it and sold," even deigning to repurpose the "beams from the roofs and floors to use for firewood," thus compounding the destruction.[219] Not all visitors were locals, of course. As early as the late 1890s, despite "the difficulties of access," US Geological Survey chief hydrographer F. H. Newell estimated in an article for *National Geographic Magazine* "that at present 75 parties a year visit the more important of these cliff-houses."[220] But while many of them—including Newell himself—were long-distance visitors, the bulk of the long-term sustained excavation and collecting of artifacts was likely performed or assisted by locals.[221]

As the example of Mesa Verde suggests, scientific quarries were often highly visible sites to local residents, sometimes even before any field work commenced and certainly once the vertical probing and excavation had begun. Local meaning is a difficult topic to approach, since the people whose experiential knowledge of quarry sites is most visible in the historical record were often atypical. Those who coveted credit for discovery in the arena of cosmopolitan science were especially eager to show the congruence between their own local meanings based on lay experience and the scientific meanings that would later be imposed on the quarries. At the Agate fossil site in Nebraska, rancher James Cook insisted that he had always considered the Miocene mammal fossils there to be ancient and significant to science rather than merely the bones of "Indians and horses," as Carnegie Museum director Holland had claimed the rancher believed, in order to undermine his discovery claim.[222] For local people who came to see themselves as collaborators in cosmopolitan science, it was imperative to close the epistemic rift between local and translocal meanings for the objects extracted from quarries.

Nevertheless, there is also scattered evidence of divergent local meanings that stood in stark contrast to those of field scientists. The American Museum's famed Bone Cabin Quarry in Wyoming, after all, was so named because a local resident had built a cabin out of the bones, seeing them as merely an abundant building material in an environment that provided little wood. In the Grand River Valley of west-central Colorado, "fossils had been locally known for fifteen years and they had been more or less collected as curios," before the arrival of Elmer S. Riggs's party from the Field Museum in 1900 to provide a "report of their presence . . . to the scientific world."[223] And the Wetherills, after digging up their first artifacts at Mesa Verde on their own before outside researchers arrived, exhibited them at the Fair Building in Durango in March 1889, attracting "several hundred Durangoans" in the "first three days," according to newspaper accounts at the time.[224] We can only guess at the whole range of diverse meanings those local residents attached to what they saw. The following winter the Wetherills excavated a second Mesa Verde collection, but as their scientific and educational aspirations increased, the public response faded. As Al Wetherill remembered it, in Durango they had "opened with a 'Hip, hip, hurray, boys! Come and see what your country produces besides 'taters, corn, and grains'" but were chagrined to discover that "the public did not particularly care about being educated."[225] Thus, while some locals readily adopted scientific perspectives on the materials, the knowledge-making value attached to quarry materials could be easily preempted, challenged, or even dismissed by local people living nearby.

An especially intriguing example of constructing alternative meanings for quarry sites comes from the Utes who lived near Mesa Verde. Historical evidence points toward a strong cultural proscription among the Ute people against disturbing the ruins and material remains found along the numerous side canyons of Mesa Verde, and particularly against digging in the ground to open an extractive quarry. When mountaineer and explorer Frederick Chapin encountered a Ute man named Wap during travels guided by Richard Wetherill, Wap "observed our spade" and "seemed dissatisfied." Although Wap "expostulated in Ute, Spanish, and English, so mixed up in vocabulary that it was difficult to comprehend," Chapin thought that he caught some key ideas, which he summed up as: "White man rich; Indian poor. White man dig up Moquis, make Ute sick: little Ute, big Ute, all heap sick." Not only was Wap implying that the cliff dwellers whose material remains and possessions were beginning to be excavated had been ancestors of the Moquis, but he "made motions indicative of the process of excavation, and with a threatening, superstitious look seemed bound to prevent any such sacrilege of the graves of the departed tribes." Wetherill then tried to convince Wap that they were merely photographing the ruins, not digging up remains—a pretense that was clearly false in a more general sense, even if true at that moment—and refused Wap's "demand for toll for the privilege of going over the mesa." Nonetheless, as they departed, the Ute man told them not to return.[226]

These crosscultural misunderstandings continued in the extended debates between government officials and Utes over the Mesa Verde cliff house ruins. At a May 1911 council meeting, officials from the Department of the Interior tried to convince the Utes to exchange some of their reservation land for land elsewhere so that the park could include the archaeological sites within its boundaries.[227] After decades of dispossession—and the Utes were especially wary given that they had once been given virtually the whole western third of Colorado in a treaty and were now left on a tiny strip of mostly desert land near Four Corners—it was difficult for the government negotiators to convince the Utes to make a land swap. The Indian negotiators offered their own commentary on the apparent absurdities of quarry field work, as it appeared to lay observers, especially those with widely divergent cultural beliefs:

> Mariano: Why do you want those old houses? Nobody is living there. The old men that owned the houses are dead. What are you going to do with them? You told us you didn't know who lived there, then why does the government want those old houses, there is no one there. Who is going to live there? Every man the government sends talks about land and that is all they talk about is the land on our reservation. . . . The government gave us this reservation long ago, now why do they want to take some of it back again?
>
> John Clark or Jawa: You want these houses so that the white men can come in and rob them. What does the government want those old houses they will do them no good. Let them alone. We might trade but we are not willing to. I know why you people want those old houses. Some people told me there was gold mines and coal mines and coal oil in there. I guess that is why you want them.[228]

As these comments indicate, for at least some Utes, the possibility of excavating the remains as a scientific quarry was inconceivable, so that conventional mining for precious metals or fossil fuels seemed a far more likely ulterior motive. This crosscultural lack of communication reminds us just how bizarre and incomprehensible quarry field work could seem to outside observers. Quarry science could perhaps more readily be assimilated into the category of "mining," and, indeed, many white observers also regarded it in those terms.

Even a hundred years later, we can see that quarry field work in science was similar to mining in the material economy. Whether in the human artifacts and remains of the spectacular Mesa Verde plateau and cliff dwellings of southwestern Colorado or the incredibly productive Dinosaur paleontological quarry of northeastern Utah, quarry scientists—and often others besides them—regarded such quarry sites with the same covetous eyes as prospectors uncovering a gold or silver vein, and they often extracted scientific materials from them just as voraciously.

In many cases, the valuable riches of fossils, artifacts, and other concentrated materials were competitively prospected by scientific institutions, and some of the most successful and well funded were based in the East, or even overseas. Tensions with regional actors were as visible in the realm of quarry science as in any mode of field science, if not more, as individuals and institutions in western states fought to concentrate the value added in the processing and interpretation of materials in the West itself. Quarry science was, however, inherently limited by its tightly narrowed geographical purview and by often shipping materials away from the site, wrenching them from their local contexts for interpretation elsewhere, though local efforts to preserve and restore quarry materials on site would provide a partial counterbalance.

Other central features of quarry field work were the necessity of long-term control of the land, often achieved through the creative manipulation of land laws, and the more intensive labor needs of excavating and removing materials buried in the ground. As one of the most distinctive and valuable environmental resources of the interior West, vertebrate fossils loomed especially large in the region's scientific landscape. While other examples such as archaeological sites and intensive field ethnography also relied on similar patterns of relating field work to geography, it was paleontology that most dominated the scene in the region. Quarry science reasserted the effective dominion of eastern scientific institutions over the environmental resources of the West, even as regional actors exercised some limited leverage to contest it. Control by eastern museums or government institutions was only possible because of the envirotechnical infrastructure, which enabled not only lines of communication, as was the case for all modes of field practice, but also especially the feasible shipment of an immense volume of physical material from quarry sites to distant centers where they could be turned into scientific knowledge. Railroad lines always grounded quarry work, as essential conduits to bring in supplies and ship out specimens, which in cases such as dinosaur and large mammal fossils could be quite bulky and heavy. The larger the quarry, the longer it took to extract, but also the more that these iron and steel veins throbbed with the flow of western riches to eastern museums.

The dominance of cosmopolitan knowledge makers over the quarry mode of practice was not absolute, however. Both in the prospecting phase, which more closely resembled the survey mode, and in the ongoing heavy labor of extraction or restoration, the experiential knowledge of local places possessed by those living around quarries enabled them to be found in the first place and made for success in the intensely physical project of vertical land excavation. And over time, field men working for distant museums on site came to occupy a middle position, identifying with local places even as they continued to work under the auspices and funding of distant authorities. They and other local actors, including both white settlers and American Indians, asserted their own meanings over quarry sites and the materials extracted from them, which more directly represented their everyday experiential knowledge of life and work.

By going vertical, quarry scientists literally carved out a new domain for field work. By revealing what was previously hidden or only superficially known, sometimes deploying new technologies of earth transformation, they overcame some of the limitations of more extensive modes of practice, such as lay networks and surveys, which remained largely on the surface. Even when the verticality was metaphorical, the intense localism of multiyear quarry-style operations in fields like ethnography meant intensifying work through novel technologies such as recording machines that began to approach the longevity through which experiential knowledge of place has been formed. As we have seen in the previous two chapters, field researchers attempted to intensify their work and increase its scientific rigor by systematically studying and mapping a geographical territory through survey, or by aggregating and standardizing the collections and observations of distributed lay people in a network. Through a quarry, by contrast, more systematic and rigorous knowledge came through simply intensifying and accelerating the collection of high-value specimens from the field, combined with their study in distant museums and laboratories. In the next chapter, we shall see that another mode of practice for intensifying work in one location also became increasingly common: situating a permanent or semipermanent knowledge production site within the place itself, at a field station.

Chapter 5

Stations

In 1904, in the semiarid western half of Nebraska, near a railroad town named North Platte, the state's first permanent branch agricultural station was born. This field station was intended to complement the main station at the University of Nebraska in Lincoln, the state's capital city, which was located in the more humid eastern half of the state. At first, settlement in western Nebraska had expanded more slowly due to the region's environmental challenges—especially semiaridity, but also temperature extremes and sandy soils in many areas. But in the first decade of the twentieth century, a land boom was underway. Government and railroad land had long been available in western Nebraska, but settlement was boosted by the Kincaid Act of 1904, which provided homesteads of 640 acres rather than the conventional 160 acres. This law made settlement in the western part of the state more attractive, as did a string of relatively rainy years. Over the next few decades, agricultural scientists at the North Platte field station launched a research program based in the environment of the semiarid Great Plains that has continued up to the present day. While simultaneously tied to the wider world of agricultural science through connections to distant institutions, station researchers at North Platte were also closely oriented to meeting the perceived demands of regional settlers for useful knowledge to cope with the environment of western Nebraska (see figure 5.1).[1]

If the branch agricultural station near North Platte, Nebraska, was typical for field stations on the Great Plains, then the University of Colorado's Mountain Laboratory in Tolland, Colorado, founded in 1909, was its characteristic counterpart in the Rocky Mountains. Located halfway between the lower foothills around the university's main campus at Boulder and the alpine heights of the Continental Divide, Tolland was a small railroad stop in a high mountain valley, surrounded by meadows, forests, and summer vacationers (see figure 5.2).[2] Francis Ramaley, an ecologist who led the University of Colorado biology department from 1900 until his retirement in 1939, established the Mountain Lab. He

BULLETIN NO. 95.
(SUBSTATION BULLETIN NO. 1.)

THE UNIVERSITY OF NEBRASKA.

BULLETIN

OF THE

AGRICULTURAL EXPERIMENT STATION

OF

NEBRASKA.

VOLUME XIX, ARTICLE II.

CROP PRODUCTION IN WESTERN NEBRASKA.

REPORT OF EXPERIMENTAL SUBSTATION, NORTH PLATTE, NEBRASKA.

W. P. SNYDER, SUPERINTENDENT.
E. A. BURNETT, DIRECTOR.

FOLLOWING THE BINDER WITH THE DISK TO SAVE MOISTURE.

DISTRIBUTED MARCH 18, 1907.

LINCOLN, NEBRASKA
U. S. A.

FIGURE 5.1. Cover of first North Platte substation bulletin, including image of farm work with horses. W. P. Snyder, *Crop Production in Western Nebraska*, Bulletin of the Nebraska Agricultural Experiment Station, no. 95 (1907), 1.

FIGURE 5.2. South Boulder Park, the valley in which Tolland, Colorado, and the University of Colorado's Mountain Lab were located. Francis Ramaley and W. W. Robbins, "A Summery Laboratory for Mountain Botany," *Plant World* 12 (1909): 105–10, on p. 106.

also published popular introductory volumes on Colorado plants, which were, in his words, "not the result of book study but of many summers' work in forest and meadow, on alpine height and level plain." The first of these was published in the same year that he founded the Mountain Lab, but already he was including photographs of the small railroad village of Tolland, as well as others of lake scenes nearby.[3]

Among Rocky Mountain field stations, the Mountain Lab was founded about a decade after the pioneers within the region. The station was closed about a decade later, but another University of Colorado mountain field station, which still exists today, was founded nearby on Niwot Ridge. Nevertheless, the Mountain Lab's founding was not only a sign of continuing interest in such lab-field hybrids but also in the appropriation of the label of "laboratory" even for sites much closer to the field end of the lab-field spectrum. The Mountain Lab and other field laboratories in the Rocky Mountains all had very low, almost nonexistent barriers with the natural world around them. Nearly all of the research done at them was place-based and—despite the bid for status implied by the use of the word "laboratory" to name and describe many of them—they functioned much more as field sites reconfigured to have lab features than as labs moved into the field. Like branch agricultural stations, they were built field sites.

These case studies from Nebraska and Colorado make it clear that, as the Euro-American settler population grew in the Great Plains and Rocky Mountains by the late nineteenth and early twentieth centuries, state and federal tax money, along with occasional private funding, became increasingly available to support more substantially built-up places that would become sites for producing science inside the boundaries of the region. The value added through the production of research results out of the region's natural resources was thus at least partially concentrated inside rather than outside the West. These sites located within the natural environment of the region were usually called stations, and I adopt that terminology for this mode of knowledge production in the field. If quarry science was like mining, then operating a field station was like farming—in fact, it often directly included the practice of farming.

The predominant type of station on the Great Plains was the agricultural experiment station—or, even more common, the substation or branch station—administered through the states, sometimes with US federal government funding, exemplified by the station in North Platte.[4] In the Rocky Mountains, on the other hand, the built landscape of science was dominated mostly by biological field stations. Other types of stations, typically operated by state universities or private colleges, such as forest stations and mining stations, were also established during the railroad era. Early precursors that shaped the field station model, and which continued to be part of the landscape of place-based built infrastructure for science, included observatories and botanical gardens. All of these stations had in common the quality of being located somewhere in the field; that is, at least partially situated within the natural environment rather than apart from it as in the case of laboratories or museums.

Yet field stations also drew heavily from the laboratory model—indeed, sometimes they had laboratories on site or were even called laboratories. Station science was a hybrid of lab and field, with a permanent, or at least long-term, built infrastructure of work spaces and dwellings, as well as established connections with transport and communication facilities of nearby cities and towns.[5] At the same time, just as was the case for all other modes of knowledge production in the field, at stations the environmental place where they were located was explicitly important to the substantive research that they performed. Most, if not all, of the knowledge produced at field stations could be linked to the environmental features of the place, such as altitude, topography, landform, rainfall, temperature, vegetation, and animal life, as well as, in many cases, the human activities there.

In the late nineteenth and early twentieth centuries, as now, the station existed in a regional context, representing a domain that was larger than a single city, town, or farm, but much smaller than the world as a whole. In contrast to the universalizing ambitions of the rhetorically placeless laboratory, the organizers of stations in the field aimed to encompass and in some cases to manage the production and circulation of knowledge in a region. Usually this regional domain was characterized by particular environmental characteristics that gave it a distinctive identity, such as the semiaridity of the Great Plains or the compression of multiple vegetation zones in a small area in the Rocky Mountains. Regions were both products of field station science and significant shapers of station-based knowledge production.

Yet despite being anchored in local and regional environments, field stations also operated as nodes in a hierarchically structured geography of knowledge production. Specimens and observations flowed into stations or were collected and produced directly around them. At stations, some of the work of turning those raw materials into systematic knowledge happened on site. In most cases, the field station was formally subordinated to another place, such as a more central agricultural station or an affiliated university (sometimes both). Strong linkages to core institutions, often at the national level, were also significant, such as with federal agencies that have supported field research. Moreover, stations faced a dual mandate to serve a regional public constituency that was concerned with the economic or practical possibilities of a specific environment and a national or even international arena of scientific standards, status, and credibility. Stations thus functioned as intermediate nodes, suspended between the two worlds and drawing from both of them.

While one could view the struggle as leading to the triumph of top-down science, it was more complicated than that. To be sure, ultimately the standards of cosmopolitan science were powerful, but local resistance persisted for an unusually long time in the Great Plains and Rocky Mountains region due to its environmental challenges and the lack of widespread public support for the notion of science for its own sake. In this regard, the political influence of Populism—stronger in most parts of this region than elsewhere in the country—shaped the context of patronage for field work.[6] Such considerations of regional political economy were important not just for agricultural stations but for other kinds as well, including those studying forests and mineral resources. Even most biological field stations had to generate a sufficient student population, from both near and far, to remain viable. Overall, as a mode of field production, the station has tended to concentrate more of the work of producing research results within the peripheral field region itself, more than lay networks, surveys, or quarries did.

Observatories, Botanical Gardens, Field Stations

Field stations may have exploded onto the scene in the late nineteenth and early twentieth centuries, but they were not the first or only kind of place-based, yet intensified and built-up, facility for doing science. Considering alternative scientific research sites with much longer historical pedigrees, such as observatories and botanical gardens, situates the field station in its longer-term trajectory and helps bring out some of its features that were becoming more prominent in the rise and spread of field stations during the railroad era. To be sure, observatories and botanical gardens themselves share many of the features of the field station—and might even, with some justification, be considered as variants of the station mode—but their distinctive characteristics are also worth noting. Moreover, situating the observatory, the botanical garden, and the field station all in relation to one another in the mixed practices of the lab-field borderlands reveals the complex cultural topography of science in which local scientific sites existed—where place was not only environmental but also technological.

One type of local site with deep historical roots was the observatory, a site for producing knowledge whose great prestige predated the rise of the laboratory and which has persisted into the twentieth century and beyond. Astronomy, the model science with which the observatory has long been associated, has long been a leading exemplar for the application of mathematical rigor to the natural world. Yet, at the same time, the observatory has also always been located somewhere—this location crucially matters for the calculations—and the siting of observatories has been shaped both by considerations of spatial location in relation to the phenomena being observed and by the practical constraints posed by the envirotechnical setting of the observatory, such as altitude, climate, accessibility for transport and communication, and remoteness from sources of light. Observatories situated in densely settled areas were increasingly compromised by industrial development, noise and light pollution, and other environmental and technological disruptions. The Paris Observatory, for example, faced challenges from urban growth that interfered with its work by the mid-nineteenth century.[7]

Mountaintops became especially attractive for observatory sites, both in astronomy and in other observatory sciences such as meteorology, for the superior environmental conditions that they provided. The Rocky Mountains proved to be no exception, although the meteorological observatory on the top of Pikes Peak in Colorado, touted as "the highest meteorological station in the world," was plagued during its fifteen-year lifespan by difficulties in maintaining its connecting telegraph line. It was eventually discontinued in 1888, when "the observations were found to be of comparatively little value for forecast purposes."[8] Moreover, its relative proximity to growing urban centers was no guarantee it would be chosen for more expensive astronomical observatory projects, and various environmental considerations led at least one observatory locating team to decide on a place outside the United States instead.[9] Meanwhile, moving in the opposite direction—toward the convenience and accessibility of the city—a donation from Henry R. Walcott funded the building of a new observatory in Colorado Springs, below Pikes Peak, at Colorado College, including "a four-inch equatorial telescope," which was supplemented by "a sidereal clock and a transit instrument" given six years later by Charles Blackman of Montreal.[10] It bears noting that the environmental context was not the only feature of the observatory that was shifting in this period. By the end of the nineteenth century, for instance, observatory work at many leading sites was also becoming more mechanized and standardized—even industrialized—with large teams of mostly female workers to process the increasing flow of information.[11] These changes were transformative, though they had their greatest impact at the more prestigious and better-funded observatories of Europe and the US East Coast.

Moreover, the rise of the laboratory, and its counterpart category of the field, also reframed how the observatory's scientific practice was perceived and evaluated by other scientists. Observatories themselves changed. The new "astrophysical observatory" brought together "elements of the traditional astronomical observatory with the chemical laboratory, physics laboratory, studio, and machine shop," writes Alex Pang. As a place, it thus came to operate "as a filter, excluding Nature's noise while letting in the signal."[12] As historical geographer Maria Lane has argued, however, by the twentieth century the lack of opportunity for laboratory-style replication in the research practices of observatories had created openings for iconoclastic amateurs to challenge scientific orthodoxy by insisting on the place-based particularity of their own findings. But while embracing the undeniable hybridity of the observatory as a research site, "in which elements of field and lab co-mingle,"[13] we should resist conflating this hybridity with a collapse of the analytical distinction between lab and field—when, in fact, it has been at such places that the differences became most salient to scientific practice. Indeed, this is one crucial way in which observatories have been very similar to field stations: in a world increasingly dominated by labs, they operate in a complex border zone of intermingling practices and ideals. The long history of mathematics and calculation in observatories may have eased the concern about the rigor of observatory science to some extent, but both stations and observatories have been subject to adverse comparison to the lab.

If observatories maintained a deep historical association with mathematical precision and rigor, the botanical garden inherited a very different past, linked with the classificatory, taxonomic tradition of natural history. The practices of botanical gardens were exacting and sophisticated in their own right, however. While they were much like museums in many respects, botanical gardens from the early modern period through the nineteenth century were also immersed more fully in the environmental conditions of the outdoors. Botanical gardens were usually located in urban areas, though more for staff accessibility and public access than any particular natural resource needs. By their location in urban and suburban places, they typically also came to include other types of scientific spaces on or nearby their grounds. The Missouri Botanical Garden in St. Louis, for instance, was proud of its "herbarium, library, and laboratories," as well as its "close relationship" with Washington University.[14] Between St. Louis and the West Coast, formal botanical gardens were exceedingly rare during the railroad era. Starting in 1916, however, Ira Cutler at the University of Denver did begin planning a "botanical and pharmaceutical garden" in conjunction with a new school of pharmacy. One news report would call it "the largest botanical and pharmaceutical garden in the west."[15]

As interlinked nodes for the movement of plants and information around the world, botanical gardens have also exhibited global structural relations analogous to those that would come to prevail at field stations, perhaps even serving as models. Consider the classic case of the satellite botanical gardens of global scientific institutions, such as Kew Gardens in London. As Lucile Brockway argues, while Kew aimed for "centralized direction of the information flow," its "satellite gardens in the colonies provided the system with multiple centers of research," which were subordinate.[16] Like the regional satellites of Kew, field stations in the Great Plains and Rocky Mountains would replicate the control of knowledge flows from and through the region (or a smaller subregion) while remaining subordinate to higher-level centers elsewhere, such as those in state university cities or the federal agency headquarters in Washington, DC. Of course, as powerful counternarratives to Brockway's center-based view remind us, we must be careful about overstating the dominance of centers, since any botanical garden was also deeply shaped by its local context.[17] But those centers *were* still centers in a very real, institutional sense, whether in the case of botanical gardens or other types of field stations.

The analogy with botanical gardens may be especially appropriate, since agricultural field stations, which also typically emphasized growing plants, predominated over all other station types in the US Great Plains and Rocky Mountains. Elsewhere the boundary line between the botanical garden and the agricultural field station could sometimes be fuzzier, especially in European colonial settings where local site-based research was closely tied to economic concerns.[18] In the United States, however, agricultural experiment stations were typically distinguished from botanical gardens by their association with practical agricultural science, their funding through a combination of state and federal money designated for agricultural research, and their affiliation with state agricultural colleges. Great Plains agricultural stations also typically differed from botanical gardens by virtue of their more complete immersion in regional and local environmental conditions. While botanical gardens were certainly place-based in their fundamental dependence on prevailing local climate and latitudinal patterns of daylight, they typically incorporated a wide range of plants from diverse origins, and they tended more often to be in urban settings rather than the rural conditions thought to be more representative of their regions. Likewise, each state's main agricultural experiment station was often in or near a city or town. Yet branch stations or substations such as the one near North Platte were almost always out in the country, and even the main state stations in the Great Plains region were usually on the edge of the countryside.

Not all field stations were oriented toward agriculture, however. Many were oriented toward the practices of teaching and the dynamics of wild vegetation in places dominated more by forestry and tourism. Most prevalent were the biological field stations in the Rocky Mountains, which followed—yet in many respects contrasted with—their station predecessors at the seashore.[19] The emerging marine stations on the East Coast "were hybrids of two distinctly American traditions": the summer school in natural history, exemplified by Louis Agassiz's 1873 venture on Penikese Island, off the coast of Massachusetts, and a research-oriented model exemplified by his son Alexander Agassiz's seaside laboratory at Newport, Rhode Island, and the Chesapeake Zoological Laboratory, a few years later.[20] We should not assume, however, that only the latter was strongly influenced by the emerging laboratory model. Even with its focus on teaching and natural history, Penikese had many laboratory features, including a strong emphasis on "dissecting instruments" and microscopes in its practices. Above all, Penikese differed from the temporary geological field camps that inspired it by its built environment, which is featured prominently in firsthand accounts of its establishment.[21]

Yet while the Penikese model contributed to the emerging model of the American hybrid teaching and research station, it was not necessarily seen that way by its leaders. Indeed, one instructor associated with Penikese, Burt G. Wilder, explicitly denied that it was a station at all, labeling it instead as a "Summer School of Natural History."[22] Wilder was responding to—and protesting—the British biologist E. Ray Lankester's adverse judgment in *Nature* that Penikese was an inferior zoological station, "a sort of holiday picnic of a few weeks," with "no money . . . to build the necessary staff or scientific or other employees," and thus, he concluded, it "was quietly and very wisely dropped." Drawing on the model of Anton Dohrn's zoological station at Naples, Lankester associated the "zoological station" with "a series of laboratories worked by a permanent staff of scientific observers" and their assistants. Rather than explicitly linking it to a summer school, then, he suggested that it "should play somewhat the same part in zoology as the State astronomical observatories do in astronomy."[23] This analogy with the observatory makes clear not only the lofty research ambitions associated with the European station ideal but also the centrality of the built environment in a purposely selected place and the hierarchical organization of scientific workers at the site.

By the time the Rocky Mountain biological field stations were founded, however, the American seaside stations "had successfully merged instruction and investigation."[24] Outside their East Coast birthplace—on the Pacific Coast, for example—that teaching had become increasingly focused on "students of colleges and universities, rather than school teachers or avocational naturalists."[25] The evidence also suggests that station scientists, while they were negotiating the complementarity of teaching and research, were also positioning their stations in the lab-field borderlands.[26] Most prominent among seaside stations was the Marine Biological Laboratory at Woods Hole, Massachusetts. Activities at Woods Hole may have been mostly on the laboratory side, but in the summer of 1900, around the same time that the Rocky Mountain biological stations were beginning to be founded, some deliberate crossing into the field—in teaching, at least—emerged. "For the first time in history," noted Thomas Montgomery, "there was started this year at the Marine Biological Laboratory a Nature-Study Course." The subject matter of this course encompassed the diversity of nature nearby, including everything from plants to insects, and from marine invertebrates to birds. For bird study, "the field work consisted of three mornings spent in the woods and fields near the laboratory, and of one day's trip to the breeding grounds of Terns at Penikese."[27] Thus, even at the Marine Biological Laboratory, where indoor research work predominated, the activities were clearly situated within the lab-field borderlands, with repeated references distinguishing "laboratory" from "field" work.

Summer vacations provided seasonal opportunities for research work by academic scientists, whether in laboratory, field, or some combination of the two. Yet not all field scientists saw such overlap between summer research and teaching as an unmitigated boon. Alfred Goldsborough Mayer, who was at the Marine Laboratory of the Carnegie Institution in Tortugas, Florida, issued a statement in 1906 calling "attention to an evil influence" that summer school exerted on college teachers: "Until within a few years the college teacher looked upon his summer vacation as a season for research and broadening study; now fully one half of this once cherished period must be sacrificed to the labor of the dissemination of superficial and elementary instruction." He worried that such activities would "kill" the "spirit of research" by allowing it to languish.[28] These concerns were real, but they did not stop many of Mayer's colleagues from seeking out productive combinations of summer teaching and research. The chances of succeeding, of course, varied depending on the type of research. Laboratory-based work depended on the existence of proper indoor facilities and the nearby supplies of natural materials. For those situated more toward the field end of the continuum, the particularities of place, such as access to scientifically productive and sufficiently diverse environments, were paramount. The main question was whether or not summer teaching could be done in a place allowing research and teaching to coexist and even support one another.

At the Rocky Mountain biological field stations, as elsewhere, summer teaching overlapped synergistically with research by professors. Often, the two activities of teaching and research were joined in the supervised research projects of students, whether the undergraduates who attended most field stations or the graduate students who dominated at a few elite stations, such as Frederic Clements's Alpine Laboratory just off the cog rail line, halfway up Pikes Peak. In contrast to the seaside stations, at least those of the East Coast, such as Woods Hole with its vast indoor laboratories using freshly obtained material, the Rocky Mountain stations were more often oriented directly toward the field. To be sure, they often had laboratory buildings on site—as did many agricultural stations—and students and professors alike moved fluidly back and forth between field and lab, but the overall effect was much more like a field site with modest partial walls raised around it than a laboratory with its walls lowered.

Other proponents of field stations—not only in the lab-field borderlands but also in the mixed zone between academic biology and practical agricultural science—focused especially on the qualities of seclusion for place-based research. In reporting his own "personal experience with small field laboratories" in 1914 to a periodical dedicated to agricultural science, David Fairchild of the US Department of Agriculture ventured to "predict that such laboratories, if properly placed, would become veritable retreats for the botanical workers—places where they could go and work quietly with their plant problems without fear of interruption and surrounded by an atmosphere of their own making, which would stimulate research in a peculiar and effective way."[29] And while most field stations were oriented toward the life sciences, a few stations established toward the end of the railroad era were permanent stations emphasizing the earth or physical sciences, such as the University of Michigan's Camp Davis, located near Hoback Junction, Wyoming, just south of the Grand Tetons. Camp Davis, which began primarily as an engineering station for teaching field surveying in the 1920s but later developed a predominantly geological focus, had a strongly field-based orientation with some modest buildings for indoor work, which were thus positioned within the lab-field borderlands much like their biological counterparts.

These biological and geological field stations were predated by forest stations on the forested slopes of the Rocky Mountains, as well as the outlying satellite uplands or planted forests of the Plains. Some of these, like most biological stations, were linked with colleges and universities. An early forest station in the Rockies was established in Manitou Park, Colorado, associated with the school of forestry at Colorado College, on "a tract of 13,000 acres of yellow pine forest in the heart of the Rocky Mountain Reserve Region."[30] There were also official state forestry stations on the Great Plains, as early as 1887, when Kansas, in cooperation with its agricultural college, created two stations in the western part of the state for forest experimentation, along with a forest commissioner, to promote planting trees where they were sparse.[31] By distributing tree seedlings and setting up experiments on site, these prospective forest stations near Ogallah and Dodge City were intended to move beyond the state's previous efforts to gather information through lay networks that relied on circulars. However, if the deeply disappointed report of a May 1904 visit to Dodge City by members of the Kansas State Horticultural Society is any indication, the station failed to do much more than freely distribute seedlings. Even after a subsequent reorganization, as one retrospective assessment noted, the two forestry stations "never received the financial support needed for creditable work."[32]

Still, the ideals of rigorous experimentation and laboratory science loomed large, if only unevenly realized, in these state and local precursors. The trope of the natural laboratory, for example, characterized the work projected at that Manitou Park forest research station: "The laboratory of the forester is the field—the forest, the logging camp, the saw mill, the lumber yard, and generally the field that uses forest products of any kind for raw material."[33] Stations operated by the US Forest Service were a few more years in coming, but between 1909 and 1914, the service established nine of what became known as "forest experiment stations," including several in the Rocky Mountain states, explicitly modeled after the agricultural experiment stations, such as the Fremont Experiment Station in Colorado.[34] Scattered examples of other types of field stations also existed, including mining experiment stations—two in the field were operated by the Colorado School of Mines at Idaho Springs and Wild Horse Park, near Pueblo, and another in the 1910s by the US Bureau of Mines on the school's campus itself—and engineering experiment stations at agricultural colleges in Kansas (1912) and Colorado (1917).[35] Since mining stations in the field were essentially laboratories located in proximity to mining sites, and many on-campus stations were virtually indistinguishable from conventional laboratories, they will receive relatively little attention in this chapter. Those occupying more mixed positions in the lab-field borderlands, however—the branch agricultural, biological, and forestry stations, especially—presented interesting challenges for place and practice in science.

Placing the Station

The place of the station was simultaneously defined in relation to the figurative lab-field borderlands and in the specific material, environmental settings of the Plains and Rockies. In many cases, place was also shaped through the practical activities undertaken in particular locations, so that the environmental shaping of stations was often mediated through utilitarian concerns. Straightforwardly economic activities based on environmental resources—widespread ways of making a living such as ranching, dryland agriculture, irrigated agriculture, mining, and forestry—elicited a demand for knowledge relevant to those activities. People in these sectors became constituencies for supporting state patronage of research work related to them, performed in the natural environments comparable to those in which those economic activities were conducted. In addition, a practical concern with education shaped the development of stations, especially the biological field stations that emerged for summer teaching and research in the Rocky Mountains, in high mountain valley or lakeside settings that were seen as the most appropriate places to accomplish such purposes. While interesting research problems without direct applications might beckon, the raison d'être of the station was often to satisfy a perceived demand for useful knowledge, or for transmission of knowledge through education.[36]

Such practical concerns were often articulated explicitly. When the Nebraska legislature first authorized a western Nebraska substation in 1903, appropriating $15,000 for acquiring land and for its first operational year, they stipulated that it be located somewhere west of the 100th meridian, recognizing the demand for agricultural knowledge specific to the conditions in that part of the state. The enabling legislation outlined the main purposes of the station, emphasizing the practical environmental knowledge that would be gained by having access to a different kind of place for intensive field work than was already available in Lincoln. It would be designed "to determine the adaptability of the arid and semi-arid portions of Nebraska to agriculture, horticulture and forest tree growing, such as the producing of grain, grasses, root crops and fruits of kinds commonly grown in the same latitude in other states, also the most economical means of producing such crops without irrigation."[37] After receiving offers from several towns, the University of Nebraska board of regents selected North Platte as the place to fit these purposes.

Nebraska was not alone. Across the Great Plains, all the states established branch field stations in the early twentieth century to produce knowledge suited to the region's distinctive environmental conditions, especially for farming without irrigation.[38] In the emerging economic hub of Garden City in western Kansas, there was considerable interest in dryland agriculture in the first decade of the twentieth century.[39] Within its first few years of existence, the new agricultural experiment station there included 148 one-tenth-acre plats for "experimental work in 'dryland agriculture' in cooperation with the United States Department of Agriculture." The Garden City branch station's 1908 report made this perceived demand for knowledge about dryland agriculture the basis for increasing its request for funding. "Information in regard to crop production under very dry conditions is in demand," the report noted, "and it is important that this Station be equipped for making more extensive investigations along this line. For this, men and more horses are needed."[40] The true extent of actual demand from local farmers is difficult to gauge.[41] However, it is a safe bet that appeals for patronage to produce useful knowledge about dryland agriculture were more likely to elicit support than appeals for work in areas that were not so clearly such a central part of the local economy, such as physiology or taxonomy.

Once a station could be projected on the basis of regional economic activity demanding the knowledge that would be produced, it had to be located in the right place. This task of placing a new station involved a delicate balance between finding the right natural conditions and mobilizing local support, thus involving both environmental and social factors.[42] The right natural conditions included not only climate—especially temperature, rainfall, and altitude—but also the particular characteristics related to the kind of research work going on there. Biological stations thrived on access to diverse environments available on mountain slopes. Agricultural stations depended on the type and quality of soils, while forest stations relied on the presence of what were judged to be the right kinds of forests, with the exception of the Nebraska National Forest and its counterparts on the High Plains, where forests were planted de novo in locations thought to be suitable.[43] Most forest stations, as Samuel Trask Dana observed, were "located on the particular forest that is most typical of the region and that presents, so far as possible, all of the important problems to be found in that region."[44] Whether in forests, farm lands, ranch country, mining districts, or high mountain valleys, station locations were thought of in relation to the regions over which their knowledge might plausibly be extended.

Despite this obvious environmental variation, places for all types of field stations were broadly similar in their dependence on a combination of natural features, technological infrastructure, and links with related human activities. Consider the biological field stations of the Rocky Mountains. While they obviously lacked the tight connections with practical economic activities in their knowledge products that were so important to agricultural, forestry, and mining stations, nonetheless, few of them could be justified and funded for research alone. For most Rocky Mountain biological stations (and their occasional counterparts in the earth and physical sciences), the key justification and mechanism of funding, as noted above, was the provision of summer teaching on site. Accordingly, they were located with an eye toward the idealized landscapes of field-based pedagogy, which meant not only a premium for easy accessibility by rail—as was true of the University of Colorado's Mountain Laboratory in the mountains above Boulder (see figure 5.3)—but also the daily ease of movement through diverse life zones ascending from the plains and foothills to the alpine heights, making the high mountain meadow location highly desirable for exposing students to the best range of environmental variability—not to mention also comfortable outdoor temperatures in the summer. If they were to entice students, it helped a great deal for them to be in attractive, yet accessible places.[45] To put it more simply: they had to be wild, but not too wild.

The placement of agricultural and forestry field stations on the landscape did often involve a stronger emphasis on potential for environmental resource development, but the criteria were quite similar to those of the mountain biological stations. Reflecting on the first five years of the US Forest Service's experience in locating forest experiment stations—and his own service as director of the Fort Valley Experiment Station north of Flagstaff, Arizona, during this period—G. A. Pearson identified the "primary requisite" as "ready access to typical stands of exploitable timber," but also ideally including "both virgin and cut-over stands in different forest types or life zones," a range of variation much like that found near biological field stations. Other considerations for placing stations included proximity to "important forest activities, particularly timber sales," as well as the availability of "facilities at the

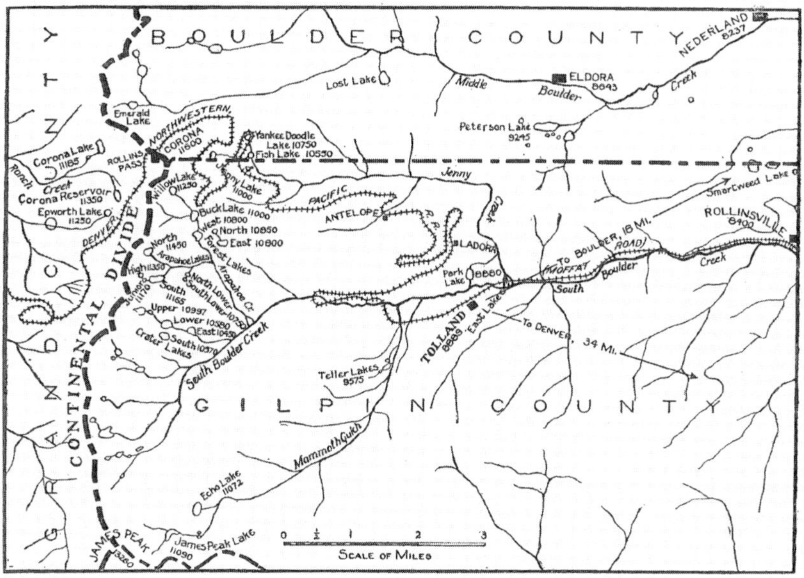

FIGURE 5.3. Map of area surrounding Tolland, Colorado, site of the University of Colorado's Mountain Lab, showing its location along the railroad line as well as midway between the Continental Divide and the plains below. Francis Ramaley and W. W. Robbins, "A Summer Laboratory for Mountain Botany," *Plant World* 12 (1909): 105–10, on p. 105.

station headquarters for at least a small forest nursery."[46] Even forest stations that were not under the control of the US Forest Service were preferentially located in proximity to that federal agency's increasingly vast system of working forest reserves. Colorado College's forest station at Manitou Park, for example, was "surrounded on three sides by the Pike National Forest," thus making it possible to maintain close ties with its active operations and facilities.[47] Whether for a mountain station oriented toward teaching or a forestry station aiming to use science to rationalize commercial development, these pragmatic purposes in locating stations embraced a combination of technological infrastructure and useful environmental characteristics.

For agricultural station scientists, too, finding the right place involved a complex combination of environment, infrastructure, and related human activities. Technological features such as accessibility by railroad intersected with natural features that placed researchers in close contact with important environmental questions. And since stations were evaluated as working farms, success in growing crops was central to forging a supportive identity among regional residents. Unlike the mountain biological stations and forest stations, however, some institutional factors related to proximity of staff also played an important role at first. To be specific, the first station in each state usually had to be near the state agricultural college because of the personnel shared between the two institutions. Over time, though, branch stations, or substations, were also established in more diverse environmental contexts, away from each state's central experiment station. Indeed, as states moved to establish arrays of branch agricultural stations, they sought out particular environmental conditions to develop research lines, varieties, and techniques for different parts of their territories.[48]

To better understand how these diverse factors in agricultural station placement—and others not yet mentioned—played out on the ground, let us consider the location decisions for branch agricultural field stations, especially given how predominant they were across the Great Plains and Rocky Mountain states. The agricultural field station in North Platte, while chosen in part for its central location and excellent rail connections, also importantly provided some diversity of land types within its bounds. The field station's 1,920-acre tract of land three miles south of the town was purchased for $16,107—half of which was contributed by local donors and the town itself—thus forming a "rectangular block 2 miles in north-south length and 1½ miles east-west width" for the station. This land consisted of two main areas of land that could be cultivated, referred to respectively as "bench land" and "table land" (see figure 5.4). At the north end, nearer to the South Platte River, were "280 acres of level bench land" situated "30 to 40 feet above the water table" and used for the main farm plots and the field station headquarters. Moving southward, the land rose "through a series of canyons to the table land, having an elevation of 3,000 feet and lying 180 feet above the bench land," with the other 1,920 acres remaining as "pastureland broken by canyons."[49]

As the case of North Platte suggests, in most cases the land for an agricultural field station was donated by civic boosters eager to attract such an institution near their communities, often as a formal condition. The branch station near Colby, Kansas, likewise, "was purchased by the County Commissioners of Thomas County for $10,000 for land and buildings and was deeded to the State of Kansas for as long as it was used for experimental purposes."[50] At Lander, Wyoming, in 1891, too, the Fremont County Business Men's Club agreed that it would "furnish, or cause to be furnished, forty acres of land in this vicinity, to be donated

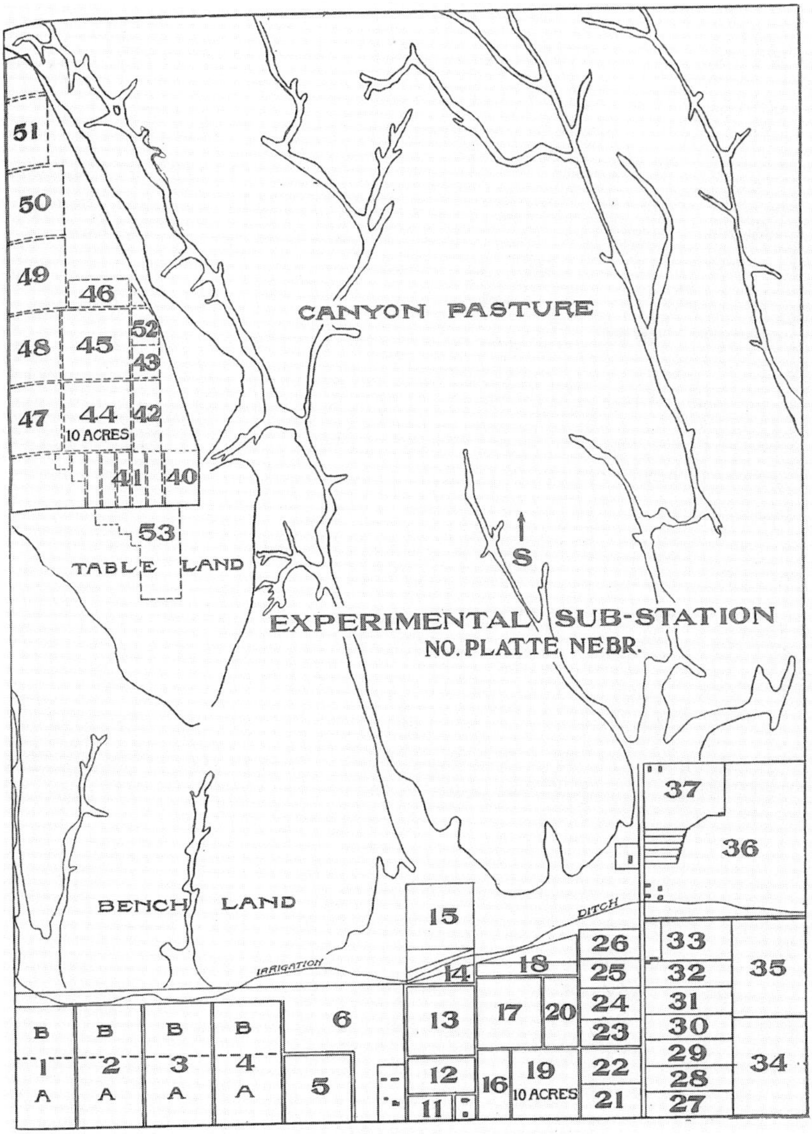

FIGURE 5.4. Map of the branch agricultural field station near North Platte, Nebraska, with diverse topography, field plats, and buildings all indicated. W. W. Burr, *Storing Moisture in the Soil,* Bulletin of the Nebraska Agricultural Experiment Station, no. 114 (1916), 3.

to the state for the purpose of establishing an Agricultural Experiment Station." Crucially for an environment with such water scarcity, along with the land deed would come the "accompanying water right," and the Club also suggested the president of the Fremont County Farmers Club, Mr. Jacob Meyer, as a possible director of the station.[51] In rare instances, a lone civic booster might singlehandedly donate the land, as did lawyer George L. Reid of Tribune, Kansas.[52] Donated land was usually better than the marginal land that remained around settled communities in the West, although even then its quality could be variable. Nevertheless, station scientists took what they could get.

Dryland farming was often the focus of the substations set up in the states of the Great Plains and Rocky Mountains, and the locations were critical to their identity and research utility. In Kansas, for example, where a small group of branch stations was established, researchers sought out sites in the more arid western half of the state. They chose to locate one substation in the small town of Tribune, "as near the central part of extreme Western Kansas as is possible to locate it"—that is, as far west as possible, within fifteen miles of the state's border with Colorado.[53] "The branch stations have been so located that they cover all of the climatic conditions of central and western Kansas, and practically all of the types and classes of soil," summarized the Kansas station's leaders.[54] Similarly, when Nebraska's legislature in 1904 established the state's first branch station, it chose the town of North Platte west of the 100th meridian in order to investigate the environment of the state's semiarid western half. In South Dakota, the equivalent station was Belle Fourche, north of the Black Hills. In Colorado and Wyoming, the main stations in Fort Collins and Laramie, respectively, were in the semiarid region already, but an equivalent move eastward, especially in the case of Colorado, with its large High Plains region covering much of the eastern half of the state, resulted in efforts to locate a branch station at Cheyenne Wells. Agricultural researchers in North Dakota and Montana were especially active in setting up a whole array of experiment farms and branch stations in the semiarid Plains region, especially during the early twentieth-century dry farming boom, which was centered in those two states.

Most agricultural stations were located not just anywhere on the landscape but close to towns on railroad lines, thereby ensuring access to personnel and supplies. Yet although they were almost always located *near* towns, as farm operations requiring extensive land areas, stations were in the adjacent countryside rather than in the towns themselves. Most stations were within close walking distance. A typical example was the Garden City station in the Arkansas River valley of Kansas, which was four and a half miles northeast of the town.[55] And in Montana, the director of the state agricultural experiment station, aware that a new railroad being constructed near its Judith Basin branch station, queried a mutual connection in nearby Lewistown who knew the agents of the Great Northern Railroad. "Do you think that a siding could be made a regular stop for our experiment farm, so that we could get off and on there without difficulty[?]" he asked, pleading that "it would be a very great convenience in visiting the experiment station to have the railway stop close to our track."[56]

Locations more distant from railroads were usually rejected for field stations or were abandoned once the disadvantages became evident. This was often true even for minimally developed station sites, such as the Bruner ranch in the Nebraska Sandhills, which served as an early experimental site for reforestation in the region during the 1890s. After about seven years of operation, forester Charles A. Keffer explained why they were not going to develop the site further, noting that "owing to the distance . . . from the railway (25 miles), and the consequent difficulties of inspection, it was thought best to abandon the plantation."[57] Nevertheless, field stations were occasionally situated in more remote locations, such as the Belle Fourche station in South Dakota, which was itself a full twenty-five miles from the town that functioned as the "post office and only railroad and principal shopping center at the time." Very few people lived near the station, and on-site director Samuel Salmon later remembered the cumbersome journey into town, which suggests by counterexample why most stations were located much closer to the communication and transport infrastructure. "A trip to Bellefourche with a farm team and light wagon," he recalled, "was made about every week or 10 days to pick up mail, and buy groceries and other supplies." This "two-day trip" took "about 4½ hours each way" and required "13 wire gates to open and close and a river to ford." Dismounting the "skittish" horses to open and close the gates meant that "one could never be quite certain they would wait," and the river was a "potential hazard": "about 3 feet deep" in good conditions and "impossible" to cross "during a flood stage," though this was rare on the semiarid plains of western South Dakota.[58] A few years later, the railroad was extended to a new town named Newell, just a few miles from the station site, but before that it was clearly an isolated place. The difficulties encountered at Belle Fourche demonstrate clearly why more proximate locations to railroads and post offices were preferred.

Biological stations were also located with an eye toward easy access by railroad. The University of Colorado's Mountain Lab was located in Tolland, "a small hamlet in a mountain valley" established on the east side of the impressive Rollins Pass crossing of the Front Range and the Continental Divide by the Denver and Salt Lake Railway ("The Moffat Road").[59] Fifty miles from Denver by rail, the Mountain Lab was thus accessible not just to university people but to outsiders as well.[60] If a laboratory's regular users and visitors might use a local transportation network to arrive or depart within the scope of daily life, so too, albeit on a somewhat more extended scale, might those who sought to access the Mountain Lab and other similarly situated field stations. At least, it was rather different from an outback field site that usually required more arduous or complex travel arrangements. Tolland was located in a high mountain valley, but it was also a railroad town. Indeed, the railroad ran up and down the valley in which station scientists did much of their work (see figure 5.5). Likewise, ecologist Frederic Clements's Alpine Laboratory was situated partway up the cog rail line climbing Pikes Peak near the resort center of Colorado Springs, located about midway between the base at Manitou Springs and the peak's summit.[61] It is striking that these two prominent Colorado mountain field stations were located along the two of the most dramatic yet accessible railroad lines ascending the Rockies directly from the Great Plains. To be sure, not all Rocky Mountain field stations were located directly on railroad lines—the University of Montana's station at Flathead Lake, for example, required further travel by stagecoach after arriving at the nearest stops along the Great Northern or Northern Pacific railways—but all were relatively easy to reach using long-distance transportation networks.

The same was true of forest experiment stations. G. A. Pearson of the US Forest Service directly emphasized the importance of accessibility to the transportation grid. "Proximity to a railroad station, while less essential than natural facilities for research," he estimated, "is important from the standpoint of economy and general efficiency."[62] Likewise, the Kansas legislature's 1887 act creating forest experiment stations specified that they be located "not more than three miles from a station on the Union Pacific and on the Santa Fe Railways, respectively."[63] Moreover, for forest stations, rail access was important not just to towns and cities but also to the exurban built infrastructure of forest operations. For example, a published article touting the early forestry station associated with Colorado College noted that "within a half-hour's ride by rail from Colorado Springs the Forest Service maintains one of the largest

FIGURE 5.5. Railroad and vegetation under study near Tolland, Colorado. Ecologists like Francis Ramaley commented on the disturbance caused by human activities near the field station, such as railway fills, but (ironically) the grassland formation visible here, despite being near a railroad line, was described as the possible end stage for vegetation "if left undisturbed." Original caption reads: "View in the *Carex stenophylla* society showing the abundance of low-growing *Antennaria*. Photograph taken in early June." Francis Ramaley, "Dry Grassland of a High Mountain Park in Northern Colorado," *Plant World* 19 (1916): 249–70, on p. 258.

forest nurseries in the region."[64] Thus, even if biological and forestry stations tended to be more distant from towns—or at least to be located close to small villages rather than larger cities—they still derived key advantages from their placement in relation to the existing railroad lines.

Complementing the transportation infrastructure in the selection of field station sites were certain environmental characteristics that mattered to station science. At agricultural stations, one such feature was the very ground in which crops were grown. While easy to overlook in the selection of a station site, the soil could prove to be as pivotal as anything else in the western environment. The centrality of good soils can most easily be made clear by considering a few places where soils were troublesome. Problems with soil at the Belle Fourche station in western South Dakota made getting favorable results there almost impossible. As Salmon described it, "This soil is known as Pierre clay by soil specialists, but locally and colloquially as gumbo. It is heavy, sticky, and impervious to water when wet, and difficult and expensive to plow at any time."[65] Alkali soils were farmed at North Platte—not at the station itself, though such land was available nearby—but it proved difficult to grow crops successfully on this alkali land.[66] Even where soils were not "gumbo" or alkali, the existence of diverse soils and subsoils on station land posed important questions, since the knowledge produced needed to match local conditions as much as possible. At the Dickinson branch station in North Dakota, the land included both "clayey and sandy subsoils" so that the superintendent, L. R. Waldron, had to choose which land type to conduct experiments on. The advice from back at the central station in Fargo was to engage in field work on smaller plats representing both subsoils, if they were "represented in the farming regions" around the station, or to choose the soil type that was most prevalent, if that was the case. Either way, in the likely absence of geological or soil survey knowledge, the decision had to be made by those in Dickinson who had the requisite local knowledge based on their experiences living in the area.[67]

Besides soil, altitude was another important environmental factor, for agricultural stations as well as many others. As a whole, the Great Plains and Rocky Mountains are generally high-altitude regions, although elevations vary from as low as two or three thousand feet above sea level in the river valleys of the Plains to over fourteen thousand feet in the case of some Rocky Mountain peaks in Colorado. Altitude was an important characteristic of a station, defining its features of place as much as rainfall or vegetation. Stations dedicated to purposes such as Rocky Mountain biology or forestry were typically located at high altitude, and the trees, grasses, and other organisms studied there were adapted accordingly. The Rocky Mountain biological stations were not only much higher than most agricultural stations but also intended in part to study conditions at a wide range of altitudes. The University of Colorado's station at Tolland, for instance, was located at 8,889 feet above sea level in the "montane zone," with the Continental Divide accessible six miles to the west at several thousand feet higher, and thousands of feet of lower elevations to the east.[68] From human-built infrastructure such as buildings and railroads to the natural endowments of place such as soil and altitude, placing the station meant situating it deliberately within a wide range of environmental variability.

At agricultural stations, on the other hand, high altitude meant more difficult growing conditions for typical field crops, trees, and grasses, as well as for raising animals. For an especially striking example, consider the difficulty of the growing conditions at 7,200 feet above sea level in Laramie, Wyoming, at the central state agricultural field station there. In 1909, the director remarked on how "peculiar conditions of soil, climate and environment due to the scanty rainfall and high elevation" forced those engaged in field work "who are more or less new to the local conditions, to grope with considerable care in order to size up the various situations":

> An experimenter coming from the lower altitudes and humid climates to an elevation of 7,200 feet, where the seasons are necessarily short, the nights cold, and many of the conditions met with in soil and animal experimentation entirely new, must study with considerable care before he can outline intelligently any experimental work, and, once begun, must carry on his investigations somewhat blindly and with very little precedent or regard of the experience of others to guide him.

Only after living and working in Laramie for long enough to develop enough local experiential knowledge to appreciate the place's "unique character" could they incorporate such insights into field practice and thereby succeed in field station science.[69] To be sure, like the North Platte station in western Nebraska at three thousand feet (or 180 feet less for the station's "bench land"), most Great Plains stations were somewhat lower—typical examples included the branch agricultural stations in western Kansas at Tribune (3,623 ft.), Colby (3,150 ft.), and Fort Hays (2,000 ft.)—though most were still moderately high compared to elsewhere in the United States.[70] The necessity for integrating experiential knowledge of living and working at altitude with the demands of cosmopolitan science was important at all elevations, if especially acute at Laramie.

Seasons and Cycles

Whatever their relatively static environmental contexts, which could to some extent be selected or occasionally even altered, all of the stations experienced fluctuating seasonal changes. Some field stations were occupied only seasonally. Most biological field station researchers, much like survey scientists, returned home from the fall to the spring. Other station researchers, however, did not usually return "home" during the winter and early spring, for the station was located at or near their usual residence—although the winter months did often mean lengthy trips by scientific staff members to state universities and to Washington, DC, for meeting, consultation, and collaboration. Overall, the number of personnel on site typically dwindled during the winter months, especially for stations that depended on summer student labor. Despite the apparent year-round occupancy, then, their patterns of human residency still displayed a marked seasonal tendency, much like Rocky Mountain field stations. Moreover, for agricultural stations, seasonality even more strongly shaped the annual rhythms of plowing, planting, weeding, and harvesting. It may be, as US Department of Agriculture chemist Ervin Ewell suggested in 1897, that all farms could become experiment stations, but in any event, all agricultural experiment stations were already farms.[71] Seasonal cycles also affected forest stations, since access to high-altitude forests was restricted by winter weather conditions. Field stations in the Plains and Rockies were products of highly seasonal environments, even as they were also knowledge-producing institutions created by human artifice.

Environmental conditions varied not just during the year but also over the longer periods of climatic cycles. Across the interior West, the climate was rarely predictable from one year to the next, going through periods of relatively wet and dry years that alternately brought hope and despair to pioneering farmers and station scientists alike on the Great Plains. Expansion during the wet part of the cycle gave way to crop failure during the ensuing dry years. In Oklahoma, for example, "unusual and prolonged" droughts in 1893 "completely demoralized many of the more delicate experiments, and were a source of great disappointment to all at the Station." Meanwhile, forage and grass tests "were ruined by the extreme heat," which resulted in "grievous failure" of their work.[72] As this experience suggests, erratic temperatures could also create problems—not only the extreme heat that often accompanied drought but also as unusually cold temperatures slowed the anticipated growth patterns of field crops.

These changes over time formed a major portion of the official record of any experiment station, especially those concerned with the possibilities of agriculture. Dating all the way back to 1868 because of its former status as a US Army fort, the station at Hays in central Kansas experienced alternating periods of drought and abundant rainfall over a sixty-seven-year period through the 1930s. Rainfall ranged widely between wet and dry portions of the cycle, from a high five-year average of 30.64 inches from 1874 to 1878 to a low of 17.87 between 1891 and 1895, or only 58 percent of the former period.[73] Individual years could also be noteworthy for their exceptional drought or humidity. Farther west than Hays on the High Plains in Tribune, Kansas, for instance, the station—only one year after its first crop season in 1912 and two years after the first work started in 1911—reported an "extremely dry season" in 1913, almost threatening its continuing viability as a research outpost.[74] By contrast, the 1919 season provided relatively good rainfall in western Kansas. At the Colby station in the northwest part of the state, it was "one of the most favorable crop seasons experienced in North Western Kansas," partly due to slightly above average rainfall and partly due to the storage of moisture in the subsoil.[75]

Similarly, the North Platte field station in Nebraska also faced the seemingly erratic swings of the Great Plains climate. One challenging period lasted from 1909 to 1911, when a particularly cold year with late rains was followed by two years of exceptionally low rain. During these years, the actual experiences of station researchers differed considerably from what would later be reported in publications and retrospective accounts. (Not all knowledge of experience—probably very little—passes unfiltered into cosmopolitan science!) Their letters reveal a much richer and more varied set of experiences on site than could possibly be summed up in precipitation totals or temperature averages. One month's experiences may be illustrative. In July 1909, the on-site US Department of Agriculture assistant in crops and soils, William W. Burr, lamented that "the records on the oats and barley will be practically useless as the stand is so poor on certain plats that the yield does not mean anything." Some fields had to be replanted, and even those were not doing too well.[76] A few days later, the farm experienced a "three inch rain" that knocked the grain "down quite badly." The erosion harmed the numbered fields, though "not nearly as badly as the land on the bench."[77] Then, a few weeks later, when they began harvesting, many of the fields were not yet "sufficiently ripe to cut." But the late rains did help them come through all right in the end.[78] "Everything is going along nicely," wrote Burr, "but the season is dreadfully late. I will not get thru harvesting before sometime about August 1st."[79] At one point in the following summer, the "extremely hot weather," with a temperature of 102 degrees and wind that felt like it was "coming thru an oven," the unbearable dry heat was so intense that Burr projected human responses onto the crops. "The young alfalfa is still holding on," he wrote, "but it seems that this afternoon I can almost hear it gasping."[80] Yet the 1911 season turned out to be even worse. Serious relief would not come until 1912.

Such cyclical climatic variations were endemic not only on the Great Plains but also in the Rocky Mountains and strongly affected field work at every station. At forest stations in the western mountains, for instance, variations in rainfall, freezing, fire, and human activities such as animal grazing could strongly influence the forest in different ways from one year to the next. Samuel Trask Dana conjectured, "One man may visit the yellow pine region during a good seed year, and the next season notice that the young seedlings making their appearance abundantly in favorable situations, and come out with the statement that yellow pine is a prolific seeder with excellent reproduction," while "another man may visit the same region two or three years later in a poor seed year and after these seedlings have been killed by drought, frost, or grazing, and come out with the equally positive statement that yellow pine produces but small quantities of seed and that its reproduction is exceedingly difficult to secure."[81] This hypothetical, yet realistic, example suggests one of the most profound epistemic advantages of field stations over more extensive field practices such as surveys: they enabled environmental experience to extend not only over the seasonal variations of a single year—already beyond the reach of the survey—but to capture the longer-term variations extending over multiple years.

Environmental conditions varied not just over time but also within the Great Plains and Rocky Mountains region, and some of those variations had implications, positive or negative, for the bodily experiences of field station scientists. Much of the southern Rocky Mountains region of New Mexico and Colorado, in particular, was viewed favorably as a health resort area, especially for those suffering from respiratory ailments. Thus, when one of the earliest known proposals for a biological station in the region was issued in 1896 by T. D. A. Cockerell—then at the agricultural college in Las Cruces, New Mexico—the announcement drew attention to New Mexico's standing as a "health and holiday resort for scientific persons, teachers, and kindred spirits" that made it a highly desirable location. "Three years experience in this country" had given Cockerell "the highest opinion of the value of the climate" for health treatment, "while the abundance of new and interesting forms of life, especially among the insects, is remarkable."[82]

Other environmental constraints, however, conspired to make field work unappealing and unproductive, thus leading researchers to move elsewhere to further their research ambitions. Consider the assessment by Salmon of his brief time at the lonely branch agricultural station at Belle Fourche in western South Dakota. Consistently unhappy with his output, despite hard work and "some contributions of minor importance," Salmon lamented that he "never developed any new variety or cultural method of practical value for the farmers of the Bellefourche or any other area." He insisted that this disappointing outcome "was not in any way related to support or lack of support from Washington headquarters or cooperation at Bellefourche" but was, rather, due to "the very low and variable rainfall and the intractable soil." The experiments he conducted, he thought, "were of value," but only "in a negative way in showing that no variety or modification of rate or time of seeding known at the time was capable of avoiding the effects of the climate and the intractable soil."[83] Few station scientists described their experiences quite this starkly, perhaps because few stations were located in places that were so environmentally challenging: the "gumbo" soils of western South Dakota, compounded by the general semiaridity of the region as a whole, simply made the place very difficult for agriculture. It may be true that scientific knowledge can be produced anywhere, but certain places were more easily productive than others for solving important research questions in the Rocky Mountains and Great Plains.

Indeed, another thing that Salmon's account suggests is how difficult it was for station researchers to separate the scientific question of rigorous testing in search of objective results (no matter how "negative" they might be) and the economic and social question of how science could serve the public by helping settlers find a way to farm successfully. Even the mountain biological stations, most of which did not face quite the same pressures for practical results, confronted the equally significant constraint of practical viability for attracting and teaching students, which of course depended on suitable environmental conditions too. In particular, these mountain stations promised access to diverse life zones for making collections and learning from observation.[84] Regardless of the type of science being pursued—and including both the "pure" and "applied" sciences—field stations lived or died according to the lived experience of place. Yet we must also remember that place was not just given—it was made.

The Built Environment: Making Stations

So far I have treated the natural environment of the field station as something of an external factor that strongly influenced the development of station science—and, up to a point, this makes sense. But what distinguished the station from other modes of field practice was the intensity with which humans transformed the local environment to make the desired kind of ongoing knowledge production possible. They built or reoccupied permanent structures, they modified landscapes (to varying degrees), and they installed features of laboratory, campus, and human settlement. To be sure, quarry science was also locally transformative in its own way, and in fact the extraction of environmental resources not only often left large pits, waste piles, and other traces but sometimes, as we have seen, involved the erection of temporary living quarters and adoption of property control measures. Yet it was at the field station that the purposeful transformation of the natural landscape for science itself was most fully realized as a long-term strategy. Key to the infrastructure that constructed the place of field stations as more lablike was the availability of housing for researchers, laborers, students, and families in close proximity. To a great extent, the environment of station science was not just found but built.

Field stations may have differed greatly in how much their human organizers wanted to change the land around them—ranging from mountain biological stations, where they sought to study nature in situ, to the aggressive plowing and planting efforts at agricultural stations, with other types such as forest stations somewhere in the middle—but, by design, all centered on significant clusters of human-built structures. Living quarters and indoor spaces for work were common to nearly all field stations, and many of them also had eating places, common gathering spaces, storage facilities, and other buildings on site. Collectively, these clusters of buildings formed little villages that defined and constituted a zone of human occupation. In some cases, field stations were deliberately located where villages already existed, such as the University of Colorado's Mountain Lab at the small mountain resort hamlet of Tolland, in order to make use of existing structures (and the village's railway station) rather than building them anew. For most stations, however, their requisite location in the midst of rural grasslands or forests meant that those clusters of buildings had to be constructed from the ground up.

In fact, it is fair to say that buildings typically were the greatest start-up expense for funding field stations, even more than acquiring or gaining the rights to use the land itself. The capital required to establish a station was often a limiting factor in determining whether a station would come to exist at all. Constructing or renovating buildings on site meant a serious commitment. While lay networks, surveys, and quarries typically operated largely out of the established built environment of museums, government offices, universities, and laboratories of scientific institutions in cities, the field station as a mode of knowledge production required major, and often specially authorized, appropriations that signaled a long-term commitment to scientific research at a particular location in the field.

Scientists establishing field stations had two basic options for assembling a built environment to house laboratories, meeting rooms, equipment, and printed materials, as well as living and dining quarters. First, they could build their own structures. This option required more capital investment up front, usually along with a parcel of land, but it could pay off in the long run and give the sponsoring institution more control over the site. It was the only realistic option available when establishing a field site in a remote, uninhabited location. The University of Montana's new field station site at Yellow Bay—which was moved down the eastern side of the lake from its initial location at the north end—was deliberately sited away from the existing built environment of the village of Bigfork. Accordingly, a "new two story brick building for a laboratory" had to be constructed.[85] Some agricultural field stations started out with a few preexisting buildings, such as a house and barn, if the land had already been occupied as a farm, but even then improving the built environment was a priority. At the North Platte station, for example, even in the bad weather year of 1911, its leaders asked the board of regents for a variety of capital improvements to the station, such as a boarding house, sewer system, corn silo, and irrigation wells.[86] Whether in the Rockies or on the Plains, field stations in their early years, if not longer, must have seemed like perpetual construction sites.

The second option, avoiding new construction, was to rent facilities from an existing property owner, if the site was already lightly occupied. (Heavily occupied areas were not usually appealing for field station sites at all.) This strategy was pursued by the University of Colorado in establishing its Mountain Lab in 1909. The small mountain village of Tolland, located in a glaciated valley below treeline, was the last stop on the railroad line before it climbed steeply over the Continental Divide—or, later, plunged into the mountain range itself through the Moffat Tunnel once it was built. As a long-established railroad town, Tolland had some vacant nonresidential buildings available. Ramaley arranged with the owner of a "store building" to use the property in exchange for paying the property taxes on it. However, such an arrangement left the

Mountain Laboratory vulnerable to the future decisions of the owner, a man named K. W. Toll. He wrote Ramaley in October 1911 declaring that the store would no longer be available, since Toll himself planned to occupy it while erecting a cottage and renovating some nearby buildings. He offered Ramaley a "small building across the track" instead. He would rent out its two rooms for the three summer months at a rate of $20 ("less than the taxes on the other place"). He also offered to allow the field station equipment to remain until the summer.[87] While it might be cheaper to rent a building, that also could make it necessary to change locations more frequently.

Nevertheless, whenever feasible station organizers typically took advantage of existing built infrastructure. At the Fort Hays branch agricultural station in Kansas established in 1901, for example, organizers envisioned repurposing a former federal military reservation outside the town. As the legislation requesting the donation of the land and buildings to the state—at first for a branch state college and later also for an agricultural experimentation—noted: "The buildings upon said military reservation, formerly used as residences for officers and their families, barracks for troops, storehouses, etc., are large and commodious, but cannot be moved without destruction of their value, but in their position are of great value, and could be used, with little additional repairs." Ultimately, the station organizers did move several buildings a short distance to use on the station grounds. They turned a "frame building" into a house for the superintendent, and subsequently for the resident dryland agriculturist, followed by the cereal grain specialist. Three other buildings became a barn, tool shed, and grain storehouse, respectively.[88] Beyond this, workers at the station disassembled several of the old fort's structures, using the lumber as a material for constructing "additional farm buildings and a 14-room house for workmen."[89] Yet even with all the buildings and disassembled lumber available at the former military post, a hefty appropriation was still required for initial capital improvements on site (and for replacing buildings destroyed in a structural fire in the summer of 1912 that destroyed the west barn).[90]

Rocky Mountain biological stations, too, despite their remote locations, were sometimes established in places that already had extant built structures. Though usually quite small, mountain villages in the Rockies were nonetheless clearly known as destinations for summer vacationing. One visiting researcher at the Mountain Lab described the village of Tolland as "largely a summer resort . . . [with] numerous cottages to rent to summer dwellers."[91] As a site, Tolland presented an ideal mix of modern infrastructure with access to diverse uninhabited and undeveloped environments: it was on an "inner frontier" of American settlement, combining the advantages of both worlds.[92] Participants in the Mountain Lab's summer activities, from the director to the students, took advantage of the existing infrastructure for housing and food. During the inaugural season of 1909 Ramaley himself rented a house that was offered for sale to the Colorado professor—along with two and a half other lots, at a "bargain" according to the selling agent at the Tolland Mercantile Company—the following spring after the owner decided to move to southern Colorado.[93] Students might also room and board with locals, who would write Ramaley with offers.[94] Alternatively, students had the options of staying in a local hotel or renting tents.[95] At the Alpine Lab on Pikes Peak, too, Clements and his associates occupied mountain resort cottages, which they apparently owned outright. Probably most of these structures were already part of the small village's built environment, even before the field scientists arrived. An early announcement of the Alpine Lab noted that a "small cabin [had] been acquired" there, which, it was hoped, might be transformed "into a laboratory ultimately."[96] Whether for residential dwellings or for laboratory facilities, the existing buildings of high mountain villages often provided the architectural basis for field stations at their founding.

Likewise, even when newly established stations did not have local predecessor campuses that could provide whole buildings and reusable building materials, their organizers took advantage of whatever modest on-site structures were there. Just as the mountain biological station at Tolland, Colorado, utilized existing village buildings as houses and work spaces, so too did many other stations, including branch agricultural stations. Some of these stations were converted from existing farms, such as the Kansas branch station at Colby. At the time this site was purchased, it included "a small six-room cottage," as well as barn, shed, granaries, silo, and "several other small buildings." Despite being in "a fair state of repair," however, these various buildings still required considerable capital investment to render them fit for the purposes of the field station. In particular, one of the granary buildings, relocated on site, was "remodeled into an office, laboratory and exhibit building."[97]

The on-site buildings where scientific work took place were often improvised from existing structures, as the examples of the first cabin at Clements's Alpine Lab near Pikes Peak and the lab facilities at branch agricultural stations such as Colby both show. Likewise, at the University of Colorado's Mountain Lab in Tolland, the "laboratory building" was a converted commercial building in the small mountain village, "admirably adapted" for its new use as a summer field station headquarters. Ramaley emphasized its essential features that made it work as a field laboratory: being "well-lighted from the north," providing "ample shelf room, counters, and tables," and having ample storage rooms, as well as "a good photographic darkroom." Such a premises could provide year-round storage for "jars, bottles and chemicals, plant driers, insect nets, water bath, drying oven, soil cans and soil borers" and a temporary home for items brought up for the summer, including "reference books, microscopes, microtomes, forestry instruments and cameras," all transported from the Boulder campus.[98] Such an array of instruments enabled students and instructors alike to engage in more elaborate original research at Tolland itself.

Another noteworthy example of repurposing the existing built environment was at the Rocky Mountain Biological Laboratory, which was established at the site of a ghost town named Gothic, Colorado. "A number of abandoned cabins . . . dating from the boom times of the 1880's were repaired for living quarters for staff and students," and "an old abandoned hotel was converted into the first laboratory, library, office room, storage room, living quarters for students, dining room, and kitchen," noted the station's longtime director.[99] While many other, more conventional examples of field work relied solely on camping in tents, often far away from towns and villages, most field stations were grounded in a natural environment engineered for clustered human occupation, often mixing scientific and other uses. None of the mountain resorts occupied by field stations had a scale anything like the most well-known contemporary Rocky Mountain vacation destinations such as Colorado Springs or Glenwood Springs—to say nothing of their sprawling counterparts of the present day throughout the region—but they did offer advantages that were both experiential and practical, as long as they were not too thickly settled.

In some cases, however, there were no preexisting villages or even single buildings to adapt for station infrastructure. This was especially true of forest stations, which were deliberately located inside forest reserves. If no buildings existed at all, then research projects might have to wait. Station organizers faced the dilemma of whether to delay research in order to concentrate all efforts on building construction or to proceed more gradually with new building activity. In this respect the challenges faced on forest reserves were not wholly different from those in other types of stations, since even stations in less remote locations, such as branch agricultural stations, often required numerous additional, often special-purpose, buildings in order to become fully functional. So should building projects be prioritized over research projects? Forest station director G. A. Pearson thought so, contending that "the most economical procedure is to construct all the buildings for which there is present need within a short period. The ideal way would be to devote the greater part of the first year to building and acquiring equipment." Only afterward, he contended, could "the station force . . . be free to devote the greater portion of its time to scientific work."[100] Similar initial emphases on building projects characterized nearly all stations during the first year or two of operation, at least when sufficient preexisting structures were not available. The reasoning for this focus on building was driven not so much by the scientific research agenda itself but instead by a concern for economy. Put simply, it could be more expensive in the long run to stretch out a building program over an extended period. To Pearson, the amount of "supervision and office routine incident to the construction of one building is almost as great as for several erected during the same season." Yet even Pearson, with his strongly articulated preference for an initial focus on building structures, recognized that such a plan was "dependent upon the availability of funds" and that changing future needs would likely require additional buildings, so that "the plans must always provide for expansion with the growth of the station."[101]

In the case of Pearson's generalized forest station, the basic assumption was that it would almost certainly be located away from existing settlements by default. Likewise, there were some Rocky Mountain biological field stations located away from settled villages, but sometimes these were the result of deliberation and choice. One such example was the University of Montana's station at Flathead Lake. There, the small village of Bigfork developed around the initial location of the field station, which was at the north end of the lake, where the Swan River flowed into it. Although Bigfork did not exist when the field station was established, within a few years it had "a postoffice with daily mail, several stores, electric light, telephone, and many of the modern conveniences."[102] Subsequently, after the opening of the Flathead Indian Reservation to settlement, the field station moved to a new eighty-seven-acre site midway along the east side of the lake, at Yellow Bay, along with two other parcels of forty acres each. Morton J. Elrod believed Yellow Bay to be "one of the most charming spots on the shore of Flathead Lake," praising especially the "wild and rugged" terrain, which was "almost untouched by man." The advantage of this more remote and less settled location was that "those attending the summer sessions of the laboratory are in the center of a collecting field extending in every direction for miles."[103] It would seem, then, that the clustering of settlement found in mountain villages could be a double-edged sword: too much settlement could interfere with field work around the station.

Nevertheless, in many respects, field stations themselves thus tended over time to increasingly resemble other small settlements: an agricultural field station was something like a farm, if not exactly. A forest or biological experiment station was like a small village in the woods, lakeshore, or high mountain meadow, overlapping in its built environment with practical activities that ranged from tourism to heavy-duty forest extraction, but to the careful eye it bore a recognizable countenance. More broadly, we might say that the particular needs of a scientific field station, while sharing some similarities with the practical activities to which they were associated, had their own distinctive built infrastructures that enabled them to become nodes of knowledge production in the field, connected to both the urban culture of labs, museums, libraries, and offices, and the field culture of local materials and environmental conditions.

Division of Labor at the Station

While the natural and built environmental context of station science in the Great Plains and Rocky Mountains was crucial to field practice, so too was its work organization. Even with ample facilities and equipment, the station could not function without people, and organizing personnel were at least as important as arranging the buildings. While nearly all field parties of any size were characterized by hierarchies that organized their work and everyday interactions, the typical social structure of the field station bore at least a family resemblance to the on-campus laboratory. Both scientific researchers and assisting manual laborers had to be recruited, supervised, and retained in order to produce the deep place-based knowledge that was so characteristic of the field station. At most stations with a practical orientation, recruiting a director was a crucial first step, whether for a state's main station or one of its substations. For their directors, these station authorities hired men with scientific training. On the other hand, a few sites, especially among the summer biological stations in the mountains, were established by a director who founded the site. Many stations also needed reliable foremen or trusted technicians, who supervised the manual labor. Nearly all such stations sought to hire manual labor under the control of the foreman, with the main criteria for choosing local men being some combination of reliability, obedience to instruction, and a good work ethic. Many of the sites, ranging from Rocky Mountain biological stations to Great Plains agricultural branch stations, also functioned based on seasonal or postgraduate student labor, which could blur the distinction between college-educated research scientists and manual laboring assistants.

First, consider the directors and other scientific staff. The North Platte field station in Nebraska was set up under a full-time director—as was typical for most field stations, not a local person but a university-trained researcher—named W. P. Snyder, who had already done some work at the main station in Lincoln.[104] During its first few years, the work of the North Platte field station was heavily guided by E. A. Burnett, the director of Nebraska's main station at Lincoln. The station chose Snyder to run the North Platte branch station on site under Burnett's instructions at an annual salary of $1,000 paid out of the state's fund for holding farmers' institutes. (As part of his job, Snyder was expected to return to Lincoln each winter to supervise the institutes.) An additional $200 was paid out of the state's research funds.[105] This close relationship between field station research and practical use knowledge for a larger public constituency was typical for branch stations and other outposts situated more deliberately in their environmental contexts. In effect, Snyder was a dependent employee of Burnett at Lincoln, who early on treated the North Platte field station as a satellite of the main station. At first, every detail was carefully orchestrated from Lincoln.

Over the next few years, however, Burnett gained enough trust in Snyder—and freed up enough funding for his salary—to give him greater authority and status. In the spring of 1906 Snyder was formally promoted to department head, placing "him in full charge of all work at the North Platte Station." Faculty members in Lincoln were still expected to relate to the North Platte station, but now less to provide startup guidance and more to coordinate their research programs as Snyder's colleagues. Burnett suggested to them "that in any work which you may have under consideration which is likely to be duplicated at North Platte you discuss with Superintendent Snyder the methods which you are to pursue at this Station so that when results are secured each Station may be able to compare its work with that done at the other plant."[106] At first, the only research staff member was Snyder, who was responsible for supervising the scientific work of the station on site. Alone among those who were physically present in North Platte, Snyder had the opportunity and obligation to publish the results of research done there, sometimes in collaboration with scientists in Lincoln.

Gradually, however, more research staff members were added. The other regular scientific researcher at North Platte during the station's early years was William W. Burr, who was appointed as the US Department of Agriculture agent, with the title of assistant in crops and soils. Educated initially in Indiana and Virginia schools, Burr had completed his college degree at the University of Nebraska.[107] At branch agricultural stations on the Great Plains, assisting scientific staff were quite often supported by and officially affiliated with the US Department of Agriculture. Burr worked under the joint direction of Snyder and the department's national headquarters in Washington, DC, which sometimes created conflicts on the ground. Although by paying his salary the national department would seem to have held ultimate authority over Burr, his physical proximity to the station director leveled things out somewhat. As this example shows, a field station's scientific staff—and this was also true of most Rocky Mountain field stations—would often consist of researchers with affiliations to multiple institutions, all of whom (if unevenly) had an interest in the place-based environmental knowledge produced at the station.

At some field stations, especially the mountain biological stations, the outside experts came not from federal science agencies but from other universities and colleges and were enticed by the Rocky Mountain location of the field station site. At the University of Wyoming Summer Science Camp that Samuel H. Knight founded, for example, scientists on site were not just from the university itself but also included a geologist from Columbia University—with which the station had a close collaborative relationship from the camp's founding—to help in teaching the geology graduate course and, at least one year, a botanist from the University of Oklahoma, for a total of nine instructors. These staff members, including those from Wyoming and elsewhere, tended to be specialists in the plants, animals, and geology of the Rocky Mountain region, such as the University of Wyoming's own Aven Nelson, touted as "the foremost authority on the flora of the Rocky Mountain Region."[108] Prof. J. F. Kemp, the Columbia geologist who collaborated with Knight in the founding of the camp, also had expertise in the Rocky Mountains, despite his East Coast institutional base.[109] Likewise, the University of Montana's biological field station at Flathead Lake often featured scientists from other places. In 1914, for instance, director Morton J.

Elrod invited Arthur G. Vestal of Illinois to teach ecology at the station for $150, with small classes and "time to do work yourself."[110]

These institutional offshoots of campus academic departments became organized in forms modeled after them, giving them an appearance that was much more similar to that of the laboratory than other modes of field work. This was true not only at more exclusively teaching-oriented field stations such as Wyoming's but also at those with greater research ambitions. The Alpine Lab near Pikes Peak, for example, was styled as a more specialized "graduate school of ecology" and had a staff of four in 1913, drawn at first mostly from the University of Nebraska, where director Frederic Clements was based during the early years of the station. The staff also included Raymond J. Pool as assistant director, Clements's wife, Edith—a credentialed scientist in her own right—as botany instructor, and H. L. Shantz as special lecturer.[111] Elsewhere, such as at Wyoming's summer station and others, there would typically be a more senior scientist acting as head instructor for each field covered there—e.g., botany, zoology, or geology—along with one or more assistant instructors, who might be advanced graduate students or junior faculty.

In addition to the regular faculty and staff researchers, nearly all field stations had short-term visitors. At branch agricultural field stations on the Great Plains, many of those short-term visiting scientists were from the affiliated institutions at the state agricultural college or the US Department of Agriculture, although occasionally others visited as well. The North Platte field station in Nebraska was typical. E. A. Burnett, the main station director in Lincoln was, of course, a frequent visitor, but he sometimes encouraged others to join him. In mid-July of 1905, for instance, he initially planned to bring along an agronomist and a botanist from the University of Nebraska, though neither ultimately got away from their duties to join the trip.[112] The botanist, Charles Bessey, decided to make up for his absence by coming out later that month, combining his site visit with some botanical field work in the area. He would "bring out [his] collecting apparatus and hope to collect grasses and other wild plants on the Sub-Station farm rather freely."[113] While few of these scientific visitors stayed for long, their appearance for visits ranging from part of one day to multiple days became an expected part of station life.

Just as at the Great Plains field stations, though often staying longer, scholars came to visit Rocky Mountain field stations from other institutions, providing for a convivial mixing of scientists from different places, much like seaside scientific resorts such as Woods Hole.[114] These stations were widely advertised, and their seasonal nature made it possible for outsiders to join in research without being absent from regular teaching obligations at their home institutions. At the Mountain Laboratory, visiting scientists from near and far would stop by a field station when in the area, usually writing Ramaley in advance to announce their intentions. In 1910, for instance, W. M. Davis of Harvard University was doing field work in Colorado, and his party of four planned to stop at Tolland during their travels, stopping between a visit to the foothills at Morrison, Colorado, and the high alpine zone.[115] That same year, Frederic and Edith Clements also planned to visit "for a day or two" in late July. They not only would study "the alpine formations" but also indicated an interest in "see[ing] the summer laboratory in session."[116]

Given Clements's international fame as a pioneer in ecological methods and ideas, his own Alpine Laboratory near Pikes Peak was a frequent destination for visiting scientists from around the world. Clements kept a log of these visitors, who included foreign scholars from Germany, Belgium, Switzerland, Denmark, England, and South Africa.[117] The 1913 International Phytogeographical Excursion to the United States made a major stop at Minnehaha, and the Alpine Laboratory served as "headquarters" for "almost daily excursions" to other places nearby.[118] Some field stations even hosted the meetings of scientific societies, such as the Summer Science Camp of Wyoming, which was said to have entertained three hundred guests attending three separate meetings of scientific organizations in a single summer.[119]

Scientific researchers, whether residing there short-term or long-term, were not the only workers at field stations. Throughout the productive existence of stations in the Great Plains and Rocky Mountains, on-site hired labor was often crucial to their work organization. For agricultural field stations, such as the one at Sundance, Wyoming, in the northeast corner of the state near the Black Hills, this meant finding reliable local men to do the routine farm work. Superintendent Thomas Dunn—himself a part-time supervisor due to his work running the Sundance Dry Goods and Clothing Company—thought he was getting a late start in arranging for hired labor on the experiment farm during the 1891 summer season. "I was not appointed until late in July and could do nothing in the way of *contracting* prior to that time," Dunn commented.[120] Yet as it turned out, the on-site farm manager already had made "an agreement with several men to work on farm whenever he calls on them," an arrangement put into place before Dunn took over the superintendent role.[121]

Beyond the permanent roles of the superintendent and foreman or manager, then, field station work needs were often met through casual and impermanent wage labor. "A common mistake of technical men," G. A. Pearson declared, "is to devote too much of their time to work which can be performed more economically by a common laborer." It might occasionally be "necessary" for station researchers to engage in manual labor, but "clearly poor economy for a $1,200 technical man to spend much of his time building fences" or "digging ditches," if "the same work can be done better by a laborer employed at $2.50 per day." He exempted student laborers, despite their possessing some scientific training, from this advice, however, noting that they "generally receive no more pay than a laborer, and who can often be employed temporarily at manual labor when not needed on scientific work."[122] Pearson's calculations applied to varying degrees at all field stations, though the relative composition of local hired labor vs. student labor differed noticeably from agricultural stations with robust local farm labor markets to remote mountain biological stations, where students usually greatly outnumbered hired workers. Sometimes students might go on to have successful scientific careers, such as future Colorado state botanist Hazel Schmoll, who worked at the Mountain Lab in Tolland in the summer of 1912 during her student years at the University of Colorado.[123] Whether students or local wage laborers, most of the people working at field stations at any given time were not the researchers with highly visible publication records.

For a more detailed example, let us consider once again the North Platte agricultural field station in Nebraska. Its ongoing functioning depended utterly on the labor of the foreman and farmhands. While researchers might need to be away from the station for consultations with other scientists, visits to others farms and stations, or extension work with members of the public, the foreman and his laborers kept the farm going day after day. And not only did hired farmhands have to perform the usual tasks required on any farm—plowing, planting, weeding, harvesting, fence building, animal tending—but they also had to adhere to the special requirements of experimental protocol and design.[124] When scientists were present on site, they would normally be expected to supervise any measuring instruments, but this could not always be assured, and farm laborers had to be trustworthy enough to follow guidelines at all times, to ensure the rigor of the knowledge that was produced. They were farmhands, to be sure, but they were also assistants in scientific practice

Because the foreman and the on-site researchers usually worked at the same physical place and communicated face-to-face, little written record of their interactions has survived. We do have some evidence of the instructions that foremen were given, such as in Utah, where they were directed to quote to visitors "only such results as appear in the regular Experiment Station publications," which suggests a limitation on the sharing of experiential knowledge unmediated by scientific protocols.[125] Other evidence is available from brief time periods when station scientists were away from branch stations like North Platte while consulting with colleagues in Lincoln or Washington, DC. Handwritten letters from the station's foreman Ward F. Weakley (sometimes spelled "Weakly") to Snyder, the field station's superintendent, in the winter of 1906 indicate a special interest in communicating the variable conditions on the station farm. In one letter shortly after New Year's Day, for example, Weakley commented, "We have had some winter since you were out here it stormed for three days but has cleared up nicely we had every thing under shelter during the storm we had the sows with little pigs in the cow barn and the old sows in the sow pens we have not got the new hog house built yet," also adding that the "sick steer is all right."[126] Such letters provide us with a glimpse of the kinds of responsibilities that field station scientists were willing to entrust to their foremen, as well as the knowledge of experience they possessed and contributed to the scientific work at the station.

Other documentation of the foreman's role comes from letters written *about* him from one scientist to another. Sometimes those communications indicated tensions in the relationship, as when the main station director in Lincoln, E. A. Burnett, wrote in September 1906 giving his opinions about recent difficulties with Weakley: "I have about made up my mind that you will have to change Foremen before you get a satisfactory man. Weakley will almost certainly cause you more and more trouble as time goes on. This will not be a question of salary but due to his wrong point of view relative to the work."[127] This lack of a properly scientific "point of view" was a common problem in relations with manual laborers, who were often accustomed to other standards of skill and work valuation. Yet despite these privately expressed reservations about Weakley, Burnett and Snyder agreed to keep him on staff. They continued to trust him, although they did insist on the need to give him "specific instructions" from time to time.[128] When Snyder was later searching for another foreman, he recruited a man named Raymond Rhodes, outlining the job duties and compensation. "It will be your duty to handle the labor on the farm," Snyder wrote, noting that he would "be allowed sufficient help in the way of teamsters of laborers to do the work but the responsibility of having the work done will rest on you." Compensation would include a $75 a month salary along with a house on site, although for the first summer he would have to live in "the three attic rooms in the 'Assistant's Residence,' with the use of the bath room on the floor below, with other occupants of the house."[129] Rhodes would end up staying as foreman at the North Platte field station for over four decades, finally retiring in 1956.[130]

Other farm workers at the North Platte field station, as at others, included a mixture of temporary and long-term farmhands. The central dilemma for selecting manual laborers, as far as the station scientists were concerned, was judging in advance their character and work ethic. Sometimes farm workers came from the University in Lincoln, as when Burnett in the spring of 1906 suggested recent graduate Fred Pochon ("a Swiss from Scotts Bluff Co." in western Nebraska) to take care of livestock at the North Platte station, noting that he was "a good faithful boy, not very fast, but I think perhaps as good as Searle was, maybe better."[131] A few months later, Burnett informed Snyder about a man named George Felt from Ogallala, about fifty miles west of North Platte. Although a colleague "liked the looks of the man," Burnett "did not make any arrangements regarding wages," leaving that to Snyder himself. "If he is a good man," Burnett advised, "I would certainly pay him as much as he can get anywhere in the neighborhood."[132] But not all farm workers came from western Nebraska. While the hired labor market was often tight in this sparsely populated region, farmhands could be recruited in the more densely settled eastern part of the state. Burnett in Lincoln once wrote to Snyder, "we might get a little better class than you would out there."[133]

Keeping good workers around required finding ways to keep them employed despite seasonal variation in labor demand. After the spring planting season in 1906, for example, Snyder told Burnett that he might not have enough work to occupy all of the hired men. "We will see if we cannot so arrange our finances so as to keep all these men at work," Burnett promised in reply, suggesting that Snyder "cut down practically all of the expense for improvements except the fencing and the hay sweeps, and a mower if absolutely necessary, but keep the labor sufficient to do the work at least until after harvest." But if expenses were unmanageably large, the station might be compelled to "discharge some of the men."[134] The same labor dilemma occurred after the harvest season that fall.

Instead of "cutting the men down to three during the winter," Burnett again advised Snyder to "keep four men as long as [you can] use them to advantage, putting in the fence if you can get the posts set before it freezes up, weaning the calves, and dehorning the steers right away."[135] As a research site that was also a farm, the North Platte field station was thus subject to the vagaries of the often tight agricultural frontier labor market. Just when they needed the most hired labor to help operate the station—during planting and harvest times—such men were the scarcest. Despite these ongoing labor challenges, the North Platte station managed to keep expanding its operations, so that by the end of the decade the workforce had tripled or quadrupled, with twelve to sixteen hired laborers during the 1910 season, many of whom were University of Nebraska students working during the summer vacation.[136]

For those station directors and researchers who relied on student assistants to make up part of their labor force, recruitment efforts made this combination of both educational and practical capabilities clear. Connections with college faculty could play a crucial role in recruiting such workers, and their recommendations could sometimes encompass both types of qualifications. A US Forest Service station director in California in 1913, for example, wrote to University of Nebraska biologist Raymond J. Pool asking him to recommend a student assistant with education in plant physiology and ecology, recognizing that Nebraska was one of the few places offering such training. The forest station director not only asked Pool to "put me next to a man whose work in these two subjects has been promising, and whom you think would pan out in experiment station work," but also with a practical, hands-on background. "Men who were raised on a farm are to be preferred," he declared. "If a man comes out of Nebraska whose work as a student is entirely satisfactory, I will agree to use every influence to get him a permanent appointment at the Station after he takes the Civil Service Examination for Forest Assistant."[137] Students were an attractive compromise offering relatively cheap labor while having some advanced training that made them more reliable and knowledgeable about science than most other manual laborers.

In most cases, however, station research required at least a few employees who were prepared to do more than just manual labor. While station work in agriculture, forestry, or mining might be a lot like running a regular farm, forestry operation, or mine—and much biological field station work was quite similar in surface appearance to tourism and recreation—science at field stations also had its own distinctive requirements. For experiments to become publishable scientific knowledge, they had to be rigorously tracked and reported; for observations to be creditable in scientific publication, they had to be recorded systematically and reliably. Forest station director Pearson expressed this imperative when he recommended "frequent inspections on the ground" of the work done by station assistants, in order to "check their judgment and to observe facts which may have escaped their notice." Even when assistants did work that was "accurate and in accordance with instructions," no set of instructions could perfectly "foresee all the factors which may enter into an experiment," he thought. "The highest type of scientific work is done by scientists who, if they can not actually perform all the work themselves, personally supervise all important details."[138] Such an imperative was even more obvious at the Rocky Mountain biological stations, whose leaders typically brought in their entire research and teaching staff from their own universities and other similar institutions, adapting their existing on-campus laboratory supervision procedures to the field station context.

Station Work: Between Two Worlds

Making field stations work during the long phase of their normal operations did not just mean a division of labor between director, scientific staff members, and manual labor force, with some logistical support from the surrounding human social community, but also required deliberate attention to how customary techniques and practices could be adapted to local environmental conditions. The field station was thus a site for producing knowledge under the conditions of long-term experience, much like those engaged in practical activities in the same area, such as farming, ranching, forestry, mining, or recreation. Yet, at the same time, scientists working at field stations sought to import key aspects of cosmopolitan knowledge production, such as standard experimental protocols, which they also adapted to field conditions. That is one reason why, whether a new station was set up as a substation or branch of a central agricultural station or as a summer educational outpost of a university campus, its establishment was heavily guided by outside forces. The histories of the North Platte field station in Nebraska and the Mountain Lab of the University of Colorado, as narrated above, bear out this generalization in the early twentieth century. Both sites constituted successful hybrids of lab and field.

By contrast, over a decade earlier, Wyoming had launched several "experiment farms" around the state, which collectively illustrate more acutely the potential dilemmas of setting up a field station that lacked this full hybridity, due to the lack of funding for a career scientist to be in charge on site. At Lander, located in central Wyoming, for example, Professor McLaren of the state university in Laramie instructed the local organizer, Fred. T. Wright, to plow, harrow, and plant eighteen acres of land in the spring of 1891. However, Wright was a local businessman attempting to run the substation "at odd moments, taken from his other regular business," and was thus hard pressed to get the work done on time. "If this is not done in the best possible manner and at the earliest possible date, the chances for a crop or for any returns at all from the station are greatly diminished," he acknowledged. While he was "ready and anxious to do all that I possibly can to make the station a success," he warned the authorities in Laramie that they "must not expect such results as could be obtained if the director was on the ground, attending to matters for himself."[139] A local director, he thought, would ensure the success of the substation as a working farm, according to the criteria that local station supporters had in mind.

The Lander booster's idea of qualification for directing the field station, however, was based largely on the value of local experiential knowledge, rather than waiting for funds to hire someone with university education in science. J. D. Woodruff, president of the Lander Mercantile Company and apparently chair of the local civic committee intent on attracting the field station, recommended a local man who "fully understands the climate and soil," Jacob Meyer, as the "best qualified person . . . to take charge" of the substation.[140] By late May, Woodruff was upset about the lack of support from the State University Commission. "I was very much disappointed and somewhat discouraged," he wrote to them, "to find on my arrival home that so little had been done and so lukewarm an interest had been taken by your Commission in our Experimental Station." Woodruff believed they had been misled about how much management and operational responsibility the locals would have to bear, even after they had "exerted" themselves "to the utmost to secure a good tract of land and in a good location." "Discouraged and disgusted," he accused the Commission of trying to "hold us back, and to not allow our Station to show the results it is justly entitled to." Pleas to hire Meyer as supervisor had gone unheeded, with a "request . . . to hire someone temporarily by the day" instead. "If you want us to run the Station, please say so," he declared, "and we will do it, and will secure the best results of any section of this State."[141] This vision was almost entirely based on the standard of valuation associated with experiential knowledge: local use value above all other concerns.

From the local perspective of community leaders in Lander, hosting the substation, eager to show the promise of farming in its vicinity through a credible state authority, Laramie was failing to hold up its end of the bargain, but this was not true at all of the Wyoming experiment farms. At another Wyoming site, at Sheridan in the north-central part of the state, the setup seems to have gone somewhat better, due to the appointment of a full-time superintendent, James A. Baker, who was apparently the son (or another relative) of a local furniture dealer. By late June, Baker had "broken about thirty acres," almost finished the bridges and irrigation ditches, and was reporting that "the crops are looking nicely, especially the corn." Things were going so well that he asked the authorities in Laramie for permission to plant "about half an acre of strawberry plants" as another local experiment, at what was now proudly called the "Sheridan Experiment Farm."[142] Nevertheless, even at Sheridan, these "experiments" were not responding to cosmopolitan scientific standards for producing knowledge to circulate more widely among scientists but, rather, to local standards of valuation based on practical experience.

Over time, however, station scientists discovered that, even when local support was strong, success in their research ultimately depended on the presence of properly trained staff members, who could navigate the delicate balance between local practical problems based on the knowledge of experience and universalized experimental rigor based on cosmopolitan standards. One great advantage of bringing in trained assistants and student labor was their increased familiarity with the expectations for systematic practice and recording techniques. While variety testing and model farming may have dominated much early station research in the interior West, even branch agricultural stations eventually developed more systematic lines of research. Special problems investigated by agricultural stations in the Plains and Rockies—based on its distinctive environmental concerns—included irrigation, dry farming, and alkali soils. The problems might be regional and local, but the means for solving them meant transcending a solely experience-based and practical approach by bridging the epistemic rift with cosmopolitan science.

Guidance from researchers at central stations or home universities would be crucial to implementing cosmopolitan experimental protocols, but they were not alone. More systematic lines of research were pushed even more aggressively by national disciplinary and institutional forces, whether emanating from academic ecologists touting quadrats or national agricultural science officers encouraging more systematic research lines in order to receive funding. In the case of Colorado, for example, A. C. True, head of the US Department of Agriculture's Office of Experiment Stations, urged the Colorado Experiment Station in the summer of 1899 to focus on irrigation problems as its own distinctive research line.[143] True favored this strategy over having substations scattered throughout the state. "The Station should work for the general interests of the agriculture of Colorado and should carry on its investigations wherever they can best be prosecuted," he acknowledged, "but should be free to move its field work from point to point as the requirements of the work may demand. It is not fair to the farmers of the State to maintain expensive substations in two or three favored localities."[144] This problem was not unique to Colorado, and True wrote to other stations with similar sentiments, including the director at Oklahoma, to whom he wrote a few years earlier that he "would use every means to avoid the establishment of permanent substations."[145]

True reiterated this argument in a letter to the Colorado station's board of trustees, following a visit to the state, during which he had told them "that in many respects the field work of the station was too much in the nature of farming and that its experimental features were relatively superficial." Now he reminded them that the promise of the station "to do a great work for the agriculture of the State and indeed of the entire West" depended on finding a specialized line of work to concentrate on. "By making irrigation problems the basis of its work and undertaking special investigations on a plan which shall involve the co-operation of different departments of the station and intelligent farmers in different parts of the State, results of much importance may be obtained," he predicted.[146] The subtext, of course, was that the station's wide-ranging and diverse existing agenda was not meeting high standards of rigor. As advocates for cosmopolitan agricultural research, US Department of Agriculture scientists provided a countervailing force to a practical, regional orientation favoring studies that simply tested varieties and techniques for adaptability to local environmental conditions. By identifying larger problems and issues that were both related to the region and relevant to larger agricultural disciplinary agendas, researchers could attempt to make station science viable in both contexts.

Nevertheless, the Colorado station, like those in other states in the Great Plains and Rocky Mountains, did not adhere strictly to the Department of Agriculture's favored policy of pursuing particular research lines rather than dispersed substations. Researchers supplemented their

Fort Collins headquarters in the north-central part of the state with other stations in such places as the Arkansas River valley in the southeast part of the state. At these sites, a variety of traditional station activities were pursued in accordance with perceived local demands. The field agent of the Rocky Ford substation reported in 1905 that the work consisted of "field observations, co-operation with farmers, and some small plat work by the field agent on the station property; the efforts have been directed largely to problems relative to cantaloups, beets and alfalfa, the three principle [*sic*] crops in this section of the state."[147] Some of this research might eventually attract some wider attention among scientists, as did the work of P. K. Blinn on the resistance of cantaloupe to leaf blight, noted as ultimately significant in a recent disciplinary history of plant pathology.[148] However, these lines of research were built on top of a predominant commitment to regional problems and the local use value that was more characteristic of experiential knowledge, even if sometimes that work happened to generate some national or international resonance.

Likewise, the opening of an agricultural station at Colby, Kansas, in 1914 demonstrated the continuing commitment of regional scientists to focus on such work as basic variety testing in diverse environmental contexts within the western part of the state. After the site was selected in December 1913, the first superintendent, S. P. Clark, began in March 1914 to administer the station, located a half mile southwest of the town. Clark and his colleague G. E. Thompson reported that "the station expects to put in actual practice on the experimental farm the best known methods of handling soils and crops under existing conditions," along with "variety tests of any crops known to be adapted to this part of the state as well as variety tests of any crops which give even a remote possibility of proving profitable for this section." In addition, a small orchard and irrigation system were intended to "insure against total crop failure."[149] Station reports for Colby and other substations focused much of their attention on reports of successes and failures, weather conditions, future plans to adapt to past events—the staples of everyday experiential knowledge of life and work.[150]

Over time, though, most agricultural field stations received guidance in designing research protocols and designing their cropping plans in accordance with cosmopolitan standards through collaboration with federal science agencies. Many of these Great Plains substations—for example, station scientists in Colby—also undertook some cooperative work with the US Department of Agriculture's Office of Dry Land Agriculture, for which J. B. Kuska served as local agent, under more strictly standardized experimental protocols.[151] In this way, despite continuing to engage in some field work in the surrounding area and basic variety testing under local environmental conditions, which satisfied the demand for more immediate and practical local use value for the knowledge produced, on-site research at field stations increasingly brought in laboratory standards and experimental techniques, adapting them to the place-bound field station. By then, other branch stations across the Great Plains had already been engaged in such formal research collaboration for as long as fifteen years. The practice of setting up cooperative work with the US Department of Agriculture's Division of Agrostology (range grasses), for example, had been pioneered by the Highmore branch station in South Dakota back in the late 1890s, even as the Office of Experiment Stations was working hard to close substations across the region.[152] In sum, working as a scientist at an agricultural field station involved not only adapting to local environmental conditions and practical knowledge needs but also connecting it through cooperative agreements and increasing rigor in publication standards to the larger world of cosmopolitan science.

Similarly, at biological field stations, university scientists bridged the rift between the knowledge of experience and the standards of cosmopolitan science by combining practices of lab and field, though in this case as often for visiting university students and schoolteachers as for local collaborators. By doing indoor lab work on local materials collected in the field, both researchers and their students could integrate lab and field—though still with a strong leaning toward the field—in a very satisfying and productive way. This meant not only undertaking field observation and collecting, which were the predominant research activities, but also undertaking kinds of scientific research that would not have been possible in the field before, practices associated more with labs. "Since almost nothing has been done with microscopic anatomy of these mountain forms," one article advertising for prospective participants at the University of Colorado's Mountain Lab noted, "the student can have the satisfaction of examining plants whose anatomical structure has never been studied before."[153] It was arguably the scientific equipment itself that enabled such productive work integrating lab and field to be undertaken, though on the whole scientific work at Tolland continued to lean decidedly toward the field end of the spectrum, where experiential and cosmopolitan knowledge closely coexisted.

The same was true of other Rocky Mountain field stations, such as the Flathead Lake station of the University of Montana, whose on-site laboratory was "a small but substantial structure, built for outdoor work."[154] Similarly, at the University of Wyoming's Summer Science Camp, the "camp laboratory," like those at seaside stations, enabled students and researchers to work on material "while [it was] still fresh."[155] Even at Clements's Alpine Lab near Pikes Peak, where attempts to bring laboratory precision to the field were at their most intense, the site was viewed as being entirely of the field, rather than in any way a placeless lab. No additional non-site-related work was supposed to be carried on there. "It is the purpose of the alpine laboratory," claimed an early announcement of the station, almost certainly written by Clements himself, "to carry on investigations under field conditions alone, and the building will be equipped only with such books, microscopes and other instruments as are necessary to field experiment."[156]

Even as the leaders of Rocky Mountain field stations emphasized the centrality of the field for all of their work, as the anecdote about microscopic anatomy at the Mountain Lab indicates, they also sometimes freely mixed in laboratory methods and tools. Another interesting piece of evidence in this regard is a 1901 article written by Morton J. Elrod about the recently established University of Montana Biological Station, one of the first field stations in the Rocky Mountains. What makes this particular article noteworthy is that it appeared in a periodical called the *Journal of Applied Microscopy and Laboratory Methods*. Despite the raging contemporary debate over the strengths and weaknesses of laboratory research—especially using the microscope—in biology, and as advocates of the "new natural history" urged a return to the field, Elrod apparently saw no conflict between these two cultures of scientific practice. Indeed, in describing and picturing the apparatus used at the Montana station, he interchanged the words "laboratory" and "field," so that what was described in the text as a "laboratory table of the microscopist" was illustrated by an image of a "field table at Sin-Yale-a-Min Lake, for microscopical work," on the very same page (see figure 5.6). (We might conjecture that while the text could emphasize the placeless qualities of the work, the visual illustration of the apparatus made its field character more visible.) Furthermore, the structure for supporting the apparatus was described as "two sticks nailed to a fir tree at the desired height, and a couple of rough boards nailed to the top of these sticks" and "located in the shade, so that it is always comfortable."[157]

This was truly a laboratory in the field, although it would perhaps be more accurate to say that the field site had been transformed into a makeshift laboratory of sorts, in that a lakeside forest had been ren-

FIGURE 5.6. Bringing the laboratory outdoors. Original caption reads: "Field table at Sin-Yale-a-Min Lake, for microscopical work." Morton J. Elrod, "The University of Montana Biological Station," *Journal of Applied Microscopy and Laboratory Methods* 4 (1901): 1269–78, on p. 1271.

dered into a place suitable for carrying on lab work without even establishing a building to divide the indoors from the outdoors. Investigators working at the Flathead Lake station would be wholly immersed in the field, yet by virtue of their equipment and its arrangement, they would also be in a laboratory. It wasn't that lab and field had blended indistinguishably but that the site was both lab and field at the same time. While most Rocky Mountain field stations relied on actual laboratory buildings—and this was true even of the Montana station, which had indoor facilities (see figure 5.7) where overcrowding and overflow had led to the improvised outdoor lab—this situation is suggestive of the ways that lab and field overlapped and coexisted at the Rocky Mountain stations.

FIGURE 5.7. Inside the indoor laboratory at a field station. Morton J. Elrod, "The University of Montana Biological Station," *Journal of Applied Microscopy and Laboratory Methods* 4 (1901): 1269–78, on p. 1275.

Facilities at other biological stations functioned in similar ways. At Wyoming's Science Summer Camp, the built infrastructure included "four large log laboratory buildings," two twelve-person rooms for "graduate students and visiting scientists" who might want "to carry on independent studies" and two thirty-person classrooms for undergraduates.[158] Similarly, at the Science Lodge (successor to the Mountain Lab), the University of Colorado provided "modern laboratories" with "almost any equipment" supplied from the Boulder campus. For school teachers, likewise, attending the summer sessions at the Rocky Mountain field stations that offered such courses, the equipment at the station provided the necessary linkage between the field and future indoor teaching needs. The Mountain Lab, for example, would furnish "necessary apparatus for general botany, ecology, taxonomy, nature study and general zoology," along with a variety of special-purpose equipment. Participants could use "botanical driers" for collecting "from the various zones" in the field for later demonstration in the classroom, cameras for photographing "general landscapes, ecological formations and . . . individual plants."[159] In effect, the station functioned as an important node not just for original research intended for scientific publication but also for materially extending more advanced ecological ideas into the school classrooms of Colorado and other states.

Whether based on dry farming or irrigation agriculture, biological field work, or something else, stations shared certain similarities in their ongoing operations. Stations needed to have some way to connect their own work, even while it was in the midst of being undertaken, to key scientific work elsewhere. In particular, setting up a new station often meant providing researchers there with access to scientific publications to consult on site. While funds were often tight for providing such materials, officials improvised. At most biological field stations, for example, a rudimentary library emphasizing publications closely related to field work on site made it possible to consult significant scientific works without leaving the field station, a distinct advantage for thinking through research problems and envisioning their broader significance. The on-site library at the University of Colorado's Mountain Laboratory, for instance, was described as "very complete for the biology of the area," and supplemented by items brought up temporarily from Boulder.[160] It was the availability of library materials for reference, at least as much as the laboratory equipment and classroom space, that enabled high-level pedagogy and research work to be undertaken at Rocky Mountain field stations.[161] In the mountain valleys and on the mountainsides, as much as at agricultural stations down on the Great Plains, field stations could function most effectively with a full array of lab space, equipment, and library materials.

Regardless of its specific topical focus, the level of funding needed to maintain the built environment of a field station and conduct research work could be substantial. Of course, the specific amounts varied widely depending on the purpose and relationship of the station to other entities. Stations frequently operated on shoestring budgets. The branch station in Tribune, Kansas, one of four in the western half of that state, for example, received an annual state appropriation of only $2,500 in 1914 for repairs and experimental work.[162] At the same time, appropriations from state funds at Great Plains agricultural stations were often supplemented by federal funding for cooperative research and to pay the salaries of appointed agents from the US Department of Agriculture. Rocky Mountain biological stations occasionally received support from national institutions as well. For instance, the Carnegie Institution of Washington furnished $6,000 to Frederic Clements in 1918, "approximately half" of which was intended "for intensive studies at the Alpine Laboratory on Pikes Peak."[163] With those external funds, whether from government or private foundations, important strings were attached, connecting them to distant centers of scientific authority, through not only the flow of money but also flows of power and supervisory force.

Stations as Satellites

While local and regional forces shaped the emergence and development of field stations, so too did larger national and international forces. Agricultural substations across the Great Plains, though at first discouraged as an inappropriate use of federal funds for state science, were later subsidized through the placement of agents under the control of federal science agencies. A few agricultural branch stations were even run directly by the US Department of Agriculture, and its forestry bureau controlled the majority of the forest stations. Likewise, by the late 1910s the US Bureau of Mines had opened eleven mining stations, sometimes in cooperation with state mining schools, including in Salt Lake City, Utah; Golden, Colorado; and Bartlesville, Oklahoma, among other places, which operated much like urban laboratories.[164]

Biological field stations, on the other hand, tended to be funded regionally, but a few of them, such as the Alpine Laboratory of Frederic Clements (and its counterpart, the Desert Lab in Tucson, Arizona), as noted above, received substantial funds from the Carnegie Institution in Washington, DC—though this made them subject to the shifting research priorities of distant authorities. In general, however, national foundations were reluctant to fund most biological field stations, except for a few favored coastal sites, which the denial of funds for the Rocky Mountain Biological Laboratory in the 1920s makes clear.[165] But for the Alpine Lab and its counterparts in agriculture, forestry, and mining, they did function to some extent as peripheral satellites of core institutions outside the region, even if we must be careful to avoid positing that such lines of authority were absolute or uncontested.

More broadly, then, not only were field stations sites of scientific research and work in the natural environment of the Great Plains and Rocky Mountains, they were also situated in a much larger geography of knowledge production and circulation, for which they were effectively satellites of metropolitan centers of science. To be sure, stations were unique among all the modes of field science in that they constituted an enduring, year-round physical and human presence on the landscape that allowed both the eventual publication of research results by people within the region itself and the definite influence of regional priorities and imperatives on its work. Station work, whether in applied sciences of agriculture, forestry, ranching, and mining, or at biological field stations dependent on summer teaching, did sometimes persist through the determination of field scientists within the Plains and Rockies to use them as outposts within their own regional environments. Yet, at the same time, more distant scientific centers and disciplines exercised significant influence—regional control was far from complete.[166]

The most important national center for most stations, especially those in agriculture, forestry, and range management, was the Department of Agriculture headquarters in Washington, DC. Scientists and administrators there dispensed patronage to state operations, which in turn provided them with place-based loci for knowledge production that they needed at dispersed environmental sites. Midwestern agricultural scientist Eugene Davenport outlined an intellectual argument for having stations spread across the country. In his view, there were "local problems over the vast extent of this great country that could not be successfully solved by an organization centered perhaps two thousand miles away, problems that require the service of local organizations impelled by local initiative, backed by local financial support, and responsible to local authority."[167] At its most grandiose, the Washington headquarters was viewed by its research staffers as a "great experiment station," as E. W. Allen of the Office of Experiment Stations declared, "with probably the largest personnel and the most liberal appropriations of any institution of its kind." As a "central experiment station," the national Department of Agriculture both guided and emulated its state-level counterparts.[168]

From the perspective of the center, the greatest problem other than funding itself was to find a way to conduct field experiments in different environmental conditions. If other sciences might try "to isolate and control conditions," noted Davenport, the Department of Agriculture addressed problems of "the field or the barn, subjected to the utmost variety of internal and external conditions, chiefly climatic."[169] Or as Allen put it, the laboratory strength of Washington, DC, needed to be complemented by practical work "under field conditions" across the country, since the environmental features of the national capital "would be representative of only a small fraction of the country" anyway.[170] While the federal Hatch Act funds allowed considerable discretion to state officers and scientists, even before the more tightly regulated Adams Act, national agencies also made funds available for special projects that interested them. In this way, they were able to exercise some leverage over the research conducted at stations. At the same time, they often recognized the challenges of directing station research at a distance. In reflecting on the Great Plains stations cooperating with the US Division of Forestry, one federal forester commented that "wherever directions for field work are to be given from a central office far away," those instructions were bound to be only partially suited to the local conditions, since the stations were "all differently situated climatically," and thus, he thought, "the full measure of success can not be expected."[171] Lacking in experiential knowledge of place, it was impossible for distant federal scientists to direct everything flawlessly from afar.

The use of field stations in the Plains and Rockies by researchers from further east in order to conduct research in particular environmental conditions was not restricted just to those agricultural stations on the Plains receiving federal funding as part of the US Department of Agriculture's applied science infrastructure. At several of the biological field stations in the Rocky Mountains, outside scientists joined their regional counterparts, using these sites as bases for conducting larger research projects. For example, the New York Botanical Garden began using facilities at the Flathead Lake biological station of the University of Montana as early as 1901, returning in later years but also staying at the Mountain Lab in Tolland, Colorado, at least twice through 1916.[172] While on site, the eastern scientists benefited from cooperation and support from their colleagues in Colorado and Montana, which included not just use of on-site facilities but sometimes even accompanying them on field excursions.

Cooperation with researchers from outside the region was common to nearly all field stations, and it was often experienced as mutual. However, state agricultural scientists in particular sometimes resented Washington's imperialistic attitude toward their stations. What national officials viewed as "cooperation" could just as easily be viewed by regional actors as oppression. "There is no other insidious agency so inclined to undermine local self-government and so centralize power," remarked Texas station director H. H. Harrington, "as the acceptance by the people of money from the national government for the industrial development of their State."[173] Such views emerged not only in peripheries like the Great Plains and Rocky Mountains but also (maybe even more) among ambitious scientists in more populated and developed areas. In California, E. W. Hilgard lambasted the "insolence" of some federal agents "in the propositions they make," which treated the state as "hewers of wood and drawers of water."[174] Likewise, even so visible a national figure as Davenport expressed a desire to avoid "the cumbersome machinery of enforced cooperation with anybody," criticizing in particular the US Department of Agriculture's efforts to involve itself in local research problems. "Cooperation is a term that sounds well," he noted. "It smacks of good nature and obligingness, but in actual practice it is to be avoided wherever possible."[175]

Davenport's resistance to cooperation was rooted in his conviction that it led to conflict over the management of research. "In plain English," he declared, "it is not good business, even with a single institution, much less where two administrative systems are involved." He resented, in particular, the national agency's imperialistic ambition to control research across all national geographical space, declaring, "If there is a local question in agriculture between the two oceans and from the Mexican to the Canadian border that the Department of Agriculture through some of its thousands of local agents has not attempted to cover, or more properly speaking that it has not dipped into, then it would be interesting indeed to hear it named."[176] US government officials, on the other hand, portrayed cooperation with state stations as "mutually helpful relations" of "equal partnership," celebrating the "paternal attitude" of the Department of Agriculture in fostering the development of stations. "The Department usually has the advantage in point of funds and the broader survey of the general field," argued one, "and it can often place a larger number of specialists and assistants in the field, whereas the stations have the plant for carrying on the work, whether the experimental field, greenhouse, feeding stable, or dairy be required; and they also have a superior knowledge of local conditions."[177]

One major reason that federal researchers cultivated their collaborative relationships with state experiment station scientists was the access such regional actors had to environments that were distant from Washington. In 1901, for instance, F. Lamson-Scribner, the chief of the national Division of Agrostology, sought more samples of the grasses from Colorado, Wyoming, and other nearby states that field agents had collected in smaller quantities on previous visits. "Some of the things we are particularly anxious to get more seed of," he wrote to the Colorado station in Fort Collins, "were secured in the Arkansas Valley at Rocky Ford and other points farther west, and could perhaps be more satisfactorily grown at your substation at Rocky Ford, if you have any space there."[178] Collaboration between the US Department of Agriculture and Colorado also included such varied subjects as entomology and disease-resistant beet seeds.[179] The files of correspondence at the National Archives bulge with the mundane records of such exchanges and collaborations between national and state agricultural scientists, with letters for any given project radiating outward from Washington, DC, to collaborators in all or nearly all of the states of the Plains and Rockies, allowing them to adapt knowledge to a wide range of environmental settings and to discern patterns across geographical space.

Just as agricultural researchers at the Department of Agriculture and other federal government agencies used their connections in multiple states to conduct place-based research across an environmentally broad spectrum of regional sites, so too did field scientists based at other institutions outside the Plains and Rockies. By moving between the biological stations of Montana and Colorado, for example, the New York Botanical Garden enjoyed the built infrastructure, as well as laboratory facilities, instruments, and field equipment, at multiple sites, in their quest to collect specimens and to produce plant knowledge across the Rocky Mountain region.[180] Moreover, even more than the federal government agencies, with their far-flung research operations spanning the entire United States and sometimes beyond, the New York Botanical Garden was at that time expanding its work into a wide variety of environmental settings, not just the Rocky Mountains or even limited to the United States, sending several dozen expeditions to the West Indies in the early decades of the twentieth century, as well as field researchers to places as far away as South America and Alaska.[181] Whether in botany or in agricultural research, then, individual field stations in the states of the Plains and Rockies were just a few of the many satellites orbiting around ambitious place-based research programs in the practice of cosmopolitan scientific research institutions.

Cosmopolitan, disciplinary science exerted its influence not just in the work done in collaboration with visiting field scientists from eastern institutions but also in how it shaped the work done by scientists within the region. However much they might poke fun at the imperiousness of hierarchically imposed standards, by the 1910s researchers at the branch agricultural stations of the Great Plains, like their counterparts at the biological field stations of the Rocky Mountains, had internalized the cosmopolitan standards of rigor and proper methodology in most of their work. Thus, just as Francis Ramaley, director of the University of Colorado's mountain field station, oriented his research publications of the 1910s toward quadrat studies and other innovative, quantitatively rigorous practices of the time,[182] so too were most of agricultural scientists at branch field stations across the Plains diligently designing ever more rigorous experimental field studies.[183] Producing knowledge for local conditions and for local use value was still important, of course, but it was increasingly embedded within a framework of practices that was designed to be simultaneously place-based *and* cosmopolitan.

Many of the field stations described in this chapter still exist today. Agricultural substations such as the ones at Hays, Kansas, and North Platte, Nebraska, have grown into regional research centers with ongoing support and robust research programs. And, of course, main agricultural stations in such places as Fort Collins, Colorado; Laramie, Wyoming; and Bozeman, Montana, continue to thrive as headquarters for a whole range of agriculture-related disciplines in the states of the Great Plains and Rocky Mountains. Many other field stations, too, such as forestry stations and biological field stations, continue to exist as small villages for research and outreach. The University of Colorado still has a summer field station relocated nearby, and other sites such as the Rocky Mountain Biological Laboratory in Gothic, Colorado, and the University of Montana's field station at Flathead Lake continue to thrive. But while the life cycles of stations are more indefinite than those of living creatures, some of them have, alas, passed out of existence. Many agricultural substations or branch experiment farms and biological field stations, among others, eventually closed down due to suspension of funding. To finish this section on the work of stations, it is therefore appropriate to review the process of closure.

The first generation of agricultural branch stations closed around the turn of the century, at least in part under the impetus of US Department of Agriculture opposition to the use of so much research funding for dispersed variety testing, so they can perhaps serve collectively to illustrate the decision making and complications involved. While some substations could be reconfigured to focus on research lines, such as the Plains station in Colorado, others, like the same state's Divide and San Luis Valley stations, were instead simply closed. In February 1901, Carpenter reported that the "remains of the substation at Divide [had] been practically sold." Negotiations were undertaken to sell the land and on-site property.[184] By that time, Wyoming had already closed many of its substations. In the summer of 1894, for example, one superintendent, A. E. Hoyt, had begun to discuss selling off equipment and "implements on what has been known as the Experiment Farm at Sundance."[185] Yet in 1900 the station was still operating, now under the supervision of another local man, who purchased much of the farm's capital stock himself. "Mr. Lawenstein [*sic*] informs me that he bought the farm barn and some other things belonging to the farm," Hoyt wrote (now on US Land Office letterhead). "I think it advisable to send me [a] list of his purchases as he is liable to claim every thing arround [*sic*] there."[186] The disposal of equipment, along with advertising for the sale of the farm land itself, continued through the summer of 1900 and into 1901 under the direction of Lauenstein, who was apparently the last superintendent at Sundance.[187] The farm, still referred to as the Sundance Experiment Farm, was finally purchased for $250 in July 1902 by Swan Pearson and Joseph T. Farrell of Aladdin, Wyoming.[188] Not all stations were closed completely. The Lander Experiment Farm, also in Wyoming, continued to be used for limited horticultural experiments even while most of the land was leased.[189] The diversity of experiences reminds us that stations typically represented substantial resource investments in a purposely built (or rebuilt) landscape, so that closing out a station's life cycle was more complex than simply coming home from a field survey or collecting party, or even abandoning a scientific quarry.

Moreover, the closure of (most of) the first generation of branch agricultural stations did not signal the end of branch agricultural station science; far from it. In the first decades of the twentieth century, the US Department of Agriculture, which before that time had lobbied states vigorously against the use of federal research funds for substations, launched a vast array of branch stations scattered throughout the Great Plains region—often in collaboration with the states—under the auspices of the department's own rapidly proliferating bureaucracy. In agencies such as the Office of Western Irrigation Agriculture or the Office of Cereal Crops and Diseases, both part of the Bureau of Plant Industry, federal scientists discovered a new love for branch stations, but under their own authority rather than local and state. Most prominent of all in the Great Plains region was the Office of Dryland Agriculture, founded in 1905 under the direction of E. C. Chilcott, who had resigned as agronomist at South Dakota's agricultural experiment station in order take the position.[190] By coordinating and systematizing observations and experiments across an entire region, Chilcott and his collaborators sought to reconfigure the field station as a site within a larger spatial matrix for making regional knowledge that more fully transcended the particularities of local place (see figure 5.8).[191]

Biological field stations, too, faced changes resulting from new funding priorities in response to larger disciplinary pressures from cosmopolitan science, though a bit later. Most famously, perhaps, the Carnegie Institution of Washington's Department of Plant Biology—major patron not only for the work of Frederic Clements at various field stations around the West but also for others based at field stations, such as For-

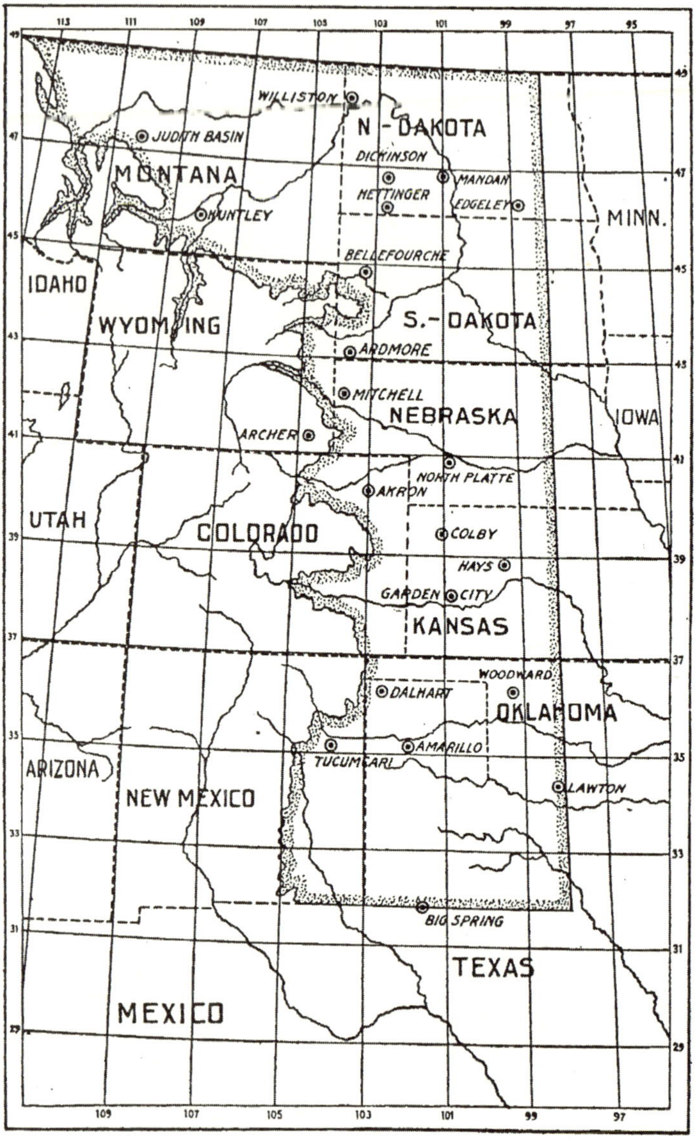

FIGURE 5.8. Regional map of various branch agricultural experiment stations scattered across the Great Plains by the mid-1910s. E. C. Chilcott, J. S. Cole, and W. W. Burr, *Spring Wheat in the Great Plains Area: Relation of Cultural Methods to Production*, Bulletin of the US Department of Agriculture, no. 214 (1915), 2.

rest Shreve at the Desert Laboratory in Tucson, Arizona—shifted emphatically away from field station science in the 1930s after the opening of its plant physiology laboratory at Stanford.[192] The Alpine Laboratory itself was formally, if unceremoniously, closed around the time of Clements's retirement in 1941, sold to Clements along with "its instruments, library material, and the 20 acres of real estate on which it sat," for just $10.[193] The more that biological field stations attracted prestigious external funding, it seemed, the more vulnerable they would be to shifts in research priorities of distant funding agencies, reflecting, if imperfectly, the ever-shifting frontiers of rigorous research in cosmopolitan science. Many other Rocky Mountain field stations, however, maintained themselves indefinitely through the practical niche of summer teaching in synergy with research, and stations focused on knowledge for agriculture, forestry, and mining maintained their own constituencies to support state and federal government funding.

Despite their obvious differences, the University of Colorado's Mountain Lab at Tolland and the branch agricultural field station in North Platte, Nebraska, were both exemplars of the mode of field practice that spanned diverse disciplines and brought together both pure and applied sciences conventionally regarded as residing in distinct categories. This chapter has outlined the rhythms, relations, and logic of the station mode of knowledge production in the field. The challenges and opportunities offered by the distinctive environmental conditions in which the stations were located were experienced not just through the technological systems of transport and communications, as were all modes of field science, but also through the extensive built environment that gave station science its distinctive contours. More than any other mode of field work, the field station was situated in the lab-field borderlands, where field observation and collecting coexisted cheek-by-jowl with indoor laboratories. Whether in the form of agricultural field experiments on dry farming methods on the Great Plains or ecological research on quadrats in high Rocky Mountain meadows, station science was rarely far removed from the culture of laboratories, libraries, and human settlement in general.

Station science was also deeply tied to the local practical activities, ranging from ranching and agriculture to forestry and summer tourism. By concentrating attention on one specific place and the area around it, scientists at field stations have partially bridged the epistemic rift between the cosmopolitan knowledge of science and the experiential knowledge of everyday life and work. With its tight intertwining of long-term residency with attempts to produce knowledge in accordance with cosmopolitan standards of rigor, and to answer questions of larger significance, the field station has formed a robust and enduring mode of practice for the field sciences, which shows no signs of abating. However, like all modes of field practice, the station was inherently limiting in its view of nature, most obviously because it was locally sited yet called upon to produce knowledge of larger regional, or even universal, significance. Even as they maintained a delicate balance between their close ties with other local practices, such as farming, and the expectations of cosmopolitan science, station scientists did not always gain easy acceptance of the relevance of the knowledge they produced to those living and working around them.[194] While it is beyond the scope of this book, the dense interactions between field scientists at stations and their local and regional publics constitute an interesting subject in their own right—and such tensions were nowhere more visible in field science than around stations, given their long-term commitment to producing knowledge in place.

The same struggles remain visible today in how station scientists relate to their regional publics. In a sociological study of present-day relations between scientific "farm advisors" and growers in California, Christopher Henke shows that the latter are often skeptical of the universalizing knowledge of station experiments and defend the irreducibly local, place-based environmental characteristics of their own land. It is thus necessary for the researchers to conduct field trials on the growers' land in order to convince them to change their practices in accordance with scientific knowledge.[195] The same could also be said of other station scientists, such as those at forest stations in their relations with logging companies, or biological field stations in their relations with other land users, such as hunters, fishermen, and tourists. Stations may help to solve the problem of place in knowledge production, but they can never completely solve it. Nevertheless, by understanding the origins of station science in the railroad era of the American West, we can lay the groundwork for thinking more deeply about how the conditions of scientific practice at field stations—as in other modes of field practice—have shaped what scientists know about the diverse and complex natural and human environments of the Earth.

Epilogue

The landscape of the interior American West was transformed between the 1860s and 1910s. Geographer W. M. Davis visited the Colorado Front Range area around Denver twice in that time, first in 1869 as a student with a party from Harvard's mining school and again in 1910 while helping lead that same college's summer field course in advanced physiography. Reflecting on his two visits, Davis noted the "strongly marked" contrasts. "In 1869," he remembered, "Cheyenne was the nearest railroad station; we went thence by stage over-night to Denver, a small frontier town, where we outfitted for a camping trip of six weeks in the surveyed mountains." By 1910, however, "Denver was a good sized city and an important railroad center, from which we could quickly enter or cross the mountains by train, to the Northwest, West, or Southwest, in the most convenient fashion for rapid work."

Yet, Davis continued, "there was another contrast," which pointed to the concurrent transformation of the practice of science over this same period: the 1869 party's "highest geographical ambitions" had been "to determine the latitude, longitude and altitude of the more important mountains, to make hachured sketch maps of the district and to write brief empirical descriptions of the scenery," while the 1910 party found that "there was no need of surveying, as fairly good governmental maps were available for much of the field." Instead, they devoted themselves "almost exclusively to the preparation of systematic explanatory descriptions of the mountain landscape, following methods that were unimagined in 1869."[1] Thus, not only did Davis describe a profound transformation over the forty-one-year span between his research visits but he also explicitly juxtaposed the parallel development of the material economy and the practice of field science.

As Davis's commentary suggests, the late nineteenth- and early twentieth-century period marks a pivotal transformation in the production and circulation of knowledge in the West. Rapid change in the organization of work and transport infrastructure had already begun even before Davis's first visit. The US Army expedition led by Raynolds in 1859–1860 had "ended an era in western exploration," in the words of one historian. During the early 1860s, when most of this activity was curtailed by the demands of the Civil War, the army's topographical engineers merged with the Corps of Engineers, eliminating the Topographical Bureau.[2] In the new era after the Civil War, expeditions were increasingly taken over by civilian leadership. A transcontinental railway line completed in 1869 had dramatically opened up access to a central corridor of the West, bringing in scientific researchers just as easily as it brought in economic fortune seekers.

These rapid changes continued through the end of the nineteenth and into the early twentieth century. In the early 1890s—as Frederick Jackson Turner then announced—the US frontier effectively "closed"—although more recent commentators have pointed out that the frontier never really closed but instead just changed its character.[3] At least as significant was that sometime around the 1890s science in the region passed the crucial fulcrum of a transition from limited funding and support to the successful regional institutionalization of research in the West. In 1901, when Denver first hosted the American Association for the Advancement of Science, the convention's president, Charles Sedgwick Minot of Harvard University, applauded the "sympathy and understanding between the practical life of Colorado and the scientific life of this Association."[4] By the end of the 1910s the establishment of permanent universities, museums, and government agencies inside the region—and in concert with their counterparts outside the region—had largely replaced the heroic, adventurous era of westward exploration. Much like the Civil War, the social demands on manpower and funding during the First World War paused the expansion of other scientific work, though of course resources related to the war were intensively studied and inventoried. Yet by around 1920 the structures for field science in the American West—as, indeed, for many other sciences in the country as a whole—were largely in place.

The greatest change in the practice of field work around the time of the First World War was probably the coming of the mass-produced automobile, whose appearance neatly marks the end of the time period covered in this book in the same way that the arrival of the railroad does the beginning. As automobiles became more widely available and affordable, many lay and amateur collaborators in geographically dispersed networks began using them. The wealthy expert amateur mammologist E. R. Warren of Colorado, for instance, gushed in 1916 about "getting to places" in his Ford that he "had no chance to get to before" since they had been "too far away" to reach.[5] So too did many field scientists working in survey mode, given their strong emphasis on systematically covering geographical territory. By the mid-to-late 1910s survey field scientists began taking up the automobile with great enthusiasm. In considering the benefits of the automobile in 1918, Stephen S. Visher of South Dakota estimated that considerable time was wasted "on foot" with "a large share of his time . . . spent in getting to and from the work. An auto continually at hand would have enabled him to cover at least three times as much territory."[6] His South Dakota colleague, the state geologist Freeman Ward, estimated that work done without an automobile was "only 25 per cent efficient."[7] The ecologist Frederic Clements, capably and expertly assisted by his wife, Edith Clements, embraced the automobile in their cross-country vegetation treks, and they negotiated with the president of the Carnegie Institution of Washington, Robert S. Woodward, about funding field expenses. When Woodward offered to allow automobile expenses if they were scrupulously separated from personal use, Clements replied that it would be "a great aid" to their field work.[8]

Similar testimonials could be found across the field sciences that were operating in survey mode. The US Department of Agriculture soil surveyor Macy Lapham noted that California and Nebraska had been the "first of the cooperating states to adopt use of the automobile in the soil survey." Despite all the problematic features of "early Ford cars" that he recalled—damage to makeshift curtains, inefficient brakes, hand cranking—they had nonetheless "welcomed this equipment and felt that we were rolling along in luxury." Cruising along at "15 to 20 miles an hour" proved to be their most remarked-upon feature, but it was perceived as both a blessing and a curse. Lapham remembered a telling conversation with US Bureau of Soils director Milton Whitney around 1913, encouraging him to come see their work, offering "that we could now show him soils along the way and still cover as much as a hundred miles a day" in the motor car. But Whitney "threw up his hands in alarm," impressed less by the distance possible to be covered than by its effects on environmental perception. "Lapham," replied Whitney, "that's too fast, you can't see anything when you go that fast."[9] Another problem, especially in the Rocky Mountains, was the poor quality of roads in steep areas. While working with the US Biological Survey, Alexander Wetmore's trip to Lake Burford in northern New Mexico, "made in a Ford truck, required four hours, a sufficient commentary," he thought, "on the state of the mountain roads, as the distance travelled was only about forty miles."[10] Despite such challenges, the automobile was adopted readily by most survey scientists working in the field.

In quarry science, too, the automobile eventually freed field researchers from the underlying railroad geography that had dominated their work since the 1860s. The automobile, in the retrospective judgment of paleontologist William Diller Matthew, had "revolutionized the fossil-hunting business." Most important, it had "greatly widened the range of practical field work." Under the ecological constraints of subsistence, especially for working animals, preautomobile field work had been restricted "to a radius of five or ten miles from water or feed." He remembered quite vividly that a "'dry camp' supplied at intervals with food and water might carry the exploration a stage further, but throughout the West . . . there were enormous areas of bad-land exposures that it was not practicable to prospect adequately for lack of water or feed." A railroad-based envirotechnical system for scientific field work may have been a dramatic change from what came before, but it still meant dependence on the limited, narrow corridors of biological productivity when away from the tracks in the arid and semiarid West.[11]

"With the automobile," Matthew noted by way of contrast, "there is probably no exposure so distant but that it can be, and will be prospected for fossils. Moreover, the range of a day's collecting . . . is so greatly increased, that most areas in the Plains can be prospected from some near-by settlement, and it hardly pays to camp out."[12] Thus, in exactly the same way that survey science was distinctly shaped by the railroad era, and then transformed again by the arrival of the automobile, so too was the prospecting phase of quarry science. For scientific quarry prospectors of all kinds, a rapid movement into the field by railroad had been followed by a season of extended dependence on the ecological resources of the arid western regions where rich deposits of fossils, artifacts, and other scientific materials were located, but now they were to be freed to a significant extent from such constraints by the automobile.

Even at field stations, despite being fixed in one place, the arrival of the automobile marked a major watershed in the 1920s. No longer would it be necessary to locate a site along a railroad line in order to provide such easy, rapid access to a field station. This was especially true for the Rocky Mountain biological field stations. The University of Colorado's Mountain Lab, for example, eventually became accessible by automobile road—in fact, already by 1917 Ramaley and his colleagues were touting the ease of automobile access from Boulder and Denver.[13] Later, in addition to the Mountain Lab, which had served as the University of Colorado's biological field station from 1909 to 1919, the institution established the "Science Lodge" for both biology and geology at an altitude of ten thousand feet on the side of Mount Niwot near the village of Nederland, located in the mountains above Boulder. Dedicated in July 1922, this new "laboratory building" was constructed next to the "summer recreational camp conducted by the associated students of the university," by which time both facilities were "accessible by automobile over good mountain roads."[14] While the rails were not abandoned completely in all cases, especially for long-distance travel to reach more remote sites—the Rocky Mountain Biological Laboratory in Gothic, Colorado, for example, continued to promote its accessibility by rail via a combination of the Denver and Rio Grande Western narrow-gage line to Crested Butte, in addition to the nearby US Highway 50 to Gunnison—the Rocky Mountain field station was no longer tied to the rails in any narrow sense.[15]

The University of Wyoming's Science Summer Camp was founded soon afterward as an auto-accessible site from its very beginning. After two initial years at temporary sites in the vicinity of Laramie, the camp moved to its permanent site along an automobile road that had just been constructed through the high mountains west of the college town.[16] Participants could by then reach the camp by means of an "improved oiled highway" (now a paved, scenic route) that extended from Laramie to Saratoga over the Medicine Bow mountain range. Automobility also reshaped the curriculum of the summer field station, making it possible in the case of the Wyoming Camp for zoology students to travel an average of 400 miles and geology students an average of 1,500 miles by car during the field season.[17] Similarly, the Rocky Mountain Biological Station near Gunnison, Colorado, had four automobiles available "for those carrying on research problems some miles from the station," charging users only for the cost of gas, oil, and tire wear.[18] These sites were part of a new postrailroad generation of field stations, sharing many characteristics with their counterparts of the railroad era but also transforming those practices and dispersing the station model much more widely through the environment. No longer tethered to the rail lifelines, field stations proliferated ever more fully into the landscape of field science. Across all modes of field practice, the end of the railroad era may not have come suddenly, but the adoption of the automobile was typically experienced as a revelation.

Field practice has continued to evolve since the 1920s. Throughout the twentieth century, undergirded by the automobile (and other internal combustion engine vehicles) along with aircraft, modes of field practice continued to evolve and proliferate, despite the continuing hegemony of the laboratory.[19] Nevertheless, even as longer-term and more quantitatively and descriptively rigorous, and sometimes even experimental, projects have proliferated, they have remained subject both to place-specific pressures that have shaped field work and the fundamental tension between the universal and the particular. As Amanda Rees has emphasized in her study of field primatology, researchers in the field have continued to "manage a number of threats to the physical and epistemological integrity of both their sites and their persons, threats rarely experienced inside the four walls of the laboratory." Moreover, while their "reported observations from a specific field site are often presented to their audiences as representative of the typical," such claims have remained recursively contestable—a conundrum she calls the "fieldworkers' regress."[20] Yet for all its challenges, the field has remained a key place for doing primatology. Likewise, in geology, as Naomi Oreskes has pointed out, while "the balance had tipped" by the second half of the twentieth century from the field to the laboratory—with its associated values of "exactitude, precision, and control"—the field tradition has also persisted, even if it was increasingly viewed as inferior.[21] More broadly, as the distinctive social role of the "field man" has persisted, its advocates have continued to feel the need to defend their place-based craft. As one self-described "field man" in mid-to-late twentieth-century archaeology put it: "We need field camps. That's where you learn, is in the field. It's where I learned. You don't learn how to use a trowel, you don't learn the textures of soils, from a book."[22]

Other historians of field science practice during the past century have stressed the role of more elaborate environmental surveillance and human surveying technologies, the contested politics of knowledge related to place, transnational and globalized field work, and the evolving role of place-based field science in environmental management, among other things.[23] In recent decades, as outside criticism of science and lay participation in science has arguably exploded, the field sciences have maintained and perhaps even increased in prominence. While the laboratory remains a somewhat forbidding location for lay involvement in science despite an architectural trend toward visible labs behind glass,[24] it is in the field that citizen scientists of all kinds have become active in monitoring, tracking, measuring, and observing the environment.

In justifying my decision to study the Great Plains and Rocky Mountains between 1860 and 1920, I have emphasized the importance of science and technology to the historical development of these regions, whose effective settlement patterns were decisively shaped by science, as much as or more than perhaps any other region of the country. While I have focused primarily on the *production* of knowledge from the Plains and Rockies, the *circulation* of knowledge to other parts of society was also important. Along the way, I have alluded many times to various examples of the political, economic, and cultural uses of science, but I have not attempted to analyze those uses comprehensively—that would take another book. Whether through educating the broader public, providing useful knowledge to farmers and industry, or influencing public policy, scientists who performed field work in the Plains and Rockies attempted to back up their frequent claims about the utility of the knowledge they were producing. For a complete account of the place of science in regional development, it would be worthwhile to direct sustained attention to these wider uses of science. Above all, it was arguably the context of capitalist economic development that most indelibly shaped the form of science in these regions during the railroad era, and probably beyond. The utility of that science was often as much symbolic as narrowly practical. The wealthy benefactors of great urban museums, whether in Denver or Pittsburgh, surely had the symbolic uses of science in mind when they sponsored such projects as the excavation and display of gigantic dinosaur skeletons.

Scientists themselves, as we have seen, had their own uses for the knowledge they produced, circulating it through disciplinary structures to researchers around the world. Indeed, it is the peculiarity of how knowledge in modern science travels and combines over great distances that often distinguishes it from other knowledge systems. But the frontier of scientific knowledge production did not simply roll unabated over a uniformly yielding natural landscape. Regional studies like this one should make clear that the particular conditions of the places where science was practiced interacted with the quasi-universal process of scientific expansion, just as distinctive environments have offered varying kinds of resistance to capitalist expansion. And getting work done in the field also involved mobilizing human workers and technologies in conjunction with natural resources, bringing together people who had different—sometimes complementary, sometimes conflicting—understandings of the regional environment. Large-scale geographical hierarchy and regional forces together shaped the practice of science and, ultimately, the production of knowledge in the Plains and Rockies.

The study of field work, here and elsewhere, opens up an alternative framework for reconstructing a global system of knowledge production and circulation that is better balanced with the particular, the local, and the regional. Universalism, the desire to attain knowledge that can be applied to all times and places, has been an overriding ambition of (and driving force behind) modern science.[25] But while high status, credibility, and legitimacy have come with knowledge aspiring to greater universality, participation in that epistemological game has not always worked to produce pragmatically useful and environmentally sustainable knowledge on the ground in particular places. Certain forms of field science might offer a model for forging a middle ground between universality of cosmopolitan science, on the one hand, and the distinctiveness and particularity of local or regional environmental and cultural context, on the other.[26] The urgency of such a task has been suggested by the commentaries of a great many historians of science, technology, and medicine. As Richard Drayton forcefully argues, "Our world is still in shock from the great 'modernizing' projects of capital, government, science, and imperialism, all the connected attempts to forge universal systems, usually without the consent of their future beneficiaries."[27]

At the same time, as I hope this book has made clear, the field sciences as they were constituted in regions such as the US Great Plains and Rocky Mountains by the early twentieth century suffered from many of the same geographical inequalities and structural tensions in mediating between the universal and the particular as the more powerful laboratory science whose shadow they inhabited. I have tried to keep squarely in view that all systems of knowledge, including Western science, represent ways of thinking about the world that are produced through human work and travel in particular places. Doing so may help us overcome what David Turnbull, one of the leading scholars of indigenous knowledge systems, identifies as a fundamental problem of modern thinking: "Our modernist predilections for displaying our knowledge as objective and universal," he observes, "denies human agency or movement in making knowledge."[28] But as field researchers negotiated with the demands—environmental and human—of localities and regions, they began at least to grasp the contours of the problem.

In the end, though, even the field sciences as constituted in the modern world, which have extended a modest bridge over the epistemic rift between the cosmopolitan knowledge of science and the experiential knowledge of everyday life, will likely not be enough. Since the railroad era, if not before, naturalists and scientists employing field practices have succeeded in producing knowledge that is impressively systematic, rigorous, and comprehensive. But when the inevitable partiality of field science practices has made compromises with the larger laboratory-dominated, universalizing, and globalizing epistemic landscape seem unavoidable—when push has come to shove—modernity has permitted too little space for the local, the regional, the particular, and the experiental. As Naomi Klein has recently written, in thinking about the enormous challenges posed by climate change, such problems are "place-based," while modern humanity is "everywhere at once":

> The problem is not just that we are moving too quickly. It is also that the terrain on which the changes are taking place is intensely local: an early blooming of a particular flower, an unusually thin layer of ice on a lake, the late arrival of a migratory bird. Noticing those kinds of subtle changes requires an intimate connection to a specific ecosystem. That kind of communion happens only when we know a place deeply, not just as scenery but also as sustenance, and when local knowledge is passed on with a sense of sacred trust from one generation to the next.[29]

Taking Klein's observation seriously, then, we need more than just a celebration of the field sciences as they were historically constructed during the railroad era—and have endured ever since. Instead, we need a clear-eyed appraisal of our predicament that pushes the field sciences toward a better balance, more evenly encompassing both the cosmopolitan perspective and the experiential perspective. The problem is not that modern science, including field science, is wrong or useless, but that it is incomplete and insufficient. The field sciences can offer a promising foundation for a new course, if its proponents become more fully aware of both the advantages and limitations inherited from its past.

Notes

Abbreviations Used in the Notes

AAP	American Antiquities Permit Files, National Anthropological Archives, Suitland, Maryland
AFEC	Alcott Farrar Elwell Collection, #1916, American Heritage Center, University of Wyoming, Laramie
ANP	Aven Nelson Papers, #400013, American Heritage Center, University of Wyoming, Laramie
CCF-NPS	Central Classified Files (NPS), P Entry 10-A, RG 79, Records of the National Park Service, National Archives II, College Park, Maryland
CHMP	C. Hart Merriam Papers, Volume II, Bancroft Library, University of California, Berkeley
CPC	Cook Papers Collection, Agate Fossil Beds National Monument, Harrison, Nebraska
EDP	Earl Douglass Papers, Ms 196, Marriott Library, University of Utah, Salt Lake City
EDP-CMNH	Earl Douglass Papers, Carnegie Museum of Natural History Diplodocus and Douglass Archives, Carnegie Mellon University Libraries, Digital Collections, http://diva.library.cmu.edu
EFCP	Edith S. and Frederic E. Clements Papers, #1678, American Heritage Center, University of Wyoming, Laramie
ESDM	Earth Sciences Department Miscellaneous, IA.050-12-26-95, Archives, Denver Museum of Nature and Science, Denver, Colorado
EXP	Experiment Station Collection, Archives/Special Collections, Morgan Library, Colorado State University, Fort Collins
FLCC	Fort Lewis College Collection, M001, Center of Southwest Studies, Fort Lewis College, Durango, Colorado

FRP	Francis Ramaley Papers, University Archives, Norlin Library, University of Colorado, Boulder
HP-LR	Hague Papers, Letters Received, Entry 141 (Old Entry 67), RG 57, Records of the Geological Survey, National Archives II, College Park, Maryland
HP-LS	Hague Papers, Letters Sent, Entry 142 (Old Entry 68), RG 57, Records of the Geological Survey, National Archives II, College Park, Maryland
JBHP	John Bell Hatcher Papers, Section of Vertebrate Paleontology, Carnegie Museum of Natural History, Pittsburgh, Pennsylvania
JDFC	Jesse Dade Figgins Correspondence, IA.501.Figgins, Archives, Denver Museum of Nature and Science, Denver, Colorado
KAESR	Agricultural Experiment Station Reports, University Archives & Manuscripts, Hale Library, Kansas State University, Manhattan
KCWB	Correspondence and Reports, Records of the Weather Bureau, Office of the Regional Director, Kansas City, Missouri, National Archives, Central Plains Region, Kansas City, Missouri
LBP	Lawrence Bruner Papers, #8/12/10, University Archives, Love Library, University of Nebraska, Lincoln
LGC	Louis George Carpenter, Archives/Special Collections, Morgan Library, Colorado State University, Fort Collins
MJEP	Morton J. Elrod Papers, MSS 486, University of Montana, Special Collections, Missoula
MVNP	Research Center, Mesa Verde National Park, Colorado
NAESR-NP	Agricultural Experiment Station Records, North Platte Experiment Station (Correspondence), #10/14/1, University Archives, Love Library, University of Nebraska, Lincoln
NDGS	North Dakota Geological Survey, UA34, Special Collections, University of North Dakota, Grand Forks
OAPP	O. A. Peterson Papers, Section of Vertebrate Paleontology, Carnegie Museum of Natural History, Pittsburgh, Pennsylvania
RBDLF	Richard "Beaver Dick" Leigh Folder, Jackson Hole Historical Society, Jackson Hole, Wyoming
RFP	Ridgway Family Papers, Caine Collection Mss 8, Tanner Reading Room, Merrill Library, Utah State University, Logan
RJPP	Raymond John Pool Papers, #12/07/12, University Archives, Love Library, University of Nebraska, Lincoln

SSD-RSS Records of the Soil Survey Division, Reports on Soil Surveys, PI-66 Entry 216, RG 54, Records of the Bureau of Plant Industry, Soils, and Agricultural Engineering, National Archives II, College Park, Maryland

SWP Samuel Williston Papers, PP32, Spencer Research Library, University of Kansas, Lawrence

USDA United States Department of Agriculture

USFS-L Medicine Wheel File, United States Forest Service, Bighorn National Forest, District Office, Lovell, Wyoming

USGS United States Geological Survey

UWAESR University of Wyoming, Agricultural Experiment Station Records, #548001, American Heritage Center, University of Wyoming, Laramie

VBP Vernon Bailey Papers, #554, American Heritage Center, University of Wyoming, Laramie

YNPA Yellowstone National Park Archives, Gardiner, Montana

Introduction

1. Goetzmann, *Exploration and Empire.* A brief account of the Raynolds expedition, as well as a map of the route followed, can be found on 310–12.

2. Hayden, "New Species of Hare," 114, 116.

3. Ramaley, "Subalpine Lake-Shore Vegetation," 57.

4. Ramaley's work also included popular introductions to Colorado's plant life for a wider audience, e.g., Ramaley, *Wild Flowers and Trees of Colorado* and *Colorado Plant Life.* On Frederic Clements and the quadrat, see Tobey, *Saving the Prairies*, 48–75; and Kohler, *Landscapes and Labscapes*, 100–108. Some of Clements's leading publications in ecology included *Research Methods in Ecology*; *Plant Succession*; and *Plant Indicators.*

5. For a helpful beginning to the regional study of science in North America, which shows how the scientific community became established in a new region, with the unusually diverse environment of California itself playing a major role, see Smith, *Pacific Visions.* Other regions of the United States have been studied from the perspective of institutional development, though not environment or work organization, including the South, Southwest, Midwest, and Mississippi Valley: see Midgette, *To Foster the Spirit of Professionalism*; Hendrickson, "Science and Culture"; Webb, *Science in the American Southwest*; and Daniel Goldstein, "Midwestern Naturalists." Some important studies of science in Victorian Britain describe the diversity of sites for knowledge production in regions with an explicitly geographical sensibility. See Alberti, "Field, Lab, and

Museum"; and Naylor, *Regionalizing Science.* On the "regionalization" of science, see Vetter, "Regionalizing Knowledge"; and Coen, "Imperial Climatographies from Tyrol to Turkestan."

6. Harvey, *Limits to Capital*, 415–16.

7. A distinct project in the "historical geography of science" has already been outlined by some geographers, most notably in Livingstone, *Putting Science in Its Place.*

8. Cunningham and Williams, "De-Centring the 'Big Picture'"; Kohler, "A Generalist's Vision"; Secord, "Knowledge in Transit."

9. The secondary literature on national science is vast. For a very selective sampling, see Harwood, *Styles of Scientific Thought*; Reingold, *Science, American Style*; Kohlstedt, *Formation of the American Scientific Community*; Pauly, *Biologists and the Promise of American Life*; Rainger, Benson, and Maienschein, *American Development of Biology*; Fox, *Savant and the State*; Morrell and Thackray, *Gentlemen of Science*; Finnegan, *Natural History Societies and Civic Culture in Victorian Scotland*; and Shen, *Unearthing the Nation.*

10. Wyatt and Balmer, "Home on the Range," 621.

11. In agreement with Lyn Schumaker, I "draw no sharp distinction between practices and methods" and view "methodology as growing out of practices, even mundane ones like choosing a tent site or hiring a cook. When practices become standardized and mythologized, scientists call them methods, but this should not obscure the fact that a variety of often contingent daily activity is what gets the work done," as noted in *Africanizing Anthropology*, 8. My conceptualization of modes of practice also draws generatively on Kohler, "Systems of Production," 88. A somewhat different taxonomy of methods or practices that cuts across disciplines, focused more on modes of cognition, can be found in Pickstone, *Ways of Knowing*; for a critique, see Barnes, "Elusive Memories of Technoscience."

12. For a taxonomic overview of these four modes, see Vetter, "Field Life in the American West." For an examination of how three of the four modes emerged in the practice of one subdiscipline during this period, see Vetter, "Capitalizing on Grass."

13. Over the past quarter century, laboratory studies have become firmly established in science and technology studies (STS). The importance of laboratories in shaping the epistemic contours of modern science certainly justifies the close scrutiny lavished on them. For a canonical example, see Latour and Woolgar, *Laboratory Life.* An early outline of this approach is Latour, "Give Me a Laboratory."

14. Kohler, "Place and Practice in Field Biology," 191. For a seminal collection of papers on scientific field work, along with a useful introduction laying

out some major conceptual issues in studying field practice, see Kuklick and Kohler, *Science in the Field.* See also Kuklick, "Personal Equations"; and Bourguet, Licoppe, and Sibum, eds., *Instruments, Travel and Science.*

15. As an alternative, historians of the laboratory have turned to laboratory notebooks for evidence about practice, which can at least give access to a researcher's ongoing account of laboratory practice if not usually communication with other participants, e.g., in Holmes, Renn, and Rheinberger, *Reworking the Bench.*

16. Vetter, "Science along the Railroad." See also Freeman, "Tracks to a New World." For a helpful analogous case of the ship for travel in the oceans, which makes the case for the ship as an actual instrument, see Sorrenson, "Ship as a Scientific Instrument."

17. White, *Railroaded,* xxix.

18. Greene, *Horses at Work,* 78.

19. Mulford, "Notes upon the Northwestern and Rocky Mountain Flora. I," 117.

20. John, *Spreading the News,* 64, 90, 92, 99, 110. See also Fuller, *American Mail.*

21. Henkin, *Postal Age,* 2–3.

22. John, *Spreading the News,* 75, 92.

23. Thompson, *Wiring a Continent.* For an insightful study of the telegraph in practice, see Downey, *Telegraph Messenger Boys.*

24. West, "Wired to the World," 78.

25. Warwick, *Masters of Theory*; Klein, *Experiments, Models, Paper Tools.*

26. te Heesen, "Boxes in Nature."

27. Rudwick, *Great Devonian Controversy,* 36.

28. Pritchard, *Confluence,* 19. See also Reuss and Cutcliffe, *Illusory Boundary.*

29. LeCain, *Mass Destruction,* 22.

30. For a useful study of the language of "head" versus "hand" in US social class history, as a conceptual relation that fostered cooperative social relations and consolidated middle-class power, see Rice, *Minding the Machine.*

31. Pratt, *Imperial Eyes,* 27. For a recent account complicating Pratt's influential analysis, see Guelke and Guelke, "Imperial Eyes on South Africa."

32. Rudwick, *Great Devonian Controversy,* 38.

33. The secondary literature on this topic is especially rich for Victorian Britain; see, for example, White, *Thomas Huxley*; Endersby, *Imperial Nature*; and Barton, "'Men of Science.'" In the United States, there seems to have been less resistance to the word *scientist*—indeed, White notes that Huxley viewed it as an Americanism—so I do not hesitate to use it. However, I do generally avoid the word *professional,* usually opting instead for the more neutral

term *career scientist*, which emphasizes the socioeconomic process of seeking to make a living through science without necessarily assuming any connection with personal identity. While I fully concur with the efforts of historians like White, Endersby, and Barton to pay attention to actors' categories, I also want to probe beneath the surface to see what those actors may or may not have perceived themselves—in other words, to use historians' categories as well. By adopting a complementary approach to the rich historical work on the emergence of identities of "philosophical" thinkers and "men of science," we can analyze how social and practical realities of individual lives structured the kinds of opportunities for identity construction that knowledge makers had. Like the generic "knowledge maker," "career scientist" is a social category imposed on the past in order to make analytical claims about the emergence and development of science in the American West. Yet while some career scientists undoubtedly saw themselves as professionals, others did not; hence, I largely avoid the latter terminology. For further skepticism about the term "professional scientist" in the nineteenth-century United States, see Lucier, "Professional and the Scientist."

34. As key actors in the world of the late nineteenth-century American collecting sciences, the specimen dealers, who bought and sold scientific objects for sale, have been insightfully discussed within the realm of natural history in Barrow, "Specimen Dealer." While the largest specimen dealers, such as Ward's Natural Science Establishment in Rochester, were in the East, there were specimen dealers in the interior West as well, including L. W. Stilwell of Deadwood, South Dakota. Moreover, Eastern specimen dealers procured specimens from Westerners, including R. C. Hills, who sold minerals to A. E. Foote of Philadelphia, and Francis Cragin, who sold at least one meteorite to Ward's in Rochester. L. W. Stillwell to Mr R. C. Hills, 5 June 1914; W. M. Foote to Mr. R. C. Hills, 25 April 1916, Richard Charles Hills Correspondence, IA.148.Hills, Archives, Denver Museum of Nature and Science, Denver, Colorado; Preston, "Franceville (El Paso County, Colorado) Meteorite," 852–57. For further details on Cragin's dealings in meteorites, see the various receipts in Francis W. Cragin Papers, MS362, Special Collections/Archives, Tutt Library, Colorado College, Colorado Springs, Colorado, folder 4. While at Colorado College in the 1890s, Cragin and his wife engaged in further dealings with Ward's, including exchanging goat skins for a llama skeleton, as noted in Catherine Cragin to Mary Ann Cragin, 26 December 1892, Francis Cragin Papers, Part 2, MS370, Special Collections/Archives, Tutt Library, Colorado College, Colorado Springs, Colorado, box 1, folder 3. Another example of a specimen dealer and "professional collector" within the region was Colorado's Charles Aiken, whose work is examined in Ruskin, "Business of Natural History."

35. For the concept of the "new middle class," see Mills, *White Collar.* See also Bledstein, *Culture of Professionalism*; Aron, *Ladies and Gentlemen of the Civil Service*; and Zunz, *Making America Corporate.*

36. Aron, *Working at Play.* See also Kohler, *All Creatures*, ch. 2; and Loefgren, *On Holiday.* A still useful classic text on tourism in the West is Pomeroy, *In Search of the Golden West*, and several other key works include Rothman, *Devil's Bargains*; Sears, *Sacred Places*; Shaffer, *See America First*; and Wrobel and Long, *Seeing and Being Seen.*

37. Kuklick, *Savage Within*, 90, 92.

38. Lucier, "Commercial Interests and Scientific Disinterestedness." See also Lucier, *Scientists and Swindlers.*

39. My thinking about this distinction has been generatively shaped by the contrast between "cosmopolitan" and "residential" knowledge mentioned in Kohler, *All Creatures*, 184. For a fuller discussion of these issues, especially the distinction between "cosmopolitan" and "experiential" knowledge, see Vetter, "Introduction: Lay Participation." The term *cosmopolitan*, especially as being in tension with the perspective of personal experience, might be associated with the classic "view from nowhere," which is discussed in philosophical terms most famously in Nagel, *View from Nowhere.* Such a philosophical point of departure has been prominently historicized in Daston and Galison, *Objectivity.* On the earlier history of cosmopolitan ideals and their partial realization, see Daston, "Ideal and Reality." Another philosophical perspective that helps to sharpen thinking about the pluralistic coexistence of different forms of knowledge, is Dupré, *Disorder of Things.* For the concept of "factification" as the process by which new domains are transformed from experiential to cosmopolitan, see Vetter, "Capitalizing on Grass."

40. The concept of an "epistemic rift" in the history of knowledge is analogous to what has been called the "metabolic rift" in social theory and historical geography, as discussed, e.g., in Moore, "Environmental Crises and the Metabolic Rift"; and Moore, "Transcending the Metabolic Rift."

41. Packard, "Introductory," 2, 4.

42. Merriam, "Biology in Our Colleges," 353.

43. Many historians of science have used these categories at various times, especially "local." Some recent works have been especially thoughtful and analytically helpful in considering the three less-frequently used words, however: On "vernacular" knowledge, see Valencius, *Lost History*, 177; Coen, "Tongues of Seismology"; Pandora, "Knowledge Held in Common"; and Tilley, "Global Histories." For further analysis of mediation between science and everyday experience as a process of translation, see Coen, *Earthquake Observers*, 11; and Pandora and Rader, "Science in the Everyday World." On "folk" knowledge in field work,

see Fan, *British Naturalists in Qing China,* 143. On "indigenous" knowledge in science, see Cooper, *Inventing the Indigenous.*

44. Moreover, we might subdivide the category of experiential knowledge still further. Historian of science Jim Endersby, for example, distinguishes three elements of local knowledge of a place, which include "topographic knowledge" (where things are), "endemic knowledge" (what is there, such as distinctive species), and "indigenous knowledge" (folk knowledge of groups long resident there), in *Imperial Nature,* 89. While no singular taxonomy should be treated as exhaustive, this tripartite division does help us to think more fully about the multiple facets of experiential knowledge as a way of appreciating its richness.

45. Many historians have noticed the leverage that came with being in the field. Richard Drayton, for example, points out that "British botanists were in a more compromised position than their mathematical and astronomical colleagues" because, in their work, "they remained dependent on the world of amateurs," in Drayton, *Nature's Government,* 138. Similarly, Endersby shows how "distance also 'tyrannized' Europeans who, for example, wanted access to Australia's flora. Unless someone like Hooker was willing to send a plant-collector halfway around the world, he had to rely on the locals to acquire plants for him." Thus colonial collectors "with exclusive access to hard-to-obtain specimens could have considerable bargaining power in their relationships with metropolitan experts," if they chose to take it. See Endersby, "A Garden Enclosed," 324, 333. For a collection of thoughtful papers on the continuing question of the role for amateurs, increasingly coming to be understood as "citizen science," see Charvolin, Micoud, and Nyhart, *Des Sciences Citoyennes?*

46. Even ethnographic field work of a year or longer might depend on the long-term stability of native field assistants, as demonstrated in Schumaker, *Africanizing Anthropology.*

47. "Denver Meeting," 233, 234.

48. On metropolitan-provincial relations in Britain and France, see Inkster and Morrell, *Metropolis and Province*; and Nye, *Science in the Provinces.*

49. "Denver Meeting," 234.

Chapter 1. Making the Field

1. For biographies of Hayden, see Cassidy, *Ferdinand V. Hayden*; and Foster, *Strange Genius.*

2. On Cockerell's botanical work, see his "Recollections of a Naturalist IV," 13; "Diverse Floras"; "Alien Clematis"; "New Variety"; "Wine-Red Sunflower"; "Suppression and Loss"; and "Sunflower Problems." A list of his zoological pub-

lications, particularly in entomology, would be enormous, but among his leading works for a wider audience was *Zoology of Colorado.*

3. For background on Cockerell's life, see Weber, *American Cockerell*; and Ewan, *Rocky Mountain Naturalists*, ch. 10.

4. The trope of the field as a natural laboratory, especially in colonial and imperial settings, has become the focus of several worthwhile historical studies, including Tilley, *Africa as a Living Laboratory*; MacLeod and Rehbock, *Darwin's Laboratory*; Fowler, *Laboratory for Anthropology*; and Cittadino, *Nature as the Laboratory*. For a related case, see Adler, "Ship as Laboratory."

5. Cockerell, "Recollections of a Naturalist II," 149.

6. Browne, "Natural History," 559. On the origins of the natural history tradition in Renaissance Europe, see Ogilvie, *Science of Describing.*

7. Jardine and Spary, "Natures of Cultural History," 8–9. The centrality of "practices" to historical accounts of natural history became well established with the publication of *Cultures of Natural History.*

8. For Linnaeus, see Koerner, *Linnaeus*; Müller-Wille, "Joining Lapland and the Topinambes"; Müller-Wille, "Walnuts at Hudson Bay"; and Dietz, "Contribution and Co-Production." For Banks, see Mackay, "Agents of Empire." For Thouin, see Spary, *Utopia's Garden*. For Hooker, see Endersby, *Imperial Nature.*

9. Fan, *British Naturalists in Qing China*, 138.

10. Endersby, *Imperial Nature*, 55; see also further discussion in chapter 2 of this volume. On the interactions between European naturalists and non-Europeans, see Green Musselman, "Plant Knowledge at the Cape."

11. For Baird, see Goldstein, "'Yours for Science'"; Deiss, "Spencer F. Baird and His Collectors"; and Rivinus and Youssef, *Spencer Baird of the Smithsonian*. A few residents of the American West participated in Baird's sprawling network during its later years, including Colorado naturalist Martha Maxwell beginning in 1869, as noted in Benson, *Martha Maxwell*, 85–86.

12. Wells, "James Hall's Amateurs."

13. Kisling, "*Naturalists' Directory* and the Evolution of Communication."

14. Burns, "Networking Ohio Valley Archaeology"; Burns, "Negotiating the Agricultural Frontier."

15. A brief general summary of natural history practices in the mid-nineteenth century, with emphasis on emerging tools and standardization, is Farber, "New Tools and Standard Practices, 1840–1859," in *Finding Order*. For highlights of the extensive literature on the natural history tradition in Britain, see Allen, *Naturalist in Britain*; Knell, *Culture of English Geology*; Larsen, "Not since Noah"; Secord, "Corresponding Interests"; Secord, "Science in the Pub"; Alberti, "Amateurs and Professionals in One County"; Alberti, "Placing Nature"; McOuat, "Cataloguing Power"; and Shteir, *Cultivating Women, Cultivating Science*. On the

local and regional natural history tradition in continental Europe, see Phillips, "Friends of Nature"; te Heesen, "From Natural Historical Investment"; and Cooper, *Inventing the Indigenous.* On American natural history, some useful works include Barrow, *Passion for Birds*, Keeney, *Botanizers*; Kohlstedt, "Nineteenth-Century Amateur Tradition"; Judd, *Untilled Garden*; Leach, *Butterfly People*; Porter, *Eagle's Nest*; Sorenson, *Brethren of the Net*; Stephens, *Science, Race, and Religion in the American South*; and Welch, *Book of Nature.*

16. Cockerell, "Recollections of a Naturalist III," 209. However, Cockerell's early forays into Rocky Mountain entomology attracted some criticism for overlooking "several hundreds of species recorded from the same region and elevation," in "Communications," 127–28.

17. Yokota, "Sowing the Seeds of Postcolonial Discontent," in *Unbecoming British*, 155; Lewis, *Democracy of Facts.*

18. For examples of conchological specimens collected in the Great Plains and Rocky Mountains states and territories, though described by distant experts, see Conrad, "Note on a Cretaceous Limestone," 5; Tryon, "Notes on Mollusca," 150–51; Gabb, "Description of a New Helix." One lonely conchologist on the Great Plains would appear in print by the mid-1880s, when Charles T. Simpson of Ogallala, Nebraska (temporarily at least—a few years later he was in Taggart, Missouri!), published in a new periodical called *Conchologists' Exchange*, which would be renamed the *Nautilus* a few years later. See Simpson, "Succinea Lineata. W. G. B. in Nebraska," 65; and Simpson, "Blacklist Them," 10. The latter article by Simpson complained bitterly about conchologists who failed to reciprocate exchanges.

19. Cockerell, "Recollections of a Naturalist IV," 13.

20. Cockerell, "Recollections of a Naturalist VI," 126. At least one topically specialized but geographically general scientific journal was based in the Great Plains: the *Journal of Mycology*, published in Manhattan, Kansas, from 1885 until 1891, when its editor moved to Ohio State University, as noted in "William Ashbrook Kellerman," 49–63.

21. Blankinship, "Century of Botanical Exploration," 3.

22. Cockerell, "Entomology," 13–14. A year later, Cockerell wrote: "We are pioneers, and the scientific work is of less than five years standing. If I were to rely on the eastern researches for the instruction of the people, I should lead them astray, and at the same time stultify myself and the station. It would amount to declaring, that in my opinion the experiment station here was unnecessary," in Cockerell, *Entomological Observations in 1894*, 49.

23. Cockerell, "New Species of Bees," 9–11; Cockerell, "Curious Case of Protective Coloration," 154.

24. "Editorial," *Muhlenbergia* 4 (1908), 15–16.

25. Weber, *King of Colorado Botany*, 3.

26. Edward Lee Greene to Thune Kumlien, 26 September 1872, reprinted in Main, "Life and Letters of Edward Lee Greene," 179–80. For background information on Greene, see Ewan, *Rocky Mountain Naturalists*, ch. 5; and McIntosh, "Edward Lee Greene."

27. For example, see Pack, "*Coleonyx vareigatus*," 101.

28. For example, in his well-known work on bison, J. A. Allen refers variously to hunters, fishermen, mountaineers, and miners as sources of information, in Allen, *American Bisons*.

29. Blankinship and Henshall, "Common Names of Montana Plants," 116.

30. See, e.g., Lovewell, "Kansas Meteorology," 612; Shear, *Field Work of the Division of Agrostology*, 12; Washburn, "Relations of the Station Entomologist," 35–36; and "Making of Forecasts by Laymen," 4–5.

31. For more details see Vetter, "Cowboys, Scientists, and Fossils."

32. Cook, *Tales of the 04 Ranch*, 8–9.

33. Cook, *Fifty Years on the Old Frontier*, 187, 196. On Native American beliefs about fossils, both in general and specifically those Cook was interacting with, see Mayor, *Fossil Legends*. The connection between thunderstorms and knowledge—a connection that has been perhaps especially strong in the thunderstorm-prone Great Plains region—resonates with the experiences of famed holy man Black Elk's account of gaining knowledge through visions, in DeMallie, *Sixth Grandfather*, 109.

34. Wilson, *Buffalo Bird Woman's Garden*, 22. The revealing preface by University of Minnesota anthropologist Albert Ernest Jenks draws attention to the compatibility of Hidatsa knowledge with science by claiming that Wilson's study "unexpectedly revealed certain varieties of maize of apparently great value to agriculture in the semi-arid areas" (n.p.). The more recent edition cited above also includes some helpful introductory material to place the account, and how it was constructed, into historical context. For further evidence of the belief that scientific agriculture in North Dakota could draw upon the experiential knowledge of the American Indians of the region, see Will and Hyde, *Corn among the Indians*, 290–91.

35. Gilmore, *Uses of Plants*, 3.

36. Alford, *Civilization, and the Story of the Absentee Shawnees*, 23. "To all who will listen and look," he commented, "nature reveals most astounding secrets" (56).

37. Cockerell, "Recollections of a Naturalist," 374–75.

38. Aiken, "Glimpse at Colorado and Its Birds," 13.

39. Cameron, "On the Migration of Phalaropes," 67–68.

40. Carpenter, *Forests and Snow*, 3–4.

41. L. G. Carpenter to W. L. Hartman, 14 October 1901, LGC, folder 2. Carpenter received a supportive reply, stating an appreciation for the value of data produced by scientists according to cosmopolitan standards, in W. L. Hartman to Hon. L. G. Carpenter, 18 October 1901, LGC, folder 2.

42. Weber, *Letters*, 154. Originals in Theodore Dru Alison Cockerell Papers, University Archives, Norlin Library, University of Colorado, Boulder.

43. Weber, *Letters*, xxii. For a few of the many examples from the letters themselves, see Weber, *Letters*, 78, 90–91, 95, 101, 108, 165.

44. Cockerell, "Classification of Slight Varieties," 270–72. This article was followed by replies from N. L. Britton (immediately afterward) and L. H. Bailey (later issue).

45. Cockerell, "Recollections of a Naturalist IV," 14–15.

46. Weber, *Letters*, 110–11. Elsewhere, Cockerell referred to Vetter (no relation to author), who worked in a town store, as "a decidedly popular man, always jovial and bouncing" and "a short little man, rather stoutly built," in Weber, *Letters*, 38.

47. Cockerell, "Entomological Ethics," 87–90.

48. Cockerell, "Recollections of a Naturalist II," 151–52.

49. Wrigley, *Continuity, Chance, and Change*. On the role of steam power in the circulation of knowledge, see also Fyfe, *Steam-Powered Knowledge*.

50. Hayden, "On the Geology of the Tertiary Formations," 9.

51. *Annual Report of the American Museum of Natural History*, no. 44 (1912), 16.

52. Andrews, *Killing for Coal*.

53. Vetter, "Science along the Railroad"; Vetter, "Field Science in the Railroad Era."

54. For some illustrative examples of the transformative effects of "steam-powered transportation" on the operations of the postal service, see John, "Private Enterprise, Public Good?," 333.

55. For an analysis of British naval expeditions in political and economic context, see Mackay, *In the Wake of Cook*. See also Robinson, *Coldest Crucible*; and Goetzmann, *Exploration and Empire*.

56. Sorrenson, "Ship as a Scientific Instrument." On exploratory scientific travel in various parts of Europe in the tradition of Humboldt, including not just collecting but also measurement with instruments, see Bourguet, "Landscape with Numbers."

57. Farber, *Finding Order in Nature*, 50.

58. Ludlow, *Report of a Reconnaissance*, 9.

59. Lass, "Missouri River Steamboating," 48–49. See also Lass, *From the Missouri to the Great Salt Lake*; and Lass, *Navigating the Missouri*.

60. Culbertson, *Journal of an Expedition*.

61. For a popular history of Hayden's work in the Dakota badlands region, including a good route map, see Stucker, "Hayden in the Badlands." For a brief overview of Hayden's contributions, see Foster, "Ferdinand Vandeveer Hayden as Naturalist." On F. B. Meek, see Nelson, "Meek at Albany."

62. Meek continued his positive relationship with Hayden but did not travel with him on future trips to the Dakotas, though the two did field work together again in late 1858 in central and northeastern Kansas, as noted in Nelson and Fryxell, "Ante-Bellum Collaboration," 187–200.

63. F. V. Hayden to Spencer Baird, 2 July 1854, in Fryxell, *Ferdinand Hayden*, 146–47. Fryxell sequentially compiled letters from this period, thus providing a quasi-archival overview of Hayden's field experiences.

64. Schubert, "Troublesome Partnership." See also Jordan, *"Happiness Is Not My Companion,"* 21. For a brief sketch of Warren's travels, see Flanagan, "Gouverneur Kemble Warren."

65. Ludlow, *Report of a Reconnaissance*, 19. For contemporaneous confirmation of Grinnell's frustration, see the diary of a fellow traveler, the journalist Fred Power, reprinted in Buecker, "'Distance Lends Enchantment,'" 214. Further photographic and documentary records of the expedition are available in Grafe and Horsted, *Exploring with Custer*; and Lubetkin, *Custer and the 1873 Yellowstone Survey*. See also Pyenson, "Athena's Retinue," who discusses Grinnell's objection to "the military discipline of a quick march" (384) in comparison to field scientists working on other colonial frontiers in Russia and Argentina.

66. Wheeler, "Western Exploration," 240.

67. Hayden, *Geological Report*.

68. An instructive comparison could be made to the shaping of oceanographic expeditions by the navy, as perceptively analyzed in Rozwadowski, "Small World." For more, see Rozwadowski, *Fathoming the Ocean*.

69. Hayden, *On the Geology and Natural History of the Upper Missouri*, 5.

70. Hayden, "Explorations under the War Department," 109–16; Hayden, "Primordial Sandstone," 68–79; and Hayden, "New Species of Hare," 113–16.

71. Leidy, "Notes of Remains," 20–29.

72. Vetter, "Science along the Railroad." Even before the Union Pacific was built, the railroad had already stimulated scientific collecting in the West, during the surveys of the 1850s for multiple alternative railway routes to the Pacific, as amply chronicled in Goetzmann, *Army Exploration in the American West*; Albright, *Official Explorations for Pacific Railroads*; and Moore, "Zoology of the Pacific Railroad Surveys." For examples of scientific work on railroad surveys, see LeConte, "New Coleoptera Collected," 49–53; LeConte, *Notes on the Geology*; and Bell, *New Tracks in North America*.

73. Jepson, "Telegraph Comes to Colorado," 4.

74. Thompson, *Wiring a Continent*, 354–55.

75. For a later example of a field scientist explicitly using a railroad line to define a corridor for scientific investigation, shortly after the completion of a transcontinental line across the northern tier of states, see Newberry, "Notes on the Geology and Botany."

76. Weber, *Letters*, 14.

77. Cockerell, "Recollections of a Naturalist III," 210.

78. Cope, "Notice of Proboscians," 580.

79. See, for example, the accounts of the 1870 and 1871 expeditions, provided in Betts, "Yale College Expedition of 1870"; Vaughan, "'Broad Are Nebraska's Rolling Plains'"; and Pankin, "Yale Scientific Expedition of 1871." Secondary accounts include McIntosh, "Route of a Sand Hills Bone Hunt"; Pankin, "George G. Lobdell, Jr."; and Pankin, "Yale Scientific Expeditions in Kansas." Other scientific travelers in the early 1870s also continued to rely on military aid, despite the existence of railroad lines that were crucial in moving in and out of the field, such as J. A. Allen of Harvard's Museum of Comparative Zoology, who received vital support and favors from Major-General John Pope, Commander of the Department of the Missouri, acknowledged in Allen, "Report on the Mammals and Birds," 11.

80. "Othniel Charles Marsh," 403, 410.

81. Reverchon, "Botanizing in Texas, I," 56. For biographical background on Reverchon, see Geiser, *Naturalists of the Frontier*, 215–24.

82. Rehn and Hebard, "Contribution to the Knowledge," 358.

83. Strecker, *Notes on the Fauna*, 1.

84. Hayden and Selwyn, *North America*, 40.

85. Hayden, "Our Great West," 196. Five years later, the Earl of Dunraven, a noted English gentleman traveler, though claiming he was "not a scientific man," made a similar point in the same venue. "There is no such thing as a chain of Rocky Mountains," he declared, then evocatively described the region instead as including "various ranges and belts of mountains which embrace within their far-stretching arms, fertile valleys, arid deserts, sunny hill slopes clothed with valuable timber, parks full of pastoral beauty, basking beneath a sun that warms them into semi-tropical life, but that never melts the virgin snow that whitens the hoary heads of the mountains that forever look down upon those smiling scenes." He added: "Rich and extensive plains, tracts of inhabitable land almost large enough to be the cradle and home of nations, are included within the Rocky Mountains," in Lord Dunraven, "Moose and Cariboo Hunting," 335–36.

86. Hayden, "Report of F. V. Hayden," 67 (in 1873 combined reprinted version of the first three annual reports).

87. Gray and Hooker, "Vegetation of the Rocky Mountain Region," 1.

88. Not everyone would agree with the precision of the terms "Great Plains" and "Rocky Mountains." In fact, one of the best-known field scientists in the region during the railroad era, John Wesley Powell, proposed renaming them the "Great Plateaus" and the "Stony" and "Park" mountains, for the northern and southern Rockies, respectively. See Powell, "Physiographic Regions," 83, 86–88. Further support for referring to the Great Plains region as a "plateau" can be found in publications of other US Geological Survey researchers, e.g., in Adams, "Physiographic Divisions of Kansas," 98.

89. For a long-term perspective, see Whitlock, Reasoner, and Key, "Paleoenvironmental History."

90. Crowley, "Ranching in the Mountain Parks of Colorado."

91. Wyckoff, *Creating Colorado*. See also Brosnan, *Uniting Mountain and Plain*; Veblen and Lorenz, *Colorado Front Range*; and Wohl, *Virtual Rivers*.

92. Axelrod, "Rise of the Grassland Biome," 163–201. See also Coupland, "Effects of Fluctuations in Weather," 273–317; and Wells, "Scarp Woodlands," 246–49.

93. Hayden and Selwyn, *North America*, 47.

94. Rice and party, "Reconnaissance Survey of Western Nebraska," 1875. For a more general discussion of the various early twentieth-century explanations, see Keyes, "Origin of the Great Plains," 352. For another account, from about a decade earlier, highlighting the dissecting forces of west-to-east-flowing streams from the Rocky Mountains in shaping the Great Plains, see Adams, "Physiographic Divisions of Kansas," 98.

95. Bessey, "Distribution of Forest Trees," 21. See also Overfield, *Science with Practice*, 138–39.

96. Larson, "Role of the Bison."

97. Borchert, "Climate of the Central North American Grassland." Using long-term climatic data, Borchert argues that the regional climate pattern during summers of drought extends eastward to the usually wetter prairies of Iowa and Illinois, thus providing an explanation of the shape of the entire grassland region. Here I am concerned primarily with the shortgrass steppe zone rather than the tallgrass prairies, which are significantly more humid during most summers.

98. A collection of writings in archaeology that emphasize the role of smaller ecological islands in human interactions with the environment is Kornfield and Osborn, *Islands on the Plains*. For an overview of the plant geography of the region as a whole, written within the railroad era itself, see Rydberg, "Plant Geography of North America," 870–73. On the deep past origins of North American nature for the whole continent, see Flannery, *Eternal Frontier*. See also Harrington and Harman, "Climate and Vegetation."

99. Brunswig, "Prehistoric Utilization of Patch Environments."

100. MacDonell and Wandsnider, "Western Niobrara River."

101. A study of prehistoric uses of the Black Hills, emphasizing the dependence of hunter-gatherers on plants and small animals more than conventional big game, is Kornfield, "Pull of the Hills." On the history of forest management, see Freeman, *Black Hills Forestry*. On conflicts over the area, see Worster, "Black Hills."

102. C. H. M., review of J. A. Allen, *List of Mammals Collected in the Black Hills Region of South Dakota*, 490.

103. On the environmental history of the Missouri River region, see Schneiders, *Big Sky Rivers*; and Schneiders, *Unruly River*.

104. Koch and Bozell, "Environmental and Cultural Variation."

105. McIntosh, *Nebraska Sand Hills*; Welsch, "Sandhill Baled-Hay Construction."

106. For an excellent and readable summary, along with some useful maps, see Trimble, *Geologic Story of the Great Plains*.

107. Worster, *Rivers of Empire*. Dan Flores makes a good case for complementing the emphasis on aridity with a focus on mountains as well, though largely for spiritual rather than material reasons, in Flores, "Islands in the Desert."

108. True and Clark, *Agricultural Experiment Stations in the United States*, 27.

109. Paul, "Colorado as a Pioneer of Science in the Mining West." Worster concurs, noting that the hard-rock miner is a possible third role that was regionally distinctive besides the two main ecological modes under white occupation: the pastoral West (cowboy and sheepherder) and the hydraulic West (irrigation and water engineer), in Worster, "New West, True West," 150–51.

110. Hollister, *Mines of Colorado*, 46.

111. As a story of human interactions with inorganic nature, the environmental history of mining in the West is now beginning to be told, e.g., in Quivik, "Smoke and Tailings"; LeCain, *Mass Destruction*; Curtis, *Gambling on Ore*; and Andrews, *Killing for Coal*.

112. Barbour, "Some Methods of Collecting," 260.

113. Barbour, "Some Methods of Collecting," 261. Like road and canal building, the increase in prairie dog activity on the Plains was probably a result of human activity, in that heavy grazing by proliferating cattle on the region's grasses inadvertently created more favorable habitat for them. On this and other aspects of the changing relationship between humans and prairie dogs, see Jones, "Becoming a Pest."

114. M'Master, "'Bad Lands' of Wyoming," 119.

115. Kingsley, "Hat Creek Bad Lands," 966.

116. The Colorado School of Mines in Golden, which opened just before statehood in the mid-1870s, a decade later had professors in chemistry and assaying, geology and drawing, physics and engineering, and mathematics, as reported in Colorado State School of Mines, *Annual Report of the State School of Mines, 1885*, 3, by which time it was also publishing reports on field work, in Colorado State School of Mines, *Annual Report of Field Work and Analyses Relating to the Economic Geology of Colorado*. For a later synopsis on the place of geological field work in the school's curriculum, see Smith, "Geological Field Work," 15–19. Junior and senior students also maintained their own journal, mostly reporting on visits to mineral processing facilities around the state but also including occasional reports on field geology, e.g., Parsons and Liddell, "Coal and Mineral Resources," 47–59. The Dakota School of Mines in Rapid City, in turn, was founded in the mid-1880s, just before the Dakotas became states, and published its first report based on field work within just a few years, in Carpenter, *Preliminary Report of the Dakota School of Mines*.

117. The first two such bulletins were *Petroleum of Salt Creek, Wyoming* (1896) and *Petroleum of the Shoshone Anticlinal* (1897). Subsequent bulletins included titles such as *The Oil Fields of Crook and Uinta Counties* and *The Dutton, Rattlesnake, Arago, Oil Mountain and Powder River Oil Fields*.

118. Cockerell, "Recollections of a Naturalist II," 151.

119. Some leading works include Nash, *Inescapable Ecologies*; Valencius, *Health of the Country*; and Mitman, *Breathing Space*.

120. Craig, "North Dakota Life," 321–22.

121. Very rarely, the Rocky Mountain region was discussed unfavorably in relation to health, as when former US Geological Survey director Clarence King wrote sympathetically to the Rocky Mountain Division head S. F. Emmons in the summer of 1881 when Emmons was suffering from health problems: "I have not failed to sympathize with you and to think very often of the terrible strain you are under. I trust that you will soon be able to come East; and that the change of air to your native sea-level may do you great good." King went on to cite his own ill health as precipitating his own recent resignation as director, though without any reference to place (and presumably King was in Washington, DC, most of the time as director), in Clarence King to Emmons, 1 July 1881, Samuel Franklin Emmons Papers, Letters Received, Entry 139 (Old Entry 65), RG 57, Records of the Geological Survey, National Archives II, College Park, Maryland, box 1.

122. Parry, "Rocky Mountain Alpine Region," 251.

123. Denison, *Rocky Mountain Health Resorts*, 100.

124. C. L. Herrick and W. R. Tipton to doctors (circular letter), 28 September 1897, JWP, folder 1. As New Mexico's leading health and climate researcher,

John Weinzirl, proclaimed: "That more equable climates can be found than that of New Mexico is certain. That warmer winter climates or cooler summer climates are available is equally true. But it will indeed be difficult to find an all-round, all-year climate which at the same time affords so many of the desirable factors required by the invalid, viz., an abundance of sunshine, a dry atmosphere, plenty of pure air, and a considerable altitude," in Weinzirl, "Availability of New Mexico's Climate," 9.

125. Health and medicine as domains of knowledge production are largely outside the scope of this book, but see Mitman, "Geographies of Hope."

126. On Waconda Lake, see Blakeslee, *Holy Ground, Healing Water.*

127. Cockerell, "Recollections of a Naturalist," 377.

128. Tight, "Clarence Luther Herrick," 1–26.

129. Langley, *New Astronomy*, 50, 55.

130. Dartt, *On the Plains*, 113. See also Benson, *Martha Maxwell*, 89.

131. Benson, *Martha Maxwell*, 104. Maxwell's location for collecting animals may have been excellent, but she did experience difficulty in affording rent for a museum nearby to display her collections. Martha to May [Mabel], 31 January 1875, folder 31; and Martha to J. L. Beverly, 23 February 1875, Martha Maxwell Collection, #762, Colorado Historical Society, Denver, folder 32. While some of Maxwell's collections ultimately made their way to Philadelphia for the Centennial Exposition in 1876, their priority was not always recognized in the cosmopolitan scientific literature, as protested in Ridgway, "*Scops flammeola* in Colorado," 185, referring to his published catalog of the birds in her collection, which hailed "the success which has crowned the enthusiastic and intelligent efforts of a 'woman Naturalist.'" He calls Maxwell "an accomplished and amiable lady," with "high attainments in the study of natural history." Ridgway, "Mrs. Maxwell's Colorado Museum," 195, 214.

132. Abbott, "Grebe Colony," 86.

133. Moorehead et al., "In Search of a Lost Race," 409.

134. Grinnell, "Old-Time Bone Hunt," 333. On the challenges of obtaining "fresh meat," at least in some parts of the West, by the early twentieth century, see Judd, *Men Met along the Trail*, 14.

135. Nelson and Sheriff, *People at War*, 218.

136. Petrick, "Feeding the Masses," 30.

137. Moorehead et al., "In Search of a Lost Race," 457.

138. Sams, "J. P. Sams Journal: Fossil Hunting in Wyoming," reprinted in Kohl, Martin, and Brinkman, *Triceratops Hunt in Pioneer Wyoming*, 63.

139. For an evocative and thorough analysis of the environmental and cultural context of provisioning of travelers at short and long distances, focusing on gold miners rather than scientific field parties, see Morse, *Nature of Gold.*

140. Gooday, "Precision Measurement and the Genesis of Physics Teaching Laboratories." An exemplary study of the development of physics laboratories in Germany in connection with the establishment of national institutions for establishing and maintaining standards is Cahan, *Institute for an Empire.*

141. Arthur, "Some Botanical Laboratories," 405. Arthur's account also suggests that the rise of the laboratory within any particular discipline might happen over a relatively brief span of time— as in the case of American botany, occurring between 1870 and 1880.

142. For a brief historiographical commentary, which argues for a periodization of lab history that would emphasize historical shifts from the early modern period to industrial society, see Kohler, "Lab History." See also the other contributions to the same focus section of *Isis* by Ursula Klein, Graeme Gooday, and Thomas F. Gieryn.

143. "Laboratory in Modern Science," 172.

144. Welch, "Evolution of Modern Scientific Laboratories," 493. The claims made in Welch's address, and even specific passages of writing, are similar enough to the aforementioned article in *Science* to suggest that either Welch was the source of that article or he borrowed some of its ideas and language.

145. Arthur, "Some Botanical Laboratories," 395.

146. "Laboratory in Modern Science," 173–74. For another example of laboratory work organization, consider the scientific practice of *Drosophila* genetics in the early twentieth century, as elucidated by Kohler, *Lords of the Fly*. Of course, while a hierarchical division of labor persisted, important changes occurred over time. During the middle of the twentieth century, for example, a gendered division of labor prevailed in *Drosophila* laboratories, which closely correlated with the hierarchy of faculty, research associates, and technical assistants. A clear majority of women was concentrated in the last category throughout this period. See Dietrich and Tambasco, "Beyond the Boss and the Boys."

147. "Laboratory in Modern Science," 173.

148. Nicholson, "Chemical Laboratory of the University of Nebraska," 82–84.

149. Arthur, "Some Botanical Laboratories," 404.

150. Bigelow, "Growth and Function of the Modern Laboratory," 645. For an example of expansionary institution and laboratory building in the Great Plains region, focused on physics at the University of Nebraska, see Cahan and Rudd, *Science at the American Frontier.*

151. Williston, "Progress of Science," 54.

152. Williston, "Progress of Science," 54. See also MacDougal, "Can Research Work Be Accomplished," 188–89.

153. An especially evocative account situating the laboratory in its urban

environmental and labor context is Dierig, "Engines for Experiment." Dierig's essay appeared in a thematic volume on "Science and the City," and the other articles help to varying degrees in establishing the urban context of scientific practice, including Mendelsohn, "Microscopist of Modern Life." On the campus laboratory environment in relation to the outdoors, see Kohler, *Lords of the Fly.*

154. A few perceptive institutional histories from before the rise of "envirotech" did lay some of the groundwork for situating laboratories in material context of technological systems and infrastructures. For a leading example of institutional history highlighting electricity and communications networks, see Cahan, "Institutional Revolution." The same modern technological systems were also available for indoor offices. One US Reclamation Service officer in Lawton, Oklahoma, "made it a point to install electric lights sufficient to enable [his] assistants to study evenings," and "the various textbooks owned by the different men formed an excellent nucleus of a library of engineering knowledge," as noted in Matthes, "Training Assistants in Office and Field," in Newell, comp., *Proceedings of Second Conference of Engineers of the Reclamation Service*, 105.

155. Some historians have even likened laboratories to factories, thereby drawing out insights relevant to their material histories. See, for example, Galison, *Image and Logic*; and Todes, *Pavlov's Physiology Factory*. On the other hand, wealthy landowners could extend the urban electrical and gas lines out to their private country house laboratories, as discussed in Schaffer, "Physics Laboratories and the Victorian Country House," 170–71.

156. Cahan, "Geopolitics and Architectural Design." See also Cahan, *Institute for an Empire*, 168–72; and Gooday, "Premisses of Premises."

157. "Personal and Scientific News," 61. In addition, deaths could occur in the laboratory due to hazards brought in from the field, as happened in the late 1910s during investigations of Rocky Mountain spotted fever in the Bitterroot Valley of western Montana, as noted in Wolbach, "Rocky Mountain Spotted Fever," 55; and in the section on "Heroic Sacrifice" mentioning the death of state bacteriologist Arthur Howard McCray in *Biennial Report of the Montana State Board of Entomology* 4 (1919–20), 13.

158. Powell, *Report on the Geological and Geographical Survey of the Rocky Mountain Region*, 18; "Drowned in Sweetwater Lake," 144.

159. *Annual Report of the United States Geological Survey* 33 (1911–12), 8.

160. *Annual Report of the United States Geological Survey* 33 (1911–12), 9.

161. *Annual Report of the United States Geological Survey* 33 (1911–12), 9–10.

162. *Annual Report of the United States Geological Survey* 33 (1911–12), 13.

163. For a laboratory at the other end of the spectrum, far from the urban centers, consider Los Alamos, whose secret atomic bomb construction phase is brilliantly and provocatively framed as materialist environmental history by

Mark Fiege in "Atomic Sublime," in *Republic of Nature*, 281–317. For the longer term historical context of Los Alamos, including this early secret laboratory period and beyond, see Rothman, *On Rims and Ridges*.

164. Mody, *Instrumental Community*. See also Rabinow, *Making PCR*. Of the many studies on laboratory organisms, some leading examples include Creager, *Life of a Virus*; Rader, *Making Mice*; and Kohler, *Lords of the Fly*.

165. Wilson, "Aims and Methods of Study," 20. For similar concerns expressed nearly a decade earlier, see Merriam, "Biology in Our Colleges."

166. Coffey, "Development of Soil Survey Work," 115.

167. "Second Decennial of the Botanical Seminar," 629–31.

168. Clements, "Scope and Significance of Paleo-ecology," 369–74.

169. Taylor, *Suggestions for Field Studies of Mammalian Life-Histories*, 2. For more discussion of these issues in the European context, see Nyhart, *Modern Nature*; and de Bont, *Stations in the Field*.

170. For an example—just one of a great many, chosen essentially at random—which happens to come from the Pikes Peak region of Colorado, see Hillebrand, "Mineralogical Notes," 48.

171. On the importance of sending in soil samples to the bureau's laboratories, see Whitney, *Work of the Bureau of Soils*, 3.

172. Secord, "Crisis of Nature," 450. See also Nyhart, *Modern Nature*.

173. Kohler, *All Creatures*; Rader and Cain, *Life on Display*.

174. Nyhart, "Natural History and the 'New' Biology," 442.

175. Hayden, *Sketch of the Origin*, 3, 5.

Chapter 2. Lay Networks

1. Bailey, *Prairie Ground Squirrels*, 11–12. On Bailey's shift to a scientific career, see Kohler, "From Farm and Family to Career Naturalist." A new interdisciplinary study of Bailey has recently appeared, which critiques the unreliability of Bailey's field data in part because of his reliance on what I have called lay networks: Nagy, "Vernon Bailey and Wolf Extermination."

2. Vetter, "Lay Observers." The predecessor of the US Weather Bureau operated out of the Department of Agriculture and had been the US Army's Signal Service, which had its own network of lay observers, to whom it directed its *Instructions for Voluntary Observers*.

3. For more details, see Vetter, "Lay Observers," 265–69. On the wide variety of businesses receiving the forecasts, see Connor to Chief of Weather Bureau, 20 October 1909, KCWB, vol. 3. On the particular interest taken by the boards of trade, in Kansas City and elsewhere, see Connor to Officials in Charge, 2 May 1905, KCWB, vol. 1; Connor to Chief of Weather Bureau, 17 June 1908, KCWB,

vol. 2; and Whitnah, *History of the United States Weather Bureau*, 99. In addition to the weather forecasts themselves, Connor also published brief meteorological articles based upon his vantage point at the center of a distributed lay network across the region, such as Connor, "Floods in the Missouri River," 399; and Connor, "Loss by Floods," 287–88; as well as articles based solely on his own data collected in Kansas City itself, such as Connor, "Notes on the Heat," 1441; and Connor, "Heavy Rainstorm," 546–47.

4. See, for example, Cooper, *Inventing the Indigenous*; Cook, *Matters of Exchange*; Carey, "Compiling Nature's History"; and Harris, "Long-Distance Corporations."

5. For a vivid example of how contentious the setting of common standards and rules of nomenclature could be, see Barrow, *Passion for Birds.*

6. Recent histories of nineteenth-century earthquake knowledge—a quintessentially lay-oriented field science, given both its practical significance for environmental risk and the impossibility of stationing expert observers ahead of time—have not only emphasized the centrality of translation as a project but also analyzed social relationships using a term from the study of language: the vernacular. Whether emphasizing a language "simultaneously scientific and vernacular" or conceiving of "vernacular science" as a meeting ground for the "local, embodied, and specific" with the "highly theoretical and interconnected across vast distances," these historical studies have raised the standing and significance of the vernacular within the history of science. See Coen, *Earthquake Observers*; Valencius, *Lost History of the New Madrid Earthquakes.* Earthquakes have been relatively rare in the Great Plains and Rocky Mountains, but one example from the railroad era was documented in Reid, "Remarkable Earthquakes," 10–16. For a broad-ranging reflection spanning into the twentieth century, see Pandora and Rader, "Science in the Everyday World." On the contest between vernacular oil prospectors and petroleum geologists, see Frehner, *Finding Oil,* which, rather than focusing on vernacular or experiential knowledge, deploys the related concept of "mētis" knowledge, deriving from Scott, *Seeing like a State.*

7. For further justification of lay observation in the history of science as a subject to examine over the long term, see Vetter, "Introduction." See also the various articles in the topical issue that this essay introduces for a variety of examples spanning the early modern period to the twentieth century.

8. For the US Entomological Commission's official investigation of this problem, which included knowledge gathered in lay network mode through "classified replies to circulars," see *First Annual Report of the United States Entomological Commission for the year 1877 relating to the Rocky Mountain locust and the best methods of preventing its injuries and of guarding against its invasions, in pursuance of*

an appropriation made by Congress for this purpose (1878). Subsequent reports were issued in 1880 and 1883. See also *Destruction of the Young or Unfledged Locusts*, Bulletin of the US Entomological Commission, no. 1 (1877); and Lockwood, *Locust.* On potato beetles and ticks, see Chittenden, *Colorado Potato Beetle*; and Bishop, *Distribution of the Rocky Mountain Spotted-Fever Tick.*

9. The precise limits of these periods differed from place to place, but for an accessible summary of their occurrence on the Texas High Plains, see Carlson, *Deep Time and the Texas High Plains.*

10. Reeves, "Concept of an Altithermal Cultural Hiatus."

11. Holliday, "Middle Holocene Drought."

12. Sears, "Pollen Profile."

13. For evidence of diverse hunting-gathering economies only occasionally supplemented with bison hunting, see Ritterbush, "Drawn by the Bison"; and Bozell, "Culture, Environment, and Bison Populations." For a more comprehensive overview of this region, see Wedel, *Central Plains Prehistory.* For archaeological data on the northwestern Plains region, including Wyoming and nearby areas, see Frison, *Prehistoric Hunters of the High Plains.*

14. Bryson, "Chinook Climates and Plains People." For macro-scale analyses based on models from population ecology, see Bamforth, *Ecology and Human Organization*; and (on the issue of horse herd size) Osborn, "Ecological Aspects of Equestrian Adaptations."

15. As Russell Spafford pointed out in 1916: "In the early day[s] people of western Nebraska and similar regions were not inclined to look upon climate and soil as fixed. They expected them to be greatly modified as a result of breaking new land and growing cultivated crops. This idea lasted only so long as they lacked actual experience with the country. A few years of effort made it plain that to do profitable farming it was necessary to conform to the natural conditions of the region"; Spafford, "Effect of Climate and Soil upon Agriculture," 92. On the environmental history of the Great Plains, see Allen, "Garden-Desert Continuum"; Blouet and Lawson, *Images of the Plains*; Blouet and Luebke, *Great Plains*; De Bres, "Come to the 'Champagne Air'"; Fleharty, *Wild Animals and Settlers*; Opie, *Ogallala*; and Sherow, *Watering the Valley.*

16. Lawson, *Climate of the Great American Desert*, 95, 97. On changing ideas about the "Great American Desert," see Bowden, "Great American Desert in the American Mind."

17. Courtwright, *Prairie Fire*; Pyne, *Fire in America.*

18. For an argument against the conventional belief that American Indians did not occupy the High Plains year round until the advent of the horse, with an interpretation of archaeological evidence regarding water supply, see Wedel, "High Plains and Their Utilization by the Indian."

19. On the environmental history of Mormon settlement, see Alexander, "Stewardship and Enterprise"; and Arrington and May, "'A Different Mode of Life.'"

20. The shift from Mormon communal to scientific environmental management is sketched out in Flores, "Zion in Eden."

21. A study of three counties at different longitudes in that area demonstrates that the extent of adjustment to the environment increased after 1890 with each step west. See Baltensberger, "Agricultural Adjustments." Another example of adaptation comes from Sheridan County in northwest Nebraska, where, by the end of the nineteenth century, "patterns of population and land use" would become "roughly similar on Box Butte Tableland and in the Sandhills," as one historical geographer has found, noting that "in the generally even but light sprinkling of population and heavy stress on livestock grazing, the strong areal distinctions" of a few years earlier would be "largely obscured," in Bowen, "Environmental Perception," 334.

22. For a compelling discussion of the importance of an environmental history of ranchlands, including the issue of public vs. private ownership, see Worster, "Cowboy Ecology."

23. On the stability of such land-use practices during the twentieth century, see Cunfer, *On the Great Plains.*

24. Bailey, *Prairie Ground Squirrels,* 10; Jones, "Becoming a Pest."

25. Anderson, "Black-Headed Grosbeak," 98.

26. Garrett, "Honey Ants," 22–23.

27. Vetter, "Lay Observers," 269.

28. On the mammal collecting of Coues, Mearns, and others, see Brodhead, "Of Mice and Mastodons," 366. On Coues, see also Cutright and Brodhead, *Elliott Coues.* Another collector who began as a US Army surgeon was the botanist Edward Palmer, who forwarded specimens to Spencer Baird at the Smithsonian while serving as a volunteer surgeon with the 2nd Colorado Regiment in Denver during the Civil War, with later posts in Arizona, American Indian Territory, and elsewhere, as discussed in McVaugh, *Edward Palmer,* 19. For a synopsis of field collecting by army men, along with army explorers, artists, and American Indian observers, see Tate, "Discoverers," in *Frontier Army,* 3–28. On bird collecting, see Hume, *Ornithologists of the United States Army Medical Corps.*

29. Useful overviews of the post surgeon's general role include Olch, "Medicine in the Indian-Fighting Army"; and Tate, "Hippocrates in Blue," in *Frontier Army,* 174–92. For a discussion of the experiences of a contract civilian army surgeon, see Riley, "Dr. David Franklin Powell."

30. Gibson, *Soldier in White,* 37. On Sternberg's fossil collecting work, see also Sternberg, *George Miller Sternberg,* 49.

31. Tate, *Frontier Army*, 18–19. On the Pawnee case, see Echo-Hawk and Echo-Hawk, "Repatriation, Reburial, and Religious Rights."

32. Tate, *Frontier Army*, 12; Herman, *Hunting and the American Imagination*, 216.

33. Tate, *Frontier Army*, 15.

34. "Remarks on Fossil Vertebrates from Wyoming," 664.

35. Scott, *Some Memories*, 166–67.

36. Higham, "Saviors and Scientists."

37. Pugsley, *Alfalfa Inoculation Tests*, 3.

38. Cooper, *Inventing the Indigenous*; Knell, *Culture of English Geology*.

39. Jankovi , *Reading the Skies*.

40. See especially Latour, "Visualization and Cognition"; and Latour, *Science in Action*, 215–57. Historical accounts applying Latour's model include Miller, "Joseph Banks, Empire, and 'Centers of Calculation'"; and Spary, *Utopia's Garden*.

41. Shapin, "Placing the View from Nowhere," 7. See also Shapin, "Following Scientists Around."

42. Lawrence Dritsas further complicates this picture of metropolitan centrality by emphasizing how metropolitan institutions like the British Museum and Kew Gardens acted as gatekeepers for materials collected in the field, sending them out to qualified specialists in other places. Dritsas, "From Lake Nyassa to Philadelphia."

43. Rusnock, "Correspondence Networks."

44. Spary, *Utopia's Garden*, 53.

45. For popular historical works on the "fossil war" between Marsh and Cope, see Jaffe, *Gilded Dinosaur*; and Wallace, *Bonehunters' Revenge*.

46. Marsh's collecting network is discussed in Schuchert and LeVene, *O. C. Marsh*, 172–205.

47. A brief discussion of this shift can be found in Rainger, *Agenda for Antiquity*, 16. For a valuable recent contribution to our understanding of fossil collecting in relationship to eastern centers, see Rieppel, "Dinosaurs."

48. Cope, "Synopsis of the Extinct Batrachia, Reptilia, and Aves of North America," 2; Almy, "Thof's Dragon."

49. Mrs. Charles K. Winne to her father and brother, 31 August 1879, 21 September 1879, in Buecker, "Letters from a Post Surgeon's Wife," 53, 54.

50. Farrington, *Catalogue of the Meteorites of North America*, 11–12.

51. Numerous examples of widespread exchange of meteorites in the Great Plains states especially, such as Nebraska and Kansas, can be found in the general science periodicals of the late nineteenth and early twentieth centuries, including national periodicals as well as state scientific and natural history asso-

ciations. For examples, see Patrick, "Waconda Meteorite," 12–13; Barbour, "Discovery of Meteoric Iron," 275–79; and Howell, "Two New Meteorites," 27–28.

52. Wistar and Nolan, "Postage," 267.

53. Wister and Nolan, "Postage," 267.

54. Harrington, "Postage," 348–49. For a further reply by another Philadelphia naturalist, with information from the US Post Office, see Calvert, "Postage," 49–50.

55. Pilsbry, Woolman, and Calvert, "Reduced Rates of Postage," 402–3.

56. C. Hart Merriam to E. R. Warren, 23 April 1912, CHMP, reel 7. Even as Warren deferred in this instance to the more expert judgment of career scientist Merriam, the Colorado gentleman amateur clearly possessed more knowledge, credibility, and authority than the typical lay collaborator, and thus the uncertainty of identification is even more striking.

57. "Second Decennial of the Botanical Seminar," 630.

58. "Second Decennial of the Botanical Seminar," 630. This did not mean, however, that Bessey was unwilling to use cosmopolitan scientific knowledge to overrule the experiential knowledge of local people. When told in the early 1890s "while collecting in the Black Hills of South Dakota . . . that Maiden Hair ferns grew at Cascade," he doubted the report based on his knowledge of the region's flora, traveling there again in 1898 with Frederic Clements to confirm himself that the plants in question were of a different species. Bessey, "One Thousand Miles," 2.

59. Pool, "Summer Botany for Forestry Students," 21.

60. Eastwood, "Two Species of Aquilegia," 560. Elsewhere, she described Al Wetherill as "my guide," who "planned the route, managed everything about the camp and horses, helped me greatly in collecting, and, altogether, was as good a friend and as efficient an aid as any botanist could desire," in Eastwood, "General Notes of a Trip," 355. See also Wilson, *Alice Eastwood's Wonderland*, 39–44; and Robertson, *Magnificent Mountain Women*, 105–8.

61. "Second Decennial of the Botanical Seminar," 630.

62. Connor to Corn and Wheat Region Observers, 13 March 1905, KCWB, vol. 1.

63. Connor to Corn and Wheat Region Observers, 25 September 1905, KCWB, vol. 1; and Connor to J. B. Loughran, 8 April 1908, KCWB, vol. 2.

64. For comparative examples of seeking information from lay networks in the British Empire, through what was initially called *Notes and Queries in Anthropology for the Use of Travellers and Residents in Uncivilized Lands* (1874), and updated five times through 1951; see Urry, "*Notes and Queries on Anthropology* and the Development of Field Methods"; Stocking, "Ethnographer's Magic"; and Stocking, "Reading the Palimpsest of Inquiry."

65. Henkin, *Postal Age*, 154.

66. Raymond, *Statistics of Mines and Mining*, 9, 13.

67. Henkin, *Postal Age*, 154–55.

68. For an example of the initial solicitation of interest in participating in a lay network through newspapers, see Hansen, *Co-operative Tests of Alfalfa*, 36.

69. For an example of a circular sent to members of a state horticultural society, see Card, *Windbreaks*.

70. *Annual Report of the Wyoming Agricultural Experiment Station* 5 (1895), 4.

71. *Annual Report of the Nebraska Agricultural Experiment Station* 12 (1898), xxxv. Even national attempts to gather information through lay networks were plagued by problems with incompleteness, errors, and imperfect data, as noted in Lyon et al., *Catalogue of Fruits*, 5–6.

72. *Annual Report of the Montana Agricultural Experiment Station* 9 (1901–02), 57.

73. *Annual Report of the Montana Agricultural Experiment Station* 10 (1902–03), 25.

74. Nowell, "Report of the Irrigation Engineer," 76.

75. In at least one example, the regions of a state to be studied more carefully using lay networks for agricultural experiments had been defined by previous botanical survey work. See Lyon, *Cooperative Variety Tests*, 3.

76. Neal, *Oklahoma Weeds*, 4.

77. Howe and Stingley, "Great Fireball of December 7, 1900," 394.

78. Udden, "Texas Meteor," 616–17.

79. On Kansas, see Jordan, "Kansas State Board of Agriculture."

80. Drake, "Waving 'A Bough of Challenge,'" 25–26.

81. Vinall, *Forage Crops*, 3.

82. *Annual Report of the Montana Agricultural Experiment Station* 1 (1893–94), 135.

83. *Forest Planting in the Sand-Hill Region of Nebraska*, 1.

84. Nowell, "Report of the Irrigation Engineer," 74.

85. Hansen, *Co-operative Tests of Alfalfa*, 36.

86. "Memorandum of Understanding between the President of the Colorado Agricultural College, the Principal of the Fort Lewis School of Agriculture, the Director of the Agricultural Experiment Station, and the Heads of the Departments of Agronomy, Animal Husbandry and Horticulture, Relative to Cooperation in the Carrying on of Experimental Work upon College Lands at Fort Lewis," unpublished typescript (1920); [E. H. Bader] to Prof. Alvin Keyes [*sic*], 8 April 1920, FLCC, RG 21, ser. 4, box 1.

87. Alvin Kezer to Mr. E. H. Bader, 14 April 1920, FLCC, RG 21, ser. 4, box 1.

88. 9 October 1891, in Bartholomew, *Pioneer Naturalist*, 239–40.

89. 8 April 1892, in Bartholomew, *Pioneer Naturalist,* 240.

90. 2 August 1893, in Bartholomew, *Pioneer Naturalist,* 241.

91. 7 June 1907, 12 June 1907, in Bartholomew, *Pioneer Naturalist,* 56.

92. 29 April 1908, in Bartholomew, *Pioneer Naturalist,* 56.

93. 10 May 1909, in Bartholomew, *Pioneer Naturalist,* 57.

94. Lawrence Bruner to Mr. Bassett, 23 May 1893, LBP, box 15.

95. Lawrence Bruner to D. F. Trunkey, 1 June 1893, LBP, box 15.

96. Lawrence Bruner to Chancellor [Snow], 22 January 1894, LBP, box 15. For an overall synopsis of Snow's contemporaneous work in neighboring Kansas, see Snow, "Work in Economic Entomology," 140–44.

97. Vetter, "Lay Observers," 265.

98. For similar collaborative dynamics in the British Empire, see Endersby, "'From Having No Herbarium.'"

99. Vetter, "Lay Observers," 269–71. Weather observers submitting paper forms only were typically unpaid volunteers, whereas those sending daily reports by telegraph were paid twenty cents per day, as noted in Vetter, "Lay Observers," 266. So the discipline of modest cash payment did transform the relationship slightly, but this was still far from a full-time job.

100. Vetter, "Lay Observers," 261.

101. Rainger, *Agenda for Antiquity,* 15.

102. Schuchert and LeVene, *O. C. Marsh,* 172–73.

103. Schuchert and LeVene, *O. C. Marsh,* 175.

104. Hinsley, "Museum Origins of Harvard Anthropology," 127; and Burns, "Networking Ohio Valley Archaeology."

105. "Co-operation in Science," 30–31.

106. Endersby, "'From Having No Herbarium,'" 356.

107. 16–19 January 1889, 22 February 1889, in Bartholomew, *Pioneer Naturalist,* 236.

108. Gould, *Covered Wagon Geologist,* 47.

109. Vanorny, *History of the South Dakota Geological Survey,* 3.

110. Nelson actively sent out circulars soliciting plant collectors across the state, including Aven Nelson and Elias Nelson, "To Collectors and Others Having Plants for Exchange" (Laramie, Wyo., Nov. 1899), typescript circular, ANP, box 10, folder 12. In the early twentieth century, Nelson thoroughly revised and updated the leading botanical guide on the Rocky Mountain region: Coulter and Nelson, *New Manual of Botany of the Central Rocky Mountains.* The earlier version was Coulter, *Manual of the Botany (Phaenogamia and Pteridophyta) of the Rocky Mountain Region.*

111. Aven Nelson to Mr. Edwin E. Payson, 13 October 1912, University of Wyoming, Botany Records, #545001, American Heritage Center, University of

Wyoming, Laramie, box 29, folder 13. See also Williams, *Aven Nelson of Wyoming*, 176–77.

112. Connor to Royal E. Dunham, August 1906, KCWB, vol. 1.

113. Connor to Royal E. Dunham, 7 July 1906, KCWB, vol. 1.

114. Smith, "On the Magnetic Declination," 46–48.

115. Smith, "On the Magnetic Declination."

116. Bailey, *Prairie Ground Squirrels*, 11.

117. Bailey, *Prairie Ground Squirrels*, 11.

118. Vasey, "Report of the Botanist" (1881), 233–34. Over the next several years, Vasey also received direct queries from many of his correspondents, such as one from Fremont County, Wyoming, that requested a grass suited to the soil, altitude, and climate there, in Vasey, "Report of the Botanist" (1886), 77.

119. Blankinship and Henshall, "Common Names of Montana Plants," 115.

120. *Biennial Report of the Oklahoma State Board of Agriculture* 2 (1909–10), 93.

121. Farrington, *Catalogue of the Meteorites*, 10.

122. Cook, *Section of Seed and Plant Introduction*, 5–6.

123. Patrick, "Great Spirit Spring," 22.

124. C. Hart Merriam to L. L. Dyche, 23 February 1904, L. L. Dyche Papers, Spencer Research Library, University of Kansas, Lawrence, box 2.

125. For more recent examples of how experts in fields as diverse as ornithology and statistics have dealt with the problem of credibility and large numbers, see Charvolin, "Une Science Citoyenne?"; and Didier, "Counting on Relief."

126. Bailey, *Prairie Ground Squirrels*, 11–12.

127. Card, *Windbreaks*, 69–70.

128. Connor to "All River Observers," 7 April 1905, KCWB, vol. 1.

129. P. Connor to Mr. Collar, 2 March 1905, KCWB, vol. 1.

130. Connor to Chief of Weather Bureau, 27 March 1905, KCWB, vol. 1.

131. Connor to Jared A. Collar, 4 April 1905; Connor to Jesse K. Thompson, 27 March 1905; Connor to Jesse K. Thompson, 4 April 1905; Connor to Chief of Weather Bureau, 11 April 1905; and Connor to Jesse K. Thompson, 18 April 1905, KCWB, vol. 1.

132. Connor to John H. Sherman, 20 April 1905, KCWB, vol. 1.

133. Connor to Manager, Western Union Telegraph Office, 28 April 1905, KCWB, vol. 1.

134. Vetter, "Lay Observers," 270.

135. P. Connor to John H. Sherman, 3 April 1905, KCWB, vol. 1.

136. The best analysis of this trend is Porter, *Trust in Numbers*.

137. Wilcox, *Directions for Collecting*, 1.

138. Pugsley, *Alfalfa Inoculation Tests*, 3.

139. Pugsley, *Alfalfa Inoculation Tests*, 3.

140. Toumey, *Practical Tree Planting in Operation*, 9, 11. Further cooperative tree plantation work by the same agency, with "seventy-six landholders in Oklahoma and adjacent regions," is discussed in Clothier, *Advice for Forest Planters in Oklahoma*.

141. Pugsley, *Alfalfa Inoculation Tests*, 4.

142. Dickson, "Report of Dickens County Experiment Association," 274–75. Another similar association about fifty miles west in Lubbock was also organized around the same time, but its report does not discuss cooperative research, and it was apparently more solely focused on circulating knowledge to farmers rather than producing new knowledge as well. Burford, "Report of the Lubbock County Experiment Station," 275.

143. Bates, "Additional Observations," 73.

144. Lovewell, "Kansas Weather Service," 83.

145. Shepperd and Richards, *Report of Four Years Work*, 95, 114.

146. "Dr. Joel Lunell," 244.

147. Lunell, "Enumerantur Plantae Dakotae Septentrionalis Vasculares," 155.

148. Snow, "Catalogue of the Birds of Kansas," 21–29. On Snow's life, see Hyder, *Snow of Kansas*.

149. Popenoe, "List of Kansas Coleoptera," 21–40.

150. Carruth, "Report on Botany, 1874," 122–23; Carruth, "Centennial Catalogue," 40, 44.

151. Snow, "Douglas County Additions," 75.

152. Forty years later, Snow's era—and his organizing role in collective collecting—would be viewed through a lens of nostalgia: "It is a thousand pities that we might not have in Kansas, some Nestor like the late Professor Snow, to criticize, formulate and re-tabulate the records made by men less learned or less critical; thus giving the world of Ornithological Science a sound and accurate survey of the *ornis* of Kansas," in Peabody, "Review of 'Migration Records for Kansas Birds,'" 149.

153. "Colorado Biological Association," 17–18. Even before the Colorado Biological Association, another organization for collectively produced science was the Colorado Meteorological Association (or Society), founded in January 1885 "entirely for the purpose of observing, collecting, recording and publishing the meteorological phenomena, occurrences and changes within the State of Colorado"; in "Colorado Meteorological Society," 462. By publishing "periodical summaries of weather records from the state at large, drawn from the observations of a few gentlemen, mostly members of the Association," they reasoned, "the notice of other persons interested in the same subject would be secured, and observers found in every part of the state to join the association,

or to aid in its work for the public welfare"; *Bulletin of the Colorado Meteorological Association*, no. 2 (May 1886), 1. Founded even earlier was the Colorado Scientific Society, which started in 1882 but functioned during its early years mainly to connect the US Geological Survey's Rocky Mountain Division in Denver with the regional mining industry. As its president Thomas L. Wilkinson said in 1907 in commemorating its first quarter century of existence, the Colorado Scientific Society's "record bears testimony to the appreciation among Coloradans of the benefits and mining industry has received from scientific research," "Celebration of the Quarto Centennial," 391. There was also a State Historical and Natural History Society in Colorado, which had been initially encouraged by state legislation as early as 1879 but does not seem to have actually become organized until about 1889, as noted in *Report of the State Historical and Natural History Society of Colorado*, 5.

154. Morrison, "List of the Birds of Colorado," 145.

155. "Colorado Biological Association," 17–18. Another society called the Colorado Ornithological Association surfaced around 1899 or 1900, met a few times, then "seemed to fall asleep," according to Fred Dille, as cited in Trostler, "History of Ornithology in Nebraska," 16.

156. Morrison, "List of the Birds of Colorado," 146.

157. Weber, *Letters*, 172–74, 177–78. As Cockerell wrote in a letter back to England after one meeting: "Vetter read a paper on 'Placer mining,' and panned out some gold before the audience. I spoke about the mountain summits and their fauna and flora, partly reading from manuscript and partly speaking as I drew on the blackboard or exhibited butterflies. I went into the Glacial Period and the origin of the boreal fauna" (177). The West Cliff meetings of the Colorado Biological Association, including Cockerell's participation, were reported in at least one national scientific journal, in Cockerell, "Doings of Societies," 31.

158. Hapeman, "Fresh-Water Algae," 13–14.

159. Hussong, "Flora of Franklin County," 9; Hussong, "Yellow Pine in the Republican Valley," 7–8; Hussong, "Data as to Wind-Distribution of Seeds," 253–57.

160. Pound, "Progress of the Botanical Survey," 7.

161. "Report of Collections Made in 1892," *Botanical Survey of Nebraska* 2 (1893), prefatory note, 4. Acknowledgment to volunteer contributors continued in later volumes, including in reports for 1893, 1894–95, and the late 1890s, in vol. 3 (1894), 4; vol. 4 (1896), 4; and vol. 5 (1901), 4.

162. C. Hart Merriam, untitled circular, Circular of the Biological Survey, no. 41 (1904). This national coordination effort had ramped up much more fully by the mid-1910s, as discussed in Cooke, *Preliminary Census*.

163. Bruner, "Ornithology in Nebraska," 8.

164. Bates, "Additional Notes and Observations," 15.

165. Cary, "Some Bird Notes," 21–29. See also Cary, "Additional Notes on Birds," 99.

166. Peabody, "Review," 149. For more on this controversy, see Barrow, *Passion for Birds.*

167. Wolcott, "Suggestions as to an Accurate and Uniform Method of Recording Observations," 41–44.

168. Wolcott, "On Migration Records," 89.

169. *Proceedings of the Nebraska Ornithologists' Union* 2 (1901), 4–5.

170. Graves, "Notes on Birds from Western Nebraska," 84–85.

171. Zimmer, "Some Notes on the Winter Birds," 19–30.

172. For example, see Crawford, "Results of a Collecting Trip," 76–79.

173. Wolcott, "Analysis of Nebraska's Bird Fauna," 25–55.

Chapter 3. Surveys

1. Joseph Paxon [*sic*] Iddings, "Recollections of a Petrologist," 64, Joseph Paxson Iddings, Correspondence, Geological Studies, and Recollections, No. 7879, US Geological Survey, Field Records Library, Denver, Colorado.

2. For an overview history of the Yellowstone Division, focusing mainly on the advocacy of its leaders, especially Arnold Hague, for greater conservation protection, see Manning, *Government in Science,* 151–67. Hague's popular writings on the geology of Yellowstone National Park include Hague, "Yellowstone National Park"; and Hague, *Geological History of the Yellowstone National Park.* The full, technical report of the Yellowstone Division was published as Hague, Iddings, Weed, Walcott, Girty, Stanton, and Knowlton, *Geology of the Yellowstone National Park, Part II, Descriptive Geology, Petrography, and Paleontology.* The first volume was never published but is available in manuscript form in YNPA, box K-127; various drafts and working materials are in Hague Papers, Working Papers and Reference Material, Entry 146 (Old Entry 62), RG 57, Records of the Geological Survey, National Archives II, College Park, Maryland. At least as important as the written reports was the geological atlas of the park, first published as Hague, Weed, and Iddings, *Yellowstone National Park, Wyo.,* Folio of the Geological Atlas of the United States, no. 30 (1896), then in an updated version a decade later: Hague, *Atlas to Accompany Monograph 32 on the Geology of the Yellowstone National Park* (1904). "No such beautiful or complete atlas of any one region has perhaps ever been issued," raved *Forest and Stream,* and "with the volumes which have preceded it, is a lasting monument to those who took part in a labor which was long and difficult—though delightful—and which was finally so creditably completed." "Maps of the Yellowstone Park," 547.

3. "Editorial," *Botanical Gazette* 22 (1896): 57–58. For histories of the US Biological Survey, see Cameron, *Bureau of Biological Survey*; and Sterling, "Builders of the U.S. Biological Survey."

4. Zahniser, "Vernon Orlando Bailey," 6–7.

5. Allen, review of C. Hart Merriam, *Monographic Revision of the Pocket Gophers*, 242.

6. For a forceful analysis of survey science as "governmentality," which, like this study, focuses on the American West in the nineteenth century, see Kirsch, "John Wesley Powell"; see also Kirsch, "Regions of Government Science."

7. Leonard, "Purpose and Value," 8.

8. Elrod, "Resources of Montana," 782.

9. Newcombe, "Scope and Method," 615–22. For further discussion of the educational aims of natural history surveys, as articulated by Newcombe and others, see Kohler, *All Creatures*, 94–107.

10. Bailey, *Birds of New Mexico*, 1. This book includes not only Vernon Bailey's "Map of Life Zones in New Mexico" (6–7) but also his wonderful map of "Routes and Collecting Localities of Biological Survey Parties in New Mexico" (14), which offers one of the best representations available of the geographical pattern of field survey work. The Baileys also applied the Bureau of Biological Survey's (i.e., C. Hart Merriam's) "life zone" framework to smaller territories, as when they mapped the life zones in Glacier National Park of Montana during a two-month visit there in the summer of 1917, following a previous visit there by Vernon alone in 1895, as noted in Bailey and Bailey, *Wild Animals of Glacier National Park*, 25, 103. Others sometimes adopted the "life zone" concept, such as in Wheeler, "Mountain Ants of Western North America," 455–569.

11. For an overview of the first quarter century of the US Geological Survey, see Manning, *Government in Science*. On the predecessor surveys, see Bartlett, *Great Surveys of the American West*.

12. Even in mineral-rich Colorado, the "state geologist" was at first an unpaid position, and its occupant in 1889 and 1890 entreated in his report "for the honor of Colorado, that the coming Legislature will dignify the office of State Geologist by providing funds with which to carry on investigations and collate information for the inquiring public"; Kedzie, *Report of the State Geologist*, 3. Most state geological surveys have had their own internal histories, but occasionally more widely contextualized histories of the frontier period have been written. For example, on the Oklahoma Geological Survey, see Frehner, *Finding Oil*, 126–33. And on the series of geological surveys in Texas, see Ferguson, *Geology and Politics in Frontier Texas*. Relevant internal histories include Bluemle, *First 100 Years*; Bryans, *History of the Geological Survey of Wyoming*; Buchanan, *'To Bring Together, Correlate, and Preserve'*; Ham, *History of the Oklahoma Geological Sur-*

vey; and Vanorny, *History of the South Dakota Geological Survey*. See also Socolow, *State Geological Surveys*.

13. For an example of survey science that was deployed to improve management of American Indians, see Powell, *Report of Special Commissioners*.

14. Bailey, *Birds of New Mexico*, 1–4, 15–36.

15. The most comprehensive account of topographical surveying by the US Army in the West is Goetzmann, *Army Exploration in the American West*. See also Allen, "Patterns of Promise"; and Garver, "Practical Military Geographers."

16. For a helpful taxonomic overview of traditional types of surveys, as viewed from the later part of the nineteenth century, see Baker, "Surveys, Their Kinds and Purposes," 254–59. On railroad survey practice, based on the work of the Atchison, Topeka, and Santa Fe Railroad in 1886 and 1887, see Newbrough, "Notes on Preliminary and Location Railroad Surveying," 1–11.

17. "Much of the work can be done with a horse and buggy, but it is practically always necessary to do more or less on foot," according to one account of soil survey practice. "In unimproved timbered sections it is impossible to keep one's location when walking, but where the country is under a high state of cultivation, it is less difficult to keep one's bearings." See Worthen, "Methods of Soil Surveying," 187.

18. For a late example of survey work by boat along the Missouri River, see Ward, *Report of the State Geologist*, 3.

19. Baker, "Surveys," 254.

20. Baker, "Surveys," 255.

21. Bartlett, *Great Surveys*. Biographies of the four "great survey" leaders include Cassidy, *Ferdinand V. Hayden*; Foster, *Strange Genius*; Moore, *King of the 40th Parallel*; Sandweiss, *Passing Strange*; Wilkins, *Clarence King*; Wilson, *Explorer King*; Stegner, *Beyond the Hundredth Meridian*; Worster, *River Running West*; and Dawdy, *George Montague Wheeler*. For a useful collection of Powell's writings, many of them first appearing in popular magazines, see DeBuys, *Seeing Things Whole*.

22. Whitney, "Geographical and Geological Surveys," 38. For another example from the 1870s, see Allen, "Notes of an Ornithological Reconnaissance," 113–83.

23. Emmons, "Little Cottonwood Granite Body," 139–40. For further expression of these sentiments, see Emmons, "Uinta Mountains," 287–302.

24. Marsh, "On the Geology," 191–92.

25. On the early US Geological Survey, see Manning, *Government in Science*. On the early Bureau of Ethnology—renamed the Bureau of American Ethnology in the mid-1890s—as driven by aspirations for a more highly structured "anthropological survey," see Hinsley, *Savages and Scientists*, ch. 10. See also Judd, *Bureau of American Ethnology*; Woodbury and Woodbury, "Rise and Fall"; and Darnell, *And along Came Boas*.

26. Detailed accounts of US Biological Survey field work in New Mexico and Texas are provided in Shaw and Weisenberger, *Twelve Hundred Miles*; and Schmidly, *Texas Natural History*. Detailed practical instructions and regulations for US Department of Agriculture cooperative soil surveys field parties are in USDA Bureau of Soils, *Instructions to Field Parties*; and USDA Bureau of Soils, *Soil Survey Field Book*. For a history of the soil survey within one Great Plains state, see Thompson, *History of Soil Survey*.

27. The grand expeditionary tradition did occasionally appear again elsewhere, including the 126-member Harriman Alaska Expedition at the turn of the century, narrated in Goetzmann and Sloan, *Looking Far North*.

28. Baker, "Botanizing in the La Plata Mountains," 29–32.

29. "Field School of Geology," 706; "New Field School of Geology for Harvard University," 237–38.

30. "Preliminary: The Plan and Scope of the Survey," 6–7. The survey was officially organized on August 24, 1892, with the first significant field work beginning in June 1893, as noted in "Minutes and Proceedings, 1886–1900," unpublished record book, box 1, Sem Bot [Botanical Seminar] Club Records, #12/7/5, University Archives, Love Library, University of Nebraska, Lincoln. The back pages of this book include coats of arms for members and scripts for club rituals, among other things. Also in the Sem Bot Club Records are an internal history suggesting that the group traced its roots back to field investigations of Nebraska flora from 1886 onward and expressing the group's pride that it was supported entirely by its own resources rather than public funds. Some members of the Botanical Seminar undertook more specialized studies based on the collective work of the group's botanical survey, as in Beattie, "Nebraska Ferns and Fern Allies," 72–74. The Botanical Seminar itself traced its origins back several years earlier, to October 1886, when it began as an "exclusive secret society" known as "the 'Sem. Bot.,'" for "advanced students in botany." Then, in "about 1891" the club "became a serious, scientific organization, aiming to promote research and, in particular, to organize the study of the vegetation of the state," as noted in "Second Decennial of the Botanical Seminar," 629.

31. Nelson, *First Report on the Flora of Wyoming*, 45–218. See also Aven Nelson, "Plants of Wyoming" (Laramie, Wyo., Apr. 1900), typescript circular, ANP, box 10, folder 12. Another example was at the agricultural experiment station of North Dakota, where a "survey of the native seed-bearing plants" included county-level surveys beginning in the summer field seasons of 1906 and 1907, as noted in *Annual Report of the North Dakota Agricultural Experiment Station* 18 (1908), 45, 53. A few years earlier, however, a North Dakota Agricultural College botanist had conceded to an out-of-state colleague that the state's earlier work had been "rather sporadic and not in the nature of a survey," thus dif-

ferentiating future work from earlier practices, in L. R. Waldron to Dr. J. W. Blankinship, 10 May 1901, Lawrence R. Waldron Collection, Mss 36, Institute for Regional Studies, North Dakota State University, Fargo, folder 2.

32. Likewise, the University of New Mexico undertook what they called a "voluntary geological and natural history survey" in the late 1890s, but with apparently the same modest results that Cragin would obtain in Kansas in the 1880s following a voluntary model. See Herrick, Terry, and Herrick, "Notes on a Collection of Lizards," 117.

33. Visher, "Biogeography of the Northern Great Plains," 89–115. Visher himself had started out back in 1906 as an earnest young prospective survey assistant pleading to join E. C. Perisho on a field survey trip through the Badlands to the Black Hills. Visher's early ambitions were centered on systematically studying the ornithology of South Dakota. Stephen S. Visher to Mr. Perisho, 24 May 1906, ECPC, box 5, folder 36; Stephen S. Visher to Mr. Perisho, 3 June 1906, E. C. Perisho Collection, USD91-0005, Arts and Sciences, Deans Series, Special Collections, University of South Dakota, Vermillion, box 5, folder 37.

34. Wolcott, "Faunal Survey," 791–92. Likewise, for an example of a delimited area study of seventy-five small lakes in the Sandhills, see Anderson and Walker, "Ecological Study of the Algae," 51–85. "At first it was hoped to cover the entire group of lakes," they wrote, "but after a few preliminary trips through the region, the work was limited to a few localities so situated that they could be frequently visited" (52).

35. Eastwood, "Report on a Collection," 270.

36. Scheffer, *Prairie-Dog Situation*, 2.

37. *Annual Report of the Oklahoma Agricultural Experiment Station* 27 (1918), 23; *Annual Report of the Oklahoma Agricultural Experiment Station* 28 (1919), 34.

38. This "agricultural survey" of Nebraska was conducted by Joseph Allen Warren as his PhD thesis and reported in Condra, "Reports of the Geographer and the Conservation Commission," 267. Another example of "farm management survey" was in Utah, in Stewart, "Can the Farms of the United States Pay for Themselves?," 177–93.

39. *University Geological Survey of Kansas* 1 (1896), title page.

40. Gould, "General Geology of Oklahoma," 39. At the time, Oklahoma had not yet achieved statehood, so the survey was funded by the territorial government. After achieving statehood in 1907, the state also experienced a fossil fuel boom that reshaped the orientation of survey field work in Oklahoma. While the first state geological survey publication, *The Oklahoma Geological Survey: Its Origin, Scope and Purpose*, still emphasized developing mineral resources for import substitution, within a few years much of what was now described as the "large amount of field work" undertaken was increasingly involving examina-

tions of oil and gas fields, as noted in Shannon, *Director's Biennial Report to the Governor of Oklahoma, 1914*, 17–18. For more details on the Oklahoma Geological Survey during the state's fossil fuel boom era, see Frehner, *Finding Oil.*

41. For the portion of this trip that crossed the very southern edge of the Great Plains and Rocky Mountains regions, see Rehn and Hebard, "Orthopterical Reconnoissance," 111–75. For a rare example of designating such preliminary, wide-ranging work as "reconnaissance" two decades earlier, see Hicks, "Soils of Nebraska," 38.

42. Means, "Reconnoissance in Sanpete, Cache, and Utah Counties, Utah," 115–20; Means, "Reconnoissance in the Cache a la Poudre Valley, Colorado," 121–24.

43. The first example was Macy Lapham and party, "Soil Survey of Western North Dakota," 1155–228. Others were published in subsequent years.

44. Marbut, Bennett, Lapham, and Lapham, *Soils of the United States*, 9.

45. Whitney, "General Review of the Work" (1909), 40. A similar justification—rapid influx of new settlers—justified a "reconnoissance" approach in the San Luis Valley of south-central Colorado, as noted in Lapham, *Soils of the San Luis Valley, Colorado.*

46. Clark Wissler, ed., *Societies of the Plains Indians*, Anthropological Papers of the American Museum of Natural History, vol. 11 (1916), vi. See also Wissler, *North American Indians of the Plains*; and Wissler, *American Indian*, which helped to successfully define the Plains as one of the major native "culture areas" of North America. See also Freed and Freed, "Clark Wissler."

47. Wissler, "Distribution and Functions of Tribal Societies," 402.

48. On the history of thematic mapping in the United States, see Schulten, *Mapping the Nation.* On the development of land use–mapping in the United States, with some attention to topographical and soil surveys, see Checkovich, "Mapping the American Way."

49. For a helpful discussion of the continuum of variation between mapping- and non-mapping-oriented surveys, as it was viewed in the late nineteenth century, see Baker, "Surveys," 254.

50. The terminology can be confusing. Some early work that was already labeled as "survey" nevertheless became more systematic over time. For example, the shift from "survey for exploration to survey for settlement, exploitation and government" in the context of the British Empire is discussed in Collier and Inkpen, "Royal Geographical Society," 106. Conversely, some natural history expeditions not labeled as surveys were transformed through their practices into surveys *avant la lettre.* For an evocative description of Humboldtian-era natural history expeditions as they began to become more systematic and similar to surveys, see Pratt, *Imperial Eyes*, 30.

51. Kohler, *All Creatures*, 17–46.

52. Davis, "Education of the Topographer," 546.

53. For example, see Vetter, "Wallace's *Other* Line."

54. On the longue-durée history of collecting in science, see Kohler, "Finders, Keepers."

55. J. B. Hatcher to Dr. Holland, 31 May 1900, JBHP.

56. J. B. Hatcher to Dr. Holland, 7 June 1900, JBHP.

57. It might not require moving very far from the railhead, however. Field travelers routinely collected during brief railroad stops, and survey parties assembling near towns with railroads often began the collecting right away, in the vicinity of the railroad town itself.

58. "Editorial," *Botanical Gazette* 21 (1896): 303. Field researchers at the Colorado Museum routinely referred to their "biological survey of Colorado" in print, without explicitly defending using state boundaries, as in Lincoln, "Notes on Some Species New to the Colorado List of Birds," 236. The opposite point of view, especially during the earlier territorial period, was implied by one ornithologist in the Bighorn Mountains, who declared that he was "so near the Montana-Wyoming boundary line that in some cases it is impossible to say in which of these Territories they were located," in Merrill, "Oölogical Notes from Montana," 203.

59. On the longue-durée history of observation in science, see Daston and Lunbeck, *Histories of Scientific Observation*.

60. Lowie, *Robert H. Lowie*, 76.

61. Rudwick, "Emergence of a Visual Language"; Hineline, "Visual Culture of the Earth Sciences."

62. Kelsey, *Archive Style*. See also Trachtenberg, "Naming the View," in *Reading American Photographs*, 119–63; and Davis and Aspinwall, *Timothy H. O'Sullivan*.

63. Jackson, *Time Exposure*, 189.

64. The roots of this impulse—particularly in the influential German context—through which advocates of a "biological perspective" advocated for greater "appreciation of living animals in their natural environments" and "observation of living animals in their natural surroundings," are fruitfully explored in Nyhart, *Modern Nature*, 35–36.

65. Lowie, *Robert H. Lowie*, 76.

66. Preble, "Vernon Bailey Passes."

67. Chesnut, "Vernon Bailey," 232.

68. Chesnut, "Vernon Bailey," 232.

69. Preble, "Vernon Bailey Passes," 329. Bailey himself appreciated the value of experiencing nature over multiple seasons and across a wide geographical

range, as he noted in "Old and New Horns of the Prong-Horned Antelope," 128–29.

70. Powell, *Report of Special Commissioners.*

71. Hague himself participated with field parties in Yellowstone National Park from July to October each year from 1883 to 1890, 1893, and 1897, as well as shorter visits in September to October 1891 and August 1902. Others from the Yellowstone Division conducted field work in Yellowstone, without Hague, in some other years, including 1892 and 1898, all according to the tabulations in Mark W. Pangborn, "Memorandum for the Chief Geologist," 21 November 1944, YNPA, box H-2, file 101, which are derived from the division's annual reports.

72. Bailey, "Identity of the Bean Mouse," 71.

73. C. Hart Merriam to Vernon Bailey, 18 October 1887, VBP.

74. There were exceptions, of course, especially in the southern part of the Plains and Rockies. See, for example, the winter and early spring travels reported in Elliot, "List of Mammals Obtained by Thaddeus Surber," 289–303.

75. For example, see Whitney, "General Review of the Work" (1902), 25.

76. Norman J. Colman to Vernon Bailey, 1 November 1888, VBP.

77. C. Hart Merriam to Vernon Bailey, 17 December 1888, VBP.

78. Minot, "Distribution of Vacations," 441–44.

79. E. B. Branson to Williston, 22 June 1905, SWP, box 1.

80. Cassidy, *Ferdinand V. Hayden*, 170.

81. Lunell, "Enumerantur Plantae Dakotae Septentrionalis Vasculares," 155–56.

82. J. Lunell, review of *The Sixth Biennial Report*, 29.

83. Allen, "Notes on the Mammals," 44.

84. Carleton, "Observations on the Native Plants," 220.

85. Allen, "Notes on the Mammals," 43–44.

86. Allen, "Notes on the Mammals," 58.

87. Ridgway, "Notes on the Bird-Fauna," 168. See also Lewis, *Feathery Tribe*, 151.

88. Dodge, *Plains of North America*, 105.

89. Lawrence Bruner to Marcia, 18 September 1881, LBP, box 11, folder 5.

90. Kimball, "Granites of Carbon County," 203.

91. Dumble, *Third Annual Report of the Geological Survey of Texas* (1891), lvii.

92. Kimball, "Granites of Carbon County," 203.

93. Macy H. Lapham to Prof. Milton Whitney, 27 June 1902, SSD-RSS, box 7.

94. Macy H. Lapham to Prof Milton Whitney, 5 July 1902; and Milton Whitney to [Lapham], 9 July 1902, SSD-RSS, box 7.

95. Walter H. Weed to Arnold Hague, 15 August 1890, HP-LR, box 1.

96. Barbour, "Account of the Nebraska Geological Survey to 1912," 12.

97. Marsh, *Odontornithes*, 2.

98. Hague, "Early Tertiary Volcanoes," 4.

99. Marsh, "On the Geology," 191.

100. King, "Report of Clarence King," 1030.

101. Despite Arnold Hague's well-known public activism for "preservation of wild animals within the park" and not himself being a sportsman—saying "upon one occasion that he had never killed any game or fish in his life"—the Yellowstone Division leader "was not averse, however, to his assistants keeping the camp table well supplied," as noted in Iddings, "Biographical Memoir of Arnold Hague," 35. References to obtaining animal food in the field are scattered throughout the primary sources, e.g., in Albert Peale's Journal, 23 August 1871, in Merrill, *Yellowstone and the Great West*, 168–69; and have also merited occasional mention in secondary literature, e.g., Pyne, *Grove Karl Gilbert*, 168.

102. Bailey, "Unforgettable Meals." The items listed as "articles of the ration" under Schedule D ("Field subsistence") of the *Regulations of the U.S. Geological Survey* for 1882 were as follows: beans, bread, coffee, flavoring extracts, flour, dried fruit, fresh fruit, ginger, hominy, lard, lime juice, cured meats, fresh meats, cornmeal, oatmeal, milk, mustard, pepper, pickles, potatoes, rice, salt, spices, sugar, tea, vegetables, vinegar, cracked wheat, and yeast powder (5).

103. Bailey, *Revision of the Pocket Gophers*, 13.

104. Bailey, *Biological Survey of Texas*, 92.

105. Kimball, "Granites of Carbon County," 203.

106. Savage, "On the Bite of the Rattlesnake," 36.

107. Brons, "Notes on the Habits," 564. For another example from Dakota Territory, see Garman, "Rattle of the Rattlesnake," 259–68.

108. Snow, "Amblychila cylindriformis Say," 31–32.

109. Snow, "Hunting Ambylchila," 732.

110. Cassidy, *Ferdinand V. Hayden*, 176.

111. Henshaw, "Autobiographical Notes" (1920), 56.

112. Denison, *Rocky Mountain Health Resorts*, 162, 164–65. On the perceived benefits of field work on the US Coast Survey compared to "unhealthy" indoor laboratory work, see Slotten, *Patronage*, 173.

113. Field work might sometimes even begin while the camp was still being assembled, as several members of an 1871 Hayden Survey party did while "encamped at Fort Ellis" in what would develop into the city of Bozeman, Montana, to begin their Yellowstone exploration during that field season, as noted in Knowlton, "Fossil Flora of the Bozeman Coal Field," 153.

114. Hague to Penfield, 12 June 1886, HP-LS.

115. Hague to Joseph P. Iddings, 13 June 1887, HP-LS.

116. Hague to J. W. Powell, 29 June 1889; Hague to G. K. Gilbert, 30 June 1890; HP-LS. On Pirsson, see also Iddings, "Louis Valentine Pirsson," 530–32.

117. Carl Barus to Arnold Hague, 28 June 1883, HP-LR, box 1; Carl Barus to Mr. Arnold Hague, 16 June 1883, HP-LR, box 1.

118. Arnold Hague to Prof. W. A. Setchell, 25[?] July, 1898, HP-LR, box 1.

119. These generalizations are based on reading the internal reports submitted by Hague to his superiors at the US Geological Survey, such as Hague to J. W. Powell, 29 June 1889, HP-LS, box 2, vol. 2E. A helpful, though incomplete, synopsis of total expenditures for each year, along with scientific field party members, is provided in "Yellowstone National Park Work" (n.d.), Hague Papers, Records Relating to Yellowstone Park and Other Matters, Entry 147 (Old Entry 73), RG 57, Records of the Geological Survey, National Archives II, College Park, Maryland. Annual expenditures dropped to around $2,000 to $4,500 most years later on, while the printed reports were being prepared.

120. Hague to Joseph P. Iddings, 13 June 1887, HP-LS, box 2, vol. 2D.

121. For an account of the survey's failure, told from the side of the rising experts who found amateur contributions lacking, see Urry, "Englishmen, Celts, and Iberians."

122. Parker, "Kansas Scientific Survey," 643.

123. Cragin, "Prospectus of the Biological Survey of Kansas," 2–3.

124. Cragin, "On the Washburn Biological Survey of Kansas," 577.

125. *Bulletin of Washburn Laboratory of Natural History* 1–2 (1884–90).

126. F. W. Cragin to Board of Trustees of Washburn College, 24 June 1891, Washburn University Archives, Mabee Library, Topeka, Kansas.

127. Ludlow, *Report of a Reconnaissance of the Black Hills of Dakota*, 8. The next year, Ludlow accompanied another sizeable US Army field party, accompanied again by a geologist, a paleontologist, and topographers, as reported in Ludlow, *Report of a Reconnaissance from Carroll, Montana Territory*, 9. The lower limit on field survey parties under threat of armed violence still present in the 1870s was probably the Texas geological survey party led by Samuel Buckley, who believed that "a small party of five or six men, well armed, can go over the route traveled by us with perfect safety," despite the presence of "hostile, thieving Indians," as noted in Buckley, *Second Annual Report of the Geological and Agricultural Survey of Texas* (1876), 6.

128. USDA Bureau of Soils, *Instructions to Field Parties*, 14. On two as "the ideal number" for soil survey parties, see also Worthen, "Methods of Soil Surveying," 187.

129. Scott, *Some Memories*, 171.

130. Among the many archival letters tracing the deteriorating relationship between the Colorado Museum's Ray Hersey and H. H. Sheldon in the field, see

Ray to Mr. Figgins, 27 May 1913, JDFC, box 3; H. H. Sheldon to Mr. Figgins, 29 May 1913, JDFC, box 3; Ray Hersey to Mr. Figgins, 1 June 1913, JDFC, box 3; and Ray Hersey to Mr. Figgins, 1 June 1913, JDFC, box 3.

131. Hague to Gilbert, 5 June 1890, HP-LS. Shifting reliance for day-to-day field party management to Iddings and Weed also allowed Hague himself to spend more time traveling with park officials. In 1889, for example, he assisted park superintendent F. A. Boutelle in investigating whether it was wise to allow a lease for building "an elevator or incline" for visitors at the Grand Canyon of the Yellowstone, as noted in Boutelle, *Report of the Superintendent*, 10.

132. Hague to Gilbert, 30 June 1892, HP-LS.

133. Hague to Director of the USGS [Powell], 4 January 1893, HP-LS.

134. Hague to Powell, 30 June 1893, HP-LS.

135. Robt Ridgway to Mother, 5 May 1867, RFP, box 1, folder 2.

136. Robert [Ridgway] to Father, 20 January 1869, RFP, box 1, folder 7.

137. Kohler, "From Farm and Family to Career Naturalist," 28–56.

138. C. Hart Merriam to Vernon Bailey, 1 November 1888, VBP.

139. Patton et al., *Geology of the Grayback Mining District*; and Patton et al., *Geology and Ore Deposits*. For another example elsewhere—not coincidentally, it was also a state with a school of mines—see Todd, *First and Second Biennial Reports, 1893–6*, 16.

140. "Preface," in *University Geological Survey of Kansas* 1 (1896), 1.

141. Haworth, "Geology of Underground Water in Western Kansas," 50.

142. Jackson, *Time Exposure*, 190.

143. For "quartermaster," see Cassidy, *Ferdinand V. Hayden*, 171. For "disbursing clerk and secretary," see Arnold Hague to the Director of the US Geological Survey [Powell], 4 April 1887, HP-LS, box 2, vol. 2D.

144. Hague to J. W. Powell, 29 June 1889, HP-LS, box 2, vol. 2E.

145. For an example from soil surveys, where lodging in towns or on farms was much more common, but referring to staying in private houses to avoid high hotel rates in Belgrade, Montana, see Macy H. Lapham to Prof. Milton Whitney, 6 August 1905, SSD-RSS, box 36. For another example of delays encountered, because of a full hotel in Lexington, Nebraska, see J. O. Martin to Prof. Milton Whitney, 25 June 1904, SSD-RSS, box 27. This is not to say, however, that camping out did not continue in some field survey work. See, for example, Perisho, *Biennial Report*, 6.

146. Lapham, *Crisscross Trails*, 80.

147. [Hague] to Mr. C. W. Hayes, 10 April 1902, HP-LR, box 1.

148. Pyne, *Grove Karl Gilbert*, 48.

149. For seminal works on laboratory technicians, see Barley and Bechky, "In the Backrooms of Science"; and Shapin, "Invisible Technician."

150. Besides the examples from soil survey work near settled areas noted above, other rare examples of women serving with field parties, primarily as cooks, were wives of survey party members, typically the party's leader. One example was Freeman Ward's wife, who accompanied a South Dakota state geological survey party in 1918, along with the Wards' daughter, while traveling in the western part of the state. See Ward, *Report of the State Geologist*, 3.

151. Henshaw, "Autobiographical Notes" (1919), 179.

152. Rhoda, "Report on the Topography," 459.

153. Southworth, "New State of Colorado," 266.

154. Henshaw, "Autobiographical Notes" (1919), 180.

155. Rhoda, "Report on the Topography," 459.

156. Hired laborers with field parties mostly came from within the region. A man named John Wilson wrote to F. V. Hayden from Boulder, Colorado, asking to be rehired as cook for another field season, in John Wilson to Prof. Hayden, 19 December 1878, box 1, Hayden Survey, Miscellaneous Letters Received, Entry 19, RG 57, Records of the Geological Survey, National Archives II, College Park, Maryland. With very few of the cooks and packers even named, it is hard to gauge their social backgrounds with any precision, although some were described through racial categories. Rhoda referred in his account to their "huge black cook," in Rhoda, "Report on the Topography," 459. Secondary sources have also referred to a black woman named Sarah Campbell, or "Aunt Sally," who accompanied the Custer expedition to the Black Hills in 1874, e.g., Grafe and Horsted, *Exploring with Custer*, 5.

157. Henshaw, "Autobiographical Notes" (1919), 179.

158. Rhoda, "Report on the Topography," 460.

159. Lawson, *Log of a Timber Cruiser*, 50–51.

160. Judd, *Men Met along the Trail*, 4.

161. Rhoda, "Report on the Topography," 480.

162. Rhoda, "Topographical Report," 326, 328.

163. Various explanations have been offered to explain how Leigh acquired the name "Beaver Dick." Two prominent ones are that Brigham Young gave him the nickname because of his "prowess in trapping beaver" and that "his two upper front teeth protruded" like a beaver's, as noted in Esther R. Allan, "Beaver Dick's Family to Visit Jackson," *Jackson Hole News*, 30 July 1970, 3, in RBDLF. Sometimes mistakenly referred to as a "half-breed" because he married American Indian women and lived in a hybrid cultural style, Leigh was actually born in England in 1831, then came over to the United States and became a trapper in the Teton Range area. His first wife, a Shoshone woman named Jenny (after whom the popular Jenny Lake in Grand Teton National Park is named), died of smallpox in 1876, as noted in "Beaver Dick Leigh," *Jackson Hole Guide*, April

8, 1965, in RBDLF. Her death was a "hard blow," but in 1879 he married a Bannock chief's daughter and had three children, whom he referred to as "Girl 16, Boy 11 [and] Girl 6 years old" in a letter, Richard Leigh to N. P. Langford, 1 January 1896, RBDLF. For further details on his life, see Norviel, "The Life of Richard Leigh."

164. Bradley, "Report of Frank H. Bradley, Geologist of the Snake River Division," 213.

165. [Leigh] to Josiah Curtis, 1 April 1877, CHMP, roll 90. All quotes *sic.*

166. Florence A. Merriam to Mr. Leigh, 23 April 1877, CHMP, roll 90. For more details on the rest of Leigh's life, see Thompson, "Richard (Beaver Dick) Leigh," also available at Jackson Hole Historical Society, Wyoming.

167. Large sections of Elwell's diary are published in Kilander, "Harvard Cook in the Wyoming Badlands"; and in Elwell, "Alcott Farrar Elwell: His Diary."

168. "Alcott Farrar Elwell," [typescript with that full title] AFEC, folder 1.

169. Mrs. Alcott Farrar Elwell to Gene M. Gressley, 12 April 1966, AFEC, folder 1.

170. Mrs. Alcott Farrar Elwell to Gene M. Gressley, 12 April 1966, AFEC, folder 1. On his contributions to drawing illustrations and making maps, see examples on 20 August, 25 August in "Alcott Farrar Elwell, His Diary, Wyoming 1908, as Camp Cook," AFEC, folder 2.

171. Another example of career mobility from hired labor to scientist is in the autobiography of geologist Charles Newton Gould, who started out as cook for Charles Prosser in 1896 but two years later was leading his own field party and hiring his own laborers: "I employed my old friend, J. E. Broadie, who furnished a covered wagon, team, and three saddle horses. As cook, I hired my schoolmate, Mark White," in Gould, *Covered Wagon Geologist*, 58, 66.

172. Scott, "Early Days," 7.

173. Scott, "Early Days," 3.

174. Scott, "Early Days," 8–9.

175. Scott, "Early Days," 7–8.

176. Lowie, "Notes on Shoshonean Ethnography," 191.

177. Jones, *Report Upon the Reconnaissance of Northwestern Wyoming*, 40.

178. Lipsey, "Light-Hearted Princetonians 'Survey' Colorado in 1877," 350–51.

179. Scott, *Some Memories*, 153–54.

180. Kiener, "P. A. Rydberg's 'Report of a Botanical Exploration,'" 316.

181. Wm. Rush Taggart to Sister Clem, n.d. [approx. 6 June 1873], William Rush Taggart Collection, #712, American Heritage Center, University of Wyoming, Laramie, folder 2.

182. Scott, *Some Memories*, 165.

183. Vernon Bailey to "Friends," 10 June 1888, VBP.

184. C. Hart Merriam to Vernon Bailey, 1 August 1888, VBP.

185. C. Hart Merriam to Vernon Bailey, 6 November 1888, VBP.

186. Lawrence Bruner to Marcia, 7 August 1881, LBP, box 11, folder 5. On the other hand, local people did not always recognize traveling field scientists for what they were, as related in Kiener, ed., "P. A. Rydberg's 'Report of a Botanical Exploration,'" 309: "The people, I saw, were just as much surprised when seeing men sent out to that arid region for the purpose of picking flowers and gathering weeds."

187. Savage, "Stone Implements in Trego County," 57–58.

188. Wallace, "English and American Flowers II," 799–800. See also Wilson, *Alice Eastwood's Wonderland*, 24.

189. S. F. Emmons to Prof. Spencer F. Baird, 9 December 1882, Samuel Franklin Emmons Papers, Letters Sent, Entry 140 (Old Entry 66), RG 57, Records of the Geological Survey, National Archives II, College Park, Maryland, vol. 1. Emmons's letter went on to request publications from the Smithsonian to add to their collections in Denver, which was, in fact, his main purpose in writing to Baird.

190. "Introductory," *American Geologist* 1 (1888), 2–3.

191. Even late in the railroad era, see *First Annual Report of the Secretary of the [Idaho] State Bureau of Mines and Geology* (1919), 9; and Umpleby and Livingston, *Reconnaissance in South Central Idaho*, v. For a contrasting case of cooperation between Idaho and the US Geological Survey, which "went forward without a hitch and with the most friendly and cordial spirit," see Livingston and Laney, *Copper Deposits*, 3.

192. C. Hart Merriam to Vernon Bailey, 18 October 1887, VBP.

193. C. Hart Merriam to Vernon Bailey, 17 December 1888, VBP.

194. H. H. Sheldon to Mr. Figgins, 24 May 1913, JDFC, box 3.

195. A. K. Fisher to Vernon Bailey, 8 August 1888, VBP.

196. C. Hart Merriam to Vernon Bailey, 10 January 1889 [second letter], VBP, emphasis Merriam's.

197. Mudge, "Annual Report of the Committee of Geology," 4.

198. Patrick, "Kansas Chalk," 13–15; Patrick, "Analysis of Kansas Soils," 15–16.

199. Brous, "Habits of the Prairie Dog," 10–11; Brous, "Habits of Amblychila cylindriformis," 11–12.

200. Snow, "On the Dermal Covering," 54–58; Snow, "Insects of Wallace County, Kansas," 61–70; Snow, "List of Coleoptera," 75–77.

201. Thompson, "Relation of a State Geological Survey," 10–11. Another point of view was that newly developing states such as New Mexico and Okla-

homa needed more focus on practical geology, as explained in Mathews, "Relation of Scientific to Practical Work in State Surveys," 181–87.

202. In Wyoming, for example, see F. F. Hintze to Mr. L. W. Trumbull, 27 February 1914, University of Wyoming, Department of Geology and Geophysics Records, #545004, American Heritage Center, University of Wyoming, Laramie, box 31, folder 2. Moreover, some Wyoming reports, though focused on economic geology of specific areas, were hampered by the lack of general geology maps, since the US Geological Survey had not yet produced them, as noted in Trumbull, "Atlantic City Gold Mining District, Fremont County," 73. Indeed, Wyoming had long aspired to producing its own statewide geological map, described as a "very important" project, in Ricketts, *Annual Report*, 3. Also, as one Oklahoma geologist observed, "Work done by the Federal Survey and by the state surveys is likely to take on both scientific and economic character, for the men at the head of these bureaus soon learn, and often to their sorrow, that the public in general has no use for purely scientific information, and unless economic results are not only promised, but actually delivered, there is no public money for the carrying on of geologic surveys." Gould, "Geological Work in the Southwest," 20.

203. Barbour, "Biennial Report," 380.

204. Barbour, "Account of the Nebraska Geological Survey," 12.

205. Leonard, "Lignite Deposits of North Dakota," 234.

206. A. J. Collier to Dr. A. G. Leonard, 4 January 1918, NDGS, box 4, folder 33. Leonard's reply was favorable and offered some additional information, in [Leonard] to Mr. A. J. Collier, 18 January 1918, NDGS, box 4, folder 33. From 1901 through the early 1910s, however, a rival "Agricultural College Survey of North Dakota" was operated out of the North Dakota Agricultural College in Fargo, with an express purpose of cooperating with federal agencies in survey work and including four divisions: hydrographic survey, soil survey, topographic survey, and biological survey, as noted in *Biennial Report of the Agricultural College Survey of North Dakota* 1 (1901–02), 8. The North Dakota Geological Survey in Grand Forks, however, fought from the beginning to have the cooperative survey work brought under their authority, as noted in *Biennial Report of the North Dakota Geological Survey* 2 (1902), 23.

207. Milton Whitney to L. G. Carpenter, 20 February 1901; Milton Whitney to L. G. Carpenter, 11 February 1902, EXP.

208. Milton Whitney to L. G. Carpenter, 24 April 1903, EXP.

209. Botany field notes, 1901 season, pt. 1, RJPP, box 8, folder 5; pt. 2, RJPP, box 8, folder 6.

210. Peterson, "Albert F. Potter's Wasatch Survey, 1902: A Beginning for Public Management of Natural Resources in Utah," 238–39. On Powell's earlier irri-

gation survey, see Sterling, "Powell Irrigation Survey, 1888–1893," 121–34; and Alexander, "John Wesley Powell, the Irrigation Survey, and the Inauguration of the Second Phase of Irrigation Development in Utah," 190–206, who judges it to be "cursory and incomplete" as a scientific study (194).

211. E. E. Free to Raymond J. Pool, 6 July 1912, RJPP, box 1, folder 1.

212. E. E. Free to Raymond J. Pool, 13 August 1912, RJPP, box 1, folder 1. For the final disappointing results, see Geo. J.[?] Young to Raymond J. Pool, 19 October 1912, RJPP, box 1, folder 1.

213. Ramaley, "What Is Biology," 60–61.

214. "New Forest Reserve," 12. Hague began advocating for protection of the park's forests, game animals, and thermal features after the Yellowstone Division's first season in the park, e.g., in a 28 December 1883 letter to US Senator George G. Vest, reprinted in *Forest and Stream*, March 13, 1884, 125. *Forest and Stream* had several years earlier deemed Hague's "acquaintance with the Park" to be "so extensive as to entitle his opinion to the greatest weight," in "Care of the National Park." Other examples of Hague's public advocacy for conservation in Yellowstone include Hague, "Yellowstone Park as a Forest Reserve"; and Hague, "Yellowstone Park as a Game Reservation."

Chapter 4. Quarries

1. A good starting point for the early history of Dinosaur National Monument is Harvey, "Utah, the National Park Service, and Dinosaur National Monument." For an overview of field work at the dinosaur quarry, see Chure and McIntosh, "Stranger in a Strange Land." A collection of Douglass's later work has recently been edited and published by his son and granddaughter, in Douglass, *Speak to the Earth*. For a literary rendition of Douglass's field work, see Stegner, "Notes on a Life." For a closer examination of Douglass's personal family life in relation to the field site, see Neel, "Love among the Fossils."

2. W. D. Matthew to Professor Osborn, 6 July 1922, Division of Paleontology, Field Correspondence, American Museum of Natural History, New York City, box 4, folder 13.

3. Lister, *Troweling through Time*; Smith, *Mesa Verde National Park*; and Torres-Reyes, *Mesa Verde National Park*. Some shorter accounts include Breternitz, "Mesa Verde National Park"; and Melinda Elliott, "Cliff Palace." For McClurg's own account of the roles she and others played, including the funding of quarry field work, see McClurg, "Making of Mesa Verde into a National Park."

4. Bell, "Fossil-Hunting in Wyoming," 267.

5. Wilson, "Visit to the Zeolite Locality," 29. Another example of a scientific quarry that was already serving as a conventional economic quarry was the

limestone quarry operated by the Northwestern Quarry Company near Rapid City, South Dakota, which also yielded fossil fish, as discussed in Hussakof, "Note on a Palaeoniscid Fish," 347.

6. Gilder, "'Spanish Diggings,' Wyoming," 3. See also Dorsey, "Aboriginal Quartzite Quarry," 231–43.

7. A few university museums were also important in vertebrate paleontology, including the University of Chicago and the University of California at Berkeley; see Rainger, "Just before Simpson," 454. However as Rainger points out elsewhere, in the case of Chicago, the departmental structure of universities made fossil work fall uneasily between biology and geology, thus hindering its development in the university context even in a city that had a major museum, in Rainger, "Biology, Geology, or Neither, or Both."

8. On the especially close relationship between capitalist wealth accumulation and the funding of dinosaur paleontology, see Rieppel, "Dinosaurs."

9. On East-West tensions in American archaeology, see Snead, *Ruins and Rivals.*

10. For an extended case study of collaboration with locals, based on a study of the Cook family at Agate Fossil Beds in northwestern Nebraska, see Vetter, "Cowboys, Scientists, and Fossils."

11. For an example of returning to a "biological quarry" in order to mine plant or animal materials for a museum's habitat diorama, consider the repeated field trips by the Colorado Museum of Natural History to Barr Lakes, a rich site that was located to the northeast of Denver, as mentioned in *Annual Report of the Colorado Museum of Natural History* (1913), 20, 24. Besides the more or less professional ethnographers mentioned below, another excavator of cultural quarries was George Bird Grinnell, who visited with the Cheyennes every year from 1890 onward and gradually gained "their confidence" after a few years, thus allowing him partly "to penetrate into the secrets of their life," as noted in Grinnell, *Cheyenne Indians,* v.

12. Riggs, *Dakota-English Dictionary*; Riggs, *Dakota Grammar, Texts, and Ethnography.* Riggs also occasionally operated in a more geographically extensive exploratory mode, as reported in Riggs, "Journal of a Tour." Another example of cultural quarrying, in which anthropologist Clark Wissler made long-term use of a local collaborator—in this case a physician who lived on the Pine Ridge Indian Reservation for nearly two decades—to extract and submit "notes, manuscripts, and texts in Indian languages to New York, where he and his staff could edit them for publication," is noted in a much later reprinted version of these materials, in Walker, *Lakota Belief and Ritual,* 14. Other Walker material was subsequently published in other books, including *Lakota Society* (1982) and *Lakota Myth* (1983). For a classic account of the shift in anthropology toward

"intensive study of limited areas," see Stocking, "Ethnographer's Magic."

13. Emmons, *Geology and Mining Industry of Leadville, Colo.*; Emmons, "Investigation of Metalliferous Ores," 14–19; Emmons and Irving, *Downtown District of Leadville, Colo.* Similar repeated, multiyear vertical probing was done at other sites in the region, such as Butte, Montana, in Weed, *Geology and Ore Deposits.* Admittedly, mapping the underground was still often labeled as a form of surveying, as in Trumbull, *Manual of Underground Surveying,* but its local intensity and verticality nevertheless mark it off as somewhat distinct. On the recursive production of knowledge in vertical space, see Braun, "Producing Vertical Territory." On verticality in science, see also Reidy, "From the Oceans to the Mountains"; and Reidy, "John Tyndall's Vertical Physics."

14. Osborn, "Hunting Prehistoric Animals," 552.

15. M'Master, "'Bad Lands' of Wyoming," 121, 125–26.

16. Wieland, *American Fossil Cycads,* 2:52.

17. W. D. Matthew to wife, 13 June 1906, in Camp, "Letters of William Diller Matthew," 269.

18. Willey, "Bone Cabin Quarry," 558. For another popular account from a few years earlier, see Osborn, "Fossil Wonders of the West." For an example of a scientific article explicitly referring to the Bone Cabin Quarry as the source of study material, see Osborn, "Fore and Hind Limbs," 161–72.

19. Fletcher, "Hako," 13–15. On the problems of differentiating researchers and American Indian informants in anthropology, using the case of Francis La Flesche, who worked with Alice Fletcher, see Liberty, "Native American 'Informants'"; and Mark, "Francis La Flesche." For a full biography of Fletcher, see Mark, *Stranger in Her Native Land.* Fletcher herself deeply appreciated La Flesche's "intimate knowledge of the customs of his tribe, together with his intelligent appreciation of the need of preserving an accurate account of the rites and ceremonies already becoming obsolete," in Fletcher, "Wa-wan, or Pipe Dance," 308.

20. Wieland, *American Fossil Cycads,* 2:6. As a recipient of funding from the Carnegie Institution of Washington for studying "living and fossil cycads"—$1,500 in 1903, $2,300 in 1904, $1,500 in 1905, and beyond—Wieland had benefited from external funding support before, as noted in *Carnegie Institution of Washington Yearbook* 2 (1903), xxxvii–xxxviii; *Carnegie Institution of Washington Yearbook* 3 (1904), 123–24; *Carnegie Institution of Washington Yearbook* 4 (1905), 241.

21. Schuchert, "Fossil Field's Expedition to Wyoming," 727. See also *Fossil Fields of Wyoming.*

22. Yarrow, "On the Exploration," 185–86. Later in the article, Yarrow crudely justified this grave robbing not only in the name of science but also by

claiming that "poetic and retributive justice was to a certain extent fulfilled as the band to which their owners belonged murdered Lt. Gunnison and his party on the Sevier River some years previously" (187).

23. Snow, "On the Dermal Covering," 54.

24. Snow, "On the Dermal Covering," 56.

25. Sternberg, *Hunting Dinosaurs*, 5. Sternberg's field teams often worked with great rapidity, as they did at one "Eldorado" in western Kansas, where they instructed their cook "to parch a kettleful of corn," then "filled our pockets with it and lived on it for three days, eating of it most of the time to keep ourselves sufficiently nourished," not leaving until they had "loaded [their] wagon with eight hundred pounds of fossil vertebrates." "Paleontologist on Campus Tells of His Life as a Fossil Hunter," *State College Leader*, February 8, 1939, in "Scrapbook, 1926–1938," George Sternberg Collection, Archives/Special Collections, Forsyth Library, Fort Hays State University, Fort Hays, Kansas. For more background on the Sternberg family, see Rogers, *Sternberg Fossil Hunters*.

26. Troxell, "Fossil Ruminant from Rock Creek," 479; Troxell, "Fossil Hunting in Texas," 82, 89.

27. Elliott, "Cliff Palace," 10. For a more detailed secondary account of Nordenskiöld's field methods, see Lister, *Troweling through Time*, 19–23. Nordenskiöld's published book, which includes some information about those methods, is *Cliff Dwellers of the Mesa Verde*. For translations of Nordenskiöld's writing about visiting southwestern Colorado and diaries from his 1891 trip, see Arrhenius, *Stones Speak and Waters Sing*.

28. Loper, "Trip to the Tertiary Formations," 751–52; Shor, *Fossils and Flies*, 222.

29. Osborn, "Skeleton of Diplodocus," 191; Hatcher to Holland, 30 November 1900, JBHP.

30. Cockerell, "Recollections of a Naturalist V," 51; Cockerell, "Florissant: A Miocene Pompeii," 112–26; Cockerell, "Fossil Hymenoptera," 31–58; and Cockerell, "New Fossil Insects," 251–56. Scudder credited Colorado field scientist Arthur Lakes with discovering another fossil insect and plant quarry in 1882 in the same general area of South Park, as noted in Scudder, "Insects of the Triassic Beds," 433.

31. Lee, "Type Section of the Morrison Formation," 183.

32. Knowlton, "Dicotyledonous Flora," 189–94.

33. W. D. Matthew to wife, 7 July 1906, in Camp, "Letters," 274.

34. W. D. Matthew to wife, 7 July 1906, in Camp, "Letters," 274.

35. W. D. Matthew to wife, 20 July 1906, in Camp, ed., "Letters," 282.

36. Brown, "Hunting Big Game of Other Days," 415.

37. Loomis, *Evolution of the Horse*, 3–4.

38. Fletcher, *Wetherills of Mesa Verde*, 95. See also Topping, "Wetherills: Explorer-Archaeologists," in *Glen Canyon and the San Juan Country*, 94–117.

39. Another exceptionally vivid example of experiential knowledge of a local person used to help locate a hidden quarry site, though complicated by the fact that it was a quarry for human remains and thus occupied very briefly, as well as the outright sale of the guiding service, is in Yarrow, "On the Exploration of some Indian Graves," 185–86.

40. Holland, "Earl Douglass," 282–83. An earlier version of this account from Holland's point of view is in Holland, "Skull of Diplodocus," 382–83. Douglass's own diary entries for August 1909 (see diary 25, EDP), which make far less mention of Holland's contributions—though they do reveal his ongoing frustration with Holland's imperious style—reveal the experience of discovery through the field man's own eyes. The diary entries also reinforce how cosmopolitan science and experiential knowledge were intertwined in the discovery process.

41. Later, at least one Carnegie Museum staff member would claim in print that another local man named Elder George A. Goodrich had initially discovered the fossil site and had led Douglass to it, in Coggeshall, "Patsy of Dinosaur Monument," 59. Douglass's son, however, would vigorously dispute this article as a "misleading account" filled with "rumors and half-truths," in Douglass, *Speak to the Earth*, xiv–xv. If Goodrich had played a role in helping Douglass locate the site, it would only reinforce the importance of local, experiential knowledge, but it remains a point of contestation.

42. J. B. Hatcher to Dr. Holland, 2 July 1901, JBHP; J. B. Hatcher to Dr. W. J. Holland, 31 December 1901, JBHP; Hatcher to Holland, 27 May 1903, JBHP.

43. On early collections at Como Bluff, see Ostrom and McIntosh, *Marsh's Dinosaurs*.

44. Menke, "From a Cabin Window," 14. Menke published this account in an ornithological journal because he spent those winter days photographing birds around the makeshift cabin.

45. Wortman, "Restoration of *Oxyaena lupine* Cope," 140.

46. W. D. Matthew to wife, 18 June 1906, in Camp, "Letters," 270.

47. W. D. Matthew to wife, 18 June 1906, in Camp, "Letters," 270.

48. Peterson to Hatcher, 9 May 1902, OAPP.

49. Harrell, "'We Contacted Smithsonian.'"

50. Gustaf Nordenskiöld to his father, 2 July 1891, translated by Daniel M. Olson, in Diamond and Olson, *Letters of Gustaf Nordenskiöld*, 30. Interestingly, according to Charles C. Mason, who married into the Wetherill family and helped to begin excavation at the Cliff Palace, in a written account that the Wetherills also signed, the "Wetherill boys" not only provided the expertise for locating

the sites but also "made a collection" for Nordenskïold. Described in Mason, "Story of the Discovery," 323–29. As we shall see below in the section on field labor at quarries, Nordenskïold himself considered them to be hired workers under his own guidance.

51. This field work, while mostly brief and observational compared to later, more intense excavations, first appeared in an archaeological journal, in an article that made a point of mentioning the field party's efforts to make ground plans, measurements, and notes. Chapin, "Cliff-Dwellings of the Mancos Cañons," 193–210.

52. Chapin, *Land of the Cliff-Dwellers*, 105–6.

53. Birdsall, "Cliff Dwellings of the Cañons of the Mesa Verde," 587–88, 611.

54. Lister and Lister, *Earl Morris and Southwestern Archaeology*, 58.

55. Hatcher to Holland, 8 July 1901, JBHP.

56. Brinkman, "Henry Fairfield Osborn and Jurassic Dinosaur Reconnaissance," 172.

57. W. D. Matthew to wife, 21 June 1906, in Camp, "Letters," 271.

58. W. D. Matthew to wife, 7 July 1906, in Camp, "Letters," 274.

59. W. D. Matthew to wife, 18 June 1906, in Camp, "Letters," 270.

60. W. D. Matthew to wife, 8 July 1906, in Camp, "Letters," 274.

61. W. D. Matthew to wife, 11 July 1906, in Camp, "Letters," 280; W. D. Matthew to wife, 16 July 1906, in Camp, "Letters," 281.

62. W. D. Matthew to wife, 18 July 1906, in Camp, "Letters," 282.

63. O. A. Peterson to Dr. W. J. Holland, 2 July 1912, OAPP.

64. 16, 17 July 1885, in Bartholomew, *Pioneer Naturalist*, 232.

65. W. D. Matthew to mother, 3 August 1902, in Camp, "Letters," 267.

66. Hatcher to Holland, 17 April 1900, JBHP.

67. Hatcher to Holland, 21 April 1900, JBHP; Hatcher to Holland, 1 May 1900, JBHP.

68. For an extended account of the rival paleontological field campaigns of the urban museums, see Brinkman, *Second Jurassic Dinosaur Rush*. For a brief sketch, see McIntosh, "Second Jurassic Dinosaur Rush."

69. Holland to Hatcher, 24 February 1900, JBHP.

70. Hatcher to Holland, 31 March 1900, JBHP.

71. J. W. Powell to Hon. W. B. Allison, 24 June 1892, Marsh Papers, Records Concerning Othniel Charles Marsh, Entry 78, RG 57, Records of the Geological Survey, National Archives II, College Park, Maryland, box 1.

72. O. A. Peterson to Hatcher, 28 May 1901, OAPP.

73. W. J. Holland to Prof. J. B. Hatcher, 29 May 1901, JBHP.

74. Hatcher to Holland, 30 September 1901, JBHP.

75. Hatcher to Holland, 30 April 1902, JBHP.

76. Hatcher to Holland, 31 March 1900, JBHP.

77. E. B. Branson to Doctor [Williston], 22 June 1905, SWP, box 1.

78. The Fossil Fund, designated by Carnegie to bring in "a representative collection of the huge fossils of the Mesozoic period as these are represented in the deposits of our Western States," was first announced a year earlier, in *Annual Report of the Carnegie Museum* (1899), 11.

79. Rea, *Bone Wars*, 1–11, 198–206.

80. Holland to Douglass, 2 September 1910, EDP, box 18, folder 1.

81. Douglas Stewart to Douglass, 3 May 1915, EDP, box 16, folder 15.

82. Rainger, *Agenda for Antiquity*, 91–92, 96; *33rd Annual Report of the American Museum of Natural History* (1901), 26.

83. J. B. Hatcher to Dr. Holland, 9 May 1900, JBHP.

84. "Agreement" (1922), unpublished typescript, ESDM, folder 4. Another (unsigned) draft agreement dated July 25, 1920, in the same folder offered $10 per year for the privilege of excavating a quarry near Stoneham, in Weld County, Colorado. That $10 was a relatively small amount compared to the overall costs of the excavation, which occurred the following year under Philip Reinheimer, at a total amount of $510.44, including travel, food, and lodging, "hauling," and auto hiring, as noted in "Expenses Field Work Collecting Fossil Skeletons at Stoneham, Weld Co. Colo." (29 September 1921), unpublished typescript, ESDM, folder 4.

85. Sternberg, *Life of a Fossil Hunter*, 128–29.

86. Two slightly different accounts lacking claim-specific source citations are in Lanham, *Bone Hunters*, 198–99; and Schuchert and LeVene, *O. C. Marsh*, 208. According to Lanham, Hatcher hired "the owner's son to help dig," while Schuchert and LeVene claim that Hatcher hired Overton himself. For his part, Sternberg claimed to have originated the idea earlier, recalling in his memoir that he "drove up to Mr. Overton's door and offered him forty dollars a month to work for us with his team during the whole summer, with the understanding that I was to have all the fossils found," in Sternberg, *Life of a Fossil Hunter*, 129.

87. Williston, "Restoration of Aceratherium fossiger Cope," 289–90.

88. On the development of US land laws and policies, see Gates, *History of Public Land Law Development*; and Hibbard, *History of the Public Land Policies*.

89. Reynolds, "Kinkaid Act and Its Effects on Western Nebraska."

90. O. A. Peterson to Hatcher, 9 May 1902, OAPP.

91. Bruner, "Comparison of the Bird-Life," 58–63.

92. Vernon Bailey to friends, 10 June 1888, VBP.

93. Hatcher to Holland, 12 August 1901, JBHP.

94. J. B. Hatcher to Dr. Holland, 25 June 1900, JBHP.

95. "Editorial Notes," *Annals of the Carnegie Museum* 10 (July 1916), 309.

96. W. J. Holland, "Earl Douglass," 284.

97. Holland to Douglass, 12 February 1912, EDP, box 18, folder 2.

98. Rainger, *Agenda for Antiquity*, 81–82.

99. P. C. Knox to Hon. James R. Garfield, 19 March 1908; James Rudolph Garfield to Hon. P. C. Knox, 19 March 1908, Central Classified Files, A1 Entry 794-A, RG 48, Records of the Office of the Secretary of the Interior, National Archives II, College Park, Maryland, box 1982.

100. Hatcher to Commissioner of Indian Affairs, 9 May 1904, JBHP.

101. A. G. Tonner to Hatcher, 14 May 1904, JBHP.

102. W. D. Matthew to Mother, 19 July 1903, in Camp, ed., "Letters," 267.

103. W. D. Matthew to wife, 24 June 1906, in Camp, ed., "Letters," 272.

104. See below for more detailed discussion of disagreement of Utes with excavation at Mesa Verde.

105. Moses, *Indian Man*, 72.

106. Moses, *Indian Man*, 131. Mooney also spent "five years among the southern Cheyenne" nearby around the same time, which resulted in a lengthy and intensive set of "notes" from his cultural quarrying, in Mooney, "Cheyenne Indians," 361–62. Overlapping with Mooney was the Field Museum anthropologist George A. Dorsey, who observed Cheyenne rituals for several years, as noted in Dorsey, "Cheyenne I: Ceremonial Organization," 1–55; and Dorsey, "Cheyenne II: The Sun Dance," 57–186. Mention of Mooney's presence at the same sun dance is in the latter article (xiv).

107. Lakes, "Florissant Basin," 179.

108. Prudden, "Summer among Cliff Dwellings," 552.

109. Cook, "Two New Bunomastodonts from Colorado," 1–14.

110. Judd, "Report on Illegal Excavations in Southwestern Ruins," 428–29.

111. Judd, "Report on Illegal Excavations in Southwestern Ruins," 429.

112. For Hewett's original pamphlet on protecting archaeological resources, written for the General Land Office, see Hewett, *Circular Relating to Historic and Prehistoric Ruins.*

113. Rothman, *Preserving Different Pasts.* On the National Park Service, see Runte, *National Parks*; and Sellars, *Preserving Nature in the National Parks.* See also Cameron, *National Park Service.* On the dispossession of American Indians (and others) from traditional uses of national park lands, see Jacoby, *Crimes against Nature*; Keller and Turek, *American Indians and National Parks*; Burnham, *Indian Country, God's Country*; and Spence, *Dispossessing the Wilderness.*

114. Mrs. W. H. Walling to W. F. [*sic*] Mondell, 30 November 1915, USFS-L.

115. F. W. Mondell to Secretary of the Interior, 9 December 1915, USFS-L.

116. H. S. Graves to Mr. W. [*sic*] D. Walcott, 10 January 1916; C. D. Walcott to H. S. Graves, 14 January 1916, AAP, box 18, Medicine Wheel folder.

117. J. Walter Fewkes to Mr. Albright, 14 December 1918, AAP, box 17, Hovenweep National Monument folder.

118. In 1911, according to his biographer, the ethnologist James Mooney suggested to the Nebraska State Historical Society that they send out "a circular letter to teachers, missionaries, church congregations, and members of pioneer associations." The information collected and mapped would then be used as the basis for sending out teams of university students to visit the sites identified. "If the site proved to be archaeologically or ethnologically important," he proposed, "the society should seek state protection for it against possible vandalism," in Moses, *Indian Man*, 172.

119. Holland to Douglass, 12 March 1912, EDP, box 18, folder 2.

120. Holland to Douglass, 16 December 1915, EDP, box 18, folder 2.

121. Holland to Douglass, 12 January 1916, EDP, box 18, folder 3.

122. Bell, "Fossil-Hunting in Wyoming," 268.

123. Gilmore, "History of the Division of Vertebrate Paleontology," 342.

124. W. D. Matthew to wife, 5 June 1906, in Camp, "Letters," 269.

125. Rainger, *Agenda for Antiquity*, 69.

126. W. J. Holland to Mr. James H. Cook, 30 October 1907, CPC, box 33; Douglass, 19–24 April 1910 entries in Diary 27, 1910—Utah, EDP, box 6, transcript in folder 1.

127. Colbert, *William Diller Matthew, Paleontologist*, 128. For a short account emphasizing Matthew's early field experiences, see Colbert, "W. D. Matthew's Early Western Field Trips."

128. Thus while ranchers were typically strangers to the field scientists who simply offered the region's customary hospitality to travelers, on rare occasions they might be relatives, such as when they stopped at Thomson's father's ranch in 1906. W. D. Matthew to wife, 18 June 1906, in Camp, "Letters," 270.

129. Hatcher to Mr. Theodore Olcott, 2 May 1904, OAPP. For comparison, in that same year (1904), Hatcher recommended raises to other staff members as follows: $75 per month each for museum paleontologists Earl Douglass and C. W. Gilmore, as well as artist Sidney Prentice, $110 per month for highly skilled fossil preparator A. S. Coggeshall, and $65 per month for Louis Coggeshall, as noted in Hatcher to Holland, 16 January 1904, JBHP.

130. Among numerous examples, see J. B. Hatcher to Dr. Holland, 9 May 1900, JBHP.

131. Nordenskiöld, *Cliff Dwellers*, 21.

132. Fewkes, *Antiquities of the Mesa Verde National Park*, 9.

133. Hatcher to Holland, 25 July 1900, JBHP.

134. Douglass, 2 October 1914 entry in Diary 32, 1914–1918, EDP, box 7.

135. Douglass, undated entry at end of August 1917 in Diary 36, 1917–1919, EDP, box 8.

136. Douglas Stewart to Douglass, 10 November 1921, EDP, box 3, folder 2.

137. O. F. McShane to Holland, n.d. [1921], EDP, box 3, folder 2.

138. Douglass to Stewart, 20 September 1920, EDP, box 16, folder 20.

139. W. D. Matthew to mother, 27 June 1898, in Camp, "Letters," 266.

140. Holland to Douglass, 16 June 1911, EDP, box 18, folder 2.

141. Holland to Douglass, 4 February 1918, EDP, box 18, folder 3.

142. Rainger, *Agenda for Antiquity*, 79.

143. Hatcher to Holland, 7 June 1900, JBHP.

144. Holland to Hatcher, 12 June 1900, JBHP. For more on Reed's life and career, see Breithaupt, "Biography of William Harlow Reed."

145. Hatcher to Holland, 25 June 1900, JBHP.

146. Fowler, *Laboratory for Anthropology*, 284. On the experiences of student participants, see Givens, *Alfred Vincent Kidder*, 131–39; Woodbury, *Alfred V. Kidder*, 1–17; Brunhouse, *Sylvanus G. Morley*, 40–47; and Johnson, *Fierce Solitude*, 29.

147. Elliott, *Exploring Human Worlds*, 12.

148. Gifford and Morris, "Digging for Credit."

149. Fletcher, *Life Is My Song*, 30, 32–33.

150. Rickner to Secretary of the Interior, 18 December 1914, MVNP, series 12, drawer 2, folder 29.

151. Thomas Rickner to J. Walter Fewkes, 27 July 1918, MVNP, series 5, drawer 4, folder 9. Interestingly, American Indian labor appears to have been fairly common at archaeological sites in other parts of the Southwest.

152. Jesse Nusbaum, "When I couldn't recruit local labor" (n.d.), Jesse Logan Nusbaum Papers, National Anthropological Archives, Suitland, Maryland, box 6.

153. Smith, *Mesa Verde National Park*, 113.

154. J. B. Hatcher to Dr. Holland, 23 September 1900, JBHP.

155. O. A. Peterson to Hatcher, 11 May 1901, OAPP.

156. Hatcher to Holland, 2 August 1900, JBHP. See also Hatcher to Mr. E. H. Sellards, 4 September 1903, JBHP.

157. For further discussion of changing practices during this period in fossil preparation labs, see Brinkman, "Dinosaurs, Museums, and the Modernization of American Fossil Preparation."

158. Riggs, "Use of Pneumatic Tools," 747.

159. Riggs, "Use of Pneumatic Tools," 748.

160. Riggs, "Use of Pneumatic Tools," 749. For an extended discussion of laboratory methods alongside field methods in vertebrate paleontology, at the

rival American Museum of Natural History—and explaining "from practical experience what method of treatment and preparation in the field" was "most practical and reliable for giving the best results in the museum laboratory," see Hermann, "Modern Laboratory Methods in Vertebrate Paleontology," 284.

161. For one example of explicit comment on the possibility of winter work at a fossil quarry site—in this case near Cañon City, Colorado, in an Arkansas River canyon tributary—see J. B. Hatcher to Dr. Holland, 9 May 1900, JBHP.

162. Holland, "Earl Douglass," 284–85.

163. Holland, "Earl Douglass," 285.

164. E. Douglass to D. Stewart, 19 September 1909, EDP-CMNH, box 1, folder 50.

165. E. Douglass to W. J. Holland, 26 August 1909, EDP-CMNH, box 1, folder 47.

166. E. Douglass to W. J. Holland, 28 December 1909, EDP-CMNH, box 1, folder 60.

167. Mooney, *Indian Man*, 84. Mooney is perhaps most famous for his pioneer ethnographic work on the origins and spread of the Ghost Dance, which required "cultural excavation" at several sites across the region. His report appeared as Mooney, "Ghost-Dance Religion and the Sioux Outbreak of 1890," 651–1110. For a brief synopsis nearly two decades later, see Mooney, "Indian Ghost Dance."

168. Fletcher, "Hako," 15. In her report, Fletcher spells it slightly differently, calling it a "graphophone." Others before and since have tended to call it simply the "phonograph." Several years earlier, she had studied the music of the Omaha Indians, using musical notation, in Fletcher, "Study of Omaha Indian Music," 233–384.

169. Fewkes, "On the Use of the Phonograph," 268. For the larger context, see Brady, *Spiral Way*. As Brady points out (54–56), Fewkes first tried out the phonograph to record spoken language among the Passamaquoddy in Maine in 1890 and then more extensively among the Zuni of New Mexico. Some have contended that Frank Hamilton Cushing may have used the phonograph for ethnography before Fewkes, also in the American Southwest, though he did not publish about this, as discussed in Brady, *Spiral Way*, 56–57.

170. Densmore, *Teton Sioux Music*, 1.

171. Densmore, *Northern Ute Music*, 5. On how Densmore, while working with the Northern Utes at Fort Duquesne, Utah, in 1914, agreed to record a complaint message to the Commissioner of Indian Affairs about local agents, in return for singing into the phonograph, see Brady, *Spiral Way*, 92–93. Densmore followed up next with an intensive study of the music of the Mandans and Hidatsas of North Dakota, in Densmore, *Mandan and Hidatsa Music*. Another suggestive

comparison of using more intensive technologies in diverse examples of quarry science is the quite literal similarity between taking measurements of the dimensions of fossil or artifact quarries and the dimensions of the bodies of Indian children, as described in Wissler, "Measurements of Dakota Indian Children."

172. Frances Densmore to Charles Hofmann, 20 October 1943, reprinted in Hoffman, *Frances Densmore and American Indian Music*, 62.

173. W. J. Holland, "Report on the Activities of the Carnegie Museum at the National Dinosaur Monument, Utah, Uinta County, during the year 1917" (5 January 1918), AAP, box 17, Dinosaur, 1918–1950 folder.

174. W. D. Matthew to wife, 13 July 1906, in Camp, "Letters," 280.

175. Arthur Lakes, journal, undated entry after 17 October 1879, in Kohl and McIntosh, *Discovering Dinosaurs in the Old West*, 138–40.

176. W. D. Matthew to wife, 24 July 1906, in Camp, "Letters," 283.

177. Nordenskiöld, *Cliff Dwellers*, 22.

178. W. D. Matthew to wife, 22 July 1906, in Camp, "Letters," 282.

179. W. D. Matthew to wife, 11 July 1906, in Camp, "Letters," 280.

180. Holland to Hatcher, 5 July 1901, JBHP.

181. Hatcher to Holland, 13 July 1901, JHBP; Hatcher to Holland, 17 July 1901, JBHP.

182. Peterson to Hatcher, 30 June 1904, OAPP; Holland to Peterson, 8 July 1904, OAPP.

183. Hatcher to Holland, 30 November 1900, JBHP.

184. Rainger, "Collectors and Entrepreneurs." See also Brinkman, *Second Jurassic Dinosaur Rush*.

185. Ward, *Needs of the State Geological and Natural History Survey*.

186. Smith, "Removing Our Relics." For a brief synopsis, see Smith, *Mesa Verde National Park*, 29–30. At the state level, this issue had already come up earlier when, "in May 1889, a large collection of relics, taken from the Cliff Dwellings in the south-western portion of the State, was brought to Denver and put on exhibition. This was visited by a number of members of the Society and others who were interested, and the officers of the Society were strongly urged to purchase it so that it might not be taken out of the State. This seemed especially advisable, as the Smithsonian Institution was negotiating for its purchase. The collection was finally bought by this Society for $3,000," as noted in *Report of the Historical and Natural History Society of the State of Colorado for the Years 1889–90*, 3.

187. Nordenskiöld, *Cliff Dwellers*, 23.

188. Gustaf Nordenskiöld to his father, 17 September 1891, translated by Daniel M. Olson, in Diamond and Olson, *Letters of Gustaf Nordenskiöld*, 53.

189. Barbour, "Museums and the People," 8, 10. See also Arnold, "Barbours."

190. Frederick J. Pack to Hubert Work, 11 August 1923, AAP, box 17, Dinosaur, 1918–1950 folder.

191. Harold Cook to H. F. Osborn, 29 February 1916, CPC, box 42. The *Moropus* skeleton, credited as a gift from the American Museum, formed the centerpiece of one of the most prominent exhibits on the Nebraska State Museum's first floor. On the second floor was a *Dinohyus* from Agate, called the "$50,000 Nebraska Hog" by the local press. These exhibits are described in "Preliminary Report on the Nebraska State Museum," 9, 13.

192. Harold Cook to W. D. Matthew, 20 January 1928, CPC, box 38.

193. Neil M. Judd to Charles D. Walcott, 18 December 1924, AAP, box 1a, Report on Illegal Excavation in the Southwest folder.

194. W. D. Matthew to H. J. Cook, 25 January 1928, CPC, Box 38. On debates over flows of scientific quarry materials out of China, see Fan, "Circulating Material Objects."

195. Wilbur C. Knight to Mr. Hatcher, 3 February 1902, JBHP.

196. See, for example, Douglass to Holland, 12 September 1910, EDP, box 18, folder 1.

197. Peterson to Douglass, 24 November 1917, EDP, box 19, folder 2.

198. W. D. Matthew to wife, 21 June 1906, in Camp, "Letters," 271.

199. Vetter, "Cowboys, Scientists, and Fossils," 298.

200. Craig's cabin, located most of the way up the side of Fossil Butte, has been preserved and adorned with interpretive signs for visitors to the National Monument. For evidence of fossil specimens from this quarry site being sold to the Museum of Comparative Zoology in Massachusetts, see Eastman, "New Fossil Bird and Fish Remains," 54–58.

201. Ward, "Description of a New Genus," 255–56.

202. See, for example W. D. Matthew to Harold [Cook], 9 February 1912; and Harold J. Cook to Dr. Matthew, 28 January 1913, CPC, box 38.

203. Holland, "Earl Douglass," 285–86. For a brief discussion of the involvement and psychological investment of local residents in the site, followed by their disappointment that so much material was removed from the site, see Harvey, *Symbol of Wilderness*, 7–8.

204. Douglass to Stewart, 20 September 1920, EDP, box 16, folder 20.

205. F. J. Pack to Douglass, 1 April 1922, EDP, box 17, folder 1; Douglass to Pack, 25 September 1922, EDP, box 17, folder 2.

206. Wm. M. Anderson to US National Park Comm., 9 May 1922; Arno B. Cammerer to Mr. Wm. M. Anderson, 17 May 1922, CCF-NPS, box 586.

207. W. J. Holland to Mr. Arno B. Cammerer, 8 November 1921, CCF-NPS, box 586.

208. Douglass [?], untitled typescript, n.d. [mid-1920s], EDP, box 20, folder 6.

209. As his son would later write, Douglass's "dream of a natural museum at the site of discovery, where all mankind could see the fossils as they were buried millions of years ago, was finally realized when the visitor's center at quarry site was opened to the public June 1, 1958, twenty-seven years after his death," in Douglass, *Speak to the Earth*, xiii.

210. James H. Cook to Dr. Matthew, 17 December 1915, Division of Paleontology, General Correspondence, American Museum of Natural History, New York City, box 24, folder 37.

211. Sheldon, "Historical Journey to the Northwest," 75–76. For more discussion of this example, see Vetter, "Cowboys, Scientists, and Fossils," 297. The Cook family did consider their on-site collections to constitute an informal museum. Much later, Harold Cook formally enrolled the "Cook Museum of Natural History" in the American Association of Museums, as discussed in Harold J. Cook to American Association of Museums, 12 February 1960; and Joseph A. Patterson to Mr. Harold J. Cook, 17 February 1960, Harold J. and James H. Cook Collection, #558, American Heritage Center, University of Wyoming, Laramie, new box 1.

212. Fletcher, *Wetherills of the Mesa Verde*, 129.

213. "Report of the Committee on American Archaeology," 53.

214. "The Old Log Cabin Museum" (n.d.), MVNP, series 12, drawer 2, folder 21.

215. Stephen T. Mather to Thomas Rickner, 20 October 1915, MVNP, series 5, drawer 4, folder 9.

216. Harvey, "Utah, the National Park Service, and Dinosaur National Monument," 258.

217. Edgar L. Hewett to Wm H. Holmes, 12 June 1906, MVNP, series 12, drawer 1, folder 42.

218. Fewkes, *Antiquities of the Mesa Verde National Park*, 9–10.

219. Fewkes, *Antiquities of the Mesa Verde National Park*, 9, 11–12, 24.

220. Newell, "Mesa Verde," 434.

221. Local rancher Al Wetherill, however, rhetorically turned the tables on government scientists, alleging in his autobiography that "government surveyors came under the category of pothunters who would wreck an entire ruin to get one bowl they neither appreciated nor really cared for," in Fletcher, *Wetherills of the Mesa Verde*, 104.

222. William J. Holland to James H. Cook, 30 November 1908, CPC, box 33; Harold J. Cook to Henry F. Osborn, 21 January 1910, CPC, box 42. In a slight variation, the Carnegie Museum scientists thought Cook might have described the bones as the remains of "Indian horses," as claimed in Holland and Peterson, "Osteology of the Chalicotheroidea," 189–406.

223. Riggs, "Dinosaur Beds of the Grand River Valley of Colorado," 267.

224. Smith, *Mesa Verde National Park*, 23.

225. Fletcher, *Wetherills of the Mesa Verde*, 125.

226. Chapin, *Land of the Cliff-Dwellers*, 117–18. Additional evidence comes from Frank McNitt's detailed biography of Richard Wetherill, which unfortunately has no footnotes, but includes an intriguing anecdote about a Ute Indian named Acowitz. According to McNitt, Acowitz not only first hinted to Richard about the existence of a large cliff dwelling but called it "a sacred place" and warned him "not to go there": "When you disturb the spirits of the dead, then you die too." See McNitt, *Richard Wetherill*, 22. This anecdote, sourced to McNitt, is further credited and repeated in Smith, *Mesa Verde National Park*, 19. McNitt's own discussion of his sources (319–22) admits that he freely mixed documents and much later interviews with people who knew Richard Wetherill, and none of the documents he mentions having consulted seems likely to be the source. Even if we presume that McNitt has rendered the story accurately, it is difficult to judge for sure how much of this story may have been embellished by Richard Wetherill himself or one of his counterparts. But it is suggestive, at least, of white settler interlocuters' perceptions of Ute beliefs and their preexisting knowledge about the cliff house ruins.

227. A brief account based on these Ute Council minutes, though without specific source notes, is Noble, "Legacy of Distrust," 32–42. Other brief accounts of the fascinating 1911 negotiation are provided in the chapter on Mesa Verde in Keller and Turek, *American Indians and National Parks*, 30–42; and in Burnham, *Indian Country*, 53–56.

228. "Minutes of Council Held at Navajo Springs, Colorado" (5 May 1911), MVNP, series 13, drawer 2, folder 29.

Chapter 5. Stations

1. Although it was officially called the "North Platte Substation" in its early years, here I will usually refer to it as the "North Platte station" or "North Platte branch station" for the sake of simplicity. Internal histories of the North Platte station are Rumery et al., "History of University of Nebraska North Platte Station"; L. L. Zook, "Sixty Years Agronomy History of the North Platte Experiment Station" (1964), box 2, folder 4; and W. W. Burr and James C. Adams, "The Early Years," in James C. Adams et al., "50 Years of Research at the North Platte Experiment Station" (1954), Circular 93, box 2, 3–9; both in Agricultural Experiment Station Records, North Platte Experiment Station (Historical), #10/14/4, University Archives, Love Library, University of Nebraska, Lincoln.

2. Portions of this paragraph and the next one are derived from Vetter, "Labs in the Field?" Selected passages from this article are incorporated throughout this chapter's text below, as noted and credited in the Acknowledgments.

3. Ramaley, *Wild Flowers and Trees of Colorado*, v, 20. Images of the Tolland area would also be used in his later book: Ramaley, *Colorado Plant Life*, e.g., fig. 11, between pp. 26–27; and fig. 126, between pp. 256–57.

4. Few branch stations or substations have been the focus of historical scholarship, beyond the internal retrospective chronicles published by the institutions themselves. For rare exceptions within the Great Plains and Rocky Mountains region, see James, "Jujubes, Grapes, and Grass"; May, "Agricultural Science in the Texas Panhandle"; Freeman, *High Plains Horticulture*, esp. ch. 11; and Johnson, "Brief History."

5. For a general account of biological field stations among many types of site-based innovations in the lab-field borderlands, see Kohler, "Labscapes: Naturalizing the Lab." Probably the most notable very recent historical work on field stations is de Bont, *Stations in the Field*, which examines several European field stations where animals were studied, including in Naples, Wimereux, Plön, Rossitten, and Brussels, with an overall aim to "study the intricate activities that enabled the zoologist to perform science in the animal's 'dwelling place'" (2).

6. On the historical role of politics at agricultural stations, see Marcus, *Agricultural Science and the Quest for Legitimacy*; Rosenberg, *No Other Gods*; and Danbom, *"Our Purpose Is to Serve."*

7. Aubin, "Fading Star of the Paris Observatory."

8. Rotch, "Mountain Meteorological Stations," 402; Moore, *Report of the Chief of the Weather Bureau*, 263; "Meteorological Observations on Pike's Peak," 122–24; and A. L. R., "Closing of the U.S. Signal Service Stations," 284–86.

9. Nisbett, "Business Practice," 60–69. On the observatory sciences more generally, see Aubin, Bigg, and Sibum, *Heavens on Earth*.

10. Loud, "Summary of Meteorological Observations, January to June, 1904," 54.

11. Bigg, "Photography and the Labour History of Astronomy." See also Rossiter, "'Women's Work' in Science"; and Pang, "'Stars Should Henceforth Register Themselves.'"

12. Pang, *Empire and the Sun*, 127. Pang also shows how the "observatory could be rebuilt in the field" (128) during solar eclipse expeditions, thus highlighting another way of hybridizing scientific space: the field observatory. Such a history becomes more directly a part of the history of scientific fieldwork, featuring many of the same material experiences, such as railroad access, staying in communication, setting up camp, division of labor, and using local informants, all

of which Pang ably analyzes in the enabling context of the British Empire. A solar eclipse in 1878, for example, generated a significant, if temporary, flow of cosmopolitan scientists to Colorado and nearby states, including many from overseas, as reported in Ranyard, "Account of Observations"; Langley, *New Astronomy*, 50–56; and US Naval Observatory, *Reports on the Total Solar Eclipses.* For an overview of this event, see Ruskin, "'Among the Favored Mortals of Earth.'"

13. Lane, *Geographies of Mars*, 93.

14. Hill, "History and Functions of Botanic Gardens," 218.

15. "History of Establishment of Botnical [*sic*] and Pharmaceutical Garden" (n.d.), unpublished typescript; and "Dr. Cutler and the Denver University Rose" (n.d.), clipping from unidentified publication, 118–21, Ira E. Cutler Collection, Special Collections, Penrose Library, University of Denver, Colorado, box 1, folder 2.

16. Brockway, *Science and Colonial Expansion*, 189. See also McCracken, *Gardens of Empire.* For a US case study emphasizing such power relations, see Luke, "Missouri Botanical Garden." On the especially prominent New York Botanical Garden, see Mickulas, "Cultivating the Big Apple"; Mickulas, "Scientific Garden in a Public Space"; and Kingsland, *Evolution of American Ecology.*

17. See, for example, Endersby, "Garden Enclosed."

18. Brockway, *Science and Colonial Expansion*; McCook, *States of Nature*; McCook, "Managing Monocultures"; Kumar, *Indigo Plantations*; Goss, *Floracrats.*

19. For a table listing biological field stations in the Rocky Mountains, see Vetter, "Rocky Mountain High Science," 112.

20. Benson, "Laboratories on the New England Shore," 78.

21. "Anderson School," 175, 174; Wilder, "Agassiz at Penikese," 190–91. On the traveling geology summer school field trips that partly inspired the summer station idea, though not at all in the station mode of practice, see Shaler and Shaler, *Autobiography of Nathaniel Southgate Shaler*, 272–74. A regional example of a traveling summer school in the 1870s, in southwestern Nebraska, aptly described as "a moving camp for the purpose of studying the natural history and resources of the Valley of the Republican river, Nebraska," is reported in "Nebraska Summer School of Science," 109.

22. Wilder, "Penikese School," 29.

23. Lankester, "American Sea-Side Laboratory," 498. Lankester's ambitious ideal of proper station science was so strong that he elevated the term "station" itself even above "laboratory," remarking that "even more modest institutions . . . may be called 'sea-side laboratories,'"

24. Benson, "Laboratories on the New England Shore," 77.

25. Benson, "Summer Camp, Seaside Station, and Marine Laboratory," 15.

26. For a useful comparative analysis of two different marine stations in Europe oriented, respectively, toward the lab and field ends of the continuum, see de Bont, "Between the Laboratory and the Deep Blue Sea." For a more complete study, recently published, on several European field stations, see de Bont, *Stations in the Field.* For a prominent British biological field station, see Erlingsson, "Plymouth Laboratory."

27. Montgomery, "Bird Course at the Marine Biological Laboratory," 153.

28. Mayer, "Should Our Colleges Establish Summer Schools?," 703.

29. Fairchild, "Small Field Laboratory and Its Atmosphere of Research," 68.

30. "Colorado School of Forestry," 677. Also in the late 1880s, the Colorado Agricultural College authorized "four experimental stations in as many different sections of the State," to study "the science of arboriculture, in its adaptation to local conditions," as noted in Ensign, "Advantages and Limitations of the Colorado Forestry System," 26.

31. Drake, "Waving 'a Bough of Challenge,'" 24.

32. "State Forestry Station," 247; Call, *Off-Campus Research,* 28.

33. "Colorado School of Forestry," 676.

34. On the Fremont Experiment Station and other US Forest Service stations, analyzed in conjunction with ecological stations such as the Alpine Lab, see Brock, "The Dynamics of Science, Rooted in Place: Field Stations and Forest Ecology," in *Money Trees,* 38–65.

35. Hoyt, *Short History*; LeCain, "From Eastern Coal to Western Copper"; Seely, "Research, Engineering, and Science."

36. Special publications were developed "to give information about the experiment stations and their work, by collating results bearing upon special topics and putting them into brief, clear, practical form for farmers and others to read," as noted in *The What and Why of Agricultural Experiment Stations,* Farmers Bulletin no. 1 (1889). However, criticism of agricultural stations for too much data gathering at the expense of reaching "results and generalizations" that would be more useful would continue to appear in many places, e.g., in "Agricultural Experiment Stations," 401.

37. *Laws, Joint Resolutions and Memorials,* 591.

38. There were early branch experiment in farms in Colorado and Wyoming in the 1890s, most of which were closed down a few years later, and Texas had authorized establishment of a substation in the state's Panhandle region in the mid-1890s, but insufficient funding prevented station authorities from actually establishing one then, as noted in *Annual Report of the Texas Agricultural Experiment Station* 8 (1895), 751; and *Annual Report of the Texas Agricultural Experiment Station* 9 (1896), 918. Finally, in 1909, Texas established and provided satisfac-

tory funding for nearly a dozen numbered substations, spanning a wide variety of environmental regions across the state, as noted in *Annual Report of the Texas Agricultural Experiment Station* 25 (1912), 43.

39. The USDA, under botanist George Vasey, had previously operated a "Forage Station" near Garden City from the late 1880s to early 1890s, which some regarded as a "wasteful expenditure" and others regarded as having produced "valuable results." "Editorial," *Agricultural Science* 7 (1893), 276–77. See also Vasey and Galloway, *Record of Some of the Work*, 9–16.

40. H. R. Reed, "Report of the Garden City Branch Station, 1908," KAESR, Garden City box 1.

41. For one indication of the strong local support for branch stations—taking the environmental variation argument further by arguing that each *county* should have its own experiment station—was in Warner, "Should We Have County Experiment Stations," 228–32. "The wide variation in the soil, climate, and needs of different parts of the state is so great that the results obtained in one locality may be of little value to other parts, and may often be misleading," he argued (229).

42. For a comparable example from Wisconsin of the aims and expectations in establishing an agricultural experiment station, see Carstensen, "Genesis of an Agricultural Experiment Station."

43. Three human-planted forest reserves were established in Nebraska between 1902 and 1906, including the Dismal Forest Reserve of 85,123 acres, the Niobrara Reserve of 123,779 acres, and the North Platte Reserve of 347,170 acres, all of which were consolidated in 1908 into the Nebraska National Forest, with a total land area of 556,000 acres. For an insightful "envirotech" analysis of the Nebraska National Forest, see Gardner, "Constructing a Technological Forest." Meanwhile, in Kansas, the Garden City Forest Reserve was established in 1903 at 97,280 acres and enlarged to 302,387 acres in 1908, when it was renamed the Kansas National Forest. See Bates and Pierce, *Forestation of the Sand Hills of Kansas and Nebraska*, 22.

44. Dana, "Experiment Stations on the National Forests," 24.

45. Vetter, "Rocky Mountain High Science."

46. Pearson, "Administration of a Forest Experiment Station," 211.

47. "Colorado School of Forestry," 677.

48. Some prominent members of practical scientific societies also advocated branch stations in different parts of a state, even before the Hatch Act provided federal funding for state experiment stations, as in Gipson, "Experimental Stations," 111.

49. Rumery et al., "History," 1–2.

50. Call, *Fifty Years of Research*, 5.

51. Fred. T. Wright to Grace Raymond Hebard, 9 April 1891, UWAESR, series 1, box 7.

52. Call, *History of the Tribune (Kansas) Branch Experiment Station*, 4.

53. Charles E. Cassel, "Tribune Substation, Annual Report for Season 1913," KAESR, Tribune box 1, p. 6.

54. *Annual Report of the Kansas Agricultural Experiment Station* 22 (1913–14), 30.

55. M. C. Sewell, "Annual Report of the Garden City Experiment Station, 1914," KAESR, Garden City box 1.

56. Director [Linfield] to Mr. S. S. Hobson, 27 May 1907, College of Agriculture Records, Acc 00005, Special Collections, Montana State University, Bozeman, box 17, folder AF-25.

57. Keffer, *Experimental Tree Planting in the Plains*, 35.

58. Salmon, *Odyssey of an Agricultural Scientist*, 16.

59. "Mountain and Alpine Laboratories, Colorado," *Journal of Ecology* 1 (1913): 157.

60. "School of Mountain Field Biology of the University of Colorado" (n.d.), FRP, box 1, folder 26.

61. Frederick [*sic*] E. Clements, "Alpine Laboratory," 327.

62. Pearson, "Administration of a Forest Experiment Station," 211–12.

63. Call, *Off-Campus Research*, 27.

64. "Colorado School of Forestry," 677.

65. Salmon, *Odyssey of an Agricultural Scientist*, 19.

66. W. W. Burr to Geo. Coupland, 20 July 1909, NAESR-NP, box 7; W. W. Burr to T. H. Kearney, 13 April 1910, NAESR-NP, box 7; W. P. Snyder to T. H. Kearney, 27 February 1911, NAESR-NP, box 1.

67. J. H. Shepperd to Supt. L. R. Waldron, 16 October 1905, Dickinson Experiment Station, University Archives, Institute for Regional Studies, North Dakota State University, Fargo, box 1, folder 4.

68. Robbins, "Succession of Vegetation at Boulder Park, Colorado," 493.

69. *Annual Report of the Wyoming Agricultural Experiment Station* 19 (1909), 9.

70. Call, *History of the Tribune*, 6; Call, *Fifty Years of Research*, 5; Call and Aicher, *History of the Fort Hays (Kansas) Branch Experiment Station*, 3.

71. Ewell, "Every Farm an Experiment Station," 292.

72. *Annual Report of the Oklahoma Agricultural Experiment Station* 3 (1893–94), 4.

73. Great Plains Committee, *Future of the Great Plains*, 27.

74. Charles E. Cassel, "Tribune Substation, Annual Report for Season 1913," KAESR, Tribune box 1.

75. John J. Bayles, "The Sixth Annual Report of the Colby Substation, 1919," KAESR, Colby box 1, 5.

76. W. W. Burr to Prof. E. C. Chilcott, 6 July 1909, NAESR-NP, box 7.

77. W. W. Burr to E. C. Chilcott, 10 July 1909, NAESR-NP, box 7.

78. W. W. Burr to E. C. Chilcott, 20 July 1909, NAESR-NP, box 7.

79. W. W. Burr to E. A. Burnett, 28 July 1909, NAESR-NP, box 7.

80. W. W. Burr to E. A. Burnett, 25 July 1910, NAESR-NP, box 7.

81. Dana, "Experiment Stations on the National Forests," 26.

82. "Proposed Biological Station," 22. Four years later, Cockerell was still debating the merits of investing time and energy in developing a field station. At first he held back from making a "definite plan," not yet knowing the level of interest he would attract in what he was calling a "nature study camp," but a month and a half later, his enthusiasm had returned, and he wrote, "I'm feeling like putting a whole *lot* of energy into the biological station. My former pessimistic ideas were absurd," in Theo. to [Wilmatte Porter], 30 March 1900; and Theo. to [Wilmatte Porter], 24 May 1900, T. D. A. Cockerell Papers, CONS 138, Denver Public Library, Colorado, folder 1. Another partial precursor to Rocky Mountain biological field stations was the Colorado Summer School of Science, Philosophy, and Languages, "held in the city of Colorado Springs at the foot of Pike's Peak" in 1895, "within easy reach of the vegetation of the plains, the cañons, foot-hills, and the strictly alpine region," as noted in Bessey, "Summer-School Botany in the Mountains," 845–47.

83. Salmon, *Odyssey of an Agricultural Scientist*, 18–19.

84. Vetter, "Rocky Mountain High Science," 114–16, 121.

85. Morton J. Elrod, "University of Montana Biological Station" (27 November 1912), unpublished manuscript, p. 1, MJEP, box 28, folder 18.

86. W. P. Snyder to E. A. Burnett, 17 July 1911, NAESR-NP, box 1.

87. K. W. Toll to Prof. Ramaley, 14 October 1911, FRP, box 1, folder 17.

88. Call and Aicher, *History of the Fort Hays (Kansas) Branch Experiment Station*, 3, 20.

89. Phillips, *History of the Agricultural Research Center-Hays*, 8.

90. On the early morning fire of 12 February 1913, see Helder and Kiene, "Fire Probably Caused by Slaking Lime," 51–52.

91. Grout, "Mosses of Colorado from Tolland and Vicinity," 1.

92. Kohler, *Landscapes and Labscapes*, 53. For more explanation of the "inner frontier" concept, see Kohler, *All Creatures*, 17–46.

93. Bruce Clifton to Dr. Ramaley, 8 February 1910, FRP, box 1, folder 11.

94. Mrs. E. F.[?] Ballard to Dr. Francis Ramaley, 16 May 1911, FRP, box 1, folder 16.

95. "School of Mountain Field Biology of the University of Colorado" (n.d.), unpublished manuscript, FRP, box 1, folder 26.

96. "Alpine Laboratory of the Botanical Seminar," 186.

97. Call, *Fifty Years of Research*, 15–16

98. Ramaley, "University of Colorado Mountain Laboratory," 93.

99. Johnson, "Brief History," 87.

100. Pearson, "Administration of a Forest Experiment Station," 212.

101. Pearson, "Administration of a Forest Experiment Station," 212.

102. "University of Montana Biological Station," 340–41.

103. Morton J. Elrod, "University of Montana Biological Station" (27 November 1912), unpublished manuscript, pp. 1, 3, MJEP, box 28, folder 18.

104. For details of Snyder's appointments, first as assistant in animal husbandry and assistant superintendent of farmers' institutes, then as director of the North Platte station, see *Annual Report of the Nebraska Agricultural Experiment Station* 17 (1903), 8; *Annual Report of the Nebraska Agricultural Experiment Station* 18 (1904), 14–17. This movement from assistant in Lincoln suggests that a more reliable ladder of advancement through merit had emerged by the early twentieth century. For a more general commentary from the earlier years of the agricultural experiment stations, which claimed it was initially "very difficult to secure men properly trained to do the work required," since there were "not enough available men of this quality in the country," see "Management of Our Experiment Stations," 245–46.

105. E. A. Burnett to Hon. W. G. Whitmore, 11 April 1904, NAESR-NP, box 9.

106. E. A. Burnett to Members of the Experiment Station Council, 3 March 1906, NAESR-NP, box 9.

107. W. W. Burr to Prof. E. C. Chilcott, 30 October 1909, NAESR-NP, box 7.

108. "Science Summer Camp," 187–88.

109. Occasionally, a midwestern or East Coast institution might establish its own field station in the Rocky Mountains for summer school teaching purposes, such as the University of Michigan's Camp Davis near Hoback Junction, Wyoming, or Indiana University's later geological field station near Pony, Montana. Another variant, lacking the permanent built infrastructure of a field station, was the "permanent maintenance of a summer course in geological field work in the Rocky Mountains," led by Harvard's Museum of Comparative Zoology in the early twentieth century with an "anonymous gift of $5,000" for "support of a summer field course in structural or stratigraphical geology, conducted preferably in the mountain region of western North America," as noted in Ward, "Report of the Department of Geology and Geography," 20. The museum's 1911–1912 annual report indicates that such a field course was established in Montana.

110. M. J. Elrod to Arthur G. Vestal, 9 June 1914, Arthur G. Vestal Papers, 15/4/24, University Archives, University of Illinois, Champaign-Urbana, box 2.

111. "News Items," 70.

112. E. A. Burnett to Dr. C. E. Bessey, 5 July 1905, C. E. Bessey Papers, #12/7/10, University Archives, Love Library, University of Nebraska, Lincoln, box 34, folder 9; E. A. Burnett to W. P. Snyder, 7 July 1905, NAESR-NP, box 9.

113. Charles E. Bessey to W. P. Snyder, 20 July 1905, NAESR-NP, box 9.

114. For an especially vivid account, see Pauly, "Summer Resort and Scientific Discipline."

115. W. M. Davis to Ramaley, 28 June 1910, FRP, box 1, folder 13.

116. Frederic E. Clements to Professor Ramaley, 20 July [1910], FRP, box 1, folder 13.

117. "Students, Staff & Visitors to Alpine Laboratory" (n.d.), unpublished manuscript, EFCP, box 31, folder 5.

118. "International Phytogeographical Excursion (I.P.E.) in America, 1913," 36–37.

119. "Science Summer Camp," 189.

120. Thomas A. Dunn to G. R. Hebard, 10 August 1891, UWAESR, series 1, box 7.

121. Thomas A. Dunn to G. R. Hebard, 14 August 1891, UWAESR, series 1, box 7.

122. Pearson, "Administration of a Forest Experiment Station," 218.

123. Robertson, *Magnificent Mountain Women*, 112.

124. For comparable cases elsewhere, consider first this statement of the station director in Oklahoma regarding the need for "skilled labor" at field stations: "We need a class of workers capable of observing, interpreting, arranging, and summarizing the observations made. To do this they should have experience and the special education for this work," in *Annual Report of the Oklahoma Agricultural Experiment Station* 3 (1894–94), 6. Similarly, several years later the superintendent of demonstration farms in North Dakota complained that, even though most of them operated "in a workmanlike manner," they sometimes failed "to do the exact work that the station requires, as they are not in the habit of keeping strict accounts of everything they do in their farming operations," as noted in *Annual Report of the Superintendent of Demonstration Farms for North Dakota* 2 (1908), 19.

125. *Memoranda of Plans for Arid Farm Investigations*, Circular of the Utah Agricultural Experiment Station, no. 1 (1904), 2.

126. W. F. Weakly to [W. P. Snyder], 4 January 1906, NAESR-NP.

127. E. A. Burnett to W. P. Snyder, 14 September 1906, NAESR-NP.

128. E. A. Burnett to W. P. Snyder, 29 December 1906, NAESR-NP.

129. W. P. Snyder to Mr. Raymond Rhodes, 30 September 1910, NAESR-NP, box 1.

130. Rumery et al., "History," 7.

131. E. A. Burnett to W. P. Snyder, 1 Mary 1905, NAESR-NP.

132. E. A. Burnett to W. P. Snyder, 6 July 1905, NAESR-NP.

133. E. A. Burnett to W. P. Snyder, 30 May 1906, NAESR-NP.

134. E. A. Burnett to W. P. Snyder, 30 April 1906, NAESR-NP

135. E. A. Burnett to W. P. Snyder, 29 October 1906, NAESR-NP.

136. W. P. Snyder to E. A. Burnett, 8 April 1910, NAESR-NP. Severe labor shortages plagued the North Platte station, as they did nearly all domestic enterprises, during the First World War, and the station's "inability to secure farm labor in sufficient quantity and of sufficient efficiency to conduct experimental work in a satisfactory manner" was reported shortly thereafter in *Annual Report of the Nebraska Agricultural Experiment Station* 33 (1919), 24. In the spring of 1918, the USDA was offering to pay $75 per month for hired laborers in Montana for a five-month season from mid-April to mid-September, as noted in Agrostologist in Charge [C. V. Piper] to Mr. P. V. Cardon, 4 April 1918, Records of the Division of Forage Crops and Diseases, Correspondence with Other Offices of the Bureau, PI-66 Entry 67, RG 54, Records of the Bureau of Plant Industry, Soils, and Agricultural Engineering, National Archives II, College Park, Maryland, box 15, folder: Dry Land Agriculture #2. This was the same monthly pay for a "temporary assistant" in forage crop work in Montana five years earlier, as reported in "Memorandum Regarding Cooperation at the Montana Experiment Station" (letter from Agrostologist to Mr. Taylor), 20 March 1913, Records of the Division of Forage Crops and Diseases, Correspondence with the Agricultural Experiment Stations, PI-66 Entry 68, RG 54, Records of the Bureau of Plant Industry, Soils, and Agricultural Engineering, National Archives II, College Park, Maryland, box 23, folder: Montana Experiment Station #2. In 1909, an unskilled laborer at the North Platte station was allotted $50 per month for work funded by the USDA, in Physiologist in Charge [Kearney] to Dr. B. T. Galloway, 29 November 1909, Records of the Division of Fruit and Vegetable Crops and Diseases, Correspondence Relating to Investigations of Alkali and Drought-Resistance Plants, PI-66 Entry 124, RG 54, Records of the Bureau of Plant Industry, Soils, and Agricultural Engineering, National Archives II, College Park, Maryland, box 1.

137. H. H. Greenamyre to R. J. Pool, 10 March 1913, RJPP, box 1, folder 1.

138. Pearson, "Administration of a Forest Experiment Station," 216.

139. Fred. T. Wright to Grace Raymond Hebard, 10 May 1891, UWAESR, series 1, box 7.

140. J. D. Woodruff to Grace Hebard, 9 April 1891, UWAESR, series 1, box 7.

141. J. D. Woodruff to State University Commission, 27 May 1891, UWAESR, series 1, box 7.

142. Jas. A. Baker to Board of Trustees, 30 June 1891, UWAESR, series 1, box 7.

143. Federal funds allocated to the states through the Hatch Act of 1887, and later also through Adams Act of 1906, were supervised and monitored by the Office of Experiment Stations. The Adams Act authorized funds to be even more tightly supervised by the Office, which its leaders thought would "draw attention sharply to a consideration of what constitutes scientific research in agriculture as distinguished from the mere acquisition of empirical knowledge." True and Allen, "Work and Expenditures of the Agricultural Experiment Stations," 49.

144. A. C. True to P. F. Sharp, 19 August 1899, EXP. While the greatest pressure to close branch stations and instead focus on a smaller number of systematic lines of research came from True in Washington, DC, evidence shows at least some private agreement with this direction from officials in Colorado dating back a few years, in Alston Ellis to Hon. A. C. True, 23 November 1896; Alston Ellis to Hon. A. C. True, 30 November 1896; Alston Ellis to Hon A. C. True, 17 March 1897; and Alston Ellis to Hon. A. C. True, 8 April 1897; General Correspondence with State Experiment Stations and Agricultural Colleges, Records of the Office of Experiment Stations, NC-132 Entry 11, RG 164, Records of the Cooperative State Research Service, box 8, folder 2. There were also some indications of Colorado officials' public agreement (directly following a site visit from True) in *Annual Report of the Colorado Agricultural Experiment Station* 8 (1895), 95–96, but also some equivocal evidence the following year showing, on the one hand, support for the continuing importance of branch stations to investigate diverse environmental conditions around the state, but, on the other hand, with strong words against the progress of the branch stations, in *Annual Report of the Colorado Agricultural Experiment Station* 9 (1896), 83–84, 90–91; and by a few years later, evidence of growing resentment of True's overbearing guidance on the matter of branch stations, in *Annual Report of the Colorado Agricultural Experiment Station* 11 (1898), 126–27.

145. A. C. True to President G. E. Morrow, 17 December 1896, Correspondence of Alfred C. True, Records of the Office of Experiment Stations, NC-132 Entry 51, RG 164, Records of the Cooperative State Research Service, National Archives II, College Park, Maryland, box 1.

146. A. C. True to P. F. Sharp, 2 December 1899, EXP.

147. Philo K. Blinn to [Carpenter], 29 September 1905, EXP.

148. Campbell, Peterson, and Griffith, *Formative Years of Plant Pathology*, 261.

149. G. E. Thompson and S. P. Clark, "Report of Colby Experiment Station, Season of 1914," KAESR, Colby box 1. Requests for information on arid-adapted crop varieties in western Kansas had been strong enough during the dry farming boom of the previous decade that farmers there were writing to agricultural experiment stations in neighboring states such as Oklahoma. W. L. English to

Mr. R. F. Chandler, 16 January 1908, Oklahoma Agricultural and Mechanical College Correspondence, 1892–1920. #86-009, Special Collections and University Archives, Edmon Low Library, Oklahoma State University, Stillwater, box 4.

150. See, for example, S. P. Clark, "Report of the Colby Experiment Station for 1915," KAESR, Colby box 1; S. P. Clark, "Colby Branch Experiment Station Annual Report 1916," KAESR, Colby box 1.

151. G. E. Thompson and S. P. Clark, "Report of Colby Experiment Station, Season of 1914," KAESR, Colby box 1.

152. Lamson-Scribner, *Cooperative Range Growth.*

153. Ramaley and Robbins, "Summer Laboratory for Mountain Botany," 108–9.

154. Elrod, "University of Montana Biological Station at Flathead Lake," 783.

155. Giles, "Algae of the Medicine Bow Mountains," 189.

156. "Alpine Laboratory of the Botanical Seminar," 186.

157. Elrod, "University of Montana Biological Station," 1271.

158. "Science Summer Camp," 186–87, 189.

159. Ramaley and Robbins, "Summer Laboratory for Mountain Botany," 110.

160. Rodeck, "Science Lodge," 103.

161. The Mount Evans Lab, with its specialized research and no teaching (and which focused on physical research more than biological research) was a revealing exception. It consisted of only a single roof-truss building lacking side walls—"to withstand the high wind velocity"—divided into two rooms, one for a laboratory and the other for sleeping quarters; Stearns, "Mount Evans Laboratory," 246. Only the most essential and compact equipment could justify occupying the severely limited space in such a remote and forbidding location.

162. Charles E. Cassel, "Tribune Station Report, 1914," KAESR, Tribune box 1, 7.

163. [Clements] to President Woodward, 9 December 1917, EFCP, box 1, folder 1.

164. Manning, *Experiment Stations of the Bureau of Mines.* State officials in Montana tried to obtain a US Bureau of Mines station for Butte, as noted in *Montana State Bureau of Mines and Metallurgy*, 8. A few years earlier, a state mining substation in Hebron, North Dakota, also produced research that was published by the US Bureau of Mines, in Babcock, *Economic Methods of Utilizing Western Lignites.* In Salt Lake City, Utah, the engineering school of the University of Utah had maintained the "engineering station" (though this was basically an on-campus laboratory) since 1909 with research in cooperation with the US Bureau of Mines, as noted in Beckstrand, *Tests of Macadam Rock*; and Lyon,

Bradford, Arentz, Ralston, and Larson, *Metallurgical Treatment,* 5. Nevertheless, researchers at the Salt Lake City station did undertake field work in the Park City region of Utah, as noted in Lyon and Ralston, *Recovery of Zinc*; as well as studies on such field-oriented topics as roads, stream measurement, and on-site mine sampling, in Lyman, *Construction and Maintenance*; Lyman, *Measurement of Flowing Streams*; and Lewis, *Mine Sampling.*

165. George E. Vincent to Dr. Quigley, 22 September 1926, John C. Johnson Collection, Leslie J. Savage Library, Western State College of Colorado, Gunnison, box 2.

166. For an example in another part of the West of the state emphasis on applied, practical concerns in conflict with the federal emphasis on basic, original research, see Nye, "Federal vs. State Agricultural Research Policy." The exertion of USDA authority may have been easier in overseas territories, such as Hawaii, Puerto Rico, Guam, the Philippines, and Alaska; see Overfield, "Science Follows the Flag"; and Overfield, "Agricultural Experiment Station and Americanization."

167. Davenport, "Relations between the Federal Department of Agriculture and the Agricultural Colleges and Experiment Stations," 122.

168. Allen, "Some Ways," 167.

169. Davenport, "Obstacles to Progress in Agricultural Science," 11.

170. Allen, "Some Ways," 168.

171. Keffer, *Experimental Tree Planting,* 8.

172. "Notes, News, and Comment," 78–79, 142–44; MacDougal, "Report of Dr. D. T. MacDougal," 8–13; Seaver, "Conference Notes," 118–21; Seaver, "Collecting Fungi in Colorado," 247–55; Pennell, "Report on a Collecting Trip," 1–6.

173. *Report of the Director on the Establishment of New State Stations,* Bulletin of the Texas Agricultural Experiment Station, no. 134 (1910), 10. Harrington attributed the vehemence of feeling on this issue in Texas in part to lingering Southern resentment of the federal government and advocacy of states' rights, though similar sentiments were expressed by those in other states in the Plains and Rockies, if less adamantly.

174. Jenny, *E. W. Hilgard and the Birth of Modern Soil Science,* 98.

175. Davenport, "Relations," 123, 124.

176. Davenport, "Relations," 124, 125.

177. Allen, "Some Ways," 175, 176, 182.

178. F. Lamson-Scribner, 12 January 1901, EXP.

179. Philo K. Blinn to [Carpenter], 29 September 1905, EXP.

180. "Notes, News, and Comment," 78; MacDougal, "Report," 10; Seaver, "Collecting Fungi," 248.

181. Kingsland, *Evolution of American Ecology,* 80–81.

182. For examples, see Ramaley, "Amount of Bare Ground," 526–28; Ramaley, "Relative Importance," 154–57; Ramaley, "Quadrat Studies," 70–74; and Reed, "Meadow Vegetation," 97–109. Similar methodological developments were also occurring at forest stations, such as at the Fort Valley station in northern Arizona, which set up "19 permanent sample plots aggregating about 2,000 acres" at sites containing "pure western yellow-pine stands in Arizona and New Mexico." Pearson, "Studies of Yield and Reproduction," 273–93.

183. For examples of more rigorous agricultural research studies at branch field stations by the 1910s, including sites across the region—North Platte, Nebraska; Judith Basin, Montana; Akron, Colorado; and Garden City, Kansas—see Burr, *Storage and Use of Soil Moisture*; Atkinson and Donaldson, *Dry Farm Grain Tests*; Atkinson and Gieseker, *Dry Farm Crop Rotations*; Briggs and Shantz, "Hourly Transpiration Rate," 583–650; Briggs and Shantz, "Daily Transpiration," 155–212; and Miller, "Daily Variation," 11–46. While much of the impetus for adhering to increasingly rigorous cosmopolitan standards undoubtedly came from state researchers themselves, by the mid-1910s "far-reaching influence" was also attributed to the federal Adams Act of 1906, which required "outlining the purpose and general plan of the inquiry at the outset," through a "project system," as noted in Allen, Wilcox, and Schulte, "Work and Expenditures of the Agricultural Experiment Stations, 1916," 13–14. However, criticisms of the lack of rigor in research methods continued to appear, e.g., Bear and McClure, "Sampling Soil Plots," 65–75.

184. L. G. Carpenter to A. C. True, 12 February 1901, EXP.

185. A. E. Hoyt to Grance [*sic*] Raymond Hebard, 10 July 1894, UWAESR, series 1, box 7.

186. A. E. Hoyt to Grace Raymond Hebard, 16 July 1900, UWAESR, series 1, box 7.

187. W. E. Lauenstein to Grace R. Hebard, 24 August 1900, UWAESR, series 1, box 7.

188. Swan Pearson and Joseph T. Farrell to Grace Raymond Hebard, 7 July 1902, UWAESR, series 1, box 7.

189. For a one-year season from March 1907 to March 1908, 137.9 acres of the farm were leased to Mrs. Carrie B. Meyer for $75, as arranged in Mrs. J. S. Meyer to G. R. Hebard, 12 February 1908, UWAESR, series 1, box 7; and Mrs. C. B. Meyer to G. R. Hebard, 20 March 1908, UWAESR, series 1, box 7.

190. *Annual Report of the South Dakota Agricultural Experiment Station* (1905), 15.

191. Vetter, "Regionalizing Knowledge." For Chilcott's perspective on the need for "coordination, systematization, and unification," see *Dry-Land Agriculture*, 7. For a synopsis of the collaborative field practices themselves, see Chilcott, *Study of Cultivation Methods*, 12–13; Chilcott, Cole, and Burr, *Spring Wheat*;

and Chilcott, Cole, and Burr, *Crop Production*. For an example of efforts undertaken to disseminate the results to a wider audience, see Chilcott and Cole, *Growing Winter Wheat*.

192. Kohler, *Landscapes and Labscapes*, 207.

193. Craig, *Centennial History of the Carnegie Institution of Washington*, 4:98.

194. For further discussion on this issue, see Danbom, "Agricultural Experiment Station and Professionalism"; and Ferleger, "Uplifting American Agriculture."

195. Henke, "Making a Place for Science."

Epilogue

1. Davis, "Colorado Front Range," 23.

2. Schubert, *Vanguard of Expansion*, 131.

3. Long, Popper, and Popper, "'Progress of the Nation.'"

4. "Executive Proceedings," 382.

5. E. R. Warren to Will Decker, 5 March 1916, Edward R. Warren Papers, MS51, Special Collections/Archives, Tutt Library, Colorado College, Colorado Springs, Colorado, box 1, folder 11.

6. Visher, "Report of the State Geologist, 1916–1918," 188.

7. Ward, *Needs of the State Geological and Natural History Survey*.

8. R. S. Woodward to Professor Frederic E. Clements, 24 September 1917; Clements to President Woodward, 8 October 1917; both in EFCP, box 1, folder 1. See also Clements, *Adventures in Ecology*.

9. Lapham, *Crisscross Trails*, 110. See also Carter, Stroud, and Bushnell, *Investigations in Adapting the Automobile*.

10. Wetmore, "Observations on the Habits of Birds," 221.

11. Matthew, "Early Days of Fossil Hunting," 454. Perhaps generated by the perception that the railroad era was passing in favor of a new automobile-based era of fossil prospecting, Matthew's memoir of fossil hunting in the American West during the railroad era was one of several to appear in the pages of the American Museum's magazine, *Natural History*, during the 1920s. Others included Grinnell, "Old-Time Bone Hunt"; and Lull, "Early Fossil Hunting in the Rocky Mountains."

12. Matthew, "Early Days," 454.

13. Ramaley and Dodds, "University of Colorado Mountain Laboratory," 53.

14. I. E. W., "Dedication of the University of Colorado Mountain Laboratory," 162.

15. Johnson, "Rocky Mountain Biological Laboratory," 105. Agricultural field stations on the Great Plains, on the other hand, typically remained near railroad lines into the 1920s, and the impact of the automobile was less dra-

matic, though its internal combustion counterpart in agricultural labor—the tractor—did reshape station practice, as foreseen by one traveler in Montana for the USDA's Bureau of Plant Industry as early as 1908, in Jardine, *Notes on Dry Farming*, 9

16. Samuel H. Knight, "Historical Resume of the Science Camp" (1963), unpublished manuscript, p. 1, Samuel H. Knight Papers, #400044, American Heritage Center, University of Wyoming, Laramie, box 33, folder 29.

17. "Science Summer Camp," 184, 188.

18. *Rocky Mountain Biological Station: 5th Annual Season* (Gunnison: Bulletin of the Western State College of Colorado, 1926), in Rocky Mountain Biological Lab Folder, Leslie J. Savage Library, Western State College of Colorado, Gunnison.

19. For example, the initial shift to airmail by the US Post Office began on the East Coast in 1918 and then expanded all the way to Omaha and then San Francisco for the first time in 1920, as noted in Rust, *Flying across America*, 23. On the role of aircraft in later field practice, see Martin-Nielsen, *Eismitte in the Scientific Imagination*, 49, 70, 73–75.

20. Rees, *Infanticide Controversy*, 8–9. See also Montgomery, "Place, Practice and Primatology"; Montgomery, "Primates in the Real World"; and Burkhardt, "Ethology, Natural History, the Life Sciences, and the Problem of Place."

21. Oreskes, *Rejection of Continental Drift*, 290.

22. Hayden, *Field Man*, 234.

23. Benson, *Wired Wilderness*; Didier, "Counting on Relief"; Alagona, *After the Grizzly*; Rumore, "Natural Laboratory"; Gieryn, "City as Truth-Spot"; Lewis, *Inventing Global Ecology*; Raby, "Making Biology Tropical"; Howkins, "Reluctant Collaborators"; Selcer, "View from Everywhere"; Way, *Conserving Southern Longleaf*; Keiner, *Oyster Question*.

24. Gieryn, "Laboratory Design for Post-Fordist Science."

25. The same has also been true of modern science-based medicine. For the argument that maturation of American medical therapeutics also involved a shift to universalism, see Numbers and Warner, "Maturation of American Medical Science."

26. In fact, scientists have sometimes drawn on place-based, regional knowledge of varied local environments, rather than universal knowledge, in order to gain credibility and authority. For a historical example of this, see Moon, "Empirical Knowledge, Scientific Authority, and Native Development."

27. Drayton, *Nature's Government*, 274.

28. Turnbull, "Travelling Knowledge," 274.

29. Klein, "Change Within." For her more fully developed account of humanity's current predicament, see Klein, *This Changes Everything*.

Bibliography

Archival Sources

Agate Fossil Beds National Monument, Harrison, Nebraska
Cook Papers Collection
American Heritage Center, University of Wyoming, Laramie
Alcott Farrar Elwell Collection, #1916
Aven Nelson Papers, #400013
Edith S. and Frederic E. Clements Papers, #1678
Harold J. and James H. Cook Collection, #558
Samuel H. Knight Papers, #400044
University of Wyoming, Agricultural Experiment Station Records, #548001
University of Wyoming, Botany Records, #545001
University of Wyoming, Department of Geology and Geophysics Records, #545004
Vernon Bailey Papers, #554
William Rush Taggart Collection, #712
American Museum of Natural History, New York City
Division of Paleontology, Field Correspondence
Division of Paleontology, General Correspondence
Carnegie Mellon University Libraries, Digital Collections
Earl Douglass Papers, Carnegie Museum of Natural History Diplodocus and Douglass Archives, http://diva.library.cmu.edu
Carnegie Museum of Natural History, Pittsburgh, Pennsylvania
John Bell Hatcher Papers, Section of Vertebrate Paleontology
O. A. Peterson Papers, Section of Vertebrate Paleontology
Center for Southwest Research, University of New Mexico, Albuquerque
John Weinzirl Papers, MSS 293SC
Center of Southwest Studies, Fort Lewis College, Durango, Colorado
Fort Lewis College Collection, M001
Colorado College, Colorado Springs, Colorado, Special Collections/Archives, Tutt Library

Edward R. Warren Papers, MS51
Francis Cragin Papers, Part 2, MS370
Francis W. Cragin Papers, MS362
Colorado Historical Society, Denver
Martha Maxwell Collection, #762
Mesa Verde Collection, #428
Colorado State University, Fort Collins, Archives/Special Collections, Morgan Library
Experiment Station Collection
Louis George Carpenter
Denver Museum of Nature and Science, Denver, Colorado, Archives
Earth Sciences Department Miscellaneous, IA.050–12–26–95
Jesse Dade Figgins Correspondence, IA.501.Figgins
Richard Charles Hills Correspondence, IA.148.Hills
Denver Public Library, Colorado
T. D. A. Cockerell Papers, CONS 138
Fort Hays State University, Fort Hays, Kansas, Archives/Special Collections, Forsyth Library
George Sternberg Collection
Institute for Regional Studies, North Dakota State University, Fargo
Dickinson Experiment Station, University Archives
Lawrence R. Waldron Collection, Mss 36
Jackson Hole Historical Society, Jackson Hole, Wyoming
Richard "Beaver Dick" Leigh Folder
Kansas State University, Manhattan, University Archives & Manuscripts, Hale Library
Agricultural Experiment Station Reports
Mesa Verde National Park, Colorado, Research Center
Montana State University, Bozeman, Special Collections
College of Agriculture Records, Acc 00005, Special Collections
National Anthropological Archives, Suitland, Maryland
American Antiquities Permit Files
Jesse Logan Nusbaum Papers
National Archives, Central Plains Region, Kansas City, Missouri
Correspondence and Reports, Records of the Weather Bureau, Office of the Regional Director, Kansas City, Missouri
National Archives II, College Park, Maryland
Hague Papers, Letters Received, Entry 141 (Old Entry 67), RG 57, Records of the Geological Survey
Hague Papers, Letters Sent, Entry 142 (Old Entry 68), RG 57, Records of the Geological Survey

Hague Papers, Records Relating to Yellowstone Park and Other Matters, Entry 147 (Old Entry 73), RG 57, Records of the Geological Survey
Hague Papers, Working Papers and Reference Material, Entry 146 (Old Entry 62), RG 57, Records of the Geological Survey
Hayden Survey, Miscellaneous Letters Received, Entry 19, RG 57, Records of the Geological Survey
Marsh Papers, Records Concerning Othniel Charles Marsh, Entry 78, RG 57, Records of the Geological Survey
Records of the Division of Forage Crops and Diseases, Correspondence with Other Offices of the Bureau, PI-66 Entry 67, RG 54, Records of the Bureau of Plant Industry, Soils, and Agricultural Engineering
Records of the Division of Fruit and Vegetable Crops and Diseases, Correspondence Relating to Investigations of Alkali and Drought-Resistant Plants, PI-66 Entry 124, RG 54, Records of the Bureau of Plant Industry, Soils, and Agricultural Engineering
Records of the National Park Service, Central Classified Files (NPS), P Entry 10-A, RG 79
Records of the Office of Experiment Stations, Correspondence of Alfred C. True, NC-132 Entry 51, RG 164, Records of the Cooperative State Research Service
Records of the Office of Experiment Stations, General Correspondence with State Experiment Stations and Agricultural Colleges, NC-132 Entry 11, RG 164, Records of the Cooperative State Research Service
Records of the Office of the Secretary of the Interior, Central Classified Files, A1 Entry 794-A, RG 48
Records of the Soil Survey Division, Reports on Soil Surveys, PI-66 Entry 216, RG 54, Records of the Bureau of Plant Industry, Soils, and Agricultural Engineering
Samuel Franklin Emmons Papers, Letters Received, Entry 139 (Old Entry 65), RG 57, Records of the Geological Survey
Samuel Franklin Emmons Papers, Letters Sent, Entry 140 (Old Entry 66), RG 57, Records of the Geological Survey
Oklahoma State University, Stillwater, Special Collections and University Archives, Edmon Low Library
Oklahoma Agricultural and Mechanical College Correspondence, 1892–1920, #86–009
University of California, Berkeley, Bancroft Library
C. Hart Merriam Papers, Volume II
University of Colorado, Boulder, University Archives, Norlin Library
Francis Ramaley Papers
Theodore Dru Alison Cockerell Papers

University of Denver, Colorado, Special Collections, Penrose Library
Ira E. Cutler Collection
University of Illinois, Champaign-Urbana, University Archives
Arthur G. Vestal Papers, 15/4/24
University of Kansas, Lawrence, Spencer Research Library
L. L. Dyche Papers
Samuel Williston Papers, PP32
University of Montana, Special Collections, Missoula
Morton J. Elrod Papers, MSS 486
University of Nebraska, Lincoln, University Archives, Love Library
Agricultural Experiment Station Records, North Platte Experiment Station (Correspondence), #10/14/1
C. E. Bessey Papers, #12/7/10
Lawrence Bruner Papers, #8/12/10
Raymond John Pool Papers, #12/07/12
Sem Bot [Botanical Seminar] Club Records, #12/7/5
University of North Dakota, Grand Forks, Special Collections
North Dakota Geological Survey, UA34
University of South Dakota, Vermillion, Special Collections
E. C. Perisho Collection, USD91–0005, Arts and Sciences, Deans Series
University of Utah, Salt Lake City, Marriott Library
Earl Douglass Papers, Ms 196
US Forest Service, Bighorn National Forest, District Office, Lovell, Wyoming
Medicine Wheel File
US Geological Survey, Field Records Library, Denver, Colorado
Joseph Paxson Iddings, Correspondence, Geologic Studies, and Recollections, No. 7879
Utah State University, Logan, Tanner Reading Room, Merrill Library
Ridgway Family Papers, Caine Collection Mss 8
Washburn University Archives, Mabee Library, Topeka, Kansas
Western State College of Colorado, Gunnison, Leslie J. Savage Library
John C. Johnson Collection
Rocky Mountain Biological Lab Folder
Yellowstone National Park Archives, Gardiner, Montana

Published Primary Sources

A. L. R. "The Closing of the U.S. Signal Service Stations on Pike's Peak and Mt. Washington." *American Meteorological Journal* 5 (1888): 284–86.

Abbott, Gerard A. "A Grebe Colony." *Bird-Lore* 4 (1902): 86–87.

Adams, George I. "Physiographic Divisions of Kansas." *Bulletin of the American Geographical Society* (1902): 89–104.

"Agricultural Experiment Stations." *Garden and Forest* 8 (1895): 401–2.

Aiken, C. E. "A Glimpse at Colorado and Its Birds." *American Naturalist* 7 (1873): 13–16.

Alexander, Thomas G. "John Wesley Powell, the Irrigation Survey, and the Inauguration of the Second Phase of Irrigation Development in Utah." *Utah Historical Quarterly* 37 (1969): 190–206.

Alford, Thomas Wildcat. *Civilization, and the Story of the Absentee Shawnees.* Norman: University of Oklahoma Press, 1936.

Allen, E. W. "Some Ways in Which the Department of Agriculture and the Experiment Stations Supplement Each Other." *Yearbook of the U.S. Department of Agriculture* (1905): 167–82.

Allen, E. W., E. V. Wilcox, and J. I. Schulte. "Work and Expenditures of the Agricultural Experiment Stations, 1916." In *Report on the Work and Expenditures of the Agricultural Experiment Stations* (1916): 13–14.

Allen, J. A. *The American Bisons, Living and Extinct.* Memoir of the Museum of Comparative Zoology 10. Cambridge, MA: University Press, 1876.

Allen, J. A. "Notes of an Ornithological Reconnaissance of Portions of Kansas, Colorado, Wyoming, and Utah." *Bulletin of the Museum of Comparative Zoology* 3 (1872): 113–83.

Allen, J. A. "Notes on the Mammals of Portions of Kansas, Colorado, Wyoming and Utah." *Bulletin of the Essex Institute* 6 (1874): 43–52, 53–66.

Allen, J. A. "Report on the Mammals and Birds." In *Annual Report of the Museum of Comparative Zoology*, 11. 1871.

Allen, J. A. Review of C. Hart Merriam, *Monographic Revision of the Pocket Gophers, North American Fauna* no. 8 (1895). *Science* 1 (1895): 241–43.

Almy, Kenneth J. "Thof's Dragon and the Letters of Capt. Theophilus H. Turner, M.D., U.S. Army." *Kansas History* 10 (1987): 170–200.

"The Alpine Laboratory of the Botanical Seminar of the University of Nebraska." *Science* 20 (1904): 185–86.

Anderson, Emma N., and Elda R. Walker. "An Ecological Study of the Algae of Some Sandhill Lakes." *Botanical Survey of Nebraska* 4, n.s. (1920): 51–85.

Anderson, Geo. P. "Black-Headed Grosbeak (*Habia melanocephala*)." *Proceedings of the Nebraska Ornithologists' Union* 2 (1901): 98.

"Anderson School of Natural History." *Nation*, September 11, 1873, 174–75.

Arrhenius, Olof W. *Stones Speak and Waters Sing: The Life and Works of Gustaf Nordenskiold.* Edited by Robert H. Lister and Florence C. Lister. Mesa Verde National Park, CO: Mesa Verde Museum Association, 1984.

Arthur, J. C. "Some Botanical Laboratories of the United States." *Botanical Gazette* 10 (1885): 395–406.

Atkinson, Alfred, and N. C. Donaldson. *Dry Farm Grain Tests in Montana.* Bulletin of the Montana Agricultural Experiment Station, no. 110 (1916).

Atkinson, Alfred, and L. F. Gieseker. *Dry Farm Crop Rotations and Cultural Methods.* Circular of the Montana Agricultural Experiment Station, no. 81 (1918).

Babcock, E. J. *Economic Methods of Utilizing Western Lignites.* Bulletin of the US Bureau of Mines, no. 89 (1915).

Bailey, Florence Merriam. *Birds of New Mexico.* Santa Fe: New Mexico Department of Game and Fish, 1928.

Bailey, Vernon. *Biological Survey of Texas.* North American Fauna, no. 25 (1905).

Bailey, Vernon. "Identity of the Bean Mouse of Lewis and Clark." *Journal of Mammalogy* 1 (1920): 70–72.

Bailey, Vernon. "Old and New Horns of the Prong-Horned Antelope." *Journal of Mammalogy* 1 (1920): 128–29.

Bailey, Vernon. *The Prairie Ground Squirrels or Spermophiles of the Mississippi Valley.* Bulletin of the Division of Ornithology and Mammalogy, no. 4 (1893).

Bailey, Vernon. *Revision of the Pocket Gophers of the Genus Thomomys.* North American Fauna, no. 39 (1915).

Bailey, Vernon. "Unforgettable Meals." *Sierra Club Bulletin* 23, no. 2 (1938): 65–69.

Bailey, Vernon, and Florence Merriam Bailey. *Wild Animals of Glacier National Park.* Washington, DC: Government Printing Office, 1918.

Baker, C. F. "Botanizing in the La Plata Mountains." *Plant World* 2 (1898): 29–32.

Baker, Marcus. "Surveys, Their Kinds and Purposes." *Science* 12 (1888): 254–59.

Barbour, Carrie Adeline. "Some Methods of Collecting, Preparing, and Mounting Fossils." *Publications of the Nebraska Academy of Sciences* 6 (1898): 258–64.

Barbour, Erwin H. "An Account of the Nebraska Geological Survey to 1912." *Nebraska Geological Survey* 7 (1912): 7–15.

Barbour, Erwin H. "Museums and the People." *Publications of the Nebraska Academy of Sciences* 8, no. 4 (1912): 8, 10.

Barbour, Erwin Hinckley. "Biennial Report." *Nebraska Geological Survey* 2 (1907): 365–87.

Barbour, Erwin Hinckley. "Discovery of Meteoric Iron in Nebraska." *Publications of the Nebraska Academy of Sciences* 6 (1898): 275–79.

Bartholomew, David M., ed. *Pioneer Naturalist on the Plains: The Diary of Elam Bartholomew, 1871 to 1934.* Manhattan, KS: Sunflower University Press, 1998.

Bates, Carlos G., and Roy G. Pierce. *Forestation of the Sand Hills of Kansas and Nebraska.* Bulletin of the US Forest Service, no. 121 (1913).

Bates, J. M. "Additional Notes and Observations on the Birds of Northern Nebraska." *Proceedings of the Nebraska Ornithologists' Union* 1 (1900): 15–18.

Bates, J. M. "Additional Observations on the Birds of Northwestern Nebraska." *Proceedings of the Nebraska Ornithologists' Union* 2 (1901): 73–75.

Bear, Firman E., and George M. McClure. "Sampling Soil Plots." *Soil Science* 9 (1920): 65–75.

Beattie, R. Kent. "Nebraska Ferns and Fern Allies." *Fern Bulletin* 6 (1898): 72–74.

Beckstrand, E. H. *Tests of Macadam Rock.* Bulletin of the State School of Mines, University of Utah, no. 2 (1909).

Bell, Howard W. "Fossil-Hunting in Wyoming." *Cosmopolitan* 28 (1900): 265–75.

Bell, William Abraham. *New Tracks in North America: A Journal of Travel and Adventure whilst Engaged in the Survey for a Southern Railroad to the Pacific Ocean during 1867–8.* 2nd ed. London: Chapman and Hall, 1870.

Bessey, Charles E. "Distribution of Forest Trees on the Nebraska Plains." *Atlantic Slope Naturalist* 1, no. 3 (July/August 1903): 21.

Bessey, Charles E. "Summer-School Botany in the Mountains." *American Naturalist* 29 (1895): 845–47.

Bessey, Chas. E. "One Thousand Miles for a Fern." *Asa Gray Bulletin* 8 (1900): 2–6.

Betts, C. W. "The Yale College Expedition of 1870." *Harper's New Monthly Magazine* 43 (1871): 663–71.

Bigelow, S. Lawrence. "The Growth and Function of the Modern Laboratory." *Science* 19 (1904): 641–50.

Birdsall, W. R. "The Cliff Dwellings of the Cañons of the Mesa Verde." *Journal of the American Geographical Society of New York* 23 (1891): 584–620.

Bishop, F. C. *The Distribution of the Rocky Mountain Spotted-Fever Tick.* Circular of the Bureau of Entomology, no. 136 (1911).

Blankinship, J. W. "A Century of Botanical Exploration in Montana, 1805–1905: Collectors, Herbaria and Bibliography." *Montana Agricultural College Science Studies* 1 (1904): 1–31.

Blankinship, J. W., and Hester F. Henshall. "Common Names of Montana Plants." *Montana Agricultural College Science Studies* 3 (1905): 113–39.

Boutelle, F. A. *Report of the Superintendent of the Yellowstone National Park to the Secretary of the Interior.* 1890.

Bradley, Frank H. "Report of Frank H. Bradley, Geologist of the Snake River Division." In *Sixth Annual Report of the U.S. Geological Survey of the Territories, Embracing Portions of Montana, Idaho, Wyoming, and Utah, Being a Report of Progress of the Explorations for the Year 1872,* 189–271.

Briggs, Lyman J., and H. L. Shantz. "Daily Transpiration during the Normal Growth Period and Its Correlation with the Weather." *Journal of Agricultural Research* 7 (1916): 155–212.

Briggs, Lyman J., and H. L. Shantz. "Hourly Transpiration Rate on Clear Days as Determined by Cyclic Environmental Factors." *Journal of Agricultural Research* 5 (1916): 583–650.

Brous, H. A. "Habits of Amblychila cylindriformis." *Transactions of the Kansas Academy of Science* 5 (1876): 11–12.

Brous, H. A. "Habits of the Prairie Dog." *Transactions of the Kansas Academy of Science* 5 (1876): 10–11.

Brons, H. A. "Notes on the Habits of Some Western Snakes." *American Naturalist* 16 (1882): 564–67.

Brown, Barnum. "Hunting Big Game of Other Days: A Boating Expedition in Search of Fossils in Alberta, Canada." *National Geographic* 35 (1919): 407–29.

Bruner, Lawrence. "A Comparison of the Bird-Life Found in the Sand-Hill Region of Holt County in 1883–84 and in 1901." *Proceedings of the Nebraska Ornithologists' Union* 3 (1902): 58–63.

Bruner, Lawrence. "Ornithology in Nebraska." *Proceedings of the Nebraska Ornithologists' Union* 1 (1900): 8–11.

Buckley, Samuel. *Second Annual Report of the Geological and Agricultural Survey of Texas.* 1876.

Buecker, Thomas R. "'Distance Lends Enchantment to the View': The 1874 Black Hills Expedition Diary of Fred W. Power." *South Dakota History* 27, no. 4 (1997): 197–260.

Buecker, Thomas R. "Letters from a Post Surgeon's Wife: The Fort Washakie Correspondence of Caroline Frey Winne, May 1879–May 1880." *Annals of Wyoming* 53 (1981): 44–63.

Bulletin of Washburn Laboratory of Natural History. Vols. 1–2 (1884–90).

Burford, L. O. "Report of the Lubbock County Experiment Station." *Seventeenth Texas Farmers' Congress, 1914.* Bulletin of the Texas Department of Agriculture, no. 40 (1914): 275.

Burr, W. W. *The Storage and Use of Soil Moisture: Report of Experimental Substation, North Platte, Nebraska.* Research Bulletin of the Nebraska Agricultural Experiment Station, no. 5 (1914).

C. H. M. Review of J. A. Allen, *List of Mammals collected in the Black Hills Region of South Dakota and in Western Kansas by W. W. Granger, with field notes by the collector. American Museum of Natural History Bulletin* 7 (1895), 259–74. *Science* 2: 490.

Calvert, Philip P. "Postage on Natural History Specimens." *Science* 23 (1894): 49–50.

Cameron, Ewen Somerled. "On the Migration of Phalaropes in Montana." *Ibis* 42 (1900): 67–70.

Camp, Charles L., ed. "The Letters of William Diller Matthew." *Journal of the West* 8 (1969): 263–90, 454–76.

Card, Fred W. *Windbreaks.* Bulletin of the Nebraska Agricultural Experiment Station, no. 48 (1897).

"The Care of the National Park." *Forest and Stream,* 29 January, 1885, 1.

Carleton, M. A. "Observations on the Native Plants of Oklahoma Territory and Adjacent Districts." *Contributions from the U.S. National Herbarium* 1 (1890): 220–32.

Carpenter, Franklin R. *Preliminary Report of the Dakota School of Mines upon the Geology, Mineral Resources and Mills of the Black Hills of Dakota.* Bulletin of the (South) Dakota School of Mines, no. 1 (1888).

Carpenter, L. G. *Forests and Snow.* Bulletin of the Colorado Agricultural Experiment Station, no. 55 (1901).

Carruth, James H. "Centennial Catalogue of the Plants of Kansas." *Transactions of the Kansas Academy of Science* 5 (1876): 40–59.

Carruth, James H. "Report on Botany, 1874." *Transactions of the Kansas Academy of Science* 3 (1874): 122–27.

Carter, W. T., J. F. Stroud, and T. M. Bushnell. *Investigations in Adapting the Automobile to Accurate Soil Survey and Road Traverse Work.* Circular of the Texas Agricultural Experiment Station, no. 19 (1919).

Cary, Merritt. "Additional Notes on Birds of the Upper Elkhorn Valley." *Proceedings of the Nebraska Ornithologists' Union* 2 (1901): 99.

Cary, Merritt. "Some Bird Notes from the Upper Elkhorn." *Proceedings of the Nebraska Ornithologists' Union* 1 (1900): 21–29.

"The Celebration of the Quarto Centennial of the Colorado Scientific Society, 1882–1907." *Proceedings of the Colorado Scientific Society* 8 (1907): 387–91.

Chapin, Frederick H. "Cliff-Dwellings of the Mancos Cañons." *American Antiquarian and Oriental Journal* 12 (1890): 193–210.

Chapin, Frederick H. *The Land of the Cliff-Dwellers.* Boston: Appalachian Mountain Club, 1892.

Chesnut, Alma. "Vernon Bailey: A Nature Nobleman." *Nature Magazine* 14 (1929): 229–32.

Chilcott, E. C. *A Study of Cultivation Methods and Crop Rotations for the Great Plains Area.* Bulletin of the Bureau of Plant Industry, no. 187 (1910).

Chilcott, E. C., and John S. Cole. *Growing Winter Wheat on the Great Plains.* Farmers Bulletin, no. 895 (1917).

Chilcott, E. C., J. S. Cole, and W. W. Burr. *Crop Production in the Great Plains Area: Relation of Cultural Methods to Yields.* (Department) Bulletin of the US Department of Agriculture, no. 268 (1915).

Chilcott, E. C., J. S. Cole, and W. W. Burr. *Spring Wheat in the Great Plains Area: Relation of Cultural Methods to Production.* (Department) Bulletin of the US Department of Agriculture, no. 214 (1915).

Chittenden, F. H. *The Colorado Potato Beetle.* Circular of the Bureau of Entomology, no. 87 (1907).

Clarke, F. W., ed. *Report of Work Done in the Division of Chemistry and Physics, Mainly during the Fiscal Year 1886–7.* Bulletin of the US Geological Survey, no. 55 (1889).

Clements, Edith S. *Adventures in Ecology: Half a Million Miles . . . From Mud to Macadam.* New York: Pageant Press, 1960.

Clements, Frederic E. *Plant Indicators: The Relation of Plant Communities to Process and Practice.* Publication 290. Washington, DC: Carnegie Institution of Washington, 1920.

Clements, Frederic E. *Plant Succession: An Analysis of the Development of Vegetation.* Publication 242. Washington, DC: Carnegie Institution of Washington, 1916.

Clements, Frederic E. *Research Methods in Ecology.* Lincoln, NE: University Publishing Company, 1905.

Clements, Frederic E. "Scope and Significance of Paleo-Ecology." *Bulletin of the Geological Society of America* 29 (1918): 369–74.

Clements, Frederick[*sic*] E. "The Alpine Laboratory." *Science* 37 (1913): 327–28.

Clothier, George I. *Advice for Forest Planters in Oklahoma and Adjacent Regions.* Bulletin of the Bureau of Forestry, no. 65 (1905).

Cockerell, T. D. A. "An Alien Clematis in New Mexico." *Science* 10 (1899): 898–99.

Cockerell, T. D. A. "The Classification of Slight Varieties." *Bulletin of the Torrey Botanical Club* 16 (1889): 270–72.

Cockerell, T. D. A. "Communications." *Proceedings of the Entomological Society of Washington* 3 (1894): 127–28.

Cockerell, T. D. A. "A Curious Case of Protective Coloration." *Psyche* 8 (1897): 154.

Cockerell, T. D. A. "The Diverse Floras of the Rocky Mountain Region." *Science* 7 (1898): 625–27.

Cockerell, T. D. A. "Doings of Societies." *Entomological News* 1 (1890): 31.

Cockerell, T. D. A. "Entomological Ethics." *Proceedings of the 10th Annual Meeting, Association of Economic Entomology.* Bulletin of the Division of Entomology, no. 17 (1898): 87–90.

Cockerell, T. D. A. *Entomological Observations in 1894*; *Life Zones in New Mexico*; *Entomological Diary at Santa Fe.* Bulletin of the New Mexico Agricultural Experiment Station, no. 15 (1895).

Cockerell, T. D. A. "Entomology." *Annual Report of the New Mexico Agricultural Experiment Station* 5 (1893–94): 13–24.

Cockerell, T. D. A. "Florissant: A Miocene Pompeii." *Popular Science Monthly* 74 (1908): 112–26.

Cockerell, T. D. A. "Fossil Hymenoptera from Florissant, Colorado." *Bulletin of the Museum of Comparative Zoology* 50 (1906): 31–58.

Cockerell, T. D. A. "New Fossil Insects from Florissant, Colorado." *Annals of the Entomological Society of America* 2 (1909): 251–56.

Cockerell, T. D. A. "New Species of Bees." *Psyche* 7, supplement (1895): 9–11.

Cockerell, T. D. A. "A New Variety of the Sunflower." *Science* 32 (1910): 384.

Cockerell, T. D. A. "Recollections of a Naturalist." *Bios* 6 (1935): 372–85.

Cockerell, T. D. A. "Recollections of a Naturalist II: Expeditions." *Bios* 7 (1936): 149–55.

Cockerell, T. D. A. "Recollections of a Naturalist III: Snails and Slugs." *Bios* 7 (1936): 205–11.

Cockerell, T. D. A. "Recollections of a Naturalist IV: The Amateur Botanist." *Bios* 8 (1937): 12–18.

Cockerell, T. D. A. "Recollections of a Naturalist V: Fossil Insects." *Bios* 8 (1937): 51–56.

Cockerell, T. D. A. "Recollections of a Naturalist VI: Scale Insects and Mealy Bugs." *Bios* 8 (1937): 122–27.

Cockerell, T. D. A. "Sunflower Problems." *Science* 40 (1914): 708–9.

Cockerell, T. D. A. "Suppression and Loss of Characters in Sunflowers." *Science* 40 (1914): 283–85.

Cockerell, T. D. A. "A Wine-Red Sunflower." *Science* 38 (1913): 312–13.

Cockerell, T. D. A. *Zoology of Colorado*. Boulder: University of Colorado, 1927.

Coffey, George N. "The Development of Soil Survey Work in the United States with a Brief Reference to Foreign Countries." *Proceedings of the American Society of Agronomy* 31 (1911): 115–29.

Coggeshall, Arthur Sterry. "Patsy of Dinosaur Monument." *National Parks Magazine* 27 (1953): 58–61, 95.

"The Colorado Biological Association." *West American Scientist* 5 (1888): 17–18.

"The Colorado Meteorological Society." *American Meteorological Journal* 1 (1885): 461–64.

"Colorado School of Forestry." *Forestry and Irrigation* 14 (1908): 675–78.

Colorado State School of Mines. *Annual Report of Field Work and Analyses Relating to the Economic Geology of Colorado*. Denver: Rocky Mountain News, 1886.

Colorado State School of Mines. *Annual Report of the State School of Mines, 1885*. Denver: Rocky Mountain News, 1886.

Condra, G. E. "Reports of the Geographer and the Conservation Commission." In *Annual Report of the Nebraska State Board of Agriculture*, 267–351. 1909.

Connor, P. "Floods in the Missouri River above Kansas City." *Monthly Weather Review* 37 (1909): 399.

Connor, P. "Heavy Rainstorm at Kansas City, Mo." *Monthly Weather Review* 42 (1914): 546–47.

Connor, P. "Loss by Floods in Kansas River and Tributaries, June, 1915." *Monthly Weather Review* 43 (1915): 287–88.

Connor, P. "Notes on the Heat and Drought during the Summer of 1913 at Kansas City, Mo." *Monthly Weather Review* 41 (1913): 1441.

Connor, P. *Proceedings of the Convention of Weather Bureau Officials, held at Omaha, Nebr., October 13–14, 1898.* Bulletin of the US Weather Bureau, no. 24 (1899).

Conrad, T. A. "Note on a Cretaceous Limestone of Dakota." *American Journal of Conchology* 3 (1867): 5.

Cook, Harold J. *Tales of the 04 Ranch: Recollections of Harold J. Cook, 1887–1909.* Lincoln: University of Nebraska Press, 1968.

Cook, Harold J. "Two New Bunomastodonts from Colorado." *Proceedings of the Colorado Museum of Natural History* 4 (1922): 1–14.

Cook, James H. *Fifty Years on the Old Frontier as Cowboy, Hunter, Guide, Scout, and Ranchman.* New Haven, CT: Yale University Press, 1923.

Cook, O. F. *The Section of Seed and Plant Introduction.* Circular of the Division of Botany, no. 16 (1899).

Cooke, Wells W. *Preliminary Census of the Birds of the United States.* (Department) Bulletin of the US Department of Agriculture, no. 187 (1915).

"Co-operation in Science." *Science* 1 (1880): 30–31.

Cope, Edw. D. "Notice of Proboscians from the Eocene of Wyoming." *Proceedings of the American Philosophical Society* 12 (1872): 580.

Cope, Edward D. "Synopsis of the Extinct Batrachia, Reptilia, and Aves of North America." *Transactions of the American Philosophical Society* 14 (1870): 1–252.

Coulter, John M. *Manual of the Botany (Phaenogamia and Pteridophyta) of the Rocky Mountain Region, from New Mexico to the British Boundary.* New York: American Book Company, 1885.

Coulter, John M., and Aven Nelson. *New Manual of Botany of the Central Rocky Mountains (Vascular Plants).* New York: American Book Company, 1909.

Cragin, F. W. "On the Washburn Biological Survey of Kansas." *Kansas City Review of Science and Industry* 4 (1885): 576–80.

Cragin, Francis W. "Prospectus of the Biological Survey of Kansas." *Bulletin of the Washburn Laboratory of Natural History* 1 (1884): 2–3.

Craig, Wallace. "North Dakota Life: Plant, Animal, and Human." *Bulletin of the American Geographical Society* 40 (1908): 321–32, 401–15.

Crawford, J. C. "Results of a Collecting Trip to Sioux County." *Proceedings of the Nebraska Ornithologists' Union* 2 (1901): 76–79.

Culbertson, Thaddeus A. *Journal of an Expedition to the Mauvaises Terres and the Upper Missouri in 1850.* Edited by John Francis McDermott. Washington, DC: Government Printing Office, 1952.

Dana, Samuel Trask. "Experiment Stations on the National Forests." *Proceedings of the Society of American Foresters* 4 (1909): 22–29.

Dartt, Mary. *On the Plains and among the Peaks; or, How Mrs. Maxwell Made Her Natural History Collection.* Philadelphia: Claxton, Remsen & Haffelfinger, 1879.

Davenport, E. "Obstacles to Progress in Agricultural Science." *Proceedings of the Society for the Promotion of Agricultural Science* 33 (1912): 11–17.

Davenport, E. "The Relations between the Federal Department of Agriculture and the Agricultural Colleges and Experiment Stations." *Proceedings of the Association of American Agricultural Colleges and Experiment Stations* 26 (1912): 121–33.

Davis, W. M. "The Colorado Front Range: A Study in Physiographic Presentation." *Annals of the Association of American Geographers* 1 (1911): 21–83.

Davis, W. M. "The Education of the Topographer." *Science* 1 (17 May 1895): 546–47.

DeBuys, William, ed. *Seeing Things Whole: The Essential John Wesley Powell.* Washington, DC: Island Press, 2001.

DeMallie, Raymond J., ed. *The Sixth Grandfather: Black Elk's Teachings Given to John G. Neihardt.* Lincoln: University of Nebraska Press, 1984.

Denison, Charles. *Rocky Mountain Health Resorts: An Analytical Study of High Altitudes in Relation to the Arrest of Chronic Pulmonary Disease.* Boston: Houghton, Osgood, 1880.

Densmore, Frances. *Mandan and Hidatsa Music.* Bulletin of the Bureau of American Ethnology, no. 80 (1923).

Densmore, Frances. *Northern Ute Music.* Bulletin of the Bureau of American Ethnology, no. 75 (1922).

Densmore, Frances. *Teton Sioux Music.* Bulletin of the Bureau of American Ethnology, no. 61 (1918).

"The Denver Meeting." *Science* 14 (1901): 233–35.

Diamond, Irving L., and Daniel M. Olson, eds. *Letters of Gustaf Nordenskiöld Written in the Year 1891, and Articles from the Journals* Ymer *and* Photographic Times. Mesa Verde National Park, CO: Mesa Verde Museum Association, 1991.

Dickson, R. E. "Report of Dickens County Experiment Association." *Seventeenth Texas Farmers' Congress, 1914.* Bulletin of the Texas Department of Agriculture, no. 40 (1914): 274–75.

Dodge, Richard Irving. *The Plains of North America and Their Inhabitants.* Edited by Wayne R. Kime. Newark: University of Delaware Press, 1989 (1876).

Dorsey, George A. "An Aboriginal Quartzite Quarry in Eastern Wyoming." *Anthropological Series of the Field Columbian Museum* 2 (1900): 231–43.

Dorsey, George A. "The Cheyenne I: Ceremonial Organization." *Anthropological Series of the Field Columbian Museum* 9 (1905): 1–55.

Dorsey, George A. "The Cheyenne II: The Sun Dance." *Anthropological Series of the Field Columbian Museum* 9 (1905): 57–186.

Douglass, G. E. *Speak to the Earth and It Will Teach You: The Life and Times of Earl Douglass, 1862–1931.* Booksurge, 2009.

"Dr. Joel Lunell." *American Midland Naturalist* 6 (1920): 243–45.

"Drowned in Sweetwater Lake." *Nidiologist* 2 (1895): 144.

Dry-Land Agriculture: Papers Read at the Second Annual Meeting of the Cooperative Experiment Association of the Great Plains, Area, Held at Manhattan, Kans., June 26–27, 1907. Bulletin of the Bureau of Plant Industry, no. 130 (1908).

Dumble, Edwin T. *Third Annual Report of the Geological Survey of Texas.* 1891.

Dunraven, Lord. "Moose and Cariboo Hunting in Colorado and Canada." *Journal of the American Geographical Society of New York* 11 (1879): 334–68.

Eastman, Charles R. "New Fossil Bird and Fish Remains from the Middle Eocene of Wyoming." *Geological Magazine* 47 (1900): 54–58.

Eastwood, Alice. "General Notes of a Trip through Southeastern Utah." *Zoe* 3 (1893): 354–61.

Eastwood, Alice. "Report on a Collection of Plants from San Juan County, in Southeastern Utah." *Proceedings of the California Academy of Sciences* 6 (1896): 270–329.

Eastwood, Alice. "Two Species of Aquilegia from the Upper Sonoran Zone of Colorado and Utah." *Proceedings of the California Academy of Sciences* 4 (1895): 559–62.

"Editorial." *Agricultural Science* 7 (1893): 274–78.

"Editorial." *Botanical Gazette* 21 (1896): 303–5.

"Editorial." *Botanical Gazette* 22 (1896): 57–58.

"Editorial." *Muhlenbergia* 4 (1908): 15–16.

"Editorial Notes." *Annals of the Carnegie Museum* 10 (July 1916): 309.

"Executive Proceedings." *Proceedings of the American Association for the Advancement of Science* (1901): 381–402.

Elliot, D. G. "List of Mammals Obtained by Thaddeus Surber, Collector for the Museum, Chiefly in Oklahoma and Indian Territories." *Field Columbian Museum, Zoological Series* 1, no. 16 (1899): 289–303.

Elrod, Morton J. "The Resources of Montana and Their Development." *Science* 19 (1904): 777–83.

Elrod, Morton J. "The University of Montana Biological Station." *Journal of Applied Microscopy and Laboratory Methods* 4 (1901): 1269–78.

Elrod, Morton J. "The University of Montana Biological Station at Flathead Lake." *Rocky Mountain Magazine* 2, no. 4 (1901): 781–87.

Elwell, Alcott Farrar. "Alcott Farrar Elwell: His Diary, Wyoming, 1908, as Camp Cook, United States Geodetic Survey Roosevelt Lignite Conservation." *Annals of Wyoming* 38 (1966): 142–72.

Emmons, S. F. *Geology and Mining Industry of Leadville, Colo.* US Geological Survey Monograph, no. 12 (1886).

Emmons, S. F. "Investigation of Metalliferous Ores." In *Contributions to Economic Geology, 1905*, ed. S. F. Emmons and E. C. Eckel, 14–19. US Geological Survey Bulletin, no. 285 (1906).

Emmons, S. F. "The Little Cottonwood Granite Body of the Wasatch Mountains." *American Journal of Science*, 4th ser., 16 (1903): 139–47.

Emmons, S. F., and J. D. Irving. *The Downtown District of Leadville, Colo.* US Geological Survey Bulletin, no. 320 (1907).

Emmons, Samuel Franklin. "Uinta Mountains." *Bulletin of the Geological Survey of America* 18 (1907): 287–302.

Ensign, Edgar T. "The Advantages and Limitations of the Colorado Forestry System." *Proceedings of the American Forestry Congress* 7 (1888–89): 25–26.

Ewell, Ervin E. "Every Farm an Experiment Station." *Yearbook of the U.S. Department of Agriculture* (1897): 291–304.

Fairchild, David. "The Small Field Laboratory and Its Atmosphere of Research." *Proceedings of the Society for the Promotion of Agricultural Science* 35 (1914): 62–71.

Farrington, Oliver Cummings. *Catalogue of the Meteorites of North America, to January 1, 1909.* Memoir of the National Academy of Sciences 13 (1915).

Fewkes, J. Walter. "On the Use of the Phonograph in the Study of the Languages of American Indians." *Science* 15 (1890): 267–69.

Fewkes, Jesse Walter. *Antiquities of the Mesa Verde National Park: Cliff Palace.* Bureau of American Ethnology Bulletin, no. 51 (1911).

"A Field School of Geology." *Science* 34 (1911): 706.

Fletcher, Alice C. "The Hako: A Pawnee Ceremony." In *Twenty-Second Annual Report of the Bureau of American Ethnology to the Secretary of the Smithsonian Institution, 1900–1901*, 5–368.

Fletcher, Alice C. "A Study of Omaha Indian Music." *Archaeological and Ethnological Papers of the Peabody Museum* 1 (1893): 233–384.

Fletcher, Alice C. "The Wa-wan, or Pipe Dance of the Omahas." *Annual Report of the Peabody Museum* 16–17 (1884): 308–23.

Fletcher, John Gould. *Life Is My Song: The Autobiography of John Gould Fletcher.* New York: Farrar & Rinehart, 1937.

Fletcher, Maurice S., ed. *The Wetherills of Mesa Verde: Autobiography of Benjamin Alfred Wetherill.* Rutherford, NJ: Farleigh Dickinson University Press, 1977.

Forest Planting in the Sand-Hill Region of Nebraska. Circular of the US Forest Service, no. 37. N.d. (c.1906).

The Fossil Fields of Wyoming: Reports by Members of the Union Pacific Expedition. Omaha, NE: Union Pacific Railroad Co., 1909.

Fryxell, Fritiof M. *Ferdinand Hayden: A Young Scientist in the Great West, 1853–1855.* Rock Island, IL: Augustana Historical Society, 2010.

Gabb, Wm. M. "Description of a New Helix from Utah." *American Journal of Conchology* 5 (1869): 24.

Garman, Samuel. "The Rattle of the Rattlesnake." *Bulletin of the Museum of Comparative Zoology* 13 (1888): 259–68.

Garrett, A. O. "The Honey Ants." *Transactions of the Utah Academy of Sciences* 1 (1909): 22–23.

Gilder, Robert F. "The 'Spanish Diggings,' Wyoming." *Records of the Past* 8 (1909): 3–10.

Giles, George H. "Algae of the Medicine Bow Mountains, in the Vicinity of the University of Wyoming Summer Camp." *University of Wyoming Publications in Science, Botany Series* 1, no. 7 (1934): 187–218.

Gilmore, Charles W. "A History of the Division of Vertebrate Paleontology in the United States National Museum." *Proceedings of the U.S. National Museum* 90 (1941): 305–77.

Gilmore, Melvin R. *Uses of Plants by the Indians of the Missouri River Region.* Washington, DC: Government Printing Office, 1919.

Gipson, A. E. "Experimental Stations." In *Third Annual Report of the Colorado State Horticultural Society for 1886*, 111. Denver: Collier & Cleaveland, 1886.

Gould, Charles N. *Covered Wagon Geologist.* Norman: University of Oklahoma Press, 1959.

Gould, Charles N. "Geological Work in the Southwest." *Bulletin of the Southwestern Association of Petroleum Geologists* 1 (1917): 20–33.

Gould, Charles Newton. "General Geology of Oklahoma." *Biennial Report of the Oklahoma Geological Survey* 2 (1901–02): 17–74.

Grafe, Ernest, and Paul Horsted. *Exploring with Custer: The 1874 Black Hills Expedition.* Custer, SD: Golden Valley Press, 2002.

Graves, A. R. "Notes on Birds from Western Nebraska." *Proceedings of the Nebraska Ornithologists' Union* 2 (1901): 84–85.

Gray, Asa, and Joseph D. Hooker. "The Vegetation of the Rocky Mountain Region and a Comparison with That of Other Parts of the World." *Bulletin of the U.S. Geological and Geographical Survey of the Territories* 6 (1881): 1–77.

Great Plains Committee. *The Future of the Great Plains: Report of the Great Plains Committee.* Washington, DC: Government Printing Office, 1936.

Grinnell, George Bird. *The Cheyenne Indians: Their History and Ways of Life*. 2 vols. New Haven, CT: Yale University Press, 1923.

Grinnell, George Bird. "An Old-Time Bone Hunt: An Account of the Expedition Undertaken by Prof. O. C. Marsh in 1870 to the Then Wild West." *Natural History* 23 (1923): 329–36.

Grout, A. J. "Mosses of Colorado from Tolland and Vicinity." *Bryologist* 19 (1916): 1–8.

Hague, Arnold. *Atlas to Accompany Monograph 32 on the Geology of the Yellowstone National Park*. 1904.

Hague, Arnold. "Early Tertiary Volcanoes of the Absaroka Range." In *Presidential Address of Arnold Hague with Abstracts of Minutes for 1897 and 1898 and Lists of Officers and Members*, 3–25. Washington, DC: Geological Society of Washington, 1899.

Hague, Arnold. *Geological History of the Yellowstone National Park*. Washington, DC: Government Printing Office, 1912.

Hague, Arnold. "Letter to George G. Vest." *Forest and Stream*, March 13, 1884, 125.

Hague, Arnold. "The Yellowstone National Park." *Scribner's Magazine* 35 (May 1904): 513–27.

Hague, Arnold. "The Yellowstone Park as a Forest Reserve." *Nation* 46 (1888): 9–10.

Hague, Arnold. "The Yellowstone Park as a Game Reservation." In *American Big-Game Hunting: The Book of the Boone and Crockett Club*, edited by Theodore Roosevelt and George Bird Grinnell, 240–70. New York: Forest and Stream Publishing Co., 1893.

Hague, Arnold, J. P. Iddings, W. H. Weed, C. D. Walcott, G. H. Girty, T. W. Stanton, and F. H. Knowlton. *Geology of the Yellowstone National Park, Part II, Descriptive Geology, Petrography, and Paleontology*. US Geological Survey Monograph, no. 32 (1899).

Hague, Arnold, W. H. Weed, and J. P. Iddings. *Yellowstone National Park, Wyo.* Folio of the Geological Atlas of the United States, no. 30 (1896).

Hansen, N. E. *Co-operative Tests of Alfalfa from Siberia and European Russia*. Bulletin of the South Dakota Agricultural Experiment Station, no. 141 (1913).

Hapeman, H. "The Fresh-water Algae of Kearney County, Nebraska." *Publications of the Nebraska Academy of Sciences* 3 (1893): 13–14.

Harrington, W. Hague. "Postage on Natural History Specimens." *Science* 22 (1893): 348–49.

Haworth, Erasmus. "The Geology of Underground Water in Western Kansas." In *Report of the Board of Irrigation Survey and Experiment, for 1895 and 1896, to the Legislature of Kansas*, 49–114. Topeka: Kansas State Printing Company, 1897.

Hayden, F. V. "Explorations under the War Department." *Proceedings of the Academy of Natural Sciences* 9 (1857): 109–16.

Hayden, F. V. *Geological Report of the Exploration of the Yellowstone and Missouri Rivers.* Washington, DC: Government Printing Office, 1869.

Hayden, F. V. "A New Species of Hare from the Summit of the Wind River Mountains." *American Naturalist* 3 (1869): 113–16.

Hayden, F. V. *On the Geology and Natural History of the Upper Missouri.* Philadelphia: Sherman, 1862.

Hayden, F. V. "On the Geology of the Tertiary Formations of Dakota and Nebraska." *Journal of the Academy of Natural Sciences* 7 (1869): 9–21.

Hayden, F. V. "Our Great West, and the Scenery of Our National Parks." *Journal of the American Geographical Society of New York* 6 (1874): 196–211.

Hayden, F. V. "The Primordial Sandstone of the Rocky Mountains in the Northwestern Territories of the United States." *American Journal of Science and Arts,* 2nd ser., 33 (1862): 68–79.

Hayden, F. V. "Report of F. V. Hayden on the Geography of the Missouri Valley." In *Second Annual Report of the United States Geological Survey of the Territories, Embracing Wyoming,* 220–29. Washington, DC: Government Printing Office, 1868.

Hayden, F. V. *Sketch of the Origin and Progress of the United States Geological and Geographical Survey of the Territories.* Washington, DC: Darby & Duvall, 1877.

Hayden, F. V., and A. R. C. Selwyn, eds. *North America.* Stanford's Compendium of Geography and Travel. London: Stanford, 1883.

Hayden, Julian D. *Field Man: Life as a Desert Archaeologist.* Southwest Center Series. Tucson: University of Arizona Press, 2011.

Helder, Geo. K., and F. A. Kiene. "A Fire Probably Caused by Slaking Lime." *Journal of the American Society of Agronomy* 5 (1913): 51–52.

Henshaw, Henry Wetherbee. "Autobiographical Notes." *Condor* 21 (1919): 171–81.

Henshaw, Henry Wetherbee. "Autobiographical Notes." *Condor* 22 (1920): 55–60.

Hermann, A. "Modern Laboratory Methods in Vertebrate Paleontology." *Bulletin of the American Museum of Natural History* 26 (1909): 283–331.

"Heroic Sacrifice." *Biennial Report of the Montana State Board of Entomology* 4 (1919–20): 13.

Herrick, C. L., John Terry, and H. N. Herrick, Jr. "Notes on a Collection of Lizards from New Mexico." *Bulletin of the University of New Mexico, Biological Series* 1, no. 4 (1898): 117.

Hewett, Edgar L. *Circular Relating to Historic and Prehistoric Ruins of the Southwest and Their Preservation.* Washington, DC: Government Printing Office, 1904.

Hicks, L. E. "Soils of Nebraska as Related to Geological Formations." *American Geologist* 3 (1889): 36–45.

Hill, Arthur W. "The History and Functions of Botanic Gardens." *Annals of the Missouri Botanical Gardens* 2 (1915): 185–240.

Hillebrand, W. F. "Mineralogical Notes." In Clarke, *Report of Work Done in the Division of Chemistry and Physics*, 48.

Hoffman, W. J. "List of Birds Observed at Grand River Agency, Dakotah." *Proceedings of the Boston Society of Natural History* 18 (1875): 169–75.

Hoffman, W. J. "List of Mammals Found in the Vicinity of Grand River, D.T." *Proceedings of the Boston Society of Natural History* 19 (1877): 94–102.

Hofmann, Charles, ed. *Frances Densmore and American Indian Music: A Memorial Volume.* New York: Museum of the American Indian, Heye Foundation, 1968.

Holland, W. J. "Earl Douglass: A Sketch in Appreciation of His Life and Work." *Annals of the Carnegie Museum* 20 (1931): 279–92.

Holland, W. J. "The Skull of Diplodocus." *Memoirs of the Carnegie Museum* 9 (1924): 379–403.

Holland, W. J., and O. A. Peterson. "The Osteology of the Chalicotheroidea with Special Reference to a Mounted Skeleton of Moropus elatus Marsh, Now Installed in the Carnegie Museum." *Memoirs of the Carnegie Museum* 3 (1914): 189–406.

Hollister, Ovando J. *The Mines of Colorado.* Springfield, MA: Bowles, 1867.

Howe, H. A., and L. L. Stingley. "The Great Fireball of December 7, 1900." In G. A. Miller, "Section A (Mathematics and Astronomy) of the American Association," *Science* 14 (1901): 393–403.

Howell, Edwin E. "Two New Meteorites." *Science* 27 (1908): 27–28.

Hussakof, L. "Note on a Palaeoniscid Fish from a Permian Formation in South Dakota." *American Journal of Science*, 4th ser., 41 (1916): 347–50.

Hussong, E. M. "Data as to Wind-Distribution of Seeds." *Publications of the Nebraska Academy of Sciences* 6 (1898): 253–57.

Hussong, E. M. "The Flora of Franklin County, Nebraska." *Publications of the Nebraska Academy of Sciences* 5 (1896): 9.

Hussong, E. M. "The Yellow Pine in the Republican Valley." *Publications of the Nebraska Academy of Sciences* 5 (1896): 7–8.

I. E. W. "Dedication of the University of Colorado Mountain Laboratory." *Science* 56 (1922): 162–63.

Iddings, J. P. "Louis Valentine Pirsson." *Science* 51 (1920): 530–32.

Iddings, Joseph P. "Biographical Memoir of Arnold Hague, 1840–1917." *National Academy of Sciences, Biographical Memoirs* 11 (1919): 21–38.

Instructions for Voluntary Observers of the Signal Service, United States Army. Washington, DC: Government Printing Office, 1882.

"International Phytogeographical Excursion (I.P.E.) in America, 1913." *New Phytologist* 13 (1914): 30–41.

"Introductory." *American Geologist* 1 (1888): 1–3.

Jackson, William Henry. *Time Exposure: The Autobiography of William Henry Jackson*. Albuquerque: University of New Mexico Press, 1986.

Jardine, William M. *Notes on Dry Farming*. Circular of the Bureau of Plant Industry, no. 10 (1908).

Johnson, John C. "A Brief History of the Rocky Mountain Biological Laboratory." *Colorado Magazine* 38 (1961): 81–103.

Jones, William A. *Report Upon the Reconnaissance of Northwestern Wyoming, Including Yellowstone National Park, Made in the Summer of 1873*. Washington, DC: Government Printing Office, 1875.

Judd, Neil M. *Men Met along the Trail: Adventures in Archaeology*. Norman: University of Oklahoma Press, 1968.

Judd, Neil M. "Report on Illegal Excavations in Southwestern Ruins." *American Anthropologist* 26 (1924): 428–32.

Kedzie, G. E. *Report of the State Geologist of the State of Colorado for the Years 1889 and 1890*. Denver: Collier and Cleaveland, 1890.

Keffer, Charles A. *Experimental Tree Planting in the Plains*. Bulletin of the Division of Forestry, no. 18 (1898).

Keyes, Charles R. "The Origin of the Great Plains." *Science* 34 (1911): 352.

Kiener, Walter, ed. "P. A. Rydberg's 'Report of a Botanical Exploration, Made in Western Nebraska in the Summer of 1891, as a Special Agent of the Department of Agriculture.'" *Nebraska History* 32 (1951): 217–39, 306–31.

Kilander, Ginny. "A Harvard Cook in the Wyoming Badlands: The 1908 Diary of Alcott Farrar Elwell." *Annals of Wyoming* 73, no. 3 (2000): 18–26.

Kimball, James P. "The Granites of Carbon County, Montana: A Division and Glacier Field of the Snowy Range." *Journal of the American Geographical Society of New York* 31 (1899): 199–216.

King, Clarence. "Report of Clarence King, Geologist in Charge of the Geological Exploration of the 40th Parallel." In *Annual Report of the Chief of Engineers to the Secretary of War for the Year 1871*, Appendix Z, 1027–30.

Kingsley, J. S. "The Hat Creek Bad Lands." *American Naturalist* 25 (1891): 963–71.

Knowlton, F. H. "A Dicotyledonous Flora in the Type Section of the Morrison Formation." *American Journal of Science*, 4th ser., 49 (1920): 189–94.

Knowlton, F. H. "The Fossil Flora of the Bozeman Coal Field." *Proceedings of the Biological Society of Washington* 7 (1892): 153–54.

Kohl, Michael F., Larry D. Martin, and Paul Brinkman, eds. *A Triceratops Hunt in Pioneer Wyoming: The Journals of Barnum Brown and J. P. Sams: The University of Kansas Expedition of 1895*. Glendo, WY: High Plains Press, 2004.

Kohl, Michael F., and John S. McIntosh, eds. *Discovering Dinosaurs in the Old West: The Field Journals of Arthur Lakes*. Washington, DC: Smithsonian Institution Press, 1997.

"The Laboratory in Modern Science." *Science* 3 (1884): 172–74.

Lakes, Arthur. "The Florissant Basin: A Prospecting Trip in the Volcanic Region between Cripple Creek and South Park." *Mines and Minerals* 20 (1899): 179–80.

Lamson-Scribner, F. *Cooperative Range Growth and Forage Plant Experiments at Highmore, S.Dak.* Circular of the Division of Agrostology, no. 21 (1899).

Langley, Samuel Pierpont. *The New Astronomy*. Boston: Ticknor, 1888.

Lankester, E. Ray. "An American Sea-Side Laboratory." *Nature*, March 25, 1880, 497–99.

Lapham, Macy. *Crisscross Trails: Narrative of a Soil Surveyor.* Berkeley, CA: Berg, 1949.

Lapham, Macy, and party. "Soil Survey of Western North Dakota." *Field Operations of the Bureau of Soils* 10 (1908): 1155–228.

Lapham, Macy H. *Soils of the San Luis Valley, Colorado.* Bulletin of the Bureau of Soils, no. 52 (1912).

Laws, Joint Resolutions and Memorials Passed by the Legislature of the State of Nebraska at the Twenty-Eighth Session. York: Nebraska Newspaper Union, 1903.

Lawson, William Pinkney. *The Log of a Timber Cruiser.* New York: Duffield, 1915.

LeConte, John L. "New Coleoptera Collected on the Survey for the Extension of the Union Pacific Railway, E. D. from Kansas to Fort Craig, New Mexico." *Transactions of the American Entomological Society* 2 (1868): 49–53.

LeConte, John L. *Notes on the Geology of the Survey for the Extension of the Union Pacific Railway*. Philadelphia: Review Printing House, 1868.

Lee, Willis T. "Type Section of the Morrison Formation." *American Journal of Science*, 4th ser., 49 (1920): 183–88.

Leidy, Joseph. "Notes of Remains of Extinct Vertebrata from the Valley of the Niobrara River." *Proceedings of the Academy of Natural Sciences* (March 1858): 20–29.

Leonard, A. G. "The Purpose and Value of a Geological Survey." *University Bulletin of the University of North Dakota* 2, no. 2 (1905): 7–11.

Leonard, Arthur Gray. "The Lignite Deposits of North Dakota." *Quarterly Journal of the University of North Dakota* 6 (1916): 234–40.

Lewis, Robert S. *Mine Sampling and the Commercial Value of Ores.* Bulletin of the Utah Engineering Experiment Station, no. 10 (1918).

Lincoln, F. C. "Notes on Some Species New to the Colorado List of Birds." *Auk* 35 (1918): 236–37.

Livingston, D. C., and F. B. Laney. *The Copper Deposits of the Seven Devils and Adjacent Districts (Including Heath, Hornet Creek, Hoodoo, and Deer Creek)*. Bulletin of the Idaho Bureau of Mines and Geology, no. 1 (1920).

Loomis, Frederic Brewster. *The Evolution of the Horse*. Boston: Marshall Jones, 1926.

Loper, S. Ward. "A Trip to the Tertiary Formations of Wyoming and Colorado." *Science* 8 (1898): 751–52.

Loud, F. H. "Summary of Meteorological Observations, January to June, 1904." *Colorado College Studies* 11 (1904): 54–76.

Lovewell, J. T. "Kansas Meteorology." *Biennial Report of the Kansas State Board of Agriculture* 4 (1883–84): 612–18.

Lovewell, J. T. "Kansas Weather Service." *Transactions of the Kansas Academy of Science* 7 (1879–80): 83–85.

Lowie, Robert H. "Notes on Shoshonean Ethnography." *Anthropological Papers of the American Museum of Natural History* 20 (1924): 185–312.

Lowie, Robert H. *Robert H. Lowie, Ethnologist: A Personal Record*. Berkeley: University of California Press, 1959.

Lubetkin, M. John, ed. *Custer and the 1873 Yellowstone Survey: A Documentary History*. Norman, OK: Arthur H. Clark, 2013.

Ludlow, William. *Report of a Reconnaissance from Carroll, Montana Territory, on the Upper Missouri, to the Yellowstone National Park, and Return, Made in the Summer of 1875*. Washington, DC: Government Printing Office, 1876.

Ludlow, William. *Report of a Reconnaissance of the Black Hills of Dakota, Made in the Summer of 1874*. Washington, DC: Government Printing Office, 1875.

Lull, Richard Swann. "Early Fossil Hunting in the Rocky Mountains." *Natural History* 26 (1926): 455–61.

Lunell, J. "Enumerantur Plantae Dakotae Septentrionalis Vasculares/The Vascular Plants of North Dakota I." *American Midland Naturalist* 4 (1915): 152–65.

Lunell, J. Review of *The Sixth Biennial Report of the Director of the Agricultural College Soil and Geological Survey of North Dakota to the Governor of North Dakota (1911–1912)*. *American Midland Naturalist* 6 (1919): 29–32.

Lyman, Richard R. *The Construction and Maintenance of Earth Roads*. Bulletin of the Utah Engineering Experiment Station, no. 3 (1910).

Lyman, Richard R. *Measurement of Flowing Streams: A Simple, Accurate Method of Using the Weir*. Bulletin of the Utah Engineering Experiment Station, no. 5 (1912).

Lyon, D. A., R. H. Bradford, S. S. Arentz, O. C. Ralston, and C. L. Larson. *Metallurgical Treatment of the Low-Grade and Complex Ores of Utah: A Preliminary Report*. Technical Paper of the US Bureau of Mines, no. 90 (1915): 5.

Lyon, Dorsey A., and Oliver C. Ralston. *Recovery of Zinc from Low-Grade and Complex Ores.* Bulletin of the US Bureau of Mines, no. 168 (1919).

Lyon, T. L. *Cooperative Variety Tests of Corn in 1902 and 1903.* Bulletin of the Nebraska Agricultural Experiment Station, no. 83 (1904).

Lyon, T. T., et al. *Catalogue of Fruits Recommended for Cultivation in the Various Sections of the United States by the American Pomological Society.* Bulletin of the Division of Pomology, no. 6 (1897).

M'Master, John B. "The 'Bad Lands' of Wyoming and Their Fossil Remains." *Journal of the American Geographical Society of New York* 12 (1880): 109–30.

MacDougal, D. T. "Can Research Work Be Accomplished in American Laboratories?" *Botanical Gazette* 22 (1896): 188–89.

MacDougal, D. T. "Report of Dr. D. T. MacDougal, First Assistant, on Explorations in Montana in 1901." *Journal of the New York Botanical Garden* 3 (1902): 8–13.

Main, Angie Kumlien. "Life and Letters of Edward Lee Greene." *Transactions of the Wisconsin Academy of Sciences, Arts, and Letters* 24 (1929): 147–85.

"The Making of Forecasts by Laymen." *Weather Bureau Topics and Personnel* (October 1915): 4–5.

"The Management of our Experiment Stations." *Garden and Forest* 3 (1890): 245–46.

Manning, Van. H. *Experiment Stations of the Bureau of Mines.* Bulletin of the US Bureau of Mines, no. 175 (1919).

"Maps of the Yellowstone Park." *Forest and Stream,* April 7, 1906, 547.

Marbut, Curtis F., Hugh H. Bennett, J. E. Lapham, and M. H. Lapham. *Soils of the United States.* Bulletin of the Bureau of Soils, no. 96 (1913).

Marsh, O. C. "On the Geology of the Eastern Uintah Mountains." *American Journal of Science and Arts,* 3rd ser., 1 (1871): 191–98.

Marsh, Othniel Charles. *Odontornithes: A Monograph on the Extinct Toothed Birds of North America.* King Survey Final Report/Monograph, no. 7 (1880).

Mason, C. C. "The Story of the Discovery and Early Exploration of the Cliff Houses at Mesa Verde." Mesa Verde Collection, #428, Colorado Historical Society, Denver. Reprinted in McNitt, *Richard Wetherill: Anasazi,* 323–29.

Mathews, Edward Bennett. "Relation of Scientific to Practical Work in State Surveys." *Economic Geology* 6 (1911): 181–87.

Matthew, W. D. "Early Days of Fossil Hunting in the High Plains." *Natural History* 26 (1926): 449–54.

Mayer, Alfred Goldsborough. "Should Our Colleges Establish Summer Schools?" *Science* 23 (1906): 703–4.

McClurg, Virginia. "The Making of Mesa Verde into a National Park." *Colorado Magazine* 7 (1930): 216–19.

Means, Thos. H. "A Reconnoissance in Sanpete, Cache, and Utah Counties, Utah." *Field Operations of the Bureau of Soils* 1 (1899): 115–20.

Means, Thos. H. "A Reconnoissance in the Cache a la Poudre Valley, Colorado." *Field Operations of the Bureau of Soils* 1 (1899): 121–24.

Memoranda of Plans for Arid Farm Investigations. Circular of the Utah Agricultural Experiment Station, no. 1 (1904): 2.

Menke, H. W. "From a Cabin Window." *Bird-Lore* 1 (1899): 14–16.

Merriam, C. Hart. "Biology in Our Colleges: A Plea for a Broader and More Liberal Biology." *Science* 21 (1893): 352–55.

Merrill, J. C. "Oölogical Notes from Montana." *Bulletin of the Nuttall Ornithological Club* 6 (1881): 203–7.

Merrill, Marlene Deahl, ed. *Yellowstone and the Great West: Journals, Letters, and Images from the 1871 Hayden Expedition.* Lincoln: University of Nebraska Press, 1999.

"Meteorological Observations on Pike's Peak." *Science* 15 (1890): 122–24.

Miller, Edwin C. "Daily Variation of Water and Dry Matter in the Leaves of Corn and the Sorghums." *Journal of Agricultural Research* 10 (1917): 11–46.

Minot, Charles S. "The Distribution of Vacations at American Universities." *Science* 15 (1902): 441–44.

Montana State Bureau of Mines and Metallurgy. Bulletin of the Montana Bureau of Mines and Geology, no. 1 (1919).

Montgomery, Thos H. "The Bird Course at the Marine Biological Laboratory, Woods Holl, Mass., during the Summer of 1900." *Bird-Lore* 2 (1900): 153–54.

Mooney, James. "The Cheyenne Indians." *Memoirs of the American Anthropological Association* 1 (1907): 357–442.

Mooney, James. "The Ghost-Dance Religion and the Sioux Outbreak of 1890." In *Fourteenth Annual Report of the Bureau of Ethnology to the Secretary of the Smithsonian Institution, 1892–1893,* 651–1110. Washington, DC: Government Printing Office, 1896.

Mooney, James. "The Indian Ghost Dance." *Collections of the Nebraska State Historical Society* 16 (1911): 168–86.

Moore, Willis L. *Report of the Chief of the Weather Bureau* (1896): 263.

Moorehead, Warren K., et al. "In Search of a Lost Race." *Illustrated American* 10–12 (1892).

Morrison, Chas. F. "A List of the Birds of Colorado." *Ornithologist and Oölogist* 13 (1888): 145–48, 165–68, 181–83.

"Mountain and Alpine Laboratories, Colorado." *Journal of Ecology* 1 (1913): 157.

Mudge, B. F. "Annual Report of the Committee of Geology, for the Year Ending November 1, 1876." *Transactions of the Kansas Academy of Science* 5 (1876): 4–5.

Mulford, A. Isabel. "Notes upon the Northwestern and Rocky Mountain Flora. I." *Botanical Gazette* 19 (1894): 117–20.

Neal, James C. *Oklahoma Weeds.* Bulletin of the Oklahoma Agricultural Experiment Station, no. 17 (1895).

"The Nebraska Summer School of Science." *Field and Forest* 3 (1877): 109–10.

Nelson, Aven. *First Report on the Flora of Wyoming.* Bulletin of the Wyoming Agricultural Experiment Station, no. 28 (1896).

Nelson, Aven. "Notes on Some Rare Wyoming Plants." *Erythea* 6 (1898): 52–57.

"A New Field School of Geology for Harvard University." *Science* 41 (1915): 237–38.

"A New Forest Reserve." *Forest and Stream,* April 9, 1891, 12.

"News Items." *Torreya* 13 (1913): 68–72.

Newberry, J. S. "Notes on the Geology and Botany of the Country Bordering the Northern Pacific Railroad." *Annals of the New York Academy of Science* 3, no. 8 (1884): 242–79.

Newbrough, W. "Notes on Preliminary and Location Railroad Surveying in the West." *School of Mines Quarterly* 10 (1888): 1–11.

Newcombe, F. C. "The Scope and Method of State Natural History Surveys." *Science* 37 (1913): 615–22.

Newell, F. H. "Mesa Verde." *National Geographic Magazine* 9 (1898): 431–34.

Newell, F. H., comp. *Proceedings of Second Conference of Engineers of the Reclamation Service.* Water-Supply and Irrigation Papers, no. 146 (1905).

Nicholson, H. H. "Chemical Laboratory of the University of Nebraska." *Science* 10 (1887): 82–84.

Noble, Bruce J. "A Legacy of Distrust: The Ute Mountain Utes and the Boundaries of Mesa Verde National Park." *Colorado Heritage* (Summer 1995): 32–42.

Nordenskiöld, Gustaf. *The Cliff Dwellers of the Mesa Verde.* Translated by D. Lloyd Morgan. Stockholm: Norstedt, 1893.

"Notes, News, and Comment." *Journal of the New York Botanical Garden* 2 (1901): 78–79, 142–44.

Nowell, Herbert T. "Report of the Irrigation Engineer." *Annual Report of the Wyoming Agricultural Experiment Station* 18 (1908): 67–85.

The Oklahoma Geological Survey: Its Origin, Scope and Purpose. Circular of the Oklahoma Geological Survey, no. 1 (1908).

Osborn, Henry Fairfield. "Fore and Hind Limbs of Carnivorous and Herbivorous Dinosaurs from the Jurassic of Wyoming." *Bulletin of the American Museum of Natural History* 12 (1899): 161–72.

Osborn, Henry Fairfield. "Fossil Wonders of the West: The Dinosaurs of the Bone Cabin Quarry." *Century Magazine* 68 (1904): 680–94.

Osborn, Henry Fairfield. "Hunting Prehistoric Animals." *Appleton's Magazine* 8 (1906): 552–54.

Osborn, Henry Fairfield. "A Skeleton of Diplodocus." *Memoirs of the American Museum of Natural History* 1 (1903): 189–214.

"Othniel Charles Marsh." *American Journal of Science* 4th ser., 7 (1899): 403–28.

P[reble], E[dward] A. "Vernon Bailey Passes." *Nature Magazine* 35 (1942): 329.

Pack, Herbert J. "*Coleonyx vareigatus* (Baird) in Utah." *Copeia*, no. 88 (1920): 101–2.

Packard, Alpheus. "Introductory." *American Naturalist* 1 (1867): 1–4.

"Paleontologist on Campus Tells of His Life as a Fossil Hunter." *State College Leader*, February 8, 1939, in "Scrapbook, 1926–1938," George Sternberg Collection, Archives Collections, Forsyth Library, Fort Hays State University, Fort Hays, Kansas.

Pankin, Mary Faith. "The Yale Scientific Expedition of 1871: A Student's-Eye View." *Oregon Historical Quarterly* 99 (1998–99): 374–435.

Parker, J. D. "Kansas Scientific Survey." *Kansas City Review of Science and Industry* 4 (1881): 640–43.

Parry, C. C. "The Rocky Mountain Alpine Region." *Proceedings of the American Association for the Advancement of Science* 18 (1869): 248–56.

Parsons, H. F., and Charles A. Liddell. "The Coal and Mineral Resources of Routt County." *Bulletin of the Colorado School of Mines* 1, no. 4 (1903): 47–59.

Patrick, G. E. "Analysis of Kansas Soils." *Transactions of the Kansas Academy of Science* 4 (1875): 15–16.

Patrick, G. E. "The Great Spirit Spring." *Transactions of the Kansas Academy of Science* 7 (1879–80): 22–26.

Patrick, G. E. "Kansas Chalk." *Transactions of the Kansas Academy of Science* 4 (1875): 13–15.

Patrick, G. E. "The Waconda Meteorite." *Transactions of the Kansas Academy of Science* 5 (1876): 12–13.

Patton, Horace B., et al. *Geology and Ore Deposits of the Alma District, Park County, Colorado.* Colorado Geological Survey Bulletin, no. 3 (1912).

Patton, Horace B., et al. *Geology of the Grayback Mining District, Costilla County, Colorado.* Colorado Geological Survey Bulletin, no. 2 (1910).

Peabody, P. B. "Review of 'Migration Records for Kansas Birds.'" *Wilson Bulletin* 32 (1920): 139–49.

Pearson, G. A. "The Administration of a Forest Experiment Station." *Forestry Quarterly* 12 (1914): 211–22.

Pearson, G. A. "Studies of Yield and Reproduction of Western Yellow Pine in Arizona and New Mexico." *Journal of Forestry* 16 (1918): 273–93.

Pennell, Francis W. "Report on a Collecting Trip to the Central Rocky Mountain Region." *Journal of the New York Botanical Garden* 17 (1916): 1–6.

Perisho, Ellwood C. *Biennial Report of the State Geologist [of South Dakota] for 1913–1914.*

"Personal and Scientific News." *American Geologist* 32 (1903): 60–64.

Peterson, Charles S. "Albert F. Potter's Wasatch Survey, 1902: A Beginning for Public Management of Natural Resources in Utah." *Utah Historical Quarterly* 39 (1971): 238–53.

The Petroleum of Salt Creek, Wyoming. Petroleum Series. Bulletin of the University of Wyoming, School of Mines, no. 1 (1896).

The Petroleum of the Shoshone Anticlinal. Petroleum Series. Bulletin of the University of Wyoming, School of Mines, no. 2 (1897).

Pilsbry, H. A., Lewis Woolman, and Philip P. Calvert. "Reduced Rates of Postage on Specimens of Natural History in the International Mails—An Appeal." *Science* 5 (1897): 402–3.

Pool, Raymond J. "Summer Botany for Forestry Students in the Mountains of Colorado." *Forest Club Annual (University of Nebraska)* 1 (1901): 21–34.

Popenoe, Edwin A. "A List of Kansas Coleoptera." *Transactions of the Kansas Academy of Science* 5 (1876): 21–40.

Pound, Roscoe. "The Progress of the Botanical Survey of Nebraska." *Publications of the Nebraska Academy of Sciences* 4 (1894): 7–8.

Powell, John Wesley. "Physiographic Regions of the United States." In *The Physiography of the United States*, 65–100. New York: American Book Company, 1896.

Powell, John Wesley. *Report of Special Commissioners on the Condition of the Ute Indians of Utah; the Paiutes of Utah, Northern Arizona, Southern Nevada, and Southeastern California; the Go-Si-Utes of Utah and Nevada; the Northwestern Shoshones of Idaho and Utah; and the Western Shoshones of Nevada; and Report Concerning Claims of Settlers in the Mo-a-Pa Valley, Southeastern Nevada.* Washington, DC: Government Printing Office, 1874.

Powell, John Wesley. *Report on the Geological and Geographical Survey of the Rocky Mountain Region.* Washington, DC: Government Printing Office, 1877.

"Preface." In *University Geological Survey of Kansas* 1 (1896), 1.

"Preliminary: The Plan and Scope of the Survey." *Botanical Survey of Nebraska* 1 (1892): 6–7.

"A Preliminary Report on the Nebraska State Museum." *Bulletin of the Nebraska State Museum* 1 (1924): 3–19.

Preston, H. L. "The Franceville (El Paso County, Colorado) Meteorite." *Journal of Geology* 10 (1902): 852–57.

Proceedings of the Nebraska Ornithologists' Union 2 (1901): 4–5.

"Proposed Biological Station." *Nautilus* 10 (1896): 22.

Prudden, T. Mitchell. "A Summer among Cliff Dwellings." *Harper's New Monthly Magazine* 93 (1896): 545–61.

Pugsley, C. W. *Alfalfa Inoculation Tests.* Bulletin of the Nebraska Agricultural Experiment Station, no. 130 (1913).

Ramaley, Francis. "The Amount of Bare Ground in Some Mountain Grasslands." *Botanical Gazette* 57 (1914): 526–28.

Ramaley, Francis. *Colorado Plant Life.* Boulder: University of Colorado, 1927.

Ramaley, Francis. "Quadrat Studies in a Mountain Grassland." *Botanical Gazette* 62 (1916): 70–74.

Ramaley, Francis. "The Relative Importance of Different Species in a Mountain Grassland." *Botanical Gazette* 60 (1915): 154–57.

Ramaley, Francis. "Subalpine Lake-Shore Vegetation in North-Central Colorado." *American Journal of Botany* 7 (1920): 57–74.

Ramaley, Francis. "The University of Colorado Mountain Laboratory." *University of Colorado Studies* 7 (1909): 91–95.

Ramaley, Francis. "What Is Biology and What Is a 'Biological Survey'?" *Science* 35 (1912): 60–61.

Ramaley, Francis. *Wild Flowers and Trees of Colorado.* Boulder, CO: Greenman, 1909.

Ramaley, Francis, and G. S. Dodds. "University of Colorado Mountain Laboratory." *University of Colorado Studies* 12 (1917): 53–64.

Ramaley, Francis, and W. W. Robbins. "A Summer Laboratory for Mountain Botany." *Plant World* 12 (1909): 105–10.

Ranyard, A. Cowper. "Account of Observations of the Total Solar Eclipse on the 29th July 1878 Made at Cherry Creek Camp near Denver, Colorado." *Royal Astronomical Society* 46 (1881): 213–39.

Raymond, Rossiter W. *Statistics of Mines and Mining of the States and Territories West of the Rocky Mountains.* Washington, DC: Government Printing Office, 1870.

Reed, E. L. "Meadow Vegetation in the Montane Region of Northern Colorado." *Bulletin of the Torrey Botanical Club* 44 (1917): 97–109.

Regulations of the U.S. Geological Survey, 1882.

Rehn, James A. G., and Morgan Hebard. "A Contribution to the Knowledge of the Orthoptera of Montana, Yellowstone Park, Utah and Colorado." *Proceedings of the Academy of Natural Sciences* 58 (1906): 358–418.

Rehn, James A. G., and Morgan Hebard. "An Orthopterical Reconnoissance of the Southwestern United States, Part II: New Mexico and Western Texas." *Proceedings of the Academy of Natural Sciences* 61 (1909): 111–75.

Reid, Harry Fielding. "Remarkable Earthquakes in Central New Mexico in 1906 and 1907." *Bulletin of the Seismological Society of America* 1 (1911): 10–16.

"Remarks on Fossil Vertebrates from Wyoming." *American Naturalist* 5 (1871): 664.

Report of the Board of Irrigation Survey and Experiment, for 1895 and 1896, to the Legislature of Kansas. Topeka: Kansas State Printing Company, 1897.

"Report of the Committee on American Archaeology." *American Journal of Archaeology* 12, supplement (1908): 44–54.

Report of the Historical and Natural History Society of the State of Colorado for the Years 1889–90. Denver: Collier & Cleaveland, 1890.

Report of the State Historical and Natural History Society of Colorado, with Articles of Incorporation and a List of Its Property. Denver: Collier & Cleaveland, January 1889.

Reverchon, J. "Botanizing in Texas, I." *Botanical Gazette* 11 (1886): 56–59, 211–16.

Rhoda, Franklin. "Report on the Topography of the San Juan Country." In *[Eighth] Annual Report of the U.S. Geological and Geographical Survey of the Territories, Embracing Colorado and Parts of Adjacent Territories; Being a Report of Progress of the Exploration for the Year 1874*, 451–96.

Rhoda, Franklin. "Topographical Report on the Southeastern District." In *Ninth Annual Report of the U.S. Geological and Geographical Survey of the Territories, Embracing Colorado and Parts of Adjacent Territories; Being a Report of Progress of the Exploration for the Year 1875*, 302–33.

Rice, T. D., and party. "Reconnaissance Survey of Western Nebraska." *Field Operations of the Bureau of Soils* 13 (1911): 1875–1989.

Ricketts, Louis D. *Annual Report of the Territorial Geologist to the Governor of Wyoming*. 1890.

Ridgway, Robert. "Mrs. Maxwell's Colorado Museum: Catalogue of the Birds." *Field and Forest* 2 (1877): 195–98, 208–14.

Ridgway, Robert. "Notes on the Bird-Fauna of the Salt Lake Valley and the Adjacent Portions of the Wahsatch Mountains." *Bulletin of the Essex Institute* 5 (1873): 168–73.

Ridgway, Robert. "*Scops flammeola* in Colorado." *Bulletin of the Nuttall Ornithological Club* 5 (1880): 185.

Riggs, E. S. "The Use of Pneumatic Tools in the Preparation of Fossils." *Science* 17 (1903): 747–49.

Riggs, Elmer S. "The Dinosaur Beds of the Grand River Valley of Colorado." *Geological Series of the Field Columbian Museum* 1 (1901): 267–74.

Riggs, Stephen Return. *A Dakota-English Dictionary*. Edited by James Owen Dorsey. Contributions to North American Ethnology 7. Washington, DC: Government Printing Office, 1890.

Riggs, Stephen Return. *Dakota Grammar, Texts, and Ethnography*. Edited by James Owen Dorsey. Contributions to North American Ethnology 9. Washington, DC: Government Printing Office, 1893.

Riggs, Stephen Return. "Journal of a Tour from Lac-Qui-Parle to the Missouri River." *South Dakota Historical Collections* 13 (1926): 330–44.

Robbins, W. W. "Succession of Vegetation at Boulder Park, Colorado." *Botanical Gazette* 65 (1918): 493–525.

Robert DeC. Ward. "Report of the Department of Geology and Geography." In *Annual Report of the Museum of Comparative Zoology*, 20. 1908–09.

The Rocky Mountain Biological Station: 5th Annual Season. Bulletin of the Western State College of Colorado, Gunnison (1926).

Rodeck, Hugo G. "Science Lodge." *Biologist* 18 (1937): 101–4.

Rotch, A. Lawrence. "The Mountain Meteorological Stations of the United States." *American Meteorological Journal* 8 (1892): 396–405.

Rydberg, P. A. "Plant Geography of North America." *Science* 12 (1900): 870–73.

Rydberg, Per A. *Flora of the Rocky Mountains and Adjacent Plains.* Lancaster, PA: Steinman & Foltz, 1917.

Salmon, Samuel C. *The Odyssey of an Agricultural Scientist: The Memoirs of Samuel C. Salmon.* Hyattsville, MD: Salmon, n.d.

Savage, J[oseph]. "Stone Implements in Trego County." *Transactions of the Kansas Academy of Science* 7 (1879–80): 57–58.

Savage, Joseph. "On the Bite of the Rattlesnake." *Transactions of the Kansas Academy of Science* 6 (1877–78): 36–38.

Scheffer, Theo. H. *Prairie-Dog Situation.* Circular of the Kansas Agricultural Experiment Station, no. 4 (1909).

Schuchert, Charles. "The Fossil Field's Expedition to Wyoming." *Science* 10 (1899): 725–28.

"Science Summer Camp." *Biologist* 18 (1937): 183–89.

Scott, Charles A. "The Early Days: The Development of the Tree Planting Project in the Nebraska Sand Hills." In *The Early Days: The Dismal River and Niobrara Forest Reserve.* USDA Forest Service, 2002.

Scott, William Berryman. *Some Memories of a Paleontologist.* Princeton, NJ: Princeton University Press, 1939.

Scudder, Samuel H. "The Insects of the Triassic Beds at Fairplay, Colorado." In *The Pretertiary Insects of North America, Including Critical Remarks on and Description of, Some European Forms,* edited by Samuel H. Scudder, 433–48. New York: Macmillan, 1890.

Seaver, Fred J. "Collecting Fungi in Colorado." *Journal of the New York Botanical Garden* 8 (1907): 247–55.

Seaver, Fred J. "Conference Notes." *Journal of the New York Botanical Garden* 10 (1909): 118–21.

"The Second Decennial of the Botanical Seminar of the University of Nebraska." *Science* 24 (1906): 629–31.

Shaler, Nathaniel Southgate, and Sophia P. Shaler. *The Autobiography of Nathaniel Southgate Shaler, with a Supplementary Memoir by His Wife.* Boston: Houghton Mifflin, 1909.

Shannon, C. W. *Director's Biennial Report to the Governor of Oklahoma, 1914.* Bulletin of the Oklahoma Geological Survey, no. 22 (1914).

Shear, Cornelius L. *Field Work of the Division of Agrostology: A Review and Summary of the Work Done since the Organization of the Division, July 1, 1895.* Bulletin of the Division of Agrostology, no. 25 (1901).

Sheldon, Addison E. "An Historical Journey to the Northwest." *Nebraska History Magazine* 13 (1932): 68–80.

Shepperd, J. H., and W. B. Richards. *Report of Four Years Work of the North Dakota Holstein Cattle Breeding Circuit.* Bulletin of the North Dakota Agricultural Experiment Station, no. 109 (1914).

Simpson, Charles T. "Blacklist Them." *Conchologists' Exchange* 2 (1887): 10.

Simpson, Charles T. "Succinea Lineata. W. G. B. in Nebraska." *Conchologists' Exchange* 1 (1887): 65.

Smith, Chas. E. "Geological Field Work at the Colorado School of Mines." *Quarterly of the Colorado School of Mines* 2, no. 2 (1907): 15–19.

Smith, H. S. S. "On the Magnetic Declination in Kansas for October, 1880." *Transactions of the Kansas Academy of Science* 7 (1879–80): 46–48.

Snow, F. H. "Douglas County Additions to the List of Kansas Coleoptera in 1879 and 1880." *Transactions of the Kansas Academy of Science* 7 (1879–80): 75–76.

Snow, F. H. "Hunting Ambylchila." *American Naturalist* 11 (1877): 731–35.

Snow, F. H. "The Insects of Wallace County, Kansas." *Transactions of the Kansas Academy of Science* 6 (1877–78): 61–70.

Snow, F. H. "List of Coleoptera Collected near 'Dome Rock,' Platte Cañon, Colorado, by the Kansas University Scientific Expedition for 1878." *Transactions of the Kansas Academy of Science* 6 (1877–78): 75–77.

Snow, F. H. "On the Dermal Covering of a Mosasauroid Reptile." *Transactions of the Kansas Academy of Science* 6 (1877–78): 54–58.

Snow, Frank H. "Amblychila cylindriformis Say." *Transactions of the Kansas Academy of Science* 6 (1877–78): 30–32.

Snow, Frank H. "Catalogue of the Birds of Kansas." *Transactions of the Kansas Academy of Science* 1 (1872): 21–29.

Snow, Francis H. "Work in Economic Entomology at the University of Kansas for the Season of 1894." *Insect Life* 7 (1895): 140–44.

Southworth, Alvan S. "The New State of Colorado." *Journal of the American Geographical Society of New York* 6 (1874): 260–85.

Spafford, Russell R. "The Effect of Climate and Soil upon Agriculture." *University Studies of the University of Nebraska* 16 (1916): 91–113.

"The State Forestry Station." *Transactions of the Kansas State Horticultural Society* 27 (1903): 247.

Stearns, J. C. "The Mount Evans Laboratory." *Scientific Monthly* 46 (1938): 242–48.

Sterling, Everett W. "The Powell Irrigation Survey, 1888–1893." *Mississippi Valley Historical Review* 27 (1940): 421–34.

Sternberg, Charles H. *Hunting Dinosaurs in the Bad Lands of the Red Deer River, Alberta, Canada.* Lawrence, KS: Sternberg, 1917.

Sternberg, Charles H. *The Life of a Fossil Hunter.* New York: Holt, 1909.

Stewart, George. "Can the Farms of the United States Pay for Themselves?" *Journal of Farm Economics* 2 (1920): 177–93.

Strecker, John H. *Notes on the Fauna of Northwestern Texas,* Baylor University Bulletin, Vol. 13, nos. 4–5. Waco, TX: Baylor University, 1910.

Taylor, Walter P. *Suggestions for Field Studies of Mammalian Life-Histories.* (Department) Circular of the US Department of Agriculture, no. 59 (1919): 2.

Thompson, A. H. "The Relation of a State Geological Survey to the Work of the National Survey." *Transactions of the Kansas Academy of Science* 10 (1886): 9–13.

Tight, W. G. "Clarence Luther Herrick." *American Geologist* 36 (1905): 1–26.

Todd, James E. *The First and Second Biennial Reports on the Geology of South Dakota, 1893–6.* Bulletin of the South Dakota Geological Survey no. 2 (1898).

Toumey, J. W. *Practical Tree Planting in Operation.* Bulletin of the Division of Forestry, no. 27 (1900).

Trostler, I. S. "History of Ornithology in Nebraska, and of State Ornithological Societies in General." *Proceedings of the Nebraska Ornithologists' Union* 2 (1901): 13–18.

Troxell, Edward L. "Fossil Hunting in Texas." *Scientific Monthly* 4 (1917): 81–89.

Troxell, Edward L. "A Fossil Ruminant from Rock Creek, Texas, Preptoceras mayfieldi sp nov." *American Journal of Science,* 4th ser., 40 (1915): 479–82.

True, A. C., and E. W. Allen. "Work and Expenditures of the Agricultural Experiment Stations." In *Annual Report of the Office of Experiment Stations,* 49–170. 1906.

True, A. C., and V. A. Clark. *The Agricultural Experiment Stations in the United States.* Bulletin of the Office of Experiment Stations, no. 80 (1900).

Trumbull, L. W. "Atlantic City Gold Mining District, Fremont County." *Bulletin of the State of Wyoming Geologist's Office, Series B,* no. 7 (1914): 69–100.

Trumbull, Loyal Wingate. *A Manual of Underground Surveying.* New York: McGraw-Hill, 1910.

Tryon, George W. "Notes on Mollusca Collected by Dr. F. V. Hayden in Nebraska." *American Journal of Conchology* 4 (1868): 150–51.

Udden, J. A. "A Texas Meteor." *Science* 46 (1917): 616–17.

Umpleby, J. B., and D. C. Livingston. *A Reconnaissance in South Central Idaho Embracing the Thunder Mountain, Big Creek, Stanley Basin, Sheep Mountain, and Seafoam Districts.* Bulletin of the Idaho Bureau of Mines and Geology, no. 3 (1920).

"University of Montana Biological Station at Flathead Lake: Announcement for the Summer of 1904." *Bulletin of the University of Montana* 22 (1904): 331–42.

US Naval Observatory. *Reports on the Total Solar Eclipses of July 29, 1878, and January 11, 1880.* Washington, DC: Government Printing Office, 1880.

USDA Bureau of Soils. *Instructions to Field Parties and Descriptions of Soil Types: Field Season, 1904.* Washington, DC: Government Printing Office, 1904.

USDA Bureau of Soils. *Soil Survey Field Book.* Washington, DC: Government Printing Office, 1906.

Vasey, Geo., and B. T. Galloway. *A Record of Some of the Work of the Division, Including Extracts from Correspondence and Other Communications.* Bulletin of the Botanical Division, no. 8 (1990).

Vasey, George. "Report of the Botanist." *Report of the Commissioner of Agriculture* (1881): 231–55.

Vasey, George. "Report of the Botanist." *Report of the Commissioner of Agriculture* (1886): 69–93.

Vaughan, Richard. "'Broad Are Nebraska's Rolling Plains': The Early Writings of George Bird Grinnell." *Nebraska History* 83 (2002): 36–49.

Vinall, H. N. *Forage Crops for the Sand-Hill Section of Nebraska.* Circular of the Bureau of Plant Industry, no. 80 (1911).

Visher, S. S. "Report of the State Geologist, 1916–1918." Bulletin of the South Dakota Geological and Natural History Survey, no. 8 (1918): 179–89.

Visher, Stephen S. "The Biogeography of the Northern Great Plains." *Geographical Review* 2 (1917): 89–115.

W[ilder], B[urt] G. "The Penikese School." *Nation,* July 8, 1880, 29.

Walker, James R. *Lakota Belief and Ritual.* Lincoln: University of Nebraska Press, 1980.

Wallace, Alfred R. "English and American Flowers II: Flowers and Forests of the Far West." *Fortnightly Review* 56 (1891): 796–810.

Ward, Freeman. *The Needs of the State Geological and Natural History Survey,* Circular of the South Dakota Geological and Natural History Survey, no. 7 (1920).

Ward, Freeman. *Report of the State Geologist [of South Dakota], 1918–1920.*

Ward, Lester F. "Description of a New Genus and Twenty New Species of Fossil Cycadean Trunks from the Jurassic of Wyoming." *Proceedings of the Washington Academy of Sciences* 1 (1900): 253–300.

Warner, H. C. "Should We Have County Experiment Stations." *Annual Report of the South Dakota State Horticultural Society* 4 (1907): 228–32.

Washburn, F. L. "The Relations of the Station Entomologist to His Environment." *Journal of Economic Entomology* 5 (1912): 33–53.

Weber, William A., ed. *The American Cockerell: A Naturalist's Life, 1866–1948.* Boulder: University Press of Colorado, 2000.

Weber, William A, ed. *Theodore D. A. Cockerell: Letters from West Cliff, Colorado, 1887–1889.* Boulder: Colorado Associated University Press, 1976.

Weed, Walter Harvey. *Geology and Ore Deposits of the Butte District, Montana.* US Geological Survey Professional Paper, no. 74 (1912).

Weinzirl, John. "The Availability of New Mexico's Climate for Outdoor Life." *Bulletin of the University of New Mexico, Biological Series* 2, no. 6 (1903): 9.

Welch, William H. "The Evolution of Modern Scientific Laboratories." In *Annual Report of the Board of Regents of the Smithsonian Institution,* 493–504. 1895.

Wetmore, Alexander. "Observations on the Habits of Birds at Lake Burford, New Mexico." *Auk* 37 (1920): 221–47, 393–412.

Wheeler, Geo. M. "Western Exploration." *Journal of the American Geographical Society of New York* 6 (1874): 233–52.

Wheeler, William Morton. "The Mountain Ants of Western North America." *Proceedings of the American Academy of Arts and Sciences* 52 (1917): 455–569.

Whitney, J. D. "Geographical and Geological Surveys." *North American Review* 121 (1875): 37–85, 270–314.

Whitney, Milton. "General Review of the Work." *Field Operations of the Bureau of Soils* 4 (1902): 25–123.

Whitney, Milton. "General Review of the Work." *Field Operations of the Bureau of Soils* 11 (1909): 35–40.

Whitney, Milton. *The Work of the Bureau of Soils.* Circular of the Bureau of Soils, no. 13 (1904).

Wieland, G. R. *American Fossil Cycads, Volume II.* Washington, DC: Carnegie Institution, 1916.

Wilcox, E. Mead. *Directions for Collecting and Preserving Insects and Plants.* Circular of Information of the Oklahoma Agricultural Experiment Station, no. 3 (1901).

Wilder, Burt G. "Agassiz at Penikese." *American Naturalist* 32 (1898): 189–96.

Will, George F., and George E. Hyde. *Corn among the Indians of the Upper Missouri.* St. Louis, MO: William Harvey Miner Co., 1917.

Willey, Day Allen. "Bone Cabin Quarry." *Appleton's Magazine* 8 (1906): 554–59.

"William Ashbrook Kellerman." *Journal of Mycology* 14 (1908): 49–63.

Williston, S. W. "Restoration of Aceratherium fossiger Cope." *Kansas University Quarterly* 2 (1894): 289–90.

Williston, Samuel W. "The Progress of Science since the Foundation of the Kansas Academy of Science." *Transactions of the Kansas Academy of Science* 29 (1918): 53–61.

Wilson, E. H. "A Visit to the Zeolite Locality at North Table Mountain, Colorado." *American Mineralogist* 2 (1917): 29–30.

Wilson, Edmund B. "Aims and Methods of Study in Natural History." *Science* 13 (1901): 14–23.

Wilson, Gilbert L. *Buffalo Bird Woman's Garden: Agriculture of the Hidatsa Indians.* St. Paul: Minnesota Historical Society Press, 1987.

Wissler, Clark. *The American Indian: An Introduction to the Anthropology of the New World.* New York: McMurtrie, 1917.

Wissler, Clark. "The Distribution and Functions of Tribal Societies among the Plains Indians: A Preliminary Report." *Proceedings of the National Academy of Sciences* 1 (1915): 401–3.

Wissler, Clark. "Measurements of Dakota Indian Children." *Annals of the New York Academy of Science* 20 (1911): 355–64.

Wissler, Clark. *North American Indians of the Plains.* New York: American Museum of Natural History, 1912.

Wissler, Clark, ed. *Societies of the Plains Indians.* Anthropological Papers of the American Museum of Natural History, vol. 11 (1916).

Wistar, Isaac J., and Edw. J. Nolan. "Postage on Natural History Specimens." *Science* 22 (1893): 267.

Wolbach, S. B. "Rocky Mountain Spotted Fever, Pathology and Etiology; Progress." *Biennial Report of the Montana State Board of Entomology* 3 (1917–18): 55–60.

Wolcott, R. H. "On Migration Records and On Our Nebraska Records." *Proceedings of the Nebraska Ornithologists' Union* 2 (1901): 86–93.

Wolcott, R. H. "Suggestions as to an Accurate and Uniform Method of Recording Observations." *Proceedings of the Nebraska Ornithologists' Union* 1 (1900): 41–44.

Wolcott, Robert H. "An Analysis of Nebraska's Bird Fauna." *Proceedings of the Nebraska Ornithologists' Union* 4 (1908): 25–55.

Wolcott, Robert H. "A Faunal Survey of the Forest Reserves in the Sandhill Region of Nebraska and of the Lakes in That Region." *Science* 21 (1905): 791–92.

Worthen, E. L. "Methods of Soil Surveying." *Proceedings of the American Society of Agronomy* 1 (1909): 185–91.

Wortman, J. L. "Restoration of *Oxyaena lupine* Cope, with Descriptions of Certain New Species of Eocene Creodonts." *Bulletin of the American Museum of Natural History* 12 (1899): 139–48.

Yarrow, H. C. "On the Exploration of some Indian Graves in Utah." *Field and Forest* 2 (1877): 185–88.

Zahniser, Howard. "Vernon Orlando Bailey." *Science* 96 (1942): 6–7.

Zimmer, John T. "Some Notes on the Winter Birds of Dawes County." *Proceedings of the Nebraska Ornithologists' Union* 5 (1911): 19–30.

SECONDARY SOURCES

Adler, Antony. "The Ship as Laboratory: Making Space for Field Science at Sea." *Journal of the History of Biology* 47 (2014): 333–62.

Alagona, Peter S. *After the Grizzly: Endangered Species and the Politics of Place in California.* Berkeley: University of California Press, 2013.

Alberti, Samuel J. M. M. "Amateurs and Professionals in One County: Biology and Natural History in Late Victorian Yorkshire." *Journal of the History of Biology* 34, no. 1 (2001): 115–47.

Alberti, Samuel J. M. M. "Field, Lab, and Museum: The Practice and Place of Life Science in Yorkshire, 1870–1904." PhD diss., University of Sheffield/ University of Leeds, 2000.

Alberti, Samuel J. M. M. "Placing Nature: Natural History Collections and Their Owners in Nineteenth-Century Provincial England." *British Journal for the History of Science* 35 (2002): 291–311.

Albright, George Leslie. *Official Explorations for Pacific Railroads, 1853–1855.* Berkeley: University of California Press, 1921.

Alexander, Thomas G. "John Wesley Powell, the Irrigation Survey, and the Inauguration of the Second Phase of Irrigation Development in Utah." *Utah Historical Quarterly* 37 (1969): 190–206.

Alexander, Thomas G. "Stewardship and Enterprise: The LDS Church and the Wasatch Oasis Environment, 1847–1930." *Western Historical Quarterly* 25, no. 3 (1994): 340–64.

Allen, D. E. *The Naturalist in Britain: A Social History.* London: Allen Lane, 1976.

Allen, John L. "The Garden-Desert Continuum: Competing Views of the Great Plains in the Nineteenth Century." *Great Plains Quarterly* 5, no. 4 (1985): 207–20.

Allen, John L. "Patterns of Promise: Mapping the Plains and Prairies, 1800–1860." *Great Plains Quarterly* 4, no. 1 (1984): 5–28.

Andrews, Thomas G. *Killing for Coal: America's Deadliest Labor War.* Cambridge, MA: Harvard University Press, 2008.

Arnold, Lois B. "The Barbours: A Family in Paleontology." *Nebraska History* 94, no. 4 (2013): 176–87.

Aron, Cindy Sondik. *Ladies and Gentlemen of the Civil Service: Middle-Class Workers in Victorian America.* New York: Oxford University Press, 1987.

Aron, Cindy Sondik. *Working at Play: A History of Vacations in the United States.* New York: Oxford University Press, 1999.

Arrington, Leonard J., and Dean May. "'A Different Mode of Life': Irrigation and Society in Nineteenth-Century Utah." *Agricultural History* 49, no. 1 (1975): 3–20.

Aubin, David. "The Fading Star of the Paris Observatory in the Nineteenth Century: Astronomers' Urban Culture of Circulation and Observation." *Osiris* 18 (2003): 79–100.

Aubin, David, Charlotte Bigg, and H. Otto Sibum, eds. *The Heavens on Earth: Observatories and Astronomy in Nineteenth-Century Science and Culture.* Durham, NC: Duke University Press, 2010.

Axelrod, Daniel I. "Rise of the Grassland Biome, Central North America." *Botanical Review* 51, no. 2 (1985): 163–201.

Baltensberger, Bradley H. "Agricultural Adjustments to Great Plains Drought: The Republican Valley, 1870–1900." In *The Great Plains: Environment and Culture,* edited by Brian W. Blouet and Frederick C. Luebke, 43–59. Lincoln: University of Nebraska Press, 1979.

Bamforth, Douglas B. *Ecology and Human Organization on the Great Plains.* New York: Plenum Press, 1988.

Barley, Stephen R., and Becky A. Bechky. "In the Backrooms of Science: The Work of Technicians in Science Labs." *Work and Occupations* 21, no. 1 (1994): 85–126.

Barnes, Barry. "Elusive Memories of Technoscience." *Perspectives on Science* 13, no. 2 (2005): 142–165.

Barrow, Mark V., Jr. *A Passion for Birds: American Ornithology after Audubon.* Princeton, NJ: Princeton University Press, 1998.

Barrow, Mark V., Jr. "The Specimen Dealer: Entrepreneurial Natural History in America's Gilded Age." *Journal of the History of Biology* 33, no. 3 (2000): 493–534.

Bartlett, Richard A. *Great Surveys of the American West.* Norman: University of Oklahoma Press, 1962.

Barton, Ruth. "'Men of Science': Language, Identity and Professionalization in the Mid-Victorian Scientific Community." *History of Science* 41, no. 1 (2003): 73–119.

Benson, Etienne. *Wired Wilderness: Technologies of Tracking and the Making of Modern Wildlife.* Baltimore, MD: Johns Hopkins University Press, 2010.

Benson, Keith R. "Laboratories on the New England Shore: The 'Somewhat Different Direction' of American Marine Biology." *New England Quarterly* 61, no. 1 (1988): 55–78.

Benson, Keith R. "Summer Camp, Seaside Station, and Marine Laboratory: Marine Biology and Its Institutional Identity." *Historical Studies in the Physical Sciences* 32, no. 1 (2001): 11–18.

Benson, Maxine. *Martha Maxwell: Rocky Mountain Naturalist.* Lincoln: University of Nebraska Press, 1986.

Bigg, Charlotte. "Photography and the Labour History of Astronomy: The *Carte du Ciel.*" In *The Role of Visual Representations in Astronomy: History and Research*

Practice, edited by Klaus Hentschel and Axel D. Wittman, 90–106. Frankfurt am Main: Harri Deutsch, 2000.

Blakeslee, Donald J. *Holy Ground, Healing Water: Cultural Landscapes at Waconda Lake, Kansas*. College Station: Texas A&M University Press, 2010.

Bledstein, Burton J. *The Culture of Professionalism: The Middle Class and the Development of Higher Education in America*. New York: Norton, 1976.

Blondheim, Menahem. *News over the Wires: The Telegraph and the Flow of Public Information in America, 1844–1897*. Cambridge, MA: Harvard University Press, 1994.

Blouet, Brian W., and Merlin P. Lawson, eds. *Images of the Plains: The Role of Human Nature in Settlement*. Lincoln: University of Nebraska Press, 1975.

Blouet, Brian W., and Frederick C. Luebke, eds. *The Great Plains: Environment and Culture*. Lincoln: University of Nebraska Press, 1979.

Bluemle, John P. *The First 100 Years: The History of the North Dakota Geological Survey, 1895–1995*. North Dakota Geological Survey Miscellaneous Series 81. [Bismarck, ND]: North Dakota Geological Survey, 1996.

Borchert, John R. "The Climate of the Central North American Grassland." *Annals of the Association of American Geographers* 40, no. 1 (1950): 1–39.

Bourguet, Marie-Noëlle. "Landscape with Numbers: Natural History, Travel and Instruments in the Late Eighteenth and Early Nineteenth Centuries." In *Instruments, Travel and Science: Itineraries of Precision from the Seventeenth to the Twentieth Century*, edited by Marie-Noëlle Bourguet, Christian Licoppe, and H. Otto Sibum, 96–125. London: Routledge, 2002.

Bourguet, Marie-Noëlle, Christian Licoppe, and H. Otto Sibum, eds. *Instruments, Travel and Science: Itineraries of Precision from the Seventeenth to the Twentieth Century*. London: Routledge, 2002.

Bowden, Martyn J. "The Great American Desert in the American Mind: The Historiography of a Geographical Notion." In *Geographies of the Mind: Essays in Historical Geosophy in Honor of John Kirtland Wright*, edited by David Lowenthal and Martyn J. Bowden, 119–47. New York: Oxford University Press, 1976.

Bowen, Marshall. "Environmental Perception and Geographic Change in Southwest Sheridan County." *Nebraska History* 51 (1970): 319–38.

Bozell, John R. "Culture, Environment, and Bison Populations on the Late Prehistoric and Early Historic Central Plains." *Plains Anthropologist* 40, no. 152 (1995): 145–63.

Braun, Bruce. "Producing Vertical Territory: Geology and Governmentality in Late Victorian Canada." *Ecumene* 7, no. 1 (2000): 7–46.

Brady, Erika. *A Spiral Way: How the Phonograph Changed Ethnography*. Jackson: University Press of Mississippi, 1999.

Breithaupt, Brent H. "Biography of William Harlow Reed: The Story of a Frontier Fossil Collector." *Earth Sciences History* 9, no. 1 (1990): 6–13.

Breternitz, David A. "Mesa Verde National Park: A History of Its Archaeology." *Essays in Colorado History* 2 (1983): 221–34.

Brinkman, Paul D. "Dinosaurs, Museums, and the Modernization of American Fossil Preparation at the Turn of the 20th Century." In *Methods in Fossil Preparation: Proceedings of the First Annual Fossil Preparation and Collections Symposium*, edited by Matthew A. Brown, John F. Kane, and William G. Parker, 21–34. http://preparation.paleo.amnh.org/assets/FPCSvolume-Final.pdf.

Brinkman, Paul D. "Henry Fairfield Osborn and Jurassic Dinosaur Reconnaissance in the San Juan Basin, along the Colorado-Utah Border, 1893–1900." *Earth Sciences History* 24, no. 2 (2005): 159–74.

Brinkman, Paul D. *The Second Jurassic Dinosaur Rush: Museums and Paleontology in America at the Turn of the Twentieth Century*. Chicago: University of Chicago Press, 2010.

Brock, Emily K. *Money Trees: The Douglas Fir and American Forestry, 1900–1944*. Corvallis: Oregon State University Press, 2015.

Brockway, Lucile H. *Science and Colonial Expansion: The Role of the British Royal Botanic Garden*. New York: Academic Press, 1979.

Brodhead, Michael J. "Of Mice and Mastodons: Contributions to the Literature of Mammalogy by Officers and Men of the United States Army in the Nineteenth Century." *Archives of Natural History* 18, no. 3 (1991): 363–74.

Brosnan, Kathleen A. *Uniting Mountain and Plain: Cities, Law, and Environmental Change along the Front Range*. Albuquerque: University of New Mexico Press, 2002.

Browne, Janet. "Natural History." In *The Oxford Companion to the History of Modern Science*, ed. J. L. Heilbron, 559–63. Oxford: Oxford University Press, 2003.

Brunhouse, Robert L. *Sylvanus G. Morley and the World of the Ancient Mayas*. Norman: University of Oklahoma Press, 1971.

Brunswig, Robert H. "Prehistoric Utilization of Patch Environments and Culture Change in Colorado's Central High Plains." In *Islands on the Plains: Ecological, Social, and Ritual Use of Landscapes*, edited by Marcel Kornfield and Alan J. Osborn, 44–64. Salt Lake City: University of Utah Press, 2003.

Bryans, William. *A History of the Geological Survey of Wyoming*. Bulletin 65. Laramie: Geological Survey of Wyoming, 1986.

Bryson, Reid A. "Chinook Climates and Plains People." *Great Plains Quarterly* 1 (1981): 5–15.

Buchanan, Rex C. *'To Bring Together, Correlate, and Preserve': A History of the Kansas Geological Survey, 1864–1989*. Bulletin 227. Lawrence: Kansas Geological Survey, 1989.

Burkhardt, Richard W. "Ethology, Natural History, the Life Sciences, and the Problem of Place." *Journal of the History of Biology* 32, no. 3 (1999): 489–508.

Burnett, D. Graham. *Masters of All They Surveyed: Exploration, Geography, and a British El Dorado.* Chicago: University of Chicago Press, 2000.

Burnham, Philip. *Indian Country, God's Country: Native Americans and the National Parks.* Washington, DC: Island Press, 2000.

Burns, J. Conor. "Negotiating the Agricultural Frontier in Nineteenth-Century Southern Ohio Archaeology." In *Knowing Global Environments: New Historical Perspectives on the Field Sciences,* edited by Jeremy Vetter, 59–86. New Brunswick, NJ: Rutgers University Press, 2011.

Burns, J. Conor. "Networking Ohio Valley Archaeology in the 1880s: The Social Dynamics of Peabody and Smithsonian Centralization." *Histories of Anthropology Annual* 4, no. 1 (2008): 1–33.

Cahan, David. "The Geopolitics and Architectural Design of a Metrological Laboratory: The Physikalisch-Technische Reichsanstalt in Imperial Germany." In *The Development of the Laboratory: Essays on the Place of Experiment in Industrial Civilization,* edited by Frank A. J. L. James, 137–54. New York: American Institute of Physics, 1989.

Cahan, David. *An Institute for an Empire: The Physikalisch-Technische Reichsanstalt 1871–1918.* Cambridge: Cambridge University Press, 1989.

Cahan, David. "The Institutional Revolution in German Physics, 1865–1914." *Historical Studies in the Physical Sciences* 15 (1985): 1–65.

Cahan, David, and M. Eugene Rudd. *Science at the American Frontier: A Biography of Dewitt Bristol Brace.* Lincoln: University of Nebraska Press, 2000.

Call, Leland E. *Fifty Years of Research at the Colby (Kansas) Branch Experiment Station, 1914–1963.* Kansas Agricultural Experiment Station Bulletin, no. 468 (1964).

Call, Leland E. *A History of the Tribune (Kansas) Branch Experiment Station, 1911–1964.* Kansas Agricultural Experiment Station Contribution, no. 112 (1965).

Call, Leland E. *Off-Campus Research by the Kansas Agricultural Experiment Station.* Kansas Agricultural Experiment Station Contribution, no. 115 (1965).

Call, Leland E., and Louis C. Aicher. *A History of the Fort Hays (Kansas) Branch Experiment Station 1901–1962.* Kansas Agricultural Experiment Station Bulletin, no. 453 (1963).

Cameron, Jenks. *The Bureau of Biological Survey: Its History, Activities, and Organization,* Service Monographs of the US Government, no. 54. Baltimore, MD: Johns Hopkins Press, 1929.

Cameron, Jenks. *The National Park Service: Its History, Activities, and Organization,* Service Monographs of the US Government, no. 11. New York: Appleton, 1922.

Campbell, C. Lee, Paul D. Peterson, and Clay S. Griffith. *The Formative Years of Plant Pathology in the United States*. St. Paul, MN: American Phytopathological Society, 1999.

Carey, Daniel. "Compiling Nature's History: Travellers and Travel Narratives in the Early Royal Society." *Annals of Science* 54, no. 3 (1997): 269–92.

Carlson, Paul H. *Deep Time and the Texas High Plains: History and Geology*. Lubbock: Texas Tech University Press, 2005.

Carstensen, Vernon. "The Genesis of an Agricultural Experiment Station." *Agricultural History* 34, no. 1 (1960): 13–20.

Cassidy, James G. *Ferdinand V. Hayden: Entrepreneur of Science*. Lincoln: University of Nebraska Press, 2000.

Charvolin, Florian. "Une science citoyenne? Le programme Feederwatch et la politique des grands nombres." *Revue Développement Durable et Territoires* 19 (2004).

Charvolin, Florian, André Micoud, and Lynn K. Nyhart, eds. *Des Sciences Citoyennes?: La question de l'amateur dans les sciences naturalistes*. La Tour-d'Aigues: Éditions de l'Aube, 2010.

Checkovich, Alex. "Mapping the American Way: Geographical Knowledge and the Development of the United States, 1890–1950." PhD diss., University of Pennsylvania, 2004.

Chure, Daniel J., and John S. McIntosh. "Stranger in a Strange Land: A Brief History of the Paleontological Operations at Dinosaur National Monument." *Earth Sciences History* 9, no. 1 (1990): 34–40.

Cittadino, Eugene. *Nature as the Laboratory: Darwinian Plant Ecology in the German Empire, 1880–1900*. Cambridge: Cambridge University Press, 1990.

Coen, Deborah R. *The Earthquake Observers: Disaster Science from Lisbon to Richter*. Chicago: University of Chicago Press, 2013.

Coen, Deborah R. "Imperial Climatographies from Tyrol to Turkestan." *Osiris* 26 (2011): 45–65.

Coen, Deborah R. "The Tongues of Seismology in Nineteenth-Century Switzerland." *Science in Context* 25, no. 1 (2012): 73–102.

Colbert, Edwin H. "W. D. Matthew's Early Western Field Trips." *Earth Sciences History* 9, no. 1 (1990): 41–44.

Colbert, Edwin H. *William Diller Matthew, Paleontologist: The Splendid Drama Observed*. New York: Columbia University Press, 1992.

Collier, Peter, and Rob Inkpen. "The Royal Geographical Society and the Development of Surveying, 1870–1914." *Journal of Historical Geography* 29, no. 1 (2003): 93–108.

Cook, Harold J. *Matters of Exchange: Commerce, Medicine, and Science in the Dutch Golden Age*. New Haven, CT: Yale University Press, 2007.

Cooper, Alix. *Inventing the Indigenous: Local Knowledge and Natural History in Early Modern Europe.* Cambridge: Cambridge University Press, 2007.

Coupland, Robert T. "The Effects of Fluctuations in Weather upon the Grasslands of the Great Plains." *Botanical Review* 24, no. 5 (1958): 273–317.

Courtwright, Julie. *Prairie Fire: A Great Plains History.* Lawrence: University Press of Kansas, 2011.

Craig, Patricia. *Centennial History of the Carnegie Institution of Washington.* Vol. 4, *The Department of Plant Biology.* Cambridge: Cambridge University Press, 2005.

Creager, Angela N. H. *The Life of a Virus: Tobacco Mosaic Virus as an Experimental Model, 1930–1965.* Chicago: University of Chicago Press, 2002.

Crowley, John M. "Ranching in the Mountain Parks of Colorado." *Geographical Review* 65, no. 4 (1975): 445–60.

Cunfer, Geoff. *On the Great Plains: Agriculture and Environment.* College Station: Texas A&M University Press, 2005.

Curtis, Kent A. *Gambling on Ore: The Nature of Metal Mining in the United States, 1860–1910.* Boulder: University Press of Colorado, 2013.

Cunningham, Andrew, and Perry Williams. "De-centring the 'Big Picture': *The Origins of Modern Science* and the Modern Origins of Science." *British Journal for the History of Science* 26 (1993): 407–32.

Cutright, Paul Russell, and Michael J. Brodhead. *Elliott Coues: Naturalist and Frontier Historian.* Urbana: University of Illinois Press, 1981.

Danbom, David B. "The Agricultural Experiment Station and Professionalism: Scientists' Goals for Agriculture, 1887–1910." *Agricultural History* 60, no. 2 (1986): 246–55.

Danbom, David B. *"Our Purpose Is to Serve": The First Century of the North Dakota Agricultural Experiment Station.* Fargo: North Dakota Institute for Regional Studies, 1990.

Darnell, Regna. *And Along Came Boas: Continuity and Revolution in Americanist Anthropology.* Amsterdam: John Benjamins, 1998.

Daston, Lorraine. "The Ideal and Reality of the Republic of Letters in the Enlightenment." *Science in Context* 4 (1991): 367–86.

Daston, Lorraine, and Peter Galison. *Objectivity.* New York: Zone Books, 2007.

Daston, Lorraine, and Elizabeth Lunbeck, eds. *Histories of Scientific Observation.* Chicago: University of Chicago Press, 2011.

Davis, Keith F., and Jane L. Aspinwall. *Timothy H. O'Sullivan: The King Survey Photographs.* Kansas City, MO: Hall Family Foundation in association with the Nelson-Atkins Museum of Art, 2011.

Dawdy, Doris Ostrander. *George Montague Wheeler: The Man and the Myth.* Athens, OH: Swallow Press, 1993.

de Bont, Raf. "Between the Laboratory and the Deep Blue Sea. Space Issues in the Marine Stations of Naples and Wimereux." *Social Studies of Science* 39, no. 2 (2009): 199–227.

de Bont, Raf. *Stations in the Field: A History of Place-Based Animal Research, 1870–1930.* Chicago: University of Chicago Press, 2015.

De Bres, Karen. "Come to the 'Champagne Air': Changing Promotional Images of the Kansas Climate, 1854–1900." *Great Plains Quarterly* 23, no. 2 (2003): 111–26.

Deiss, William A. "Spencer F. Baird and His Collectors." *Journal of the Society for the Bibliography of Natural History* 9, no. 4 (1980): 635–45.

Didier, Emmanuel. "Counting on Relief: Industrializing the Statistical Interviewer during the New Deal." *Science in Context* 24, no. 2 (2011): 281–310.

Dierig, Sven. "Engines for Experiment: Laboratory Revolution and Industrial Labor in the Nineteenth-Century City." *Osiris* 18 (2003): 116–34.

Dietrich, Michael R., and Brandi H. Tambasco. "Beyond the Boss and the Boys: Women and the Division of Labor in *Drosophila* Genetics in the United States, 1934–1970." *Journal of the History of Biology* 40, no. 3 (2007): 509–28.

Dietz, Bettina. "Contribution and Co-production: The Collaborative Culture of Linnaean Botany." *Annals of Science* 69, no. 4 (2012): 551–69.

Downey, Gregory J. *Telegraph Messenger Boys: Labor, Technology, and Geography, 1850–1950.* New York: Routledge, 2002.

Drake, Brian Allen. "Waving 'A Bough of Challenge': Forestry on the Kansas Grasslands, 1868–1915." *Great Plains Quarterly* 23, no. 1 (2003): 19–34.

Drayton, Richard. *Nature's Government: Science, Imperial Britain, and the "Improvement" of the World.* New Haven, CT: Yale University Press, 2000.

Dritsas, Lawrence. "From Lake Nyassa to Philadelphia: A Geography of the Zambesi Expedition, 1858–64." *British Journal for the History of Science* 38 (2005): 35–52.

Dupré, John. *The Disorder of Things: Metaphysical Foundations of the Disunity of Science.* Cambridge, MA: Harvard University Press, 1993.

Echo-Hawk, Walter R., and Roger C. Echo-Hawk. "Repatriation, Reburial, and Religious Rights." In *Handbook of American Indian Religious Freedom*, edited by Christopher Vecsey, 63–80. New York: Crossroad, 1991.

Edney, Matthew H. *Mapping an Empire: The Geographical Construction of British India, 1765–1843.* Chicago: University of Chicago Press, 1997.

Elliott, Melinda. "Cliff Palace: Three Generations of Exploration at Mesa Verde." In *Great Excavations: Tales of Early Southwestern Archaeology, 1888–1939*, 3–23. Santa Fe, NM: School of American Research Press, 1995.

Elliott, Melinda. *Exploring Human Worlds: A History of the School of American Research.* Santa Fe, NM: School of American Research Press, 1991.

Endersby, Jim. "'From Having No Herbarium.' Local Knowledge versus Metropolitan Expertise: Joseph Hooker's Australasian Correspondence with William Colenso and Ronald Gunn." *Pacific Science* 55, no. 4 (2001): 343–58.

Endersby, Jim. "A Garden Enclosed: Botanical Barter in Sydney, 1818–39." *British Journal for the History of Science* 33 (2000): 313–34.

Endersby, Jim. *Imperial Nature: Joseph Dalton Hooker and the Practices of Victorian Science.* Chicago: University of Chicago Press, 2008.

Erlingsson, Steindór J. "The Plymouth Laboratory and the Institutionalization of Experimental Zoology in Britain in the 1920s." *Journal of the History of Biology* 42 (2009): 151–83.

Ewan, Joseph. *Rocky Mountain Naturalists.* Denver, CO: University of Denver Press, 1950.

Fan, Fa-ti. *British Naturalists in Qing China: Science, Empire, and Cultural Encounter.* Cambridge, MA: Harvard University Press, 2004.

Fan, Fa-ti. "Circulating Material Objects: The International Controversy over Antiquities and Fossils in Twentieth-Century China." In *The Circulation of Knowledge between Britain, India, and China: The Early-Modern World to the Twentieth Century*, edited by Bernard Lightman, Gordon McOuat, and Larry Stewart, 209–36. Leiden: Brill, 2013.

Farber, Paul Lawrence. *Finding Order in Nature: The Naturalist Tradition from Linnaeus to E. O. Wilson.* Baltimore, MD: Johns Hopkins University Press, 2000.

Ferguson, Walter Keene. *Geology and Politics in Frontier Texas, 1845–1909.* Austin: University of Texas Press, 1969.

Ferleger, Lou. "Uplifting American Agriculture: Experiment Station Scientists and the Office of Experiment Stations in the Early Years after the Hatch Act." *Agricultural History* 64, no. 2 (1990): 5–23.

Fiege, Mark. *The Republic of Nature: An Environmental History of the United States.* Seattle: University of Washington Press, 2012.

Finnegan, Diarmid A. *Natural History Societies and Civic Culture in Victorian Scotland.* London: Pickering & Chatto, 2009.

Flanagan, Vincent J. "Gouverneur Kemble Warren, Explorer of the Nebraska Territory." *Nebraska History* 51 (1970): 171–98.

Flannery, Tim. *The Eternal Frontier: An Ecological History of North America and Its Peoples.* New York: Atlantic Monthly Press, 2001.

Fleharty, Eugene D. *Wild Animals and Settlers on the Great Plains.* Norman: University of Oklahoma Press, 1995.

Flores, Dan L. "Islands in the Desert: The Rocky Mountains in Environmental History." In *The Natural West: Environmental History in the Great Plains and Rocky Mountains*, 107–23. Norman: University of Oklahoma Press, 2001.

Flores, Dan L. "Zion in Eden: Phases of the Environmental History of Utah." *Environmental Review* 7 (1983): 325–44. Reprinted in *The Natural West: Environmental History in the Great Plains and Rocky Mountains*, 124–44. Norman: University of Oklahoma Press, 2001.

Foster, Mike. "Ferdinand Vandeveer Hayden as Naturalist." *American Zoologist* 26, no. 2 (1986): 343–49.

Foster, Mike. *Strange Genius: The Life of Ferdinand Vandeveer Hayden*. Niwot, CO: Roberts Rinehart, 1994.

Fowler, Don D. *A Laboratory for Anthropology: Science and Romanticism in the American Southwest, 1846–1930*. Albuquerque: University of New Mexico Press, 2000.

Fox, Robert. *The Savant and the State: Science and Cultural Politics in Nineteenth-Century France*. Baltimore, MD: Johns Hopkins University Press, 2012.

Freed, Stanley A., and Ruth S. Freed. "Clark Wissler and the Development of Anthropology in the United States." *American Anthropologist* 85 (1983): 800–825.

Freeman, John F. *Black Hills Forestry: A History*. Boulder: University Press of Colorado, 2015.

Freeman, John F. *High Plains Horticulture: A History*. Boulder: University Press of Colorado, 2008.

Freeman, Michael. "Tracks to a New World: Railway Excavation and the Extension of Geological Knowledge in Mid-Nineteenth-Century Britain." *British Journal for the History of Science* 34 (2001): 51–65.

Frehner, Brian. *Finding Oil: The Nature of Petroleum Geology, 1859–1920*. Lincoln: University of Nebraska Press, 2011.

Frison, George C. *Prehistoric Hunters of the High Plains*. 2nd ed. San Diego: Academic Press, 1991.

Fuller, Wayne E. *The American Mail: Enlarger of the Common Life*. Chicago: University of Chicago Press, 1972.

Fyfe, Aileen. *Steam-Powered Knowledge: William Chambers and the Business of Publishing, 1820–1860*. Chicago: University of Chicago Press, 2012.

Gabler, Edwin. *The American Telegrapher: A Social History, 1860–1900*. New Brunswick, NJ: Rutgers University Press, 1988.

Galison, Peter. *Image and Logic: A Material Culture of Microphysics*. Chicago: University of Chicago Press, 1997.

Gardner, Robert. "Constructing a Technological Forest: Nature, Culture, and Tree-Planting in the Nebraska Sand Hills." *Environmental History* 14, no. 2 (2009): 275–97.

Garver, John B. "Practical Military Geographers and Mappers of the Trans-Missouri West, 1820–1860." In *Mapping the North American Plains: Essays in the History of Cartography*, edited by Frederick C. Luebke, Frances W. Kaye, and Gary E. Moulton, 111–26. Norman: University of Oklahoma Press, 1987.

Gates, Paul W. *History of Public Land Law Development.* Washington, DC: Government Printing Office, 1968.

Geiser, Samuel Wood. *Naturalists of the Frontier.* 2nd ed. Dallas: Southern Methodist University Press, 1948.

Gibson, John M. *Soldier in White: The Life of General George Miller Sternberg.* Durham, NC: Duke University Press, 1958.

Gieryn, Thomas F. "City as Truth-Spot: Laboratories and Field-Sites in Urban Studies." *Social Studies of Science* 36, no. 1 (2006): 5–38.

Gieryn, Thomas F. "Laboratory Design for Post-Fordist Science." *Isis* 99 (2008): 796–802.

Gifford, Carol A., and Elizabeth A. Morris. "Digging for Credit: Early Archaeological Field Schools in the American Southwest." *American Antiquity* 50, no. 2 (1985): 395–411.

Givens, Douglas R. *Alfred Vincent Kidder and the Development of Americanist Archaeology.* Albuquerque: University of New Mexico Press, 1992.

Goetzmann, William H. *Army Exploration in the American West, 1803–1863.* New Haven, CT: Yale University Press, 1959.

Goetzmann, William H. *Exploration and Empire: The Explorer and the Scientist in the Winning of the American West.* New York, Norton, 1966.

Goetzmann, William H., and Kay Sloan. *Looking Far North: The Harriman Expedition to Alaska, 1899.* New York: Viking, 1982.

Goldstein, Daniel. "Midwestern Naturalists: Academies of Science in the Mississippi Valley, 1850–1900." PhD diss., Yale University, 1989.

Goldstein, Daniel. "'Yours for Science': The Smithsonian Institution's Correspondents and the Shape of Scientific Community in Nineteenth-Century America." *Isis* 85, no. 4 (1994): 573–99.

Gooday, Graeme. *The Morals of Measurement: Accuracy, Irony, and Trust in Late Victorian Electrical Practice.* Cambridge: Cambridge University Press, 2004.

Gooday, Graeme. "Precision Measurement and the Genesis of Physics Teaching Laboratories in Victorian Britain." *British Journal for the History of Science* 23 (1990): 25–51.

Gooday, Graeme. "The Premisses of Premises: Spatial Issues in the Historical Construction of Laboratory Credibility." In *Making Space for Science: Territorial Themes in the Shaping of Knowledge,* edited by Crosbie Smith and Jon Agar, 216–45. Basingstoke: Macmillan, 1998.

Goss, Andrew. *The Floracrats: State-Sponsored Science and the Failure of the Enlightenment in Indonesia.* Madison: University of Wisconsin Press, 2011.

Green Musselman, Elizabeth. "Plant Knowledge at the Cape: A Study in African and European Collaboration." *International Journal of African Historical Stud-*

ies 36, no. 2 (2003): 367–92. Greene, Ann Norton. *Horses at Work: Harnessing Power in Industrial America*. Cambridge, MA: Harvard University Press, 2008.

Guelke, Leonard, and Jeanne Kay Guelke. "Imperial Eyes on South Africa: Reassessing Travel Narratives." *Journal of Historical Geography* 30, no. 1 (2004): 11–31.

Ham, Elizabeth A. *A History of the Oklahoma Geological Survey, 1908–1983*. Oklahoma Geological Survey Special Publication 83-2. Norman: University of Oklahoma, 1983.

Harrell, David. "'We Contacted Smithsonian': The Wetherills at Mesa Verde." *New Mexico Historical Review* 62 (1987): 229–48.

Harrington, John A., Jr., and Jay R. Harman. "Climate and Vegetation in Central North America: Natural Patterns and Human Alterations." *Great Plains Quarterly* 11, no. 2 (1991): 103–12.

Harris, Steven J. "Long-Distance Corporations, Big Sciences, and the Geography of Knowledge." *Configurations* 6, no. 2 (1998): 269–304.

Harvey, David. *The Limits to Capital*. Chicago: University of Chicago Press, 1982.

Harvey, Mark W. T. *A Symbol of Wilderness: Echo Park and the American Conservation Movement*. Albuquerque: University of New Mexico Press, 1994.

Harvey, Mark W. T. "Utah, the National Park Service, and Dinosaur National Monument, 1909–56." *Utah Historical Quarterly* 59, no. 3 (1991): 243–63.

Harwood, Jonathan. *Styles of Scientific Thought: The German Genetics Community, 1900–1933*. Chicago: University of Chicago Press, 1993.

Hendrickson, Walter B. "Science and Culture in the American Middle West." *Isis* 64, no. 3 (1973): 326–40.

Henke, Christopher R. "Making a Place for Science: The Field Trial." *Social Studies of Science* 30, no. 4 (2000): 483–511.

Henkin, David M. *The Postal Age: The Emergence of Modern Communication in Nineteenth-Century America*. Chicago: University of Chicago Press, 2006.

Herman, Daniel Justin. *Hunting and the American Imagination*. Washington, DC: Smithsonian Institution Press, 2001.

Hibbard, Benjamin Horace. *A History of the Public Land Policies*. Madison: University of Wisconsin Press, 1965.

Higham, C. L. "Saviors and Scientists: North American Protestant Missionaries and the Development of Anthropology." *Pacific Historical Review* 72, no. 4 (2003): 531–59.

Hineline, Mark L. "The Visual Culture of the Earth Sciences, 1862–1970." PhD dissertation, University of California, San Diego, 1993.

Hinsley, Curtis M. "The Museum Origins of Harvard Anthropology, 1866–1915." In *Science at Harvard University: Historical Perspectives*, edited by Clark

A. Elliott and Margaret W. Rossiter, 121–45. Bethlehem: Lehigh University Press, 1992.

Hinsley, Curtis M. *Savages and Scientists: The Smithsonian Institution and the Development of American Anthropology 1846–1910.* Washington, DC: Smithsonian Institution Press, 1981.

Hochfelder, David. *The Telegraph in America, 1832–1920.* Baltimore, MD: Johns Hopkins University Press, 2012.

Holliday, Vance T. "Middle Holocene Drought on the Southern High Plains." *Quaternary Research* 31, no. 1 (1989): 74–82.

Holmes, Frederic L., Jürgen Renn, and Hans-Jörg Rheinberger, eds. *Reworking the Bench: Research Notebooks in the History of Science.* Dordrecht: Kluwer, 2003.

Howkins, Adrian. "Reluctant Collaborators: Argentina and Chile in Antarctica during the International Geophysical Year, 1957–58." *Journal of Historical Geography* 34, no. 4 (2008): 596–617.

Hoyt, Mary E. *A Short History of the Colorado School of Mines.* Golden: Colorado School of Mines Library, 1949.

Hume, Edgar Erskine. *Ornithologists of the United States Army Medical Corps: Thirty-Six Biographies.* Baltimore, MD: Johns Hopkins University Press, 1942.

Hunt, Bruce J. "The Ohm Is Where the Art Is: British Telegraph Engineers and the Development of Electrical Standards." *Osiris* 9 (1994): 48–63.

Hyder, Clyde Kenneth. *Snow of Kansas: The Life of Francis Huntington Snow, with Extracts from His Journals and Letters.* Lawrence: University Press of Kansas, 1953.

Inkster, Ian, and Jack B. Morrell, eds. *Metropolis and Province: Science in British Culture, 1780–1850.* Philadelphia: University of Pennsylvania Press, 1983.

Jack, Homer A. "Biological Field Stations of the World." *Chronica Botanica* 9 (1945): 1–73.

Jackson, Catherine M. "Chemistry as the Defining Science: Discipline and Training in Nineteenth-Century Chemical Laboratories." *Endeavour* 35, no. 2–3 (2011): 55–62.

Jacoby, Karl. *Crimes against Nature: Squatters, Poachers, Thieves, and the Hidden History of American Conservation.* Berkeley: University of California Press, 2001.

Jaffe, Mark. *The Gilded Dinosaur: The Fossil War between E. D. Cope and O. C. Marsh and the Rise of American Science.* New York: Crown, 2000.

James, Louise Boyd. "Jujubes, Grapes, and Grass: The USDA Research Station at Woodward, 1913–1987." *Chronicles of Oklahoma* 65 (1988): 354–79.

Janković, Vladimir. *Reading the Skies: A Cultural History of English Weather, 1650–1820.* Chicago: University of Chicago Press, 2000.

Jardine, Nicholas, and Emma Spary. "The Natures of Cultural History." In *Cultures of Natural History*, edited by Nicholas Jardine, James A. Secord, and E. C. Spary, 1–13. Cambridge: Cambridge University Press, 1996.

Jenny, Hans. *E. W. Hilgard and the Birth of Modern Soil Science.* Pisa: Agrochimica, 1961.

Jepson, Thomas C. "The Telegraph Comes to Colorado: A New Technology and Its Consequences." *Essays in Colorado History* 7 (1987): 1–25.

John, Richard R. "Private Enterprise, Public Good? Communications Deregulation as a National Political Issue, 1839–1851." In *Beyond the Founders: New Approaches to the Political History of the Early American Republic,* edited by Jeffrey L. Pasley, Andrew W. Robertson, and David Waldstreicher, 339–42. Chapel Hill: University of North Carolina Press, 2004.

John, Richard R. *Spreading the News: The American Postal System from Franklin to Morse.* Cambridge, MA: Harvard University Press, 1995.

Johnson, Ben F. *Fierce Solitude: A Life of John Gould Fletcher.* Fayetteville: University of Arkansas Press, 1994.

Jones, Susan. "Becoming a Pest: Prairie Dog Ecology and the Human Economy in the Euroamerican West." *Environmental History* 4, no. 4 (1999): 531–52.

Jordan, Carole. "The Kansas State Board of Agriculture." In *The Rise of the Wheat State: A History of Kansas Agriculture, 1861–1986,* edited by George E. Ham and Robin Higham, 298–39. Manhattan, KS: Sunflower University Press, 1987.

Jordan, David M. *"Happiness Is Not My Companion": The Life of General G. K. Warren.* Bloomington: Indiana University Press, 2001.

Judd, Neil M. *The Bureau of American Ethnology: A Partial History.* Norman: University of Oklahoma Press, 1967.

Judd, Richard W. *The Untilled Garden: Natural History and the Spirit of Conservation in America, 1740–1840.* Cambridge: Cambridge University Press, 2009.

Keeney, Elizabeth B. *The Botanizers: Amateur Scientists in Nineteenth-Century America.* Chapel Hill: University of North Carolina Press, 1992.

Keiner, Christine. *The Oyster Question: Scientists, Watermen, and the Maryland Chesapeake Bay since 1880.* Athens: University of Georgia Press, 2009.

Keller, Robert H., and Michael F. Turek. *American Indians and National Parks.* Tucson: University of Arizona Press, 1998.

Kelsey, Robin. *Archive Style: Photographs and Illustrations for U.S. Surveys, 1850–1890.* Berkeley: University of California Press, 2007.

Kingsland, Sharon E. *The Evolution of American Ecology, 1890–2000.* Baltimore, MD: Johns Hopkins University Press, 2005.

Kirsch, Scott. "John Wesley Powell and the Mapping of the Colorado Plateau, 1869–1879: Survey Science, Geographical Solutions, and the Economy of Environmental Values." *Annals of the Association of American Geographers* 92, no. 3 (2002): 548–72.

Kirsch, Scott. "Regions of Government Science: John Wesley Powell in Washington and the American West." *Endeavour* 23, no. 4 (1999): 155–58.

Kisling, Vernon N. "*The Naturalists' Directory* and the Evolution of Communication among American Naturalists." *Archives of Natural History* 21, no. 3 (1994): 393–406.

Klein, Naomi. "The Change Within: The Obstacles We Face Are Not Just External." *Nation*, April 21, 2014.

Klein, Naomi. *This Changes Everything: Capitalism vs. The Climate.* New York: Simon & Schuster, 2014.

Klein, Ursula. *Experiments, Models, Paper Tools: Cultures of Organic Chemistry in the Nineteenth Century*. Stanford, CA: Stanford University Press, 2002.

Knell, Simon J. *The Culture of English Geology, 1815–1851: A Science Revealed Through Its Collecting*. Aldershot: Ashgate, 2000.

Koch, Amy, and John R. Bozell. "Environmental and Cultural Variation in the Nebraska Sand Hills." In *Islands on the Plains: Ecological, Social, and Ritual Use of Landscapes*, edited by Marcel Kornfield and Alan J. Osborn, 167–90. Salt Lake City: University of Utah Press, 2003.

Koerner, Lisbet. *Linnaeus: Nature and Nation.* Cambridge, MA: Harvard University Press, 1999.

Kohler, Robert E. *All Creatures: Naturalists, Collectors, and Biodiversity, 1850–1950.* Princeton, NJ: Princeton University Press, 2006.

Kohler, Robert E. "Finders, Keepers: Collecting Sciences and Collecting Practice." *History of Science* 45, no. 4 (2007): 428–54.

Kohler, Robert E. "From Farm and Family to Career Naturalist: The Apprenticeship of Vernon Bailey." *Isis* 99, no. 1 (2008): 28–56.

Kohler, Robert E. "A Generalist's Vision." *Isis* 96, no. 2 (2005): 224–29.

Kohler, Robert E. "Lab History: Reflections." *Isis* 99, no. 4 (2008): 761–68.

Kohler, Robert E. "Labscapes: Naturalizing the Lab." *History of Science* 40 (2002): 473–501.

Kohler, Robert E. *Landscapes and Labscapes: Exploring the Lab-Field Border in Biology*. Chicago: University of Chicago Press, 2002.

Kohler, Robert E. *Lords of the Fly:* Drosophila *Genetics and the Experimental Life.* Chicago: University of Chicago Press, 1994.

Kohler, Robert E. "Place and Practice in Field Biology." *History of Science* 40 (2002): 189–210.

Kohler, Robert E. "Systems of Production: *Drosophila, Neurospora*, and Biochemical Genetics." *Historical Studies in the Physical and Biological Sciences* 22 (1991): 87–130.

Kohlstedt, Sally Gregory. *The Formation of the American Scientific Community: The American Association for the Advancement of Science, 1848–60.* Urbana: University of Illinois Press, 1976.

Kohlstedt, Sally Gregory. "The Nineteenth-Century Amateur Tradition: The Case of the Boston Society of Natural History." In *Science and Its Public: The Changing Relationship*, edited by Gerald Holton and William A. Blanpied, 173–90. Dordrecht: Reidel, 1976.

Kornfield, Marcel. "Pull of the Hills: Technological Structures around Biogeographical Islands." In *Islands on the Plains: Ecological, Social, and Ritual Use of Landscapes*, edited by Marcel Kornfield and Alan J. Osborn, 111–41. Salt Lake City: University of Utah Press, 2003.

Kornfield, Marcel, and Alan J. Osborn, eds. *Islands on the Plains: Ecological, Social, and Ritual Use of Landscapes*. Salt Lake City: University of Utah Press, 2003.

Kuklick, Henrika. "Personal Equations: Reflections on the History of Fieldwork, with Special Reference to Sociocultural Anthropology." *Isis* 102, no. 1 (2011): 1–33.

Kuklick, Henrika. *The Savage Within: The Social History of British Anthropology, 1885–1945*. New York: Cambridge University Press, 1991.

Kuklick, Henrika, and Robert E. Kohler, eds. "Science in the Field." Special issue, *Osiris* 11 (1996).

Kumar, Deepak. "Problems in Science Administration: A Study of the Scientific Surveys in British India 1757–1900." In *Science and Empires: Historical Studies about Scientific Development and European Expansion*, edited by Patrick Petitjean, Catherine Jami, and Anne Marie Moulin, 269–80. Dordrecht: Kluwer, 1992.

Kumar, Prakash. *Indigo Plantations and Science in Colonial India*. New York: Cambridge University Press, 2012.

Lane, K. Maria D. *Geographies of Mars: Seeing and Knowing the Red Planet*. Chicago: University of Chicago Press, 2011.

Lanham, Url. *The Bone Hunters*. New York: Columbia University Press, 1973.

Larsen, Anne. "Not since Noah: The English Scientific Zoologists and the Craft of Collecting, 1800–1840." PhD diss., Princeton University, 1993.

Larson, Floyd. "The Role of the Bison in Maintaining the Short Grass Plains." *Ecology* 21, no. 2 (1940): 113–21.

Lass, William E. *From the Missouri to the Great Salt Lake: An Account of Overland Freighting*. Lincoln: Nebraska State Historical Society, 1972.

Lass, William E. "Missouri River Steamboating: The Supplier of the Oregon-California Trail." *Overland Journal* 30, no. 2 (2012): 44–64.

Lass, William E. *Navigating the Missouri: Steamboating on Nature's Highway, 1819–1935*. Norman, OK: Arthur H. Clark, 2008.

Latour, Bruno. "Give Me a Laboratory and I Will Raise the World." In *Science Observed: Perspectives on the Social Study of Science*, edited by Karin D. Knorr-Cetina and Michael Mulkay, 141–70. London: SAGE Publications, 1983.

Latour, Bruno. *Science in Action: How to Follow Scientists and Engineers through Society.* Cambridge, MA: Harvard University Press, 1987.

Latour, Bruno. "Visualization and Cognition: Thinking with Eyes and Hands." In *Knowledge and Society: Studies in the Sociology of Culture Past and Present.* Vol. 6. Edited by Henrika Kuklick and Elizabeth Long, 1–40. Greenwich, CT: JAI Press, 1986.

Latour, Bruno, and Steve Woolgar. *Laboratory Life: The Construction of Scientific Facts.* Princeton, NJ: Princeton University Press, 1986.

Lawson, Merlin Paul. *The Climate of the Great American Desert: Reconstruction of the Climate of Western Interior United States, 1800–1850.* Lincoln: University of Nebraska, 1974.

Leach, William. *Butterfly People: An American Encounter with the Beauty of the World.* New York: Pantheon, 2013.

LeCain, Timothy J. "From Eastern Coal to Western Copper: The Bureau of Mines and Industrial Mining in the West." *Journal of the West* 43, no. 2 (2004): 20–28.

LeCain, Timothy J. *Mass Destruction: The Men and Giant Mines That Wired America and Scarred the Planet.* New Brunswick, NJ: Rutgers University Press, 2009.

Lewis, Andrew John. *A Democracy of Facts: Natural History in the Early Republic.* Philadelphia: University of Pennsylvania Press, 2011.

Lewis, Daniel. *The Feathery Tribe: Robert Ridgway and the Modern Study of Birds.* New Haven, CT: Yale University Press, 2012.

Lewis, Michael L. *Inventing Global Ecology: Tracking the Biodiversity Ideal in India, 1947–1997.* Athens: Ohio University Press, 2004.

Liberty, Margot. "Native American 'Informants': The Contribution of Francis La Flesche." In *American Anthropology: The Early Years,* edited by John V. Murra, 99–110. St. Paul, MN: West, 1976.

Lipsey, John J. "Light-Hearted Princetonians 'Survey' Colorado in 1877." *Denver Posse of Westerners Brand Book* 18 (1962): 349–53.

Lister, Florence C. *Troweling through Time: The First Century of Mesa Verdean Archaeology.* Albuquerque: University of New Mexico Press, 2004.

Lister, Florence C., and Robert H. Lister. *Earl Morris and Southwestern Archaeology.* Albuquerque: University of New Mexico Press, 1968.

Livingstone, David N. *Putting Science in Its Place: Geographies of Scientific Knowledge.* Chicago: University of Chicago Press, 2003.

Lockwood, Jeffrey A. *Locust: The Devastating Rise and Mysterious Disappearance of the Insect That Shaped the American Frontier.* New York: Basic Books, 2004.

Loefgren, Orvar. *On Holiday: A History of Vacationing.* Berkeley: University of California Press, 1999.

Long, Robert E., Deborah Epstein Popper, and Frank J. Popper. "'Progress of the Nation': The Settlement History of the Enduring American Frontier." *Western Historical Quarterly* 26, no. 3 (1995): 289–307.

Lucier, Paul. "Commercial Interests and Scientific Disinterestedness: Consulting Geologists in Antebellum America." *Isis* 86, no. 2 (1995): 245–67.

Lucier, Paul. "The Professional and the Scientist in Nineteenth-Century America." *Isis* 100, no. 4 (2010): 699–732.

Lucier, Paul. *Scientists and Swindlers: Consulting on Coal and Oil in America, 1820–1890.* Baltimore, MD: Johns Hopkins University Press, 2008.

Luke, Timothy W. "The Missouri Botanical Garden: Reworking Biopower as Florapower." *Organization and Environment* 13, no. 3 (2000): 305–21.

MacDonell, George H., and LuAnn Wandsnider. "The Western Niobrara River: An Inter-Island Passage on the Plains." In *Islands on the Plains: Ecological, Social, and Ritual Use of Landscapes*, edited by Marcel Kornfield and Alan J. Osborn, 89–107. Salt Lake City: University of Utah Press, 2003.

Mackay, David. "Agents of Empire: The Banksian Collectors and Evaluation of New Lands." In *Visions of Empire: Voyages, Botany, and Representations of Nature*, edited by David Philip Miller and Peter Hanns Reill, 38–57. Cambridge: Cambridge University Press, 1996.

Mackay, David. *In the Wake of Cook: Exploration, Science and Empire, 1780–1801.* London: Croom Helm, 1985.

MacLeod, Roy, and Philip F. Rehbock, eds. *Darwin's Laboratory: Evolutionary Theory and Natural History in the Pacific.* Honolulu: University of Hawaii Press, 1994.

Manning, Thomas G. *Government in Science: The U.S. Geological Survey, 1867–1894.* Lexington: University of Kentucky Press, 1967.

Marcus, Alan I. *Agricultural Science and the Quest for Legitimacy: Farmers, Agricultural Colleges, and Experiment Stations, 1870–1890.* Ames: Iowa State University Press, 1985.

Mark, Joan. "Francis La Flesche: The American Indian as Anthropologist." *Isis* 73, no. 4 (1982): 496–510.

Mark, Joan. *A Stranger in Her Native Land: Alice Fletcher and the American Indians.* Lincoln: University of Nebraska Press, 1988.

Martin-Nielsen, Janet. *Eismitte in the Scientific Imagination: Knowledge and Politics at the Center of Greenland.* New York: Palgrave Macmillan, 2013.

May, Irvin M. "Agricultural Science in the Texas Panhandle: The Texas Agricultural Experiment Station." *Panhandle-Plains Historical Review* 53 (1980): 111–38.

Mayor, Adrienne. *Fossil Legends of the First Americans.* Princeton, NJ: Princeton University Press, 2005.

McCook, Stuart. "Managing Monocultures: Coffee, the Coffee Rust, and the Science of Working Landscapes." In *Knowing Global Environments: New Historical Perspectives on the Field Sciences*, edited by Jeremy Vetter, 87–107. New Brunswick, NJ: Rutgers University Press, 2011.

McCook, Stuart. *States of Nature: Science, Agriculture, and Environment in the Spanish Caribbean, 1760–1940.* Austin: University of Texas Press, 2002.

McCracken, Donal P. *Gardens of Empire: Botanical Institutions and the Victorian British Empire.* London: Leicester University Press, 1997.

McIntosh, Charles Barron. *The Nebraska Sand Hills: The Human Landscape.* Lincoln: University of Nebraska Press, 1996.

McIntosh, Charles Barron. "The Route of a Sand Hills Bone Hunt: The Yale College Expedition of 1870." *Nebraska History* 69, no. 2 (1988): 84–94.

McIntosh, John S. "The Second Jurassic Dinosaur Rush." *Earth Sciences History* 9, no. 1 (1990): 22–27.

McIntosh, Robert P. "Edward Lee Greene: The Man." In *Landmarks of Botanical History*, edited by Frank N. Egerton, 18–53. Stanford, CA: Stanford University Press, 1983.

McNitt, Frank. *Richard Wetherill: Anasazi.* 2nd ed. Albuquerque: University of New Mexico Press, 1966.

McOuat, Gordon. "Cataloguing Power: Delineating 'Competent Naturalists' and the Meaning of Species in the British Museum." *British Journal for the History of Science* 34 (2001): 1–28.

McVaugh, Rogers. *Edward Palmer: Plant Explorer of the American West.* Norman: University of Oklahoma Press, 1956.

Mendelsohn, J. Andrew. "The Microscopist of Modern Life." *Osiris* 18 (2003): 150–70.

Mickulas, Peter. "Cultivating the Big Apple: The New York Horticultural Society, Nineteenth-Century New York Botany, and the New York Botanical Garden." *New York History* 83 (2002): 34–54.

Mickulas, Peter. "A Scientific Garden in a Public Space: The New York Botanical Garden, 1900–10." *Studies in the History of Gardens and Designed Landscapes* 22 (2002): 193–213.

Midgette, Nancy Smith. *To Foster the Spirit of Professionalism: Southern Scientists and State Academies of Science.* Tuscaloosa: University of Alabama Press, 1991.

Miller, David Philip. "Joseph Banks, Empire, and 'Centers of Calculation' in Late Hanoverian London." In *Visions of Empire: Voyages, Botany, and Representations of Nature*, edited by David Philip Miller and Peter Hanns Reill, 21–37. Cambridge: Cambridge University Press, 1996.

Mills, C. Wright. *White Collar: The American Middle Classes.* New York: Oxford University Press, 1953.

Mitman, Gregg. *Breathing Space: How Allergies Shape Our Lives and Landscapes.* New Haven, CT: Yale University Press, 2007.

Mitman, Gregg. "Geographies of Hope: Mining the Frontiers of Health in Denver and Beyond, 1870–1965." *Osiris* 19 (2004): 93–111.

Mody, Cyrus C. M. *Instrumental Community: Probe Microscopy and the Path to Nanotechnology.* Cambridge, MA: MIT Press, 2011.

Montgomery, Georgina. "Place, Practice and Primatology: Clarence Ray Carpenter, Primate Communication and the Development of Field Methodology, 1931–1945." *Journal of the History of Biology* 38, no. 3 (2005): 495–533.

Montgomery, Georgina. "Primates in the Real World: Place, Practice and the History of Primate Field Studies, 1924–1970." PhD diss., University of Minnesota, 2005.

Moon, Suzanne. "Empirical Knowledge, Scientific Authority, and Native Development: The Controversy over Sugar/Rice Ecology in the Netherlands East Indies, 1905–1914." *Environment and History* 10, no. 1 (2004): 59–81.

Moore, James Gregory. *King of the 40th Parallel: Discovery in the American West.* Stanford, CA: Stanford General Books, 2006.

Moore, Jason W. "Environmental Crises and the Metabolic Rift in World-Historical Perspective." *Organization and Environment* 13, no. 2 (2000): 123–57.

Moore, Jason W. "Transcending the Metabolic Rift: A Theory of Crises in the Capitalist World-Ecology." *Journal of Peasant Studies* 38, no. 1 (2011): 1–46.

Moore, John A. "Zoology of the Pacific Railroad Surveys." *American Zoologist* 26, no. 2 (1986): 331–41.

Morrell, Jack, and Arnold Thackray. *Gentlemen of Science: Early Years of the British Association for the Advancement of Science.* Oxford: Clarendon, 1981.

Morse, Kathryn. *The Nature of Gold: An Environmental History of the Klondike Gold Rush.* Seattle: University of Washington Press, 2003.

Moses, L. G. *The Indian Man: A Biography of James Mooney.* Urbana: University of Illinois Press, 1984.

Müller-Wille, Staffan. "Joining Lapland and the Topinambes in Flourishing Holland: Center and Periphery in Linnaean Botany." *Science in Context* 16, no. 4 (2003): 461–88.

Müller-Wille, Staffan. "Walnuts at Hudson Bay, Coral Reefs in Gotland: The Colonialism of Linnaean Botany." In *Colonial Botany: Science, Commerce, and Politics in the Early Modern World,* edited by Londa Schiebinger and Claudia Swan, 34–48. Philadelphia: University of Pennsylvania Press, 2005.

Nagel, Thomas. *The View from Nowhere.* New York: Oxford University Press, 1986.

Nagy, Laura L. "Vernon Bailey and Wolf Extermination: Economy, Ecology, and Progressive Era Policy." PhD diss., University of New Hampshire, 2014.

Nash, Linda. *Inescapable Ecologies: A History of Environment, Disease, and Knowledge.* Berkeley: University of California Press, 2006.

Naylor, Simon. *Regionalizing Science: Placing Knowledges in Victorian England.* London: Pickering and Chatto, 2010

Neel, Susan Rhoades. "Love among the Fossils: Earl and Pearl Douglass at Dinosaur National Monument." *Utah Historical Quarterly* 89, no. 4 (2015): 290–307.

Nelson, Clifford M. "Meek at Albany, 1852–58." *Earth Sciences History* 6, no. 1 (1987): 40–46.

Nelson, Clifford M., and Fritiof M. Fryxell. "The Ante-bellum Collaboration of Meek and Hayden in Stratigraphy." In *Two Hundred Years of Geology in America: Proceedings of the New Hampshire Bicentennial Conference on the History of Geology,* edited by Cecil J. Schneer, 187–200. Hanover, NH: University Press of New England, 1979.

Nelson, Scott Reynolds, and Carol Sheriff. *A People at War: Civilians and Soldiers in America's Civil War, 1854–1877.* New York: Oxford University Press, 2007.

Nisbett, Catherine Elaine. "Business Practice: The Rise of American Astrophysics, 1859–1919." PhD diss., Princeton University, 2007.

Noble, Bruce J., Jr. "A Legacy of Distrust: The Ute Mountain Utes and the Boundaries of Mesa Verde National Park." *Colorado Heritage* (Summer 1995): 32–42.

Norviel, Jordan. "The Life of Richard Leigh (Beaver Dick) and the Teton Mountains." *Annals of Wyoming* 85, no. 4 (2013): 17–25.

Numbers, Ronald L., and John Harley Warner. "The Maturation of American Medical Science." In *Scientific Colonialism: A Cross-Cultural Comparison,* edited by Nathan Reingold and Marc Rothenberg, 191–214. Washington, DC: Smithsonian Institution Press, 1987.

Nye, Mary Jo. *Science in the Provinces: Scientific Communities and Provincial Leadership in France, 1860–1930.* Berkeley: University of California Press, 1986.

Nye, Ronald L. "Federal vs. State Agricultural Research Policy: The Case of California's Tulare Experiment Station, 1888–1909." *Agricultural History* 57, no. 4 (1983): 436–49.

Nyhart, Lynn K. *Modern Nature: The Rise of the Biological Perspective in Germany.* Chicago: University of Chicago Press, 2009.

Nyhart, Lynn K. "Natural History and the 'New' Biology." In *Cultures of Natural History,* edited by Nicholas Jardine, James A. Secord, and E. C. Spary, 426–43. Cambridge: Cambridge University Press, 1996.

O'Connell, Joseph. "Metrology: The Creation of Universality by the Circulation of Particulars." *Social Studies of Science* 23, no. 1 (1993): 129–73.

Ogilvie, Brian. *The Science of Describing: Natural History in Renaissance Europe.* Chicago: University of Chicago Press, 2006.

Olch, Peter D. "Medicine in the Indian-Fighting Army, 1866–1890." *Journal of the West* 21, no. 3 (1982): 32–41.

Olesko, Kathryn M. "Precision, Tolerance, and Consensus: Local Cultures in German and British Resistance Standards." In *Scientific Credibility and Technical Standards in 19th and Early 20th Century Germany and Britain,* edited by Jed Z. Buchwald, 117–56. Dordrecht: Kluwer, 1996.

Opie, John. *Ogallala: Water for a Dry Land.* 2nd ed. Lincoln: University of Nebraska Press, 2000.

Oreskes, Naomi. *The Rejection of Continental Drift: Theory and Method in American Earth Science.* New York: Oxford University Press, 1999.

Osborn, Alan J. "Ecological Aspects of Equestrian Adaptations in Aboriginal North America." *American Anthropologist* 85, no. 3 (1983): 563–91

Ostrom, John H., and John S. McIntosh. *Marsh's Dinosaurs: The Collections from Como Bluff.* 2nd ed. New Haven, CT: Yale University Press, 1999.

Overfield, Richard A. "The Agricultural Experiment Station and Americanization: The Hawaiian Experience, 1900–1910." *Agricultural History* 60, no. 2 (1986): 256–66.

Overfield, Richard A. "Science Follows the Flag: The Office of Experiment Stations and American Expansion." *Agricultural History* 64, no. 2 (1990): 31–40.

Overfield, Richard A. *Science with Practice: Charles E. Bessey and the Maturing of American Botany.* Ames: Iowa State University Press, 1993.

Pandora, Katherine. "Knowledge Held in Common: Tales of Luther Burbank and Science in the American Vernacular." *Isis* 92, no. 3 (2001): 484–516.

Pandora, Katherine, and Karen Rader. "Science in the Everyday World: Why Perspectives from the History of Science Matter." *Isis* 99, no. 2 (2008): 350–64.

Pang, Alex Soojung-Kim. *Empire and the Sun: Victorian Solar Eclipse Expeditions.* Stanford, CA: Stanford University Press, 2002.

Pang, Alex Soojung-Kim. "'Stars Should Henceforth Register Themselves': Astrophotography at the Early Lick Observatory." *British Journal for the History of Science* 30, no. 2 (1997): 177–202.

Pankin, Mary Faith. "George G. Lobdell, Jr., and the Yale Scientific Expedition of 1871 at Fort Bridger." *Annals of Wyoming* 70, no. 1 (1998): 25–44.

Pankin, Mary Faith. "The Yale Scientific Expeditions in Kansas." *Heritage of the Great Plains* 35, no. 2 (2002): 21–35.

Paul, Rodman Wilson. "Colorado as a Pioneer of Science in the Mining West." *Mississippi Valley Historical Review* 47, no. 1 (1960): 34–50.

Pauly, Philip J. *Biologists and the Promise of American Life: From Meriwether Lewis to Alfred Kinsey.* Princeton, NJ: Princeton University Press, 2000.

Pauly, Philip J. "Summer Resort and Scientific Discipline: Woods Hole and the Structure of American Biology, 1882–1925." In *The American Development of Biology*, edited by Ronald Rainger, Keith R. Benson, and Jane Maienschein, 121–50. Philadelphia: University of Pennsylvania Press, 1988.

Peterson, Charles S. "Albert F. Potter's Wasatch Survey, 1902: A Beginning for Public Management of Natural Resources in Utah." *Utah Historical Quarterly* 39, no. 3 (1971): 238–53.

Petrick, Gabriella M. "Feeding the Masses: H. J. Heinz and the Creation of Industrial Food." *Endeavour* 33, no. 1 (2009): 29–34.

Phillips, Denise. "Friends of Nature: Urban Sociability and Regional Natural History in Dresden, 1800–1850." *Osiris* 18 (2003): 43–59.

Phillips, Melba. "Laboratories and the Rise of the Physics Profession in the Nineteenth Century." *American Journal of Physics* 51, no. 6 (1983): 497–503.

Phillips, William B. *A History of the Agricultural Research Center-Hays: The First 100 Years.* Kansas Agricultural Experiment Station Contribution, no. 01-100-B (2001).

Pickstone, John V. *Ways of Knowing: A New History of Science, Technology, and Medicine.* Chicago: University of Chicago Press, 2001.

Plate, Robert. *The Dinosaur Hunters: Othniel C. Marsh and Edward D. Cope.* New York: McKay, 1964.

Pomeroy, Earl. *In Search of the Golden West: The Tourist in Western America.* New York: Knopf, 1957.

Porter, Charlotte M. *The Eagle's Nest: Natural History and American Ideas, 1812–1842.* Tuscaloosa: University of Alabama Press, 1986.

Porter, Theodore M. *Trust in Numbers: The Pursuit of Objectivity in Science and Public Life.* Princeton, NJ: Princeton University Press, 1995.

Pratt, Mary Louise. *Imperial Eyes: Travel Writing and Transculturation.* London: Routledge, 1992.

Pritchard, Sara B. *Confluence: The Nature of Technology and the Remaking of the Rhône.* Cambridge, MA: Harvard University Press, 2011.

Pyenson, Lewis. "Athena's Retinue: Nineteenth-Century Scientists Embedded in the Army." *British Journal for the History of Science* 45, no. 3 (2012): 377–400.

Pyne, Stephen J. *Fire in America: A Cultural History of Wildland and Rural Fire.* Princeton, NJ: Princeton University Press, 1982.

Pyne, Stephen J. *Grove Karl Gilbert: A Great Engine of Research.* Austin: University of Texas Press, 1980.

Quivik, Frederic L. "Smoke and Tailings: An Environmental History of Copper Smelting Technologies in Montana, 1880–1930." PhD diss., University of Pennsylvania, 1998.

Rabinow, Paul. *Making PCR: The Story of Biotechnology*. Chicago: University of Chicago Press, 1996.

Raby, Megan. "Making Biology Tropical: American Science in the Caribbean, 1898–1963." PhD diss., University of Wisconsin, 2013.

Rader, Karen. *Making Mice: Standardizing Animals for American Biomedical Research, 1900–1955*. Princeton, NJ: Princeton University Press, 2004.

Rader, Karen, and Victoria E. M. Cain. *Life on Display: Revolutionizing U.S. Museums of Science and Natural History in the Twentieth Century*. Chicago: University of Chicago Press, 2014.

Rainger, Ronald. *An Agenda for Antiquity: Henry Fairfield Osborn and Vertebrate Paleontology at the American Museum of Natural History, 1890–1935*. Tuscaloosa: University of Alabama Press, 1991.

Rainger, Ronald. "Biology, Geology, or Neither, or Both: Vertebrate Paleontology at the University of Chicago, 1892–1950." *Perspectives in Science* 1, no. 3 (1993): 478–519.

Rainger, Ronald. "Collectors and Entrepreneurs: Hatcher, Wortman, and the Structure of American Vertebrate Paleontology circa 1900." *Earth Sciences History* 9, no. 1 (1990): 14–21.

Rainger, Ronald. "Just before Simpson: William Diller Matthew's Understanding of Evolution." *Proceedings of the American Philosophical Society* 130, no. 4 (1986): 453–74.

Rainger, Ronald, Keith R. Benson, and Jane Maienschein, eds. *The American Development of Biology*. Philadelphia: University of Pennsylvania Press, 1988.

Rea, Tom. *Bone Wars: The Excavation and Celebrity of Andrew Carnegie's Dinosaur*. Pittsburgh, PA: University of Pittsburgh Press, 2001.

Rees, Amanda. *The Infanticide Controversy: Primatology and the Art of Field Science*. Chicago: University of Chicago Press, 2009.

Reeves, Brian. "The Concept of an Altithermal Cultural Hiatus in Northern Plains Prehistory." *American Anthropologist* 75 (1973): 1221–53.

Reidy, Michael S. "From the Oceans to the Mountains: Spatial Science in an Age of Empire." In *Knowing Global Environments: New Historical Perspectives on the Field Sciences*, edited by Jeremy Vetter, 17–38. New Brunswick, NJ: Rutgers University Press, 2011.

Reidy, Michael S. "John Tyndall's Vertical Physics: From Rock Quarries to Icy Peaks." *Physics in Perspective* 12, no. 2 (2010): 122–45.

Reingold, Nathan. *Science, American Style*. New Brunswick, NJ: Rutgers University Press, 1991.

Reuss, Martin, and Stephen H. Cutcliffe, eds. *The Illusory Boundary: Environment and Technology in History*. Charlottesville: University of Virginia Press, 2010.

Reynolds, Arthur R. "The Kinkaid Act and Its Effects on Western Nebraska." *Agricultural History* 23, no. 1 (1949): 20–29.

Rice, Stephen P. *Minding the Machine: Languages of Class in Early Industrial America.* Berkeley: University of California Press, 2004.

Rieppel, Lukas Benjamin. "Dinosaurs: Assembling an Icon of Science." PhD diss., Harvard University, 2012.

Riley, Paul D. "Dr. David Franklin Powell and Fort McPherson." *Nebraska History* 51 (1970): 153–70.

Ritterbush, Lauren W. "Drawn by the Bison: Late Prehistoric Native Migration into the Central Plains." *Great Plains Quarterly* 22, no. 4 (2002): 259–70.

Rivinus, E. F., and E. M. Youssef. *Spencer Baird of the Smithsonian.* Washington, DC: Smithsonian Institution Press, 1992.

Robertson, Janet. *The Magnificent Mountain Women: Adventures in the Colorado Rockies.* Lincoln: University of Nebraska Press, 1990.

Robinson, Michael F. *The Coldest Crucible: Arctic Exploration and American Culture.* Chicago: University of Chicago Press, 2006.

Rogers, Katherine. *The Sternberg Fossil Hunters: A Dinosaur Dynasty.* Missoula, MT: Mountain Press, 1991.

Rosenberg, Charles E. *No Other Gods: On Science and American Social Thought.* Baltimore, MD: Johns Hopkins University Press, 1976.

Rossiter, Margaret W. "'Women's Work' in Science, 1880–1910." *Isis* 71 (1980): 381–98.

Rothman, Hal. *Devil's Bargains: Tourism in the Twentieth-Century American West.* Lawrence: University Press of Kansas, 1998.

Rothman, Hal. *On Rims and Ridges: The Los Alamos Area since 1880.* Lincoln: University of Nebraska Press, 1992.

Rothman, Hal. *Preserving Different Pasts: The American National Monuments.* Urbana: University of Illinois Press, 1989.

Rozwadowski, Helen M. *Fathoming the Ocean: The Discovery and Exploration of the Deep Sea.* Cambridge, MA: Harvard University Press, 2005.

Rozwadowski, Helen M. "Small World: Forging a Scientific Maritime Culture for Oceanography." *Isis* 87, no. 3 (1996): 409–29.

Rudwick, Martin. "The Emergence of a Visual Language for Geological Science 1760–1840." *History of Science* 14, no. 3 (1976): 149–95.

Rudwick, Martin. *The Great Devonian Controversy: The Shaping of Scientific Knowledge among Gentlemanly Specialists.* Chicago: University of Chicago Press, 1985.

Rumery, Myron G. A., et al. "A History of University of Nebraska North Platte Station." In *University of Nebraska North Platte Station: Its Beginning and Continuation, 1904–1972,* 1–95. Misc. Pub. 29 of Nebraska Agricultural Experiment Station, n.d.

Rumore, Gina Maria. "A Natural Laboratory, a National Monument: Carving out a Place for Science in Glacier Bay, Alaska, 1879–1959." PhD diss., University of Minnesota, 2009.

Runte, Alfred. *National Parks: The American Experience.* 3rd ed. Lincoln: University of Nebraska Press, 1997.

Ruskin, Steve. "'Among the Favored Mortals of Earth': The Press, State Pride, and the Eclipse of 1878." *Colorado Heritage* (Spring 2008): 22–35.

Ruskin, Steve. "The Business of Natural History: Charles Aiken, Colorado Ornithology, and the Role of the Professional Collector." *Historical Studies in the Natural Sciences* 45, no. 3 (2015): 357–96.

Rusnock, Andrea. "Correspondence Networks and the Royal Society, 1700–1750." *British Journal for the History of Science* 32, no. 2 (1999): 155–69.

Rust, Daniel L. *Flying across America: The Airline Passenger Experience.* Norman: University of Oklahoma Press, 2009.

Sandweiss, Martha A. *Passing Strange: A Gilded Age Tale of Love and Deception across the Color Line.* New York: Penguin Press, 2009.

Schaffer, Simon. "Metrology, Metrication, and Victorian Values." In *Victorian Science in Context,* edited by Bernard Lightman, 438–74. Chicago: University of Chicago Press, 1997.

Schaffer, Simon. "Physics Laboratories and the Victorian Country House." In *Making Space for Science: Territorial Themes in the Shaping of Knowledge,* edited by Crosbie Smith and Jon Agar, 149–80. Basingstoke: Macmillan, 1998.

Schmidly, David J. *Texas Natural History: A Century of Change.* Lubbock: Texas Tech University Press, 2002.

Schneiders, Robert Kelley. *Big Sky Rivers: The Yellowstone and Upper Missouri.* Lawrence: University Press of Kansas, 2003.

Schneiders, Robert Kelley. *Unruly River: Two Centuries of Change along the Missouri.* Lawrence: University Press of Kansas, 1999.

Schubert, Frank N. "Troublesome Partnership: Gouverneur K. Warren and Ferdinand V. Hayden on the Northern Plains in 1856 and 1857." *Earth Sciences History* 3, no. 2 (1984): 143–48.

Schubert, Frank N. *Vanguard of Expansion: Army Engineers in the Trans-Mississippi West, 1819–1879.* Washington, DC: Government Printing Office, n.d.

Schuchert, Charles, and Clara Mae LeVene. *O. C. Marsh: Pioneer in Paleontology.* New Haven, CT: Yale University Press, 1940.

Schulten, Susan. *Mapping the Nation: History and Cartography in Nineteenth-Century America.* Chicago: University of Chicago Press, 2012.

Schumaker, Lyn. *Africanizing Anthropology: Fieldwork, Networks, and the Making of Cultural Knowledge in Central Africa.* Durham, NC: Duke University Press, 2001.

Scott, James C. *Seeing like a State: How Certain Schemes to Improve the Human Condition Have Failed.* New Haven, CT: Yale University Press, 1998.

Sears, John F. *Sacred Places: American Tourist Attractions in the Nineteenth Century.* New York: Oxford University Press, 1989.

Sears, Paul B. "A Pollen Profile from the Grassland Province." *Science* 134 (1961): 2038–40.

Secord, Anne. "Corresponding Interests: Artisans and Gentlemen in Nineteenth-Century Natural History." *British Journal for the History of Science* 27, no. 4 (1994): 383–408.

Secord, Anne. "Science in the Pub: Artisan Botanists in Early Nineteenth-Century Lancashire." *History of Science* 32, no. 3 (1994): 269–315.

Secord, James A. "The Crisis of Nature." In *Cultures of Natural History,* edited by Nicholas Jardine, James A. Secord, and E. C. Spary, 447–59. Cambridge: Cambridge University Press, 1996.

Secord, James A. "Knowledge in Transit." *Isis* 95, no. 4 (2004): 654–72.

Seely, Bruce. "Research, Engineering, and Science in American Engineering Colleges: 1900–1960." *Technology and Culture* 34, no. 2 (1993): 344–86.

Selcer, Perrin. "The View from Everywhere: Disciplining Diversity in Post–World War II International Social Science." *Journal of the History of the Behavioral Sciences* 45, no. 4 (2009): 309–29.

Sellars, Richard West. *Preserving Nature in the National Parks: A History.* New Haven, CT: Yale University Press, 1997.

Shaffer, Marguerite S. *See America First: Tourism and National Identity, 1880–1940.* Washington, DC: Smithsonian Institution Press, 2001.

Shapin, Steven. "Following Scientists Around." *Social Studies of Science* 18, no. 3 (1988): 533–50.

Shapin, Steven. "The Invisible Technician." *American Scientist* 77, no. 6 (1989): 554–63.

Shapin, Steven. "Placing the View from Nowhere: Historical and Sociological Problems in the Location of Science." *Transactions of the Institute of British Geographers* 23, no. 1 (1998): 5–12.

Shaw, Harley G., and Mara E. Weisenberger. *Twelve Hundred Miles by Horse and Burro: J. Stokely Ligon and New Mexico's First Breeding Bird Survey.* Tucson: University of Arizona Press, 2011.

Shen, Grace Yen. *Unearthing the Nation: Modern Geology and Nationalism in Republican China.* Chicago: University of Chicago Press, 2014.

Sherow, James Earl. *Watering the Valley: Development along the High Plains Arkansas River, 1870–1950.* Lawrence: University Press of Kansas, 1990.

Shor, Elizabeth Noble. *Fossils and Flies: The Life of a Compleat Scientist, Samuel Wendell Williston (1851–1918).* Norman: University of Oklahoma Press, 1971.

Shteir, Ann B. *Cultivating Women, Cultivating Science: Flora's Daughters and Botany in England, 1760 to 1860.* Baltimore, MD: Johns Hopkins University Press, 1996.

Slotten, Hugh Richard. *Patronage, Practice, and the Culture of American Science: Alexander Dallas Bache and the U.S. Coast Survey.* Cambridge: Cambridge University Press, 1994.

Smith, Duane A. *Mesa Verde National Park: Shadows of the Centuries.* Lawrence: University Press of Kansas, 1988.

Smith, Duane A. "Removing Our Relics: Durango versus Gustaf Nordenskiold over Mesa Verde Ruins." *Colorado Heritage,* no. 3 (1989): 2–11.

Smith, Michael L. *Pacific Visions: California Scientists and the Environment, 1850–1915.* New Haven, CT: Yale University Press, 1987.

Snead, James A. *Ruins and Rivals: The Making of Southwest Archaeology.* Tucson: University of Arizona Press, 2001.

Socolow, Arthur A., ed. *The State Geological Surveys: A History.* N.p.: Association of American State Geologists, 1988.

Sorenson, W. Conner. *Brethren of the Net: American Entomology, 1840–1880.* Tuscaloosa: University of Alabama Press, 1995.

Sorrenson, Richard. "The Ship as a Scientific Instrument in the Eighteenth Century." *Osiris* 11 (1996): 221–36.

Spary, E. C. *Utopia's Garden: French Natural History from Old Regime to Revolution.* Chicago: University of Chicago Press, 2000.

Spence, Mark David. *Dispossessing the Wilderness: Indian Removal and the Making of the National Parks.* New York: Oxford University Press, 1999.

Stegner, Wallace. *Beyond the Hundredth Meridian: John Wesley Powell and the Second Opening of the West.* Boston: Houghton Mifflin, 1954.

Stegner, Wallace. "Notes on a life spent pecking at a sandstone cliff . . ." In *Mormon Country.* New York: Bonanza Books, 1942.

Stephens, Lester D. *Science, Race, and Religion in the American South: John Bachman and the Charleston Circle of Naturalists, 1815–1895.* Chapel Hill: University of North Carolina Press, 2000.

Sterling, Everett W. "The Powell Irrigation Survey, 1888–1893." *Mississippi Valley Historical Review* 27, no. 3 (1940): 421–34.

Sterling, Keir B. "Builders of the U.S. Biological Survey, 1885–1930." *Journal of Forest History* 33 (1989): 180–87.

Sternberg, Martha L. *George Miller Sternberg.* Chicago: American Medical Association, 1920.

Stocking, George W., Jr. "The Ethnographer's Magic: Fieldwork in British Anthropology from Tylor to Malinowski." In *Observers Observed: Essays on Ethnographic Fieldwork,* edited by George Stocking, 70–120. Madison: University of Wisconsin Press, 1983.

Stocking, George W., Jr. "Reading the Palimpsest of Inquiry: *Notes and Queries* and the History of British Social Anthropology." In *Delimiting Anthropology: Occasional Essays and Reflections*, 164–206. Madison: University of Wisconsin Press, 2001.

Stucker, Gilbert F. "Hayden in the Badlands." *American West* 4, no. 1 (1967): 40–45, 79–85.

Sviedyrs, Romualdas. "The Rise of Physics Laboratories in Britain." *Historical Studies in the Physical Sciences* 7 (1976): 405–36.

Tate, Michael L. *The Frontier Army in the Settlement of the West.* Norman: University of Oklahoma Press, 1999.

te Heesen, Anke. "Boxes in Nature." *Studies in History and Philosophy of Science* 31, no. 3 (2000): 381–403.

te Heesen, Anke. "From Natural Historical Investment to State Service: Collectors and Collections of the Berlin Society of Friends of Nature Research." *History of Science* 42, no. 1 (2004): 113–31.

Thompson, Kenneth W. *A History of Soil Survey in North Dakota.* Dickinson, ND: King Speed Printing, 1992.

Thompson, Robert Luther. *Wiring a Continent: The History of the Telegraph Industry in the United States, 1832–1866.* Princeton, NJ: Princeton University Press, 1947.

Thompson, William Leigh. "Richard (Beaver Dick) Leigh: Last of the Mountain Men." MS thesis, Utah State University, 1970.

Tilley, Helen. *Africa as a Living Laboratory: Empire, Development, and the Problem of Scientific Knowledge, 1870–1950.* Chicago: University of Chicago Press, 2011.

Tilley, Helen. "Global Histories, Vernacular Science, and African Genealogies; or, Is the History of Science Ready for the World?" *Isis* 101 (2010): 110–19.

Tobey, Ronald C. *Saving the Prairies: The Life Cycle of the Founding School of American Plant Ecology, 1895–1955.* Berkeley: University of California Press, 1981.

Todes, Daniel P. *Pavlov's Physiology Factory: Experiment, Interpretation, Laboratory Enterprise.* Baltimore, MD: Johns Hopkins University Press, 2002.

Topping, Gary. *Glen Canyon and the San Juan Country.* Moscow: University of Idaho Press, 1997.

Torres-Reyes, Ricardo. *Mesa Verde National Park: An Administrative History, 1906–1970.* Washington, DC: Department of the Interior, 1970.

Trachtenberg, Alan. *Reading American Photographs: Images as History, Mathew Brady to Walker Evans.* New York: Hill and Wang, 1989.

Trimble, Donald E. *The Geologic Story of the Great Plains.* Bulletin of the US Geological Survey, no. 1493 (1980).

Turnbull, David. "Travelling Knowledge: Narratives, Assemblage and Encounters." In *Instruments, Travel and Science: Itineraries of Precision from the Seven-*

teenth to the Twentieth Century, edited by Marie-Noëlle Bourguet, Christian Licoppe, and H. Otto Sibum, 273–94. London: Routledge, 2002.

Urry, James. "Englishmen, Celts, and Iberians: The Ethnographic Survey of the United Kingdom, 1892–1899." In *Functionalism Historicized: Essays on British Social Anthropology*, edited by George W. Stocking, 83–105. Madison: University of Wisconsin Press, 1984.

Urry, James. "*Notes and Queries on Anthropology* and the Development of Field Methods in British Anthropology, 1870–1920." *Proceedings of the Royal Anthropological Institute* (1972): 45–57.

Valencius, Conevery Bolton. *The Health of the Country: How American Settlers Understood Themselves and Their Land.* New York: Basic Books, 2002.

Valencius, Conevery Bolton. *The Lost History of the New Madrid Earthquakes.* Chicago: University of Chicago Press, 2013.

Vanorny, Patricia M. *A History of the South Dakota Geological Survey.* Vermillion: University of South Dakota, 1970.

Veblen, Thomas T., and Diane C. Lorenz. *The Colorado Front Range: A Century of Ecological Change.* Salt Lake City: University of Utah Press, 1991.

Vetter, Jeremy. "Capitalizing on Grass: The Science of Agrostology and the Sustainability of Ranching in the American West." *Science as Culture* 19, no. 4 (2010): 483–507.

Vetter, Jeremy. "Cowboys, Scientists, and Fossils: The Field Site and Local Collaboration in the American West." *Isis* 99, no. 2 (2008): 273–303.

Vetter, Jeremy. "Field Life in the American West: Surveys, Networks, Stations, and Quarries." In *Scientists and Scholars in the Field: Studies in the History of Fieldwork and Expeditions*, edited by Kristian H. Nielsen, Michael Harbsmeier, and Christopher J. Ries, 225–58. Aarhus: Aarhus University Press, 2012.

Vetter, Jeremy. "Field Science in the Railroad Era: The Tools of Knowledge Empire in the American West, 1869–1916." *História, Ciências, Saúde—Manguinhos* 15, no. 3 (2008): 597–613.

Vetter, Jeremy. "Introduction: Lay Participation in the History of Scientific Observation." *Science in Context* 24, no. 2 (2011): 127–41.

Vetter, Jeremy. "Labs in the Field? Rocky Mountain Biological Stations in the Early Twentieth Century." *Journal of the History of Biology* 45, no. 4 (2012): 587–611.

Vetter, Jeremy. "Lay Observers, Telegraph Lines, and Kansas Weather: The Field Network as a Mode of Knowledge Production." *Science in Context* 24, no. 2 (2011): 259–80.

Vetter, Jeremy. "Regionalizing Knowledge: The Ecological Approach of the USDA Office of Dryland Agriculture on the Great Plains." In *New Perspec-*

tives on the History of Life Sciences and Agriculture, edited by Denise Phillips and Sharon Kingsland, 277–96. Springer, 2015.

Vetter, Jeremy. "Rocky Mountain High Science: Teaching, Research, and Nature at Field Stations." In *Knowing Global Environments: New Historical Perspectives on the Field Sciences*, edited by Jeremy Vetter, 108–34. New Brunswick, NJ: Rutgers University Press, 2011.

Vetter, Jeremy. "Science along the Railroad: Expanding Field Work in the U.S. Central West." *Annals of Science* 61 (2004): 187–211.

Vetter, Jeremy. "Wallace's *Other* Line: Human Biogeography and Field Practice in the Eastern Colonial Tropics." *Journal of the History of Biology* 39, no. 1 (2006): 89–123.

Wallace, David Rains. *The Bonehunters' Revenge: Dinosaurs, Greed, and the Greatest Scientific Fraud of the Gilded Age*. Boston: Houghton Mifflin, 1999.

Warwick, Andrew. *Masters of Theory: Cambridge and the Rise of Mathematical Physics*. Chicago: University of Chicago Press, 2003.

Way, Albert G. *Conserving Southern Longleaf: Herbert Stoddard and the Rise of Ecological Land Management*. Athens: University of Georgia Press, 2011.

Webb, George E. *Science in the American Southwest: A Topical History*. Tucson: University of Arizona Press, 2002.

Weber, William A. *King of Colorado Botany: Charles Christopher Parry, 1823–1890*. Niwot: University Press of Colorado, 1997.

Wedel, Waldo R. *Central Plains Prehistory: Holocene Environments and Culture Change in the Republic River Basin*. Lincoln: University of Nebraska Press, 1986.

Wedel, Waldo R. "The High Plains and Their Utilization by the Indian." *American Antiquity* 29, no. 1 (1963): 1–16.

Welch, Margaret. *The Book of Nature: Natural History in the United States, 1825–1875*. Boston: Northeastern University Press, 1998.

Wells, John W. "James Hall's Amateurs." *Earth Sciences History* 6, no. 1 (1987): 34–39.

Wells, Philip V. "Scarp Woodlands, Transported Grassland Soils, and Concept of Grassland Climate in the Great Plains Region." *Science* 148 (1965): 246–49.

Welsch, Roger L. "Sandhill Baled-Hay Construction." *Keystone Folklore Quarterly* 15 (1970): 16–34.

West, Elliott. "Wired to the World: The Telegraph and the Making of the West." In *The Essential West: Collected Essays*, 78–99. Norman: University of Oklahoma Press, 2012.

White, Paul. *Thomas Huxley: Making the "Man of Science."* Cambridge: Cambridge University Press, 2003.

White, Richard. *Railroaded: The Transcontinentals and the Making of Modern America.* New York: Norton, 2011.

Whitlock, Cathy, Mel A. Reasoner, and Carl H. Key. "Paleoenvironmental History of the Rocky Mountain Region during the Past 20,000 Years." In *Rocky Mountain Futures: An Ecological Perspective,* edited by Jill S. Baron, 41–57. Washington, DC: Island Press, 2002.

Whitnah, Donald R. *A History of the United States Weather Bureau.* Urbana: University of Illinois Press, 1961.

Whittlesey, Derwent. "The Impress of Effective Central Authority upon the Landscape." *Annals of the Association of American Geographers* 25, no. 2 (1935): 85–97.

Wilkins, Thurman. *Clarence King: A Biography.* 2nd ed. Albuquerque: University of New Mexico Press, 1988.

Williams, Roger L. *Aven Nelson of Wyoming.* Boulder: Colorado Associated University Press, 1984.

Wilson, Carol Green. *Alice Eastwood's Wonderland: The Adventures of a Botanist.* San Francisco: California Academy of Sciences, 1955.

Wilson, Robert. *The Explorer King: Adventure, Science, and the Great Diamond Hoax—Clarence King in the Old West.* New York: Scribner, 2006.

Wohl, Ellen E. *Virtual Rivers: Lessons from the Mountain Rivers of the Colorado Front Range.* New Haven, CT: Yale University Press, 2001.

Woodbury, Richard B. *Alfred V. Kidder.* New York: Columbia University Press, 1973.

Woodbury, Richard B., and Nathalie F. S. Woodbury. "The Rise and Fall of the Bureau of American Ethnology." *Journal of the Southwest* 41, no. 3 (1999): 283–96.

Worster, Donald. "The Black Hills: Sacred or Profane?" In *Under Western Skies: Nature and History in the American West,* 106–53. New York: Oxford University Press, 1992.

Worster, Donald. "Cowboy Ecology." In *Under Western Skies: Nature and History in the American West,* 34–52. New York: Oxford University Press, 1992.

Worster, Donald. "New West, True West: Interpreting the Region's History." *Western Historical Quarterly* 18, no. 2 (1987): 141–56.

Worster, Donald. *A River Running West: The Life of John Wesley Powell.* Oxford: Oxford University Press, 2001.

Worster, Donald. *Rivers of Empire: Water, Aridity, and the Growth of the American West.* New York: Pantheon, 1985.

Wrigley, E. A. *Continuity, Chance, and Change: The Character of the Industrial Revolution in England.* Cambridge: Cambridge University Press, 1988.

Wrobel, David M., and Patrick T. Long, eds. *Seeing and Being Seen: Tourism in the American West.* Lawrence: University Press of Kansas, 2001.

Wyatt, Sally, and Brian Balmer. "Home on the Range: What and Where Is the Middle in Science and Technology Studies?" *Science, Technology and Human Values* 32, no. 6 (2007): 619–26.

Wyckoff, William. *Creating Colorado; The Making of a Western American Landscape, 1860–1940.* New Haven, CT: Yale University Press, 1999.

Yokota, Kariann Akemi. *Unbecoming British: How Revolutionary America Became a Postcolonial Nation.* Oxford: Oxford University Press, 2011.

Zunz, Olivier. *Making America Corporate, 1870–1920.* Chicago: University of Chicago Press, 1990.

Index